VOLUME TWO HUNDRED AND SIXTEEN

ADVANCES IN IMAGING AND ELECTRON PHYSICS

VOLUME TWO HUNDRED AND SIXTEEN

ADVANCES IN IMAGING AND ELECTRON PHYSICS

Morphological Image Operators

Edited by

MARTIN HŸTCH and PETER W. HAWKES

CEMES-CNRS

Toulouse, France

ELSEVIER

ACADEMIC PRESS

An imprint of Elsevier

Cover photo credit:
Dilation and erosion by a disc (above) and a 3 × 3 square below.

Academic Press is an imprint of Elsevier
125 London Wall, London EC2Y 5AS, United Kingdom
525 B Street, Suite 1650, San Diego, CA 92101, United States
50 Hampshire Street, 5th Floor, Cambridge, MA 02139, United States
The Boulevard, Langford Lane, Kidlington, Oxford OX5 1GB, United Kingdom

ISBN: 978-0-12-821003-1
ISSN: 1076-5670

For information on all Academic Press publications
visit our website at https://www.elsevier.com/books-and-journals

Publisher: Zoe Kruze
Acquisitions Editor: Jason Mitchell
Editorial Project Manager: Shellie Bryant
Production Project Manager: James Selvam
Designer: Alan Studholme

Typeset by VTeX

to Marianne

Contents

Contributor

Henk J.A.M. Heijmans
Formerly Centre for Mathematics and Computer Science, Amsterdam, Netherlands

Preface

This volume is titled "Morphological Image Operators" for the simple reason that it deals with morphological (image) operators. This raises the obvious question: what is a morphological operator and what are they meant for? In fact, this book contains quite a few definitions, but that of a morphological operator is missing. Granted, this is somewhat peculiar, especially for a book which seeks to present a mathematically rigorous exposition on morphological operators. Any attempt to find a formal definition of a morphological operator, however, would lead inevitably to the following dilemma: either it would be too restrictive, excluding operators that should not be excluded a priori, or it would be too general, leading to a "theory of everything". For those readers who are content with a less formal approach: morphological operators are mappings on spaces of images (complete lattices in this book) which emerge in the context of mathematical morphology.

The monographs *Random Sets and Integral Geometry* by Georges Matheron (1975) and *Image Analysis and Mathematical Morphology* by Jean Serra (1982) discuss a number of mappings on subsets of the Euclidean plane (which serve as a model for continuous binary images), which have in common that they are based on set-theoretical operations (union, intersection, complementation) as well as translations. Only recently, these operations have been extended to arbitrary complete lattices, possibly endowed with some automorphism group. It is these operations which will be called "morphological operators" in this book. The adjective "morphological" is to indicate that our interest goes to these operators which are suited for the task of describing certain aspects of shape. The word "image" indicates that the operators apply to images.

Mathematical morphology is a theory which is concerned with the processing and analysis of images, using operators and functionals based on topological and geometrical concepts. During the last decade, it has acquired a special status within the field of image processing, pattern recognition, and, to a lesser extent, computer vision. This achievement is particularly due to the pioneering work of Matheron, Serra, and other workers of the Ecole Nationale Supérieure des Mines de Paris. Their work will get ample attention in this book, the main goal of which is to present a rigorous mathematical treatment of morphology, in particular morphological operators. Our exposition adopts the complete lattice framework; nevertheless we shall, quite often, restrict attention to concrete examples (e.g., subsets of the Euclidean space).

It is conceivable that the non-mathematical reader will be discouraged by the mathematical abstraction in this exposition. We want two make two comments on this. If the reader prefers, he can always keep a concrete example in mind, e.g., he may substitute "subsets of $\mathbb{R}^d$" for "complete lattice". But on the other hand, the adoption of the complete lattice framework avoids the replication of analogous concepts and results in different situations. This may sound rather obvious, but the recent morphological literature, where one encounters phrases like binary morphology, grey-scale morphology, graph morphology, vector morphology, matrix morphology, and fuzzy morphology, reveals that many authors are not aware of the existence of a unifying framework for mathematical morphology (or at least, they fail to see the range of this overall unification).

So far so good, but doesn't the comprehension of such an algebraic approach presume that the reader is an experienced mathematician with a deep knowledge of modern algebra? The answer to such a question in the preface of a book is usually to be regarded with some suspicion and disbelief. To lure the reader, the writer should suggest that hardly any prerequisites are needed (a first year course in calculus and a basic knowledge of linear algebra). Then, after digestion of the first two or three pages, the reader realizes that he'd better first become acquainted with the latest developments in algebraic topology and study some graduate books on the theory of pseudo-differential operators. Let us ease the reader's mind at this point and assure him or her that for an apprehension of this book no specific knowledge is supposed. In fact, the reader finds in Chapter 2 all basic notions concerning complete lattices which are needed in this book. In Chapters 1 and 4 he finds an introduction into mathematical morphology. Although the book is not intended as a primer on mathematical morphology, Chapters 1, 4, 9, and 11, and to a lesser extent Chapter 7, can be used as an introduction into this field. Furthermore, Chapter 12 contains a rather elementary exposition on the theory of morphological filters.

I am indebted to several colleagues in the field of mathematical morphology, in particular to Jean Serra through whose first book I became interested in this field. His amazing intuition in combination with his enthusiasm to share his ideas with other people are gratefully acknowledged. His work and that of his colleagues of the Fontainebleau school, in particular Georges Matheron, form the basis of this volume. Many thanks are also due to Christian Ronse, with whom I had the pleasure to cooperate and whose ideas can be found at several instances in this book. I am much indebted for the support which I received from my home institute, the Centre for Mathematics and Computer Science (CWI) in Amsterdam. This applies particularly to my colleagues Adrian Baddeley and Annoesjka Cabo and my ex-colleague Jos Roerdink, who commented on parts of the text, and to Adri Steenbeek (our C++ magician), whom I annoyed many times when some basic operator had to be implemented. (I quietly hope that ever the day will break when I will be able to have "Hello world" printed on my computer screen.) The figures containing "real-world" images were produced with SCILIMAGE, a software package for image processing developed by the University of Amsterdam in cooperation with the TNO Institute of Applied Physics, Delft. Special thanks are due to Peter Nacken, who went through the pain of reading the whole manuscript and correcting many mistakes. As such he solely is responsible for remaining errors (just kidding, Peter). Of this long list, Marianne, my wife, is probably the only person who understands almost nothing of what is inside this book. But she is the one and only to whom I never had to explain a single word.

Henk J.A.M. Heijmans
Formerly Centre for Mathematics and Computer Science, Amsterdam, Netherlands

Editorial Preface

Many of the Supplements to *Advances in Electronics and Electron Physics* remain of considerable interest and continue to be cited. Supplement 10 by W.O. Saxton has already been reprinted in these Advances (Volume 214) and here we republish another widely used Supplement, *Morphological Image Operators* by Henk Heijmans.

Mathematical morphology is one of the regular themes of these Advances– we recall that Jean Serra, co-founder of the subject, contributed to volume 150. Heijmans' volume is a self-contained account of all the basic theory of mathematical morphology, ranging from complete lattices, via adjunctions, dilations and erosions and hence openings and closings, to the morphology of grey-scale images and morphological filters. Although these topics are covered at length in more recent books on mathematical morphology, Heijmans' very readable account is ideal for comparative newcomers to the subject, daunted by the abstract tone of the more advanced texts.

We are particularly grateful to Professor Roerdink who has enthusiastically supported the republication of the volume. Thanks to him, the original source files were made available for this reprint.

Further Supplements that will be available in forthcoming volumes are (Jansen, *Coulomb Interactions in Charged Particle Beams*), (Hawkes, *Quadrupoles in Electron Lens Design* and *The Beginnings of Electron Microscopy*) and (Ximen, *Aberration Theory in Electron and Ion Optics*).

Martin Hÿtch
Peter W. Hawkes

CHAPTER ONE

First principles

Henk J.A.M. Heijmans
Formerly Centre for Mathematics and Computer Science, Amsterdam, Netherlands

Contents

The primary goal of this introductory chapter is to acquaint the reader with the basic principles of mathematical morphology. Rather than to present formal concepts and definitions, this chapter is intended to give the reader a flavour of what mathematical morphology is about. This is done by means of two examples in two subsequent sections. The first example concerns the extraction of connected components; the second example discusses Gaussian convolution and its "morphological analogue". These examples serve also as a mean to introduce some of the theoretical notions which play a major role in this book.

1.1. A typical example

Consider the following problem. Given a set $X \subseteq \mathbb{R}^2$ which comprises several connected components, extract all components that contain a closed disk with radius r. This situation is illustrated in Fig. 1.1.

We introduce some terminology to formalize this problem. Let α_r be the mapping which sends X to the union of all components that contain a translate of rB; here B is the unit disk centred at the origin and rB is the disk with radius r. Henceforth, α_r will be called *operator*. If A is a subset of $\mathbb{R}^2$ and h a vector in $\mathbb{R}^2$, then A_h denotes the translate of A along the vector h, i.e.,

$$A_h = \{a + h \mid a \in A\}.$$

Advances in Imaging and Electron Physics, Volume 216
ISSN 1076-5670
https://doi.org/10.1016/bs.aiep.2020.07.001

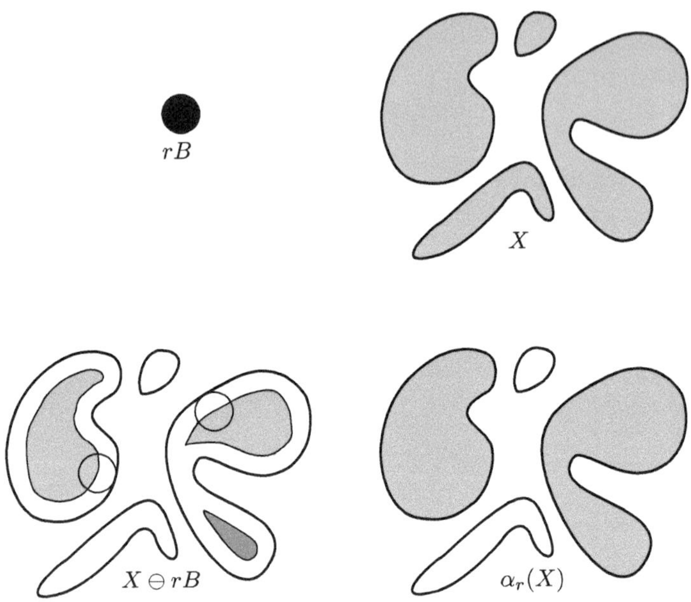

Figure 1.1 Extraction of the connected components which contain a disk with radius $r \geq 0$.

The Minkowski sum and difference of a set X and another set A are, respectively, defined by

$$X \oplus A = \bigcup_{a \in A} X_a, \tag{1.1}$$

$$X \ominus A = \bigcap_{a \in A} X_{-a}. \tag{1.2}$$

Furthermore, $\gamma_h(X)$ is the connected component of X which contains h. If $h \notin X$, then $\gamma_h(X) = \varnothing$. It is obvious that

$$X = \bigcup_{h \in \mathbb{R}^2} \gamma_h(X);$$

in other words, X is the union of all its connected components. The following identity holds:

$$\alpha_r(X) = \bigcup_{h \in X \ominus rB} \gamma_h(X). \tag{1.3}$$

To understand this, one has to verify that

$$X \ominus rB = \{h \in \mathbb{R}^2 \mid (rB)_h \subseteq X\}. \tag{1.4}$$

Figure 1.2 Hexagonal grid.

In other words, $X \ominus rB$ comprises the centres of the disks with radius r which are contained in X. If Y is a component of X which contains a disk with radius r, say $(rB)_h$, then $h \in X \ominus rB$ and $\gamma_h(X) = Y$. An illustration of the identity in (1.3) is given in Fig. 1.1.

The family of operators $\{\alpha_r \mid r > 0\}$ has several interesting properties:

- α_r *is increasing;* that is, if $X \subseteq X'$, then $\alpha_r(X) \subseteq \alpha_r(X')$.
- α_r *is translation invariant;* in symbols, $\alpha_r(X_h) = [\alpha_r(X)]_h$, for every $X \subseteq \mathbb{R}^2$ and $h \in \mathbb{R}^2$. This means that translation of X followed by application of α_r leads to the same result as application of α_r followed by translation (see also Fig. 4.1).
- $\alpha_r(X) \subseteq X$; that is, $\alpha_r(X)$ is a subset of X.
- $\alpha_r\alpha_s = \alpha_s\alpha_r = \alpha_s$, $s \geq r$. Here $\alpha_r\alpha_s$ denotes the composition of both operators: first α_s is applied and subsequently α_r. The given identities mean that the effect of α_r is overruled by α_s. We will encounter this property, which is called semigroup property or granulometry property, at several places in this book.

The problem, as well as its solution, becomes slightly different if X is not a continuous set but a subset of points on some regular grid.

First of all, we have to define what is meant by a disk with radius r in this case. And second, we have to define what a connected set is. Suppose we are interested in the hexagonal grid; see Fig. 1.2. Two points on the grid are neighbours if they are connected by an edge. A path is a sequence of points $x_1, x_2, \ldots, x_N$ such that x_k and x_{k+1} are neighbours for every $k = 1, 2, \ldots, N-1$. A subset X of a regular grid is connected if for every pair $x, y \in X$ there exists a path in X with endpoints x and y. Now $\gamma_h(X)$ can be defined as before. In this case there is a simple algorithm to compute

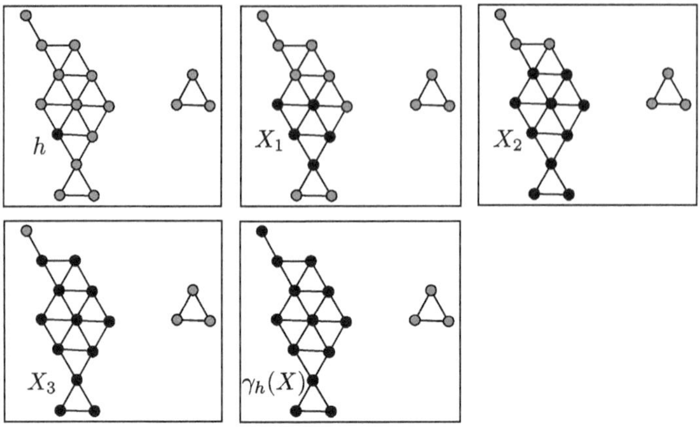

Figure 1.3 Extraction of the connected component of X (grey points) containing h. Depicted (from left to right and top to bottom) are $X_0 = \{h\}, X_1, X_2, X_3$ and $X_4 = \gamma_h(X)$.

$\gamma_h(X)$. In fact, let H be the hexagon comprising seven points, the origin and its six neighbours. Define

$$X_0 = \{h\},$$
$$X_{n+1} = (X_n \oplus H) \cap X.$$

Presuming that X is bounded, one gets $X_N = X_{N+1}$ for N large enough; now $\gamma_h(X) = X_N$. The procedure is illustrated in Fig. 1.3.

We point out that in this particular case a reasonable approximation of the disk with radius r is the set

$$H \oplus H \oplus \cdots \oplus H \qquad [r \text{ terms}].$$

The operators α_r and γ_h on $\mathcal{P}(\mathbb{R}^2)$, the subsets of $\mathbb{R}^2$, are typical examples of morphological operators. As a matter of fact, the family $\{\alpha_r \mid r > 0\}$ is a *granulometry* (Section 4.4, Section 6.7, Section 9.6) and γ_h is a *connectivity opening* (Section 9.7). To a large extent this book is devoted to a theoretical investigation of various morphological operators. As in the previous example, such operators are based on geometrical and topological concepts. Yet their properties are easiest described if one uses the language of algebra and set theory. Morphological operators can be defined in purely algebraic terms. For example, Minkowski addition $X \mapsto X \oplus A$ has the property that it distributes over unions:

$$(\bigcup_{i \in I} X_i) \oplus A = \bigcup_{i \in I} (X_i \oplus A), \tag{1.5}$$

for an arbitrary family of sets X_i. This operator is also called *dilation by A*. Analogously, Minkowski subtraction $X \mapsto X \ominus A$ distributes over intersections and is called *erosion by A*. In Chapter 3, a dilation (resp. erosion) is defined as an operator from one complete lattice into another which distributes over arbitrary suprema (resp. infima). Such definitions are purely algebraic and do not refer to geometrical or topological concepts. This book attempts to present a consistent algebraic theory of morphological operators, but with an eye on their geometrical nature.

1.2. Morphological convolution

A common method in mathematical morphology to model grey-scale images is by means of functions. For convenience we restrict ourselves to real-valued functions on $\mathbb{R}^2$. A well-known mapping in classical signal analysis is the convolution product

$$(F * K)(x) = \int_{\mathbb{R}^2} F(x-y)K(y)dy; \tag{1.6}$$

here F is the representation of an image and K the convolution kernel. In practice one often chooses for K the Gaussian function

$$K(x) = \frac{1}{\sqrt{2\pi}\sigma} \exp\left(\frac{-|x|^2}{2\sigma^2}\right).$$

The convolution operator L given by

$$L(F) = F * K \tag{1.7}$$

is a linear operator, i.e.,

$$L(aF + bG) = aL(F) + bL(G). \tag{1.8}$$

From a formal viewpoint the expression in (1.6) makes no sense if we do not prescribe to which class of functions F belongs. As it is not our intention to be very formal, we confine ourselves to pointing out that in practice one usually assumes that the underlying function space is $L^2(\mathbb{R}^2)$, the space of square integrable functions. Other choices are conceivable as well, however.

Obviously, L is invariant under translations:

$$L(F_h) = [L(F)]_h. \tag{1.9}$$

Here F_h is the translation of F along h, i.e.,

$$F_h(x) = F(x - h). \tag{1.10}$$

Fig. 1.4(b) shows the effect of a convolution by a Gaussian kernel.

In classical signal analysis linear operators, like convolution, form a major tool. This is not surprising when one realizes that acoustic signals combine linearly by superposition. This explains why one has attempted to apply linear methods, viz. Fourier transforms and convolution operators, to analyze such signals. However, visual signals do not combine linearly. Serra (1988, Introduction) writes:

> *Objects in space generally have three dimensions, which are reduced to two dimensions in a photograph or on the retina. In this projection, the luminances of the points located along a line oriented directly away from the viewer are not summed, because most physical objects are not translucent to light rays, in the way they would be to X-rays, but are opaque. Consequently, any object that is seen hides those that are placed beyond it with respect to the viewer: this self-evident property is a basic one. In fact it serves as a starting point for mathematical morphology, since, whenever we wish to describe quantitively phenomena in this domain, a set-theoretic approach must be used.*

The physical properties of visual signals can be modeled mathematically by a partial ordering such as set inclusion. Suppose one can represent a scene by a pencil drawing consisting of simple primitives like contours, line segments, blobs, shadows, etc. If object X is behind another object Y, the pencil drawing shows the contour of $X \setminus Y$, the set difference of X and Y. Similarly, the shadow of $X \cup Y$ is the union of their respective shadows. For a deeper discussion, the reader may refer to the introduction of Serra (1988), a paper by Ronse (1989b), or the book by Marr (1982).

These arguments make plausible that binary image operators should be based on set-theoretical notions. Ultimately, such an approach leads to a class of operators, called morphological operators; some typical examples have been discussed in the previous section. If one thinks about a mathematical analogy between sets and functions, one ends up automatically with a lattice-theoretical characterization of functions; such a characterization (which will be discussed in great detail later in this book) is induced by the partial ordering

$$F \leq G \qquad \text{if} \qquad F(x) \leq G(x), \ \text{for every } x \in \mathbb{R}^2.$$

The set of all functions $F : \mathbb{R}^2 \to \overline{\mathbb{R}}$, where $\overline{\mathbb{R}}$ is the extended real line $\mathbb{R} \cup \{-\infty, \infty\}$, is a complete lattice, henceforth denoted by $\mathrm{Fun}(\mathbb{R}^2)$. Note that

$\mathcal{P}(\mathbb{R}^2)$ endowed with the inclusion is also a complete lattice (see Chapter 2 for a formal definition). An arbitrary collection of functions F_i, $i \in I$, in $\mathrm{Fun}(\mathbb{R}^2)$ has a supremum, denoted by $\bigvee_{i\in I} F_i$, and an infimum, denoted by $\bigwedge_{i\in I} F_i$. If the index set I is finite, then the supremum coincides with the maximum and the infimum coincides with the minimum.

Let us return to the convolution operator $L(F) = F * K$ which, as said, is linear. Motivated by the observations above we replace the integral (a summation) by a supremum. Thus we get an operator Δ given by

$$\Delta(F)(x) = \bigvee_{y\in\mathbb{R}^2} F(x-y)K(y). \tag{1.11}$$

Instead of the supremum we can also take the infimum. For technical reasons (see (1.17) below) we replace $K(y)$ by $1/K(-y)$, and we arrive at the operator $\mathcal{E}$ given by

$$\mathcal{E}(F)(x) = \bigwedge_{y\in\mathbb{R}^2} F(x+y)/K(y). \tag{1.12}$$

Henceforth, we assume that $K(y) > 0$. In Figs. 1.4(c) and (d), respectively, the operators Δ and $\mathcal{E}$ are applied to a grey-scale image.

What kind of properties do the operators Δ and $\mathcal{E}$ have? Note, first of all, the dual character of both operators. In the sequel it will be explained that such duality relations are a universal phenomenon in mathematical morphology, known as the Duality Principle; see the next section for a general discussion.

It is evident that Δ and $\mathcal{E}$ are translation invariant, i.e.,

$$\Delta(F_h) = [\Delta(F)]_h \quad \text{and} \quad \mathcal{E}(F_h) = [\mathcal{E}(F)]_h. \tag{1.13}$$

It is also evident that Δ and $\mathcal{E}$ are not linear. This is due to the fact that the integral has been replaced by a supremum and an infimum, respectively. Instead, however, one gets the following properties:

$$\Delta(F \vee G) = \Delta(F) \vee \Delta(G), \tag{1.14}$$

$$\mathcal{E}(F \wedge G) = \mathcal{E}(F) \wedge \mathcal{E}(G), \tag{1.15}$$

for $F, G \in \mathrm{Fun}(\mathbb{R}^2)$. This implies in particular that $\Delta(F) \leq \Delta(F')$, and $\mathcal{E}(F) \leq \mathcal{E}(F')$ if $F \leq F'$. An operator with this property is called increasing. Note that the convolution operator L is also increasing if $K \geq 0$. In

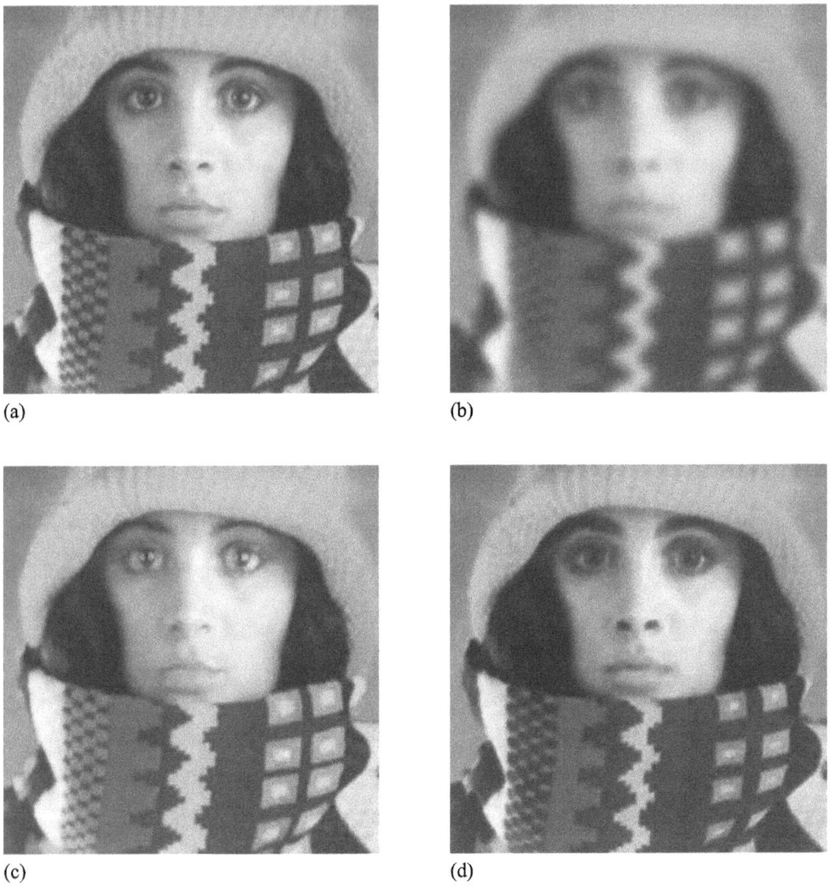

Figure 1.4 (a) Original image F, (b) convolution $F * K$, where K is a Gaussian kernel of size 3×3 pixels, (c) dilation $\Delta(F)$, and (d) erosion $\mathcal{E}(F)$, with K a 3×3 flat structuring function, i.e., K has the value 1 at a 3×3 neighbourhood of the origin.

fact, (1.14) holds for arbitrary suprema, and, dually, (1.15) holds for arbitrary infima. The operator Δ is called a *dilation*, and the operator $\mathcal{E}$ is called an *erosion*. In this particular example the operators $\Delta, \mathcal{E}$ also satisfy

$$\Delta(aF) = a\Delta(F) \qquad \text{and} \qquad \mathcal{E}(aF) = a\mathcal{E}(F), \tag{1.16}$$

if $a > 0$. If we substitute $-F$ (the negation of F) in Δ, we get

$$\Delta(-F) = -\bigwedge_{y \in \mathbb{R}^2} F(x-y)K(y).$$

This means that the mapping $F \mapsto -\Delta(-F)$ is an erosion. This observation can be expressed as follows: "the negative of a dilation is an erosion". Later we shall present a formal definition of a negation.

The operators $\Delta, \mathcal{E}$ will be treated in more detail in Section 5.7.2. There the function K will be called *multiplicative structuring function*. By far the most important relation between these two operators is

$$\Delta(F) \le G \iff F \le \mathcal{E}(G), \tag{1.17}$$

for arbitrary functions $F, G \in \mathrm{Fun}(\mathbb{R}^2)$. This relation, which can easily be verified, is called *adjunction relation*, and the pair $(\mathcal{E}, \Delta)$ is called an adjunction on $\mathrm{Fun}(\mathbb{R}^2)$.

In the literature one usually encounters a different adjunction of convolution type, namely,

$$\Delta(F)(x) = (F \oplus K)(x) = \bigvee_{y \in \mathbb{R}^2} [F(x-y) + K(y)], \tag{1.18}$$

$$\mathcal{E}(F)(x) = (F \ominus K)(x) = \bigwedge_{y \in \mathbb{R}^2} [F(x+y) - K(y)]. \tag{1.19}$$

These operators satisfy (1.13)–(1.15) as well as the adjunction relation (1.17). Instead of (1.16), however, they obey

$$\Delta(F+a) = \Delta(F) + a \qquad \text{and} \qquad \mathcal{E}(F+a) = \mathcal{E}(F) + a, \tag{1.20}$$

for $a \in \mathbb{R}$. Here the function K is called an *additive structuring function*. In Section 5.7.2 it is demonstrated that the adjunctions in (1.11)–(1.12) and (1.18)–(1.19) are related by a logarithmic mapping of the grey-values.

Finally, we point out how the convolution formulas (1.11)–(1.12) can be used to obtain the Minkowski addition $\oplus$ and subtraction $\ominus$ for sets introduced in the previous section. Identify a set $X \subseteq \mathbb{R}^2$ with its characteristic function $X(\cdot)$, given by $X(h) = 1$ if $h \in X$, and 0 elsewhere. Take $X, A \subseteq \mathbb{R}^2$, and use the convention that $0/0 = 1$. It is easy to show that $\Delta(X)(h) = \bigvee_{y \in \mathbb{R}^2} X(h-y)A(y)$ is the characteristic function of $X \oplus A$, and that $\mathcal{E}(X)(h) = \bigwedge_{y \in \mathbb{R}^2} X(h+y)/A(y)$ is the characteristic function of $X \ominus A$. This implies that the pair δ, ε given by $\delta(X) = X \oplus A$, $\varepsilon(X) = X \ominus A$ constitutes an adjunction on $\mathcal{P}(\mathbb{R}^2)$ in the sense that

$$X \oplus A \subseteq Y \iff X \subseteq Y \ominus A.$$

Of course, one can also give a direct proof of this relation.

1.3. A bird's eye view

Ask three persons with a different scientific background how they conceive an image, and it is very likely that they will give quite different answers. A physicist will tell you that an image is a graphical representation of a physical scene by some brightness function. For a computer scientist it is yet another data structure. And a statistician will regard it as a large amount of correlated data. We, however, shall adopt the mathematician's viewpoint, and consider an image as a mathematical model for the representation of certain objects. In the previous two sections two such models have been introduced. In Section 1.1 we have represented binary images as subsets of $\mathbb{R}^2$, meaning that $\mathcal{P}(\mathbb{R}^2)$ is the mathematical model. And in the previous section we have considered $\mathrm{Fun}(\mathbb{R}^2)$, the functions mapping $\mathbb{R}^2$ into $\overline{\mathbb{R}}$, as a model for grey-scale images. It is evident that the choice of the model is partially determined by the specific application. However, the kind of operations one wants to apply to the image has to be taken into account as well. If one wants to perform a convolution with a Gaussian function, it doesn't make sense to take $\mathcal{P}(\mathbb{R}^2)$ as the image space, for a set convolved with a Gaussian function is not a set. Moreover, convolution is a linear operation, and for that reason one should assume that the image space has a vector space structure. A useful choice in this case would be $L^2(\mathbb{R}^2)$, the space of square integrable functions.

Morphological operators, like dilations and erosions, are nonlinear. On the other hand, for a systematic study of such operators the intrinsic partial ordering (e.g., set inclusion for binary images) is enormously important. In the previous section it was noticed that dilation distributes over suprema, and dually, that erosion distributes over infima. This suggests that a complete lattice is the right structure for a formal theory of morphological operators. Indeed, the complete lattice framework forms the continuing thread through this book. In this section we discuss some basic notions from mathematical morphology; as much as possible, a formal discussion will be postponed until later.

An operator $\psi : \mathcal{L} \to \mathcal{M}$ between two complete lattices $\mathcal{L}, \mathcal{M}$ is said to be increasing if it preserves the ordering, i.e.,

$$X \le X' \Rightarrow \psi(X) \le \psi(X'), \text{ for all } X, X' \in \mathcal{L}.$$

If ψ distributes over suprema, i.e.,

$$\psi(\bigvee_{i\in I} X_i) = \bigvee_{i\in I} \psi(X_i),$$

where $X_i \in \mathcal{L}$ for i in some index set I, then ψ is called a dilation. And dually, if

$$\psi(\bigwedge_{i\in I} X_i) = \bigwedge_{i\in I} \psi(X_i),$$

that is, if ψ distributes over infima, then ψ is called an erosion. The pair (ε, δ), where ε is an operator from $\mathcal{L}$ to $\mathcal{M}$, and δ is an operator from $\mathcal{M}$ to $\mathcal{L}$, is called an adjunction if

$$\delta(Y) \leq X \iff Y \leq \varepsilon(X), \qquad X \in \mathcal{L},\ Y \in \mathcal{M}. \tag{1.21}$$

From a theoretical point of view, the adjunction is the most important notion in mathematical morphology, and it will show up at several places in this book. A comprehensive discussion can be found in Section 3.2 and Chapter 5. It will be shown that the operators ε and δ which constitute an adjunction are an erosion and a dilation, respectively. Moreover, to every dilation there corresponds a unique erosion such that both operators form an adjunction; dually, to every erosion can be associated a unique dilation such that the pair constitutes an adjunction. In a sense, the existence of adjunctions in mathematical morphology can be considered as a compensation for the absence of linearity.

In (1.21) we can choose $Y = \varepsilon(X)$ on the right-hand side; substitution on the left-hand side gives

$$\delta\varepsilon(X) \leq X.$$

The composite operator $\alpha = \delta\varepsilon$ has the following properties:

- α is increasing;
- α is anti-extensive, i.e., $\alpha(X) \leq X$;
- α is idempotent, i.e., $\alpha^2 = \alpha$.

An operator with these three properties is called an *opening*. Dually, the composite operator $\beta = \varepsilon\delta$ is increasing, extensive ($\beta(X) \geq X$), and idempotent, and in mathematical morphology such operators are called *closings*. Together with dilations, erosions, and adjunctions, openings and closings are the cornerstones of mathematical morphology, and we shall frequently meet them in this book.

Besides the existence of adjunctions, there is yet another compensation for the absence of linearity, namely, the Duality Principle. This principle is based on the trivial observation that given a complete lattice $\mathcal{L}$ with partial ordering $\leq$, one gets another complete lattice $\mathcal{L}'$ by simply reversing the ordering. In other words, $X \leq' Y$ iff $Y \leq X$. The lattice $\mathcal{L}'$ is called the opposite lattice. If, for instance, ψ is an erosion mapping one complete lattice $\mathcal{L}$ into another complete lattice $\mathcal{M}$, then the same ψ, but now considered as an operator from $\mathcal{L}'$ into $\mathcal{M}'$, is a dilation. The main implication of the Duality Principle is that to every class of morphological operators there corresponds an opposite class; the opposite of a dilation is an erosion, and the opposite of an opening is a closing. The Duality Principle associates with every statement an opposite statement. For example, Theorem 3.28(a) states that an arbitrary supremum of openings is an opening. The opposite of this statement says that an arbitrary infimum of closings is a closing. In most (but not all) cases the opposite of a statement is different; in such cases the opposite of the statement in (a) will be given in (a′). For example, that an infimum of closings is a closing is the content of Theorem 3.28(a′). In such cases a proof of (a) in combination with the Duality Principle yields automatically a proof of (a′).

A second, completely different, type of duality may emerge in the theory of morphological operators. This duality is based on the observation that images often have a unique "negative". We refer to the operator which maps an image onto its negative (if it exists) as a *negation*. It depends exclusively on the image space whether a negation exists or not. On $\mathcal{P}(\mathbb{R}^2)$, for example, the operator $X \mapsto X^c$, where X^c is the ordinary set complement, is a negation. On $\mathrm{Fun}(\mathbb{R}^2)$ the operator $F \mapsto -F$ is a negation. Generally speaking, an operator ν on a complete lattice $\mathcal{L}$ is a negation if it is a bijection, reverses the ordering ($X \leq Y \Rightarrow \nu(Y) \leq \nu(X)$), and satisfies $\nu^2 = \mathrm{id}$, the identity operator. If $\mathcal{L}$ is a Boolean lattice, then the operator mapping an element to its complement is a negation. Hereafter we write $X^* = \nu(X)$ if ν is a negation. Now, if ψ is an operator mapping $\mathcal{L}$ into itself, and if $X \mapsto X^*$ is a negation, then the operator ψ^* given by

$$\psi^*(X) = (\psi(X^*))^*, \qquad X \in \mathcal{L},$$

is called the *negative* of ψ. It will become clear in the sequel that this procedure transforms dilations into erosions, openings into closings, etc. Unfortunately, there are quite a few image spaces without negations. A typical example is the complete lattice comprising all closed subsets of $\mathbb{R}^2$.

The discussion so far in this section suggests that mathematical morphology is an algebraic theory of image spaces and image operators, and, indeed, this impression will be affirmed many times hereafter. But how can we reconcile this with the main observation of Section 1.1, that morphology is a geometrical approach in image analysis? It is clear that merely the assumption that the image space $\mathcal{L}$ is a complete lattice does not suffice. Recall that in Section 1.1, $\mathcal{L} = \mathcal{P}(\mathbb{R}^2)$; the operators introduced there utilize the topological and geometrical properties of the Euclidean space $\mathbb{R}^2$. To be specific, γ_h uses connectivity (a topological notion) and the erosion $X \mapsto X \ominus rB$ uses the vector space structure. Finally, to establish that $\alpha_r\alpha_s = \alpha_s\alpha_r = \alpha_s$ if $s \geq r$, one needs that the unit disk is convex. The additional geometrical structure of $\mathbb{R}^2$ enables the definition of a large class of morphological operators (e.g., dilations $X \mapsto X \oplus A$) which satisfy, besides the usual algebraic properties (distributivity over unions in the case of dilations), geometrical invariance properties (translation invariance) also.

In Chapters 5 and 6 the image space $\mathcal{L}$ will be endowed with additional geometrical structure; this is achieved by assuming that there exists an automorphism group $\mathbb{T}$ on $\mathcal{L}$ (in Section 1.1, $\mathbb{T}$ is the translation group). We shall give a rather complete characterization of dilations, erosions, openings, and closings which are invariant under $\mathbb{T}$. This is useful for instance if one is interested in operators invariant under, e.g., rotations and translations. Furthermore, Chapter 9 concentrates on the role of convexity and distance (including geodesic distance) in morphology. In Section 1.1 we have seen that convexity of the disk B implies that the openings α_r satisfy the semigroup property $\alpha_r\alpha_s = \alpha_s\alpha_r = \alpha_s$ if $s \geq r$. This property provides a basis for a formal definition of a size distribution. The importance of convexity in mathematical morphology can hardly be overestimated. If the reader is not convinced of this after reading Chapter 9, he should refer to Matheron's monumental work (Matheron, 1975), in particular to Chapters 4–6 where Matheron discusses the relations between random closed sets and integral geometry.

Topology, which is the main theme of Chapter 7, plays a role in mathematical morphology at two levels. First of all, topology may enter at the modeling stage. For instance, there are circumstances in which the closed sets are the most appropriate model for binary images. But on a higher level there may be a need for an image space which is itself a topological space; this allows the definition of (semi-) continuous operators. In Chapter 8, which concerns discretization problems, a topology on the image space (in casu: the hit-or-miss topology on the closed sets) will be exploited to capture the notion of approximation; furthermore, it will be shown there that (semi-) continuity properties of morphological operators are indispensable for their discretizability.

Alternatively, it is possible to introduce a notion of convergence on arbitrary complete lattices which is based on the partial ordering. This leads in a natural way to yet another class of (semi-) continuous operators. In Chapter 13 it is shown that this concept of continuity shows up naturally in a method for the construction of morphological filters by iteration. Here the term *morphological filter* refers to an increasing operator ψ which is idempotent, i.e., $\psi^2 = \psi$. It will be explained in Chapter 12 that idempotence is a sensible requirement for operators which are utilized for image cleaning. We have already met two classes of filters, the openings and the closings. If α is an opening and β is a closing, then $\alpha\beta$ as well as $\beta\alpha$ are filters. For $\alpha\beta\alpha\beta \le \alpha\beta^3 = \alpha\beta$ and $\alpha\beta\alpha\beta \ge \alpha^3\beta = \alpha\beta$, thus $(\alpha\beta)^2 = \alpha\beta$. These two filters are called *alternating sequential filters* and will be discussed in Chapter 12, along with many other examples. Again the complete lattice framework is eminently suited for a general theory of morphological filters.

In the first section we have discussed some basic operators for sets, whereas in the second section we were mainly concerned with functions. In the literature one refers to these cases by means of the terminology *binary morphology* and *grey-scale morphology*, respectively. Although the two theories are substantially different, they fit both well within the complete lattice framework. Chapter 10 presents a very general theory for the representation of functions; this theory applies to ordinary grey-scale images and (partially) to colour images. Furthermore, it applies to ordinary functions as well as to u.s.c. functions (here, "u.s.c." is an abbreviation of "upper semi-continuous"). Using this representation, which is essentially based on thresholding, we show in Chapter 11 how to extend binary image operators to grey-scale image operators.

1.4. Historical remarks

The last decade has shown an outburst of publications in the area of mathematical morphology. Likely, this is caused by the appearance in 1982 of an inspiring and highly original book written by Jean Serra, entitled *Image Analysis and Mathematical Morphology* (Serra, 1982). Usually, this work is considered as a first systematic treatment of mathematical morphology as a new approach in image analysis. The main theoretical foundations, however, were laid some years earlier in a book by Matheron entitled *Random Sets and Integral Geometry* (Matheron, 1975). Matheron as well as his colleague Serra, both researchers at the Ecole Nationale Supérieure des Mines de Paris at Fontainebleau, are very much inspired by experimental techniques of texture analysis. Although the importance of Matheron's work was recognized immediately by workers in the field of stochastic geometry, it took considerable time until his ideas found general acceptance with members of the image processing community.

Already in the early sixties researchers like Matheron and Serra realized that the study of geometrical data required radically new ideas. Their efforts ultimately led to a theory which they called "mathematical morphology", a term which shows up for the first time in two papers by Haas, Matheron, and Serra[1] (Haas et al., 1967a, 1967b). To a large extent, the breakthrough of mathematical morphology in the U.S. is due to the work of Dougherty and Giardina (Dougherty & Giardina, 1987b, Chapter 3; Giardina & Dougherty, 1988), Haralick and co-workers (Haralick et al., 1987), Maragos (1985, 1986, 1987, 1989a, 1989b) Maragos and Schafer (1987a, 1987b), and Sternberg (1982, 1986).

As mentioned, integral geometry forms an essential ingredient in Matheron's work (Matheron, 1975). In 1957 Hadwiger published an interesting book *Vorlesungen über Inhalt, Oberfläche und Isoperimetrie* (Hadwiger, 1957). Many notions which have become the major constituent parts of morphology can be traced back to Hadwiger's work. This holds in particular for Minkowski addition, an operation which originates from the work of Minkowski (1903), and Minkowski subtraction, an operation first introduced by Hadwiger (1950). Both operations play a crucial role in Hadwiger's work. All elementary properties of these two operations such as $(X \ominus A) \oplus A \subseteq X \subseteq (X \oplus A) \ominus A$, have been established by Hadwiger (1957).

Above, we have emphasized the importance of adjunctions with respect to the theory of mathematical morphology; in fact this notion will show up in every chapter of this book. Adjunctions are not an exclusive tool of the "theoretical morphologist", however, but appear as a natural concept in several mathematical disciplines. A somewhat different manifestation of this concept is the *Galois connection*, a notion originating from Galois theory; see Birkhoff (1967). A detailed description of the relation between both concepts is given in Heijmans and Ronse (1990, pp. 268ff). Adjunctions also play a role (again under a different name) in residuation theory (Blyth & Janowitz, 1972); refer also to Section 3.7.

[1] In fact, the name *mathematical morphology* was coined by these authors in 1967 in a bar in Nancy. Note that the word *morphology* stems from the Greek words $\mu o \rho \phi \eta$ and $\lambda o \gamma o \varsigma$, which mean study of forms.

CHAPTER TWO

Complete lattices

Henk J.A.M. Heijmans
Formerly Centre for Mathematics and Computer Science, Amsterdam, Netherlands

Contents

This chapter introduces several basic notions from the theory of complete lattices which are used in the sequel of the book. As much as possible new concepts will be illustrated by means of concrete examples. In some cases these examples anticipate issues, such as topology and convexity, to be considered in much greater detail in subsequent chapters. The first section presents basic definitions concerning (complete) lattices. Section 2.2 considers Boolean lattices; these play an important role in the theory of mathematical morphology. The prototypical example of a complete Boolean lattice is the power set $\mathcal{P}(E)$, where E is an arbitrary nonempty set. Another, more unusual, example of a complete Boolean lattice is the family of regular closed subsets of a topological space; this lattice is examined in Section 2.3. Finally, Section 2.4 recalls some basic facts from the theory of Boolean functions.

2.1. Basic concepts

This first section summarizes some basic definitions and results from the theory of complete lattices. For a comprehensive discussion, refer to the monograph by Birkhoff (1967).

2.1 Definition. Given a nonempty set $\mathcal{L}$, a binary relation $\leq$ on $\mathcal{L}$ is called a *partial ordering* if the following properties hold:

(O1) $X \leq X$; (reflexivity)

(O2) $X \leq Y$ and $Y \leq X$ implies $X = Y$; (anti-symmetry)

Advances in Imaging and Electron Physics, Volume 216
ISSN 1076-5670
https://doi.org/10.1016/bs.aiep.2020.07.002

(O3) $X \le Y$ and $Y \le Z$ implies $X \le Z$; (transitivity)
for every $X, Y, Z \in \mathcal{L}$. A set $\mathcal{L}$ which carries a partial ordering $\le$ is called a *partially ordered set*, or briefly *poset*, and is denoted by $(\mathcal{L}, \le)$. We say that the poset $(\mathcal{L}, \le)$ is *totally ordered* if
(O4) $X \le Y$ or $Y \le X$, for every pair $X, Y \in \mathcal{L}$.
A totally ordered poset is called a *chain*.

Instead of $X \le Y$ we also write $Y \ge X$, and we say that "X is smaller than or equal to Y." When there is no danger of confusion as to which partial ordering is meant, we write $\mathcal{L}$ instead of $(\mathcal{L}, \le)$. By $X < Y$ we mean that $X \le Y$ and $X \neq Y$. If $A \le B$, then we denote by $[A, B]$ the collection of all $X \in \mathcal{L}$ with $A \le X \le B$.

2.2 Definition. A subset $\mathcal{K}$ of a poset $\mathcal{L}$ is called an *upper set* if $A \in \mathcal{K}$ and $B \ge A$ implies that $B \in \mathcal{K}$. It is called a *lower set* if $A \in \mathcal{K}$ and $B \le A$ implies that $B \in \mathcal{K}$.

2.3 Examples.

(a) Let $\mathbb{R}$ be the set of all real numbers, and let $x \le y$ have its usual meaning; then $\mathbb{R}$ is a chain. The set $\mathbb{R}^d$ with the relation "$(x_1, x_2, \dots, x_d) \le (y_1, y_2, \dots, y_d)$ if $x_i \le y_i$ for all i" is a poset. It is a chain if and only if $d = 1$.

(b) Let $\mathbb{N}$ be the set of all nonnegative integers, and put $m \le n$ if m divides n. Then $\mathbb{N}$ is a poset.

(c) Given a set E, the power set $\mathcal{P}(E)$ comprising all subsets of E becomes a poset under the inclusion relation, that is, "$X \le Y$ if and only if $X \subseteq Y$." The empty set is denoted by $\varnothing$.

(d) Let G be a group. The set of all subgroups of G ordered by "$H \le K$ if H is a subgroup of K" is a poset.

If $\le$ is a partial ordering on $\mathcal{L}$, then the binary relation $\le'$ given by "$X \le' Y$ iff $X \ge Y$" also defines a partial ordering, called the *dual partial ordering*.

2.4 Duality Principle. *If $(\mathcal{L}, \le)$ is a poset, then $(\mathcal{L}, \le')$ is a poset too, called the dual poset. To every definition, property, statement, etc., referring to $(\mathcal{L}, \le)$ there corresponds a dual one referring to $(\mathcal{L}, \le')$, interchanging the role of $\le$ and $\le'$.*

Note that the second dual partial ordering $(\le')'$ coincides with $\le$. The Duality Principle, seemingly a trivial and rather useless observation, plays a prominent role in this book. Its major implication is that every notion and statement concerning posets has a dual counterpart. Throughout this book this fact is exploited whenever possible, often without explicit mention. If some part of a definition or statement has a dual counterpart in the sense of the Duality Principle, we will express this in our formulation. For instance, the dual of statement (b) in a given proposition is denoted by (b′). In such cases we need only give a demonstration of (b) (or (b′)), as the other statement follows from the Duality Principle.

If there is no danger of confusion concerning the partial ordering, we write $\mathcal{L}'$ instead of $(\mathcal{L}, \preceq')$.

Given a poset $\mathcal{L}$ and a subset of $\mathcal{K}$ of $\mathcal{L}$, an element $A \in \mathcal{K}$ is called a *least element* of $\mathcal{K}$ if $A \preceq X$ for all $X \in \mathcal{K}$. An element $B \in \mathcal{K}$ is called a *greatest element* of $\mathcal{K}$ if $B \succeq X$ for all $X \in \mathcal{K}$. A subset $\mathcal{K}$ can have at most one least and one greatest element. In fact, if both A and A' are least elements of $\mathcal{K}$, then $A \preceq A'$ and $A' \preceq A$, and we conclude from property (O2) that $A = A'$. An element $A \in \mathcal{L}$ is called a *lower bound* of $\mathcal{K}$ if $A \preceq X$ for every $X \in \mathcal{K}$; note that A need not lie in $\mathcal{K}$. The least element of $\mathcal{K}$, if it exists, is a lower bound of $\mathcal{K}$. If the set of lower bounds of $\mathcal{K}$, which is a subset of $\mathcal{L}$, contains a greatest element A_0, then this is called the *greatest lower bound*, or *infimum*, of $\mathcal{K}$. Note that A_0 satisfies

(i) $A_0 \preceq X$ for $X \in \mathcal{K}$ (A_0 is a lower bound of $\mathcal{K}$);
(ii) $A \preceq A_0$ for every other lower bound A of $\mathcal{K}$.

The notions *upper bound* and *least upper bound*, also called *supremum*, are defined analogously. The infimum (resp. supremum) of a subset $\mathcal{K}$, if it exists, is unique, and is denoted by $\inf \mathcal{K}$ or $\bigwedge \mathcal{K}$ (resp. $\sup \mathcal{K}$ or $\bigvee \mathcal{K}$). If $\mathcal{K}$ contains only finitely many elements $X_1, X_2, \ldots, X_n$, then we write $X_1 \wedge X_2 \wedge \cdots \wedge X_n$ instead of $\bigwedge\{X_1, X_2, \ldots, X_n\}$, and $X_1 \vee X_2 \vee \cdots \vee X_n$ instead of $\bigvee\{X_1, X_2, \ldots, X_n\}$. Furthermore, if $X_i \in \mathcal{L}$ for all i in some index set I, then we write $\bigwedge_{i \in I} X_i$ instead of $\bigwedge\{X_i \mid i \in I\}$, and $\bigvee_{i \in I} X_i$ instead of $\bigvee\{X_i \mid i \in I\}$.

In fact, infimum and supremum are dual notions in the sense of the Duality Principle. If $\mathcal{K}$ is a subset of the poset $\mathcal{L}$ with infimum A, then A is the supremum of $\mathcal{K}$ with respect to the dual poset $\mathcal{L}'$; we denote this as $\bigwedge \mathcal{K} = \bigvee' \mathcal{K}$.

2.5 Definition. A poset $\mathcal{L}$ is called a *lattice* if every finite subset of $\mathcal{L}$ has an infimum and a supremum. A lattice is called *complete* if every subset of $\mathcal{L}$ has an infimum and a supremum.

Every chain is a lattice, for every finite set of elements of a chain can be arranged in increasing order, and therefore contains a least and a greatest element. Not every lattice (or even chain) is complete; for instance, the open interval $(0,1)$ with the usual partial ordering is a chain, but it is not complete: the set $\{1/2, 1/3, 1/4, \ldots\}$ does not have a lower bound.

By definition, every complete lattice $\mathcal{L}$ must possess a least element O and a greatest element I, called the *universal bounds* of $\mathcal{L}$. The empty set seems to play an exceptional role. A moment of reflection, however, makes clear that in a complete lattice every element is both an upper bound and a lower bound of the empty set. Thus the least upper bound of $\varnothing$ is O, and the greatest lower bound of $\varnothing$ is I. In other words,

$$O = \bigwedge \mathcal{L} = \bigvee \varnothing, \qquad I = \bigvee \mathcal{L} = \bigwedge \varnothing. \tag{2.1}$$

2.6 Definition. Let $\mathcal{L}, \mathcal{M}$ be lattices. The mapping $\psi : \mathcal{L} \to \mathcal{M}$ is called a *lattice isomorphism* if ψ is a bijection (one–one and onto), and if ψ as well as its inverse ψ^{-1} are *order-preserving*, that is,

$$X \le Y \qquad \text{if and only if} \qquad \psi(X) \le \psi(Y),$$

for $X, Y \in \mathcal{L}$.

A lattice isomorphism preserves infima and suprema, that is,

$$\psi(X \wedge Y) = \psi(X) \wedge \psi(Y),$$
$$\psi(X \vee Y) = \psi(X) \vee \psi(Y),$$

for $X, Y \in \mathcal{L}$. We prove the first relation. Since $X \wedge Y \le X$, it follows that $\psi(X \wedge Y) \le \psi(X)$. Analogously, $\psi(X \wedge Y) \le \psi(Y)$. This gives $\psi(X \wedge Y) \le \psi(X) \wedge \psi(Y)$. Assume on the other hand that $A \le \psi(X) \wedge \psi(Y)$; then $A \le \psi(X), \psi(Y)$, and hence $\psi^{-1}(A) \le X, Y$. This gives $\psi^{-1}(A) \le X \wedge Y$; hence $A \le \psi(X \wedge Y)$. Therefore, $\psi(X \wedge Y)$ is the greatest lower bound of $\psi(X), \psi(Y)$; in other words, $\psi(X \wedge Y) = \psi(X) \wedge \psi(Y)$.

If $\mathcal{L} = \mathcal{M}$, then ψ is called an *automorphism*. The lattices $\mathcal{L}$ and $\mathcal{M}$ are called *isomorphic* if there exists an isomorphism between them.

We return to the settings in Example 2.3.

2.7 Examples.

(a) The set of real numbers $\mathbb{R}$ is a lattice; even more, it is a chain. It is not complete for it does not contain a least and greatest element. To make it complete one has to add $-\infty$ (as the least element), and $+\infty$ (as the greatest element). Henceforth, the set $\mathbb{R} \cup \{-\infty, +\infty\}$ will be denoted by $\overline{\mathbb{R}}$. Analogously, $\overline{\mathbb{Z}} = \mathbb{Z} \cup \{-\infty, +\infty\}$ is a complete chain.

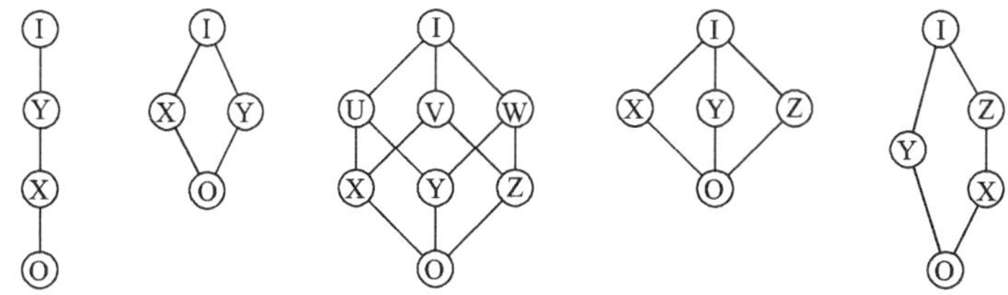

(a) (b) (c) (d) (e)

Figure 2.1 Hasse diagrams.

The sets $\overline{\mathbb{R}}_+$ and $\overline{\mathbb{Z}}_+$, comprising, respectively, the positive real numbers and positive integers including ∞, are complete chains, too.

(b) The set of nonnegative integers ordered by "$m \le n$ if m divides n" is a complete lattice with least element 1 and greatest element 0.

(c) The power set $\mathcal{P}(E)$ with the inclusion ordering is a complete lattice. The infimum is given by set intersection, and the supremum by set union. The least element is $\varnothing$, and the greatest element E.

(d) The set of all subgroups of a group G ordered by "$H \le K$ if H is a subgroup of K" is a complete lattice with $\bigwedge_{i\in I} H_i = \bigcap_{i\in I} H_i$, and $\bigvee_{i\in I} H_i$ the smallest subgroup in G containing $\bigcup_{i\in I} H_i$.

2.8 Example. (Hasse diagrams)
A finite lattice $\mathcal{L}$ can be represented $\mathcal{L}$ graphically as follows. If $A < B$, and there is no element $X \in \mathcal{L}$ with $A < X < B$, then we place B higher than A, and draw a line segment connecting the two elements. The resulting diagram is called *Hasse diagram*. Some examples are depicted in Fig. 2.1.

In (e), for example, one has $Z \ge X$, $Y \vee Z = Y \vee X = I$, $Y \wedge X = Y \wedge Z = O$. The lattice in (a) is a chain, e.g., $\{0, 1, 2, 3\}$, with the usual ordering; (b) and (c) represent the power set of a set consisting of two and three elements, respectively.

2.9 Example. (Closed and open sets)
Denote by $\mathcal{F}(\mathbb{R}^d)$ the closed subsets of $\mathbb{R}^d$; here $\mathbb{R}^d$ may be replaced by any other topological space E. Those readers not acquainted with the basic aspects of topology may consult Chapter 7, where such spaces are considered in more detail. It is evident that inclusion defines a partial ordering on $\mathcal{F}(\mathbb{R}^d)$. Moreover, with this partial ordering, $\mathcal{F}(\mathbb{R}^d)$ becomes a complete lattice. The infimum and supremum of a family X_i, $i \in I$, in $\mathcal{F}(\mathbb{R}^d)$ are,

respectively, given by

$$\bigwedge_{i\in I} X_i = \bigcap_{i\in I} X_i,$$

$$\bigvee_{i\in I} X_i = \overline{\bigcup_{i\in I} X_i}.$$

Here $\overline{X}$ denotes the closure of X. As the intersection of closed sets is always closed, it is not necessary to take the closure in the expression for the infimum.

The open subsets of $\mathbb{R}^d$, denoted by $\mathcal{G}(\mathbb{R}^d)$, are also a complete lattice under the inclusion ordering. In this case the infimum and supremum are given by

$$\bigwedge_{i\in I} X_i = (\bigcap_{i\in I} X_i)^\circ,$$

$$\bigvee_{i\in I} X_i = \bigcup_{i\in I} X_i,$$

respectively; here X° denotes the interior of X, the largest open set contained in X. In fact, it is fairly easy to show that $(\mathcal{G}(\mathbb{R}^d), \subseteq)$ is isomorphic to the dual lattice of $(\mathcal{F}(\mathbb{R}^d), \subseteq)$.

2.10 Example. (Functions)
Given a complete lattice $\mathcal{T}$ and an arbitrary nonempty set E, define the *power lattice* $\mathcal{L} = \mathcal{T}^E$ as the space of all functions mapping E into $\mathcal{T}$. The relation $\le$ given by "$F \le G$ if $F(x) \le G(x)$ for every $x \in E$" defines a partial ordering on $\mathcal{L}$. It is easy to show that $\mathcal{L}$ is a complete lattice with the infimum $\bigwedge_{i\in I} F_i$ and supremum $\bigvee_{i\in I} F_i$ of a collection F_i in $\mathcal{L}$, respectively, given by

$$(\bigwedge_{i\in I} F_i)(x) = \bigwedge_{i\in I}(F_i(x)), \qquad x \in E,$$

$$(\bigvee_{i\in I} F_i)(x) = \bigvee_{i\in I}(F_i(x)), \qquad x \in E.$$

The lattice $\mathcal{L} = \mathcal{T}^E$ inherits many properties from the lattice $\mathcal{T}$. Examples can be found in Examples 2.18(c) and 2.30(b). It is obvious that $\{0, 1\}^E$ is isomorphic to the lattice $\mathcal{P}(E)$. The lattice $\mathcal{T}^E$, henceforth denoted by $\mathrm{Fun}(E, \mathcal{T})$ (the functions from E to $\mathcal{T}$), emerges at many instances in this book; see in particular Chapters 10 and 11. This is due to the fact that functions constitute a mathematical model for grey-scale images.

2.11 Example. (Products of lattices)
Given the complete lattices $(\mathcal{L}_1, \leq_1), (\mathcal{L}_2, \leq_2), \ldots, (\mathcal{L}_d, \leq_d)$, define $\mathcal{M} = \mathcal{L}_1 \times \mathcal{L}_2 \times \cdots \times \mathcal{L}_d$; that is, $\mathcal{M}$ contains all d-tuples $(X_1, X_2, \ldots, X_d)$ where $X_k \in \mathcal{L}_k$, $k = 1, 2, \ldots, d$. Furthermore, define the relation $\leq$ on $\mathcal{M}$ by

$$(X_1, X_2, \ldots, X_d) \leq (Y_1, Y_2, \ldots, Y_d), \text{ if } X_k \leq_k Y_k \text{ for every } k = 1, 2, \ldots, d.$$

Hereafter, we refer to this ordering as the *product ordering*. Obviously, $(\mathcal{M}, \leq)$ is a complete lattice. If $\mathcal{L}_1 = \mathcal{L}_2 = \cdots = \mathcal{L}_d = \mathcal{L}$, then we write $\mathcal{M} = \mathcal{L}^d$. Note that $\mathcal{L}^d$ is isomorphic to $\mathrm{Fun}(E, \mathcal{L})$, where E is an arbitrary set comprising d elements.

2.12 Proposition. *Given a poset $\mathcal{L}$, the following three statements are equivalent:*

(i) *$\mathcal{L}$ is a complete lattice;*
(ii) *$\mathcal{L}$ has a least element O and every subset of $\mathcal{L}$ has a supremum;*
(iii) *$\mathcal{L}$ has a greatest element I and every subset of $\mathcal{L}$ has an infimum.*

Proof. It is obvious that (i) implies (ii) and (iii). We show that (ii) implies (i). The other implication follows from the Duality Principle. Assume that (ii) holds, and let $\mathcal{K} \subseteq \mathcal{L}$. Denote the set of lower bounds of $\mathcal{K}$ by $\mathcal{J}$. Then $\mathcal{J} \neq \varnothing$ since $O \in \mathcal{J}$. Let $A = \sup \mathcal{J}$; we show that A is the infimum of $\mathcal{K}$. Note first that every $X \in \mathcal{K}$ is an upper bound of $\mathcal{J}$; since A is the least upper bound, $A \leq X$. Let A' be a lower bound of $\mathcal{K}$; then $A' \in \mathcal{J}$, and so $A' \leq A$. This proves that $A = \inf \mathcal{K}$. □

The next result lists some basic properties of the infimum and supremum.

2.13 Proposition. *Let $\mathcal{L}$ be a poset and $X, Y, Z \in \mathcal{L}$.*

(a) $X \wedge X = X \vee X = X$; (idempotence)
(b) $X \wedge Y = Y \wedge X$, $X \vee Y = Y \vee X$; (commutativity)
(c) $X \wedge (Y \wedge Z) = (X \wedge Y) \wedge Z$, $X \vee (Y \vee Z) = (X \vee Y) \vee Z$; (associativity)
(d) $X \wedge (X \vee Y) = X \vee (X \wedge Y) = X$. (absorption)

The proof is easy and is left as an exercise to the reader. In fact, one can use the identities (a)–(d) as an alternative definition of a lattice. To be precise, a set $\mathcal{L}$ with two binary operations $\wedge$ and $\vee$ is a lattice if and only if it satisfies these identities. To prove the if-statement one defines a partial ordering on $\mathcal{L}$ by putting $X \leq Y$ if $X \wedge Y = X$.

2.14 Definition. A lattice $\mathcal{L}$ is *distributive* if

$$X \wedge (Y \vee Z) = (X \wedge Y) \vee (X \wedge Z), \tag{2.2}$$

$$X \vee (Y \wedge Z) = (X \vee Y) \wedge (X \vee Z), \tag{2.3}$$

for all $X, Y, Z \in \mathcal{L}$.

The lattices in Examples 2.3(a)–(c), 2.9, and 2.10 are distributive. For Example 2.3(b) distributivity is not completely trivial. A proof of this fact can be found in Birkhoff (1967, pp. 12). The lattice in Example 2.3(d) is not distributive in general.

2.15 Definition. A lattice $\mathcal{L}$ is called *modular* if it satisfies

$$X \vee (Y \wedge Z) = (X \vee Y) \wedge Z \qquad \text{if } X \leq Z, \tag{2.4}$$

for all $X, Y, Z \in \mathcal{L}$.

It is obvious that a distributive lattice is modular. The converse is not true, however. The lattices represented by the Hasse diagrams in Figs. 2.1(a)–(c) are distributive. Diagram (d) shows a modular lattice which is not distributive; e.g., $X \vee (Y \wedge Z) = X \neq I = (X \vee Y) \wedge (X \vee Z)$. Diagram (e) shows a lattice which is not modular, for $X \vee (Y \wedge Z) = X \neq Z = (X \vee Y) \wedge Z$.

We point out here that the Duality Principle is no longer valid if one makes an assumption on $\mathcal{L}$ which is not self-dual. For instance, the assumption that $\mathcal{L}$ has a least element is not self-dual, since it does not imply that $\mathcal{L}$ also has a greatest element. The reader may verify that both distributivity and modularity are self-dual constraints.

2.16 Examples. (Convex sets)

(a) A set $X \subseteq \mathbb{R}^2$ is *convex* if for every two points $x, y \in X$ the entire straight line segment between x and y is contained in X. Note that this definition includes the empty set. A comprehensive treatment of convexity is given in Section 9.1. Denote the convex subsets of $\mathbb{R}^2$ by $\mathcal{C}(\mathbb{R}^2)$. With the inclusion as partial ordering this set becomes a complete lattice. The infimum is the ordinary set intersection; here we use that an arbitrary intersection of convex sets is convex, too. This is no longer true for the union. Define the *convex hull* $\text{co}(X)$ of a set $X \subseteq \mathbb{R}^2$ as the smallest convex set that contains X. It is evident that $\text{co}(X)$ is the intersection of all convex sets which

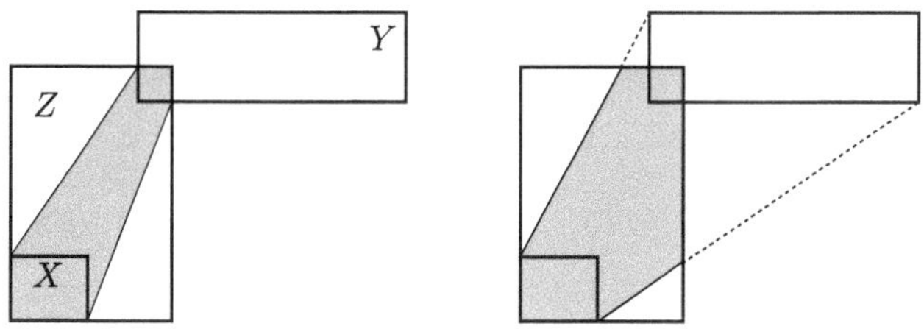

Figure 2.2 The lattice $(\mathcal{C}(\mathbb{R}^2), \subseteq)$ is not modular. The shaded region at the left represents $X \vee (Y \wedge Z)$; the shaded region at the right represents $(X \vee Y) \wedge Z$.

contain X. Now the supremum of a family $X_i \in \mathcal{C}(\mathbb{R}^2)$ is given by

$$\bigvee_{i \in I} X_i = \mathrm{co}(\bigcup_{i \in I} X_i).$$

The example depicted in Fig. 2.2 shows that $(\mathcal{C}(\mathbb{R}^2), \subseteq)$ is not modular; in particular, it is not distributive.

(b) The *closed convex sets*, denoted by $\mathcal{C}_{\mathcal{F}}(\mathbb{R}^2)$, also form a complete lattice under inclusion. The infimum is again the ordinary set intersection, but the supremum is given by

$$\bigvee_{i \in I} X_i = \overline{\mathrm{co}}(\bigcup_{i \in I} X_i).$$

Here $\overline{\mathrm{co}}(X)$ is the *closed convex hull* of X, the smallest closed convex set that contains X.

We say that $\mathcal{L}$ satisfies the *infinite distributive laws* if

$$A \wedge \bigvee_{i \in I} X_i = \bigvee_{i \in I} (A \wedge X_i), \tag{2.5}$$

$$A \vee \bigwedge_{i \in I} X_i = \bigwedge_{i \in I} (A \vee X_i), \tag{2.6}$$

for an arbitrary index set I. We call (2.5) and (2.6) the *infinite supremum distributive law* and the *infinite infimum distributive law*, respectively. One can show by induction that, in a distributive lattice, these laws hold for every *finite* index set I. But in general, they do not hold for arbitrary index sets. For example, (2.5) is not valid in the complete lattice $\mathcal{F}(\mathbb{R})$ consisting of all closed subsets of $\mathbb{R}$ (which is a distributive lattice). In fact, take $A = \{0\}$ and $X_i = [1/i, 1]$ for $i = 1, 2, \ldots$; then $\bigvee_{i \geq 1} X_i = [0, 1]$, and so the left-hand side of (2.5) yields $\{0\}$, whereas the right-hand side yields the empty set. It is easy to show that (2.6) is valid for $\mathcal{F}(\mathbb{R})$. Theorem 2.32 will show that the infinite distributive laws are valid in every complete Boolean lattice.

Let I be a nonempty index set, and let, for every $i \in I$, J_i be a nonempty index set. Let Φ be the collection of functions with domain I and with $\phi(i) \in J_i$ for $i \in I$. Consider the *extended distributive laws*:

$$\bigwedge_{i\in I}\Big(\bigvee_{j\in J_i} X_{i,j}\Big) = \bigvee_{\phi\in\Phi}\Big(\bigwedge_{i\in I} X_{i,\phi(i)}\Big), \tag{2.7}$$

$$\bigvee_{i\in I}\Big(\bigwedge_{j\in J_i} X_{i,j}\Big) = \bigwedge_{\phi\in\Phi}\Big(\bigvee_{i\in I} X_{i,\phi(i)}\Big). \tag{2.8}$$

2.17 Definition. A *completely distributive lattice* is a complete lattice for which the extended distributive laws (2.7) and (2.8) hold.

In the next section we shall see that, in contrast to the infinite distributive laws, the extended distributive laws do *not* hold in every complete Boolean lattice. This suggests that they are stronger. That this is indeed the case can be seen as follows. Substitute $J_i = \{0, 1\}$, $X_{i,0} = A$, and $X_{i,1} = X_i$ in (2.7). In that case $\Phi = \{0, 1\}^I$, the functions mapping I into $\{0, 1\}$. The only two functions $\phi \in \Phi$ that contribute to the right-hand side are $\phi \equiv 0$ and $\phi \equiv 1$. In the first case we get A, whereas in the second case we get $\bigvee_{i\in I} X_i$. Now (2.5) follows immediately.

2.18 Examples.

(a) Every complete chain is a completely distributive lattice.

(b) $\mathcal{P}(E)$ is a completely distributive lattice for every set E. In fact, Theorem 2.36 states that every complete Boolean lattice that is completely distributive is isomorphic to a power set.

(c) Consider the function lattice $\mathrm{Fun}(E, \mathcal{T})$ introduced in Example 2.10, where $\mathcal{T}$ is a complete lattice and E an arbitrary nonempty set. If $\mathcal{T}$ is (completely) distributive, then $\mathrm{Fun}(E, \mathcal{T})$ is (completely) distributive as well. Similar conclusions hold with regard to modularity and the infinite distributive laws.

We introduce the notions of a sublattice and an underlattice by means of two examples. A closed interval $[a, b]$ in $\overline{\mathbb{R}}$ is a complete lattice under the usual ordering. We say that $[a, b]$ is a *complete sublattice* of $\overline{\mathbb{R}}$. As a second example, let E be a topological space and $E' \subseteq E$. Then $\mathcal{P}(E')$ can be regarded as a subset of $\mathcal{P}(E)$; it is a complete lattice with the same infimum and supremum. On the other hand, $\mathcal{F}(E)$, the set of closed subsets of E, is a complete lattice under the partial ordering of $\mathcal{P}(E)$; it has the same infimum but a different supremum; see also Example 2.9. To distinguish between these two cases, we introduce the following definitions. To understand them, the reader should bear in mind that a subset of a poset becomes a poset when provided with the induced partial ordering.

2.19 Definition. Let $\mathcal{L}$ be a lattice and $\mathcal{M}$ a subset of $\mathcal{L}$.

(a) If $\mathcal{M}$ becomes a lattice under the partial ordering of $\mathcal{L}$, then it is called an *underlattice*. If the lattice $\mathcal{M}$ is complete, then it is called a *complete underlattice*.

(b) $\mathcal{M}$ is called a *sublattice* if for every two elements $X, Y \in \mathcal{M}$ we have $X \vee Y \in \mathcal{M}$ and $X \wedge Y \in \mathcal{M}$. If, moreover, infinite infima and suprema of elements in $\mathcal{M}$ lie in $\mathcal{M}$, then it is called a *complete sublattice*.

To get further insight into these definitions we make a few observations. In the second definition it is required that $\mathcal{M}$ inherits the infimum and supremum of $\mathcal{L}$, whereas in the first definition $\mathcal{M}$ may have a different infimum and supremum. In particular, every sublattice is an underlattice, but not conversely. For example, if E is a topological space, then $\mathcal{F}(E)$ is a complete underlattice of $\mathcal{P}(E)$. It is also a sublattice, but not one which is complete.

If $\mathcal{M}$ is a subset of a *complete* lattice $\mathcal{L}$, then we say that $\mathcal{M}$ is *inf-closed* if every subset of $\mathcal{M}$ has an infimum in $\mathcal{M}$; it follows in particular that $I = \bigwedge \varnothing \in \mathcal{M}$. Dually, $\mathcal{M}$ is said to be *sup-closed* if the supremum of an arbitrary subset of $\mathcal{M}$ is contained in $\mathcal{M}$; in particular, this gives that $O = \bigvee \varnothing \in \mathcal{M}$. For example, $\mathcal{F}(E)$ is an inf-closed subset of $\mathcal{P}(E)$; this subset is not sup-closed, however.

The following result is an immediate consequence of Proposition 2.12.

2.20 Proposition. *Let $\mathcal{L}$ be a complete lattice and $\mathcal{M} \subseteq \mathcal{L}$. Then $\mathcal{M}$ is a complete underlattice in either of the following two cases:*

(a) *$\mathcal{M}$ is inf-closed;*

(a′) *$\mathcal{M}$ is sup-closed.*

Note that in the first case the greatest element of $\mathcal{M}$ is $I = \wedge\varnothing$, whereas in the second case the least element of $\mathcal{M}$ is $O = \vee\varnothing$. We emphasize that this proposition gives only sufficient conditions for $\mathcal{M}$ to be an underlattice. These conditions are not necessary. Section 2.3 discusses an example of an underlattice of $\mathcal{P}(E)$, E a topological space, which is neither inf-closed nor sup-closed.

For future use we introduce the following notation. Let $\mathcal{L}$ be a complete lattice and $\mathcal{M}$ some subset of $\mathcal{L}$. Denote by $\langle \mathcal{M} \mid \wedge \rangle$ the smallest subset of $\mathcal{L}$ that contains $\mathcal{M}$ and is inf-closed. Dually, define $\langle \mathcal{M} \mid \vee \rangle$ as the smallest subset of $\mathcal{L}$ that contains $\mathcal{M}$ and is sup-closed. Note that, by Proposition 2.20, both sets are complete underlattices of $\mathcal{L}$.

2.21 Example. (Convex sets)
Let $\mathcal{L} = \mathcal{P}(\mathbb{R}^2)$, and let $\mathcal{M}$ be the subset comprising all closed half-spaces of $\mathbb{R}^2$, that is, sets of the form $\{(x, y) \mid ax + by \le c\}$ with $a, b, c \in \mathbb{R}$. Intersections of closed half-spaces yield closed convex sets, and conversely every closed convex set is the intersection of all closed half-spaces which contain this set; cf. Proposition 9.7. Therefore, $\langle \mathcal{M} \mid \wedge \rangle = \mathcal{C}_{\mathcal{F}}(\mathbb{R}^2)$, the space of all closed convex subsets of $\mathbb{R}^2$.

A subset $\mathcal{H}$ of $\mathcal{L}$ is called a *sup-generating family* if every element of $\mathcal{L}$ is a supremum of elements of $\mathcal{H}$. Dually, $\mathcal{H}$ is called an *inf-generating family* if every element of $\mathcal{L}$ is an infimum of elements of $\mathcal{H}$. For instance, the singletons (i.e., sets with one element) form a sup-generating family in $\mathcal{P}(E)$. Also, the open balls form a sup-generating family in $\mathcal{G}(\mathbb{R}^d)$, the open subsets of $\mathbb{R}^d$.

2.2. Boolean lattices

Given a lattice $\mathcal{L}$ with universal bounds O and I; if $X, Y \in \mathcal{L}$ are such that $X \wedge Y = O$ and $X \vee Y = I$, then Y is called a *complement* of X (and X a complement of Y). The lattice $\mathcal{L}$ is called *complemented* if all elements in $\mathcal{L}$ have a complement.

2.22 Proposition. *If $\mathcal{L}$ is a distributive lattice, then every element of $\mathcal{L}$ has at most one complement.*

Proof. Let $X \in \mathcal{L}$ and assume that Y, Z are both complements of X. Then

$$Y = Y \wedge I = Y \wedge (Z \vee X) = (Y \wedge Z) \vee (Y \wedge X) = Y \wedge Z.$$

Analogously $Z = Y \wedge Z$, and we conclude that $Y = Z$. □

2.23 Definition. A *Boolean lattice* is a complemented distributive lattice.

In a Boolean lattice every element X possesses a unique complement, denoted by X^*. It is obvious that

$$(X^*)^* = X. \tag{2.9}$$

2.24 Proposition. *In a Boolean lattice the following holds:*
(a) $X \le Y$ *if and only if* $X^* \ge Y^*$;
(b) $(X \vee Y)^* = X^* \wedge Y^*$;
(b′) $(X \wedge Y)^* = X^* \vee Y^*$.

Proof. We prove only the first assertion. Let $X \le Y$. Then $I = X \vee X^* \le Y \vee X^*$, and so $Y \vee X^* = I$. Now $X^* = X^* \vee O = X^* \vee (Y \wedge Y^*) = (X^* \vee Y) \wedge (X^* \vee Y^*) = I \wedge (X^* \vee Y^*) = X^* \vee Y^*$, which implies $X^* \ge Y^*$. This proves the only if-statement. Now the if-statement follows immediately from (2.9). □

The simplest example of a Boolean lattice is the set $\{0, 1\}$, with $0 \wedge 1 = 0$, $0 \vee 1 = 1$, $0^* = 1$, and $1^* = 0$. This trivial, but nevertheless important example plays an important role in Section 2.4, which deals with Boolean functions.

We denote the operator which maps X onto X^* by ν; that is,

$$\nu(X) = X^*. \tag{2.10}$$

ν is a bijection on $\mathcal{L}$ which reverses the partial ordering; that is,

$$X \le Y \iff \nu(X) \ge \nu(Y).$$

Mappings with such properties are called *dual automorphisms*.

2.25 Proposition. *Given a lattice $\mathcal{L}$ and a dual automorphism ψ on $\mathcal{L}$, then*

$$\psi(\bigwedge_{i \in I} X_i) = \bigvee_{i \in I} \psi(X_i),$$
$$\psi(\bigvee_{i \in I} X_i) = \bigwedge_{i \in I} \psi(X_i),$$

for every finite family X_i. If, moreover, the lattice $\mathcal{L}$ is complete, then these identities also hold for infinite families.

The proof is straightforward.

2.26 Definition. A lattice is called *self-dual* if there exists a dual automorphism on it.

2.27 Corollary. *Every Boolean lattice is self-dual.*

The next result establishes the connection between self-duality and the Duality Principle 2.4. Recall that $\mathcal{L}'$ is the dual lattice obtained by reversing the partial ordering.

2.28 Theorem. *A lattice $\mathcal{L}$ is self-dual if and only if it is isomorphic with the dual lattice $\mathcal{L}'$.*

The proof of this result follows directly from the observation that a dual automorphism on $\mathcal{L}$ is an isomorphism between $\mathcal{L}$ and $\mathcal{L}'$, and vice versa.

2.29 Definition. A dual automorphism ν on $\mathcal{L}$ which satisfies $\nu(\nu(X)) = X$ for $X \in \mathcal{L}$ or, in operator notation,

$$\nu^2 = \mathrm{id},$$

is called a *negation*; here id is the identity operator mapping every element of $\mathcal{L}$ onto itself.

On a Boolean lattice the complement operator is a negation, and we use the same notation ν for the general notion of a negation and the particular notion of a complement operator. The examples that follow show that there are lattices which are not Boolean but which do have a negation. Furthermore, they show that negations need not be unique. In fact, if ν is a negation and ψ an automorphism with inverse ψ^{-1}, then $\psi^{-1}\nu\psi$ is also a negation. In spite of this fact we use the ambiguous notation X^* to denote $\nu(X)$ if ν is a negation on $\mathcal{L}$. This notation coincides with the notation for the complement in a Boolean lattice. It will always be clear from the context which negation or complement is meant.

2.30 Examples.

(a) The power set $\mathcal{P}(E)$ with the ordinary set complement X^c is a Boolean lattice. On this lattice there exist in general many other negations. For example, on $\mathcal{P}(\mathbb{R}^d)$ the operator $\nu(X) = (-X)^c$ is a negation different from the complement operator; here $-X = \{-x \mid x \in X\}$.

(b) Consider the function lattice $\mathrm{Fun}(E, \mathcal{T})$ of Example 2.10. It is easy to verify that this lattice is Boolean if and only if $\mathcal{T}$ is Boolean. In particular, $\mathrm{Fun}(E, \{0, 1\})$ is isomorphic to the complete Boolean lattice $\mathcal{P}(E)$. Furthermore, $\mathrm{Fun}(E, \mathcal{T})$ is self-dual if and only if $\mathcal{T}$ is. Take, for example, $\mathcal{T} = \overline{\mathbb{R}}$; the mapping $t \mapsto -at + b$ with $a > 0$ and $b \in \mathbb{R}$ defines a dual automorphism on $\overline{\mathbb{R}}$. It is a negation if and only if $a = 1$; the corresponding negation on $\mathrm{Fun}(E, \mathcal{T})$ is given by $F \mapsto -F + b$. Hereafter, we will choose $b = 0$ when considering negations on this function lattice.

The following result, called *de Morgan's laws*, has its roots in set theory and logic.

2.31 De Morgan's Laws. *Consider a lattice $\mathcal{L}$ with a negation $X \mapsto X^*$. For every finite family X_i in $\mathcal{L}$,*

$$(\bigwedge_{i\in I} X_i)^* = \bigvee_{i\in I} X_i^*,$$
$$(\bigvee_{i\in I} X_i)^* = \bigwedge_{i\in I} X_i^*.$$

If $\mathcal{L}$ is complete, then these laws are also valid for infinite families.

In particular, these laws hold in every Boolean lattice.

2.32 Theorem. *In every complete Boolean lattice the infinite distributive laws* (2.5) *and* (2.6) *hold.*

Proof. Let $\mathcal{L}$ be a complete Boolean lattice. We show that (2.5) holds; then (2.6) follows by the Duality Principle. We must show that

$$A \wedge \bigvee_{i\in I} X_i = \bigvee_{i\in I}(A \wedge X_i)$$

if $A, X_i \in \mathcal{L}$. Since $A \wedge \bigvee_{i\in I} X_i \geq A \wedge X_i$, the inequality $\geq$ follows immediately.

To prove $\leq$ put $Y = \bigvee_{i\in I}(A \wedge X_i)$. Then $A \wedge X_i \leq Y$ for every i; hence

$$X_i = (A \wedge X_i) \vee (A^* \wedge X_i) \leq Y \vee A^*.$$

This gives

$$A \wedge \bigvee_{i\in I} X_i \leq A \wedge (Y \vee A^*) = A \wedge Y \leq Y.$$

This concludes the proof. □

It was noticed that there exist complete Boolean lattices for which the extended distributive laws (2.7) and (2.8) do not hold. In a sense, the only Boolean lattice for which they do hold is $\mathcal{P}(E)$. Before we can give a precise formulation of this remarkable fact we need some extra definitions.

2.33 Definition. A *field of sets* is a family $\mathcal{N} \subseteq \mathcal{P}(E)$, where E is an arbitrary set, which is closed under finite intersections and finite unions, as well as under complements.

A field of sets has a Boolean lattice structure. Theorem 2.35 states that the converse also holds.

2.34 Examples.

(a) The subsets of a set E which are finite or whose complement is finite is a field of sets.

(b) The collection of finite unions of half-closed intervals in $\mathbb{R}$, including $(-\infty, a)$ and (a, ∞), is a field of sets.

(c) Let E be a topological space. A subset of E which is simultaneously open and closed is called a *clopen set*. In particular, $\varnothing$ and E are clopen sets. If E is connected, then these are the only clopen sets. The clopen subsets of an arbitrary topological space E form a field of sets.

Section 2.3 presents another important example, the so-called regular closed sets. The proof of the following result can be found in Halmos (1963).

2.35 Theorem. *Every Boolean lattice is isomorphic to a field of sets.*

Atoms are defined to be the smallest nonzero elements in a lattice. To be more specific, a nonzero element A of a lattice $\mathcal{L}$ is called an *atom* if $X \leq A$ implies $X = O$ or $X = A$. For example, in the lattice $\mathcal{P}(E)$ the atoms are the singletons, the sets containing precisely one element. In the lattice $\mathbb{N}$ of nonnegative integers ordered by "$m \leq n$ if m divides n" the atoms are the prime numbers. The set of atoms in $\mathcal{L}$ is denoted by $\ell_{\mathcal{L}}$, or briefly ℓ, if there is no danger of confusion. Note that a lattice need not have any atoms; $\mathcal{G}(\mathbb{R}^d)$ is a typical example of an atomless lattice.

Atoms are denoted by lower case letters such as a, b, x, y. An *atomic lattice* is a lattice in which every element is a supremum of atoms. In other words, in an atomic lattice the atoms constitute a sup-generating family. Obviously $\mathcal{P}(E)$ is an atomic lattice since every set is the union of all singletons contained in it. A similar remark applies to $\mathcal{F}(\mathbb{R}^d)$, the closed sets in $\mathbb{R}^d$.

We also introduce the two dual notions. An element $B \neq I$ is called a *dual atom* if $X \geq B$ implies $X = B$ or $X = I$. A *dual-atomic lattice* is a lattice in which every element can be written as an infimum of dual atoms, i.e., a lattice in which the dual atoms constitute an inf-generating family. By the Duality Principle, a lattice $\mathcal{L}$ is atomic if and only if the dual lattice $\mathcal{L}'$ is dual-atomic. There exists the following important representation theorem for complete Boolean lattices which are atomic. The proof can be found in Halmos (1963, pp. 70).

2.36 Theorem. *In a Boolean lattice $\mathcal{L}$ the following assertions are equivalent:*

(i) *$\mathcal{L}$ is complete and atomic;*

(ii) *$\mathcal{L}$ is complete and completely distributive;*
(iii) *$\mathcal{L}$ is isomorphic to the field $\mathcal{P}(E)$ for some set E.*

An element is called a *semi-atom* if $A \neq O$ and $X \vee Y \geq A$ implies $X \geq A$ or $Y \geq A$.

2.37 Proposition.
(a) *In a distributive lattice, every atom is also a semi-atom.*
(b) *In a Boolean lattice, every semi-atom is an atom.*

Proof. (a): Let A be an atom in the distributive lattice $\mathcal{L}$, and assume that $A \leq X \vee Y$ and that neither $A \leq X$ nor $A \leq Y$. Since $A \wedge X \leq A$, we have $A \wedge X = A$ or $A \wedge X = O$. The first relation would imply that $A \leq X$, which is false by assumption, and so $A \wedge X = O$. Similarly $A \wedge Y = O$. By the distributivity of $\mathcal{L}$,

$$A = (X \vee Y) \wedge A = (X \wedge A) \vee (Y \wedge A) = O,$$

a contradiction. This proves (a).

(b): Assume that A is a semi-atom and $B < A$; we show $B = O$. Since $A \leq I = B \vee B^*$, we conclude that $A \leq B^*$. This implies $B \leq B^*$, and therefore $B = B \wedge B^* = O$. □

A lattice is called *semi-atomic* if every element can be written as a supremum of semi-atoms. In a similar way as we defined dual atoms and dual-atomic lattices, we can also define *dual semi-atoms* and *dual-semi-atomic lattices*. It follows from the Duality Principle that the analogue of Proposition 2.37 for dual (semi-) atoms also holds. Every element in a chain $\mathcal{T}$ is a (dual) semi-atom, and $\mathcal{T}$ is (dual-) semi-atomic.

If $\mathcal{L}$ is a complete lattice with a negation (or more generally, a dual automorphism), then this maps every (semi-) atom to a dual (semi-) atom, and vice versa. Moreover, if $\mathcal{L}$ is (semi-) atomic, it is also dual- (semi-) atomic. We illustrate these abstract concepts with the aid of some examples.

2.38 Examples.

(a) Consider the function lattice Fun$(E, \overline{\mathbb{R}})$. It is easy to see that this lattice does not contain any atoms. The function $f_{x,t}$ $(x \in E,\ t \in \mathbb{R} \cup \{\infty\})$ given by $f_{x,t}(y) = t$ if $y = x$ and $-\infty$ elsewhere, however, defines a semi-atom. Furthermore, the lattice is semi-atomic, as every element F can be written as

$$F = \bigvee_{x \in E} f_{x,F(x)}.$$

We call the functions $f_{x,t}$ *pulse functions*. They emerge at several instances in this book, in particular in Section 5.7. Note that the lattice $\mathrm{Fun}(E, \overline{\mathbb{R}})$ is also dual-semi-atomic.

(b) The lattice $\mathcal{C}(\mathbb{R}^2)$ is atomic because every singleton is a convex set. If $\{a\}$ is a singleton and if $x, y \in \mathbb{R}^2$ are two points distinct from a such that a lies on the segment between x and y, then $\{a\} \subseteq \{x\} \vee \{y\}$; note that $\{x\} \vee \{y\}$ is the segment between x and y. This shows that $\{a\}$ is not a semi-atom. This example does not contradict Proposition 2.37 because the lattice $\mathcal{C}(\mathbb{R}^2)$ is not distributive. In fact, one can easily verify that $\mathcal{C}(\mathbb{R}^2)$ contains no semi-atoms.

2.3. Regular closed sets

This section presents an example of a Boolean lattice which has no atoms. The elements of this lattice are subsets of some nonempty topological space E. Although a systematic treatment of topological spaces is postponed until Chapter 7, a discussion of the complete lattice of regular closed sets fits best in the present chapter. Anyhow, if he wishes the reader may skip this section altogether.

For a subset X of E, the closure is denoted by $\overline{X}$ or X^- and the interior by X°. A closed set X is called *regular* if

$$X = X^{\circ -}.$$

Since $X^- = X^{coc}$ (Y^c being the ordinary set complement of Y), it follows that X is regular if and only if

$$X = X^{\perp\perp},$$

where $X^\perp = X^{\circ c}$. In $\mathbb{R}^2$ a closed set is regular if it contains no parts of thickness zero or isolated points; see Fig. 2.3.

The family of regular closed subsets of E is denoted by $\mathcal{R}(E)$; we show that $\mathcal{R}(E)$ with the inclusion ordering is a complete Boolean lattice with $O = \varnothing$, $I = E$, and

$$\bigwedge_{i \in I} X_i = \Big(\bigcap_{i \in I} X_i\Big)^{\perp\perp}, \tag{2.11}$$

$$\bigvee_{i \in I} X_i = \Big(\bigcup_{i \in I} X_i\Big)^{\perp\perp}, \tag{2.12}$$

$$\nu(X) = X^\perp. \tag{2.13}$$

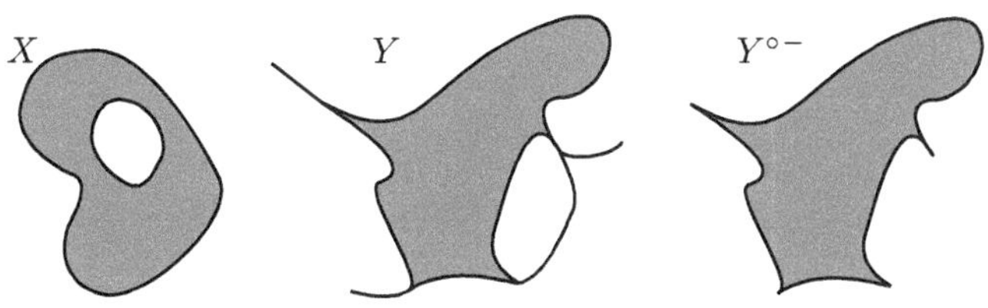

Figure 2.3 A regular set X and a set Y which is not regular; $Y^{\circ -}$ is a strict subset of Y.

Furthermore, we show

$$\bigvee_{i \in I} X_i = \bigcup_{i \in I} X_i \qquad \text{if } I \text{ is finite.} \tag{2.14}$$

To prove these assertions some preparations are necessary. First note that for every two subsets X, Y of E,

$$X \subseteq Y \Rightarrow Y^{\perp} \subseteq X^{\perp}. \tag{2.15}$$

Assume that X is closed; since $X^{\circ} \subseteq X$, we get $X^{c} \subseteq X^{\perp}$. Taking the interior on both sides and using that X^{c} is open, we find $X^{c} \subseteq X^{\perp\circ}$. Taking complements again, we find

$$X^{\perp\perp} \subseteq X \qquad \text{if } X \text{ is closed.} \tag{2.16}$$

Combining this fact with (2.15) gives

$$X^{\perp\perp\perp} = X^{\perp} \qquad \text{if } X \text{ is closed.} \tag{2.17}$$

In particular, $X^{\perp}$ is regular if X is closed. It is not difficult to give examples which show that this assertion is false for arbitrary X. Take, e.g., $X = \mathbb{R} \setminus \{0\}$; then $X^{\perp} = \{0\}$.

Given a family of regular closed sets X_i, define $A = (\bigcap X_i)^{\perp\perp}$. Then A is a regular closed set and a lower bound of X_i by (2.16). If $A' \in \mathcal{R}(E)$ is another lower bound, then $A' \subseteq \bigcap X_i$, and so

$$A' = (A')^{\perp\perp} \subseteq (\bigcap X_i)^{\perp\perp} = A,$$

meaning that A is the infimum of the X_i. In a similar way, one shows that $(\bigcup X_i)^{\perp\perp}$ is the supremum of X_i. We establish the following relation:

$$(X \cup Y)^{\perp\perp} = X^{\perp\perp} \cup Y^{\perp\perp} \qquad \text{if } X, Y \text{ are closed.} \tag{2.18}$$

It is obvious that

$$X^{\perp\perp} \cup Y^{\perp\perp} \subseteq (X \cup Y)^{\perp\perp}.$$

We prove the reverse inclusion. Using that X is closed, it is easy to show that

$$(X \cup Y)^{\circ} \subseteq X \cup Y^{\circ}.$$

Taking complements of both sides we find

$$X^{c} \cap Y^{\perp} \subseteq (X \cup Y)^{\perp}.$$

Now taking interiors at both sides, using that this operation is distributive over intersections and that X^c is open, we obtain

$$X^{c} \cap Y^{\perp\circ} \subseteq (X \cup Y)^{\perp\circ}.$$

Taking complements again, we arrive at

$$(X \cup Y)^{\perp\perp} \subseteq X \cup Y^{\perp\perp}.$$

Applying $\perp$ twice and using the previous relation once more, we find

$$(X \cup Y)^{\perp\perp\perp\perp} \subseteq (X \cup Y^{\perp\perp})^{\perp\perp} \subseteq X^{\perp\perp} \cup Y^{\perp\perp}.$$

Using (2.17), we get

$$(X \cup Y)^{\perp\perp} \subseteq X^{\perp\perp} \cup Y^{\perp\perp},$$

and thus (2.18) is proved. Note that (2.14) follows immediately from this relation.

To prove (2.13) we must show that $X \vee X^{\perp} = E$ and $X \wedge X^{\perp} = \varnothing$. The first relation is trivial, since $X^{\perp} = X^{\circ c} \supseteq X^{c}$, and so $X \vee X^{\perp} = X \cup X^{\perp} \supseteq X \cup X^{c} = E$. To prove $X \wedge X^{\perp} = \varnothing$ we observe that $X \cap X^{\perp} = X \cap X^{\circ c}$ has empty interior, and so $(X \cap X^{\perp})^{\perp\perp} = (X \cap X^{\perp})^{\circ c \circ c} = \varnothing$.

It remains to prove distributivity. Let $X, Y, Z \in \mathcal{R}(E)$. We show

$$X \wedge (Y \vee Z) = (X \wedge Y) \vee (X \wedge Z).$$

The other distributivity law is proved analogously. Using (2.14) and (2.18), we find

$$X \wedge (Y \vee Z) = X \wedge (Y \cup Z) = (X \cap (Y \cup Z))^{\perp\perp}$$

$$= \left[(X \cap Y) \cup (X \cap Z)\right]^{\perp\perp} = (X \cap Y)^{\perp\perp} \cup (X \cap Z)^{\perp\perp}$$
$$= (X \wedge Y) \vee (X \wedge Z).$$

It is obvious that $\mathcal{R}(\mathbb{R}^d)$ does not contain atoms, as every closed ball with positive radius contains a closed ball with a smaller radius. In Example 7.9 we show that the converse also holds: every complete Boolean lattice is isomorphic to the lattice of regular closed sets of some topological space.

2.4. Boolean functions

A Boolean lattice with a very simple algebraic structure is the set comprising the elements 0 and 1. A *Boolean variable* is a variable which can only take the values 0 (false) or 1 (true). If u, v are Boolean variables, then we write $u \cdot v$ or uv instead of $u \wedge v$, $u + v$ instead of $u \vee v$, and $\overline{u}$ instead of u^c. The n-fold product of $\{0, 1\}$, denoted by $\{0, 1\}^n$, becomes a Boolean lattice by the following definitions:

$$(u_1, \ldots, u_n) \wedge (v_1, \ldots, v_n) = (u_1 v_1, \ldots, u_n v_n),$$
$$(u_1, \ldots, u_n) \vee (v_1, \ldots, v_n) = (u_1 + v_1, \ldots, u_n + v_n),$$
$$(u_1, \ldots, u_n)^c = (\overline{u}_1, \ldots, \overline{u}_n).$$

We introduce the notation $\mathbf{u}$ for the Boolean vector $(u_1, \ldots, u_n)$. If $\mathbf{u}, \mathbf{v}$ are Boolean vectors, then $\mathbf{u} + \mathbf{v}$ and $\mathbf{u} \cdot \mathbf{v}$ represent the Boolean vectors $(u_1 + v_1, \ldots, u_n + v_n)$ and $(u_1 v_1, \ldots, u_n v_n)$, respectively. Furthermore, $\overline{\mathbf{u}} = (\overline{u}_1, \ldots, \overline{u}_n)$.

2.39 Definition. A mapping from $\{0, 1\}^n$ into $\{0, 1\}$ is called a *Boolean function* (of n variables).

The space of all Boolean functions of n variables is denoted by $\mathbb{B}_n$. It is easily seen that there are exactly 2^{2^n} Boolean functions of n variables. Every Boolean function can be represented by a so-called *truth table*, which gives the different outcomes of $b(\mathbf{u})$ for all vectors $\mathbf{u}$.

It is not the intention of this section to give an exhaustive treatment of Boolean functions but rather to present the basic notions and results which are needed later. In Section 4.5 Boolean functions will be used to construct a large class of translation invariant morphological operators.

2.40 Proposition. *The space $\mathbb{B}_n$ is a Boolean lattice with infimum, supremum, and complement respectively given by*

$$(b_1 b_2)(\mathbf{u}) = b_1(\mathbf{u}) b_2(\mathbf{u}),$$

$$(b_1 + b_2)(\mathbf{u}) = b_1(\mathbf{u}) + b_2(\mathbf{u}),$$
$$b^*(\mathbf{u}) = \overline{b(\overline{\mathbf{u}})}.$$

The least and greatest element are the functions identically 0 *and* 1, *respectively.*

The function identically 0 (resp. 1) is denoted by **0** (resp. **1**).

A *Boolean polynomial* of n variables is an expression consisting of the elements $0, 1$, the variables $u_1, u_2, \ldots, u_n$, complementation, and the binary operations $\cdot$ and $+$. For example, $u_1\overline{u_2} + u_3$ and $u_1 + \overline{u_2} + u_2(u_1 + u_3)$ are Boolean polynomials of three variables. Every Boolean polynomial of n variables represents an element of $\mathbb{B}_n$. Note that two seemingly different polynomials can represent the same function. For example, instead of the second expression preceding we may also write $u_1 + \overline{u_2} + u_2u_3$ because $u_1 + u_2u_1 = u_1$. Two Boolean polynomials are called *equivalent* if they represent the same Boolean function. A natural question is whether every Boolean function can be represented as a polynomial. The next result shows that the answer to this question is affirmative. We use the convention that $u^0 = \overline{u}$ and $u^1 = u$. Furthermore $\sum$ denotes a Boolean sum and $\prod$ a Boolean product.

2.41 Theorem. (Representation of Boolean functions)
Every Boolean function b of n variables can be written as

$$b(\mathbf{u}) = \sum_{e_1,\ldots,e_n=0}^{1} b(e_1, \ldots, e_n) u_1^{e_1} u_2^{e_2} \cdots u_n^{e_n}, \tag{2.19}$$

$$b(\mathbf{u}) = \prod_{e_1,\ldots,e_n=0}^{1} \big(b(e_1, \ldots, e_n) + u_1^{\overline{e_1}} + u_2^{\overline{e_2}} + \cdots + u_n^{\overline{e_n}}\big). \tag{2.20}$$

The proof of formula (2.19) follows easily from the observation that the product $u_1^{e_1} u_2^{e_2} \cdots u_n^{e_n}$ gives the outcome 1 if and only if $\mathbf{u} = (e_1, \ldots, e_n)$. A similar argument leads to (2.20). Note that (2.20) is the dual representation of (2.19) in the sense of the Duality Principle. The expressions in (2.19) and (2.20) are the canonical polynomials representing b, called *sum-of-products* and *product-of-sums* representations, respectively. The product terms $u_1^{e_1} u_2^{e_2} \cdots u_n^{e_n}$, where $e_i \in \{0, 1\}$, are the atoms of the lattice $\mathbb{B}_n$, and the first formula in the representation theorem writes b as a supremum of atoms. In particular, this means that $\mathbb{B}_n$ is atomic.

2.42 Example. There is a simple algorithm which yields both canonical polynomials. We will not describe this algorithm explicitly; we find it more instructive to give an example where we compute the sum-of-products representation. Let the 3-variable Boolean function be given by

$$b(u_1, u_2, u_3) = u_1 u_2 + u_1(\overline{u_3} + u_2 u_3).$$

First, we write b as a sum of products, simplifying the expression whenever possible. This gives

$$b(u_1, u_2, u_3) = u_1 u_2 + u_1 \overline{u_3}.$$

The term $u_1 u_2 u_3$ has been deleted since it is smaller than $u_1 u_2$. In the first term, the variable u_3 is absent, and we replace this term by $u_1 u_2(u_3 + \overline{u_3}) = u_1 u_2 u_3 + u_1 u_2 \overline{u_3}$. Following the same procedure for the second term, we get

$$\begin{aligned} b(u_1, \ldots, u_3) &= u_1 u_2 u_3 + u_1 u_2 \overline{u_3} + u_1 u_2 \overline{u_3} + u_1 \overline{u_2}\,\overline{u_3} \\ &= u_1 u_2 u_3 + u_1 u_2 \overline{u_3} + u_1 u_2 u_3. \end{aligned}$$

2.43 Definition. The Boolean function b is said to be *increasing* (or *positive*) if $\mathbf{u} \leq \mathbf{v}$ implies that $b(\mathbf{u}) \leq b(\mathbf{v})$.

Consider an increasing Boolean function b for which the term $u_1^0 u_2^{e_2} \cdots u_n^{e_n}$ occurs in the sum-of-products expansion, that is, $b(0, e_2, \ldots, e_n) = 1$; then, by the fact that b is increasing, $b(1, e_2, \ldots, e_n) = 1$ as well, and therefore also $u_1^1 u_2^{e_2} \cdots u_n^{e_n}$ occurs in the sum-of-products expansion. Both terms sum to $u_2^{e_2} \cdots u_n^{e_n}$. This argument shows that in the sum-of-products representation complemented variables $\overline{u_i}$ may be replaced by 1. This gives the following result.

2.44 Theorem. *A Boolean function b is increasing if and only if b can be represented as a sum-of-products (or dually, a product-of-sums) in which no complements appear.*

2.45 Definition. A Boolean function b is called *multiplicative* if $b(\mathbf{1}) = 1$ and $b(\mathbf{uv}) = b(\mathbf{u})b(\mathbf{v})$. It is called *additive* if $b(\mathbf{0}) = 0$ and $b(\mathbf{u} + \mathbf{v}) = b(\mathbf{u}) + b(\mathbf{v})$.

For example, $u_1 u_3$ is a multiplicative function in $\mathbb{B}_3$, whereas $u_2 + u_3$ is an additive function. It is easy to see that multiplicative and additive functions are increasing. Therefore, such functions can be written as polynomials without complemented variables. Actually, a much stronger result holds.

2.46 Proposition.
(a) *Every multiplicative Boolean function b can be represented as*

$$b(u_1, \dots, u_n) = (c_1 + u_1)(c_2 + u_2) \cdots (c_n + u_n),$$

where $c_i \in \{0, 1\}$.
(a′) *Every additive Boolean function b can be represented as*

$$b(u_1, \dots, u_n) = c_1 u_1 + c_2 u_2 + \cdots + c_n u_n,$$

where $c_i \in \{0, 1\}$.

Proof. Assume that b is multiplicative. First we observe

$$(u_1, \dots, u_n) = (u_1, 1, 1, \dots, 1) \cdot (1, u_2, 1, \dots, 1) \cdots (1, 1, \dots, 1, u_n).$$

Furthermore

$$b(u_1, 1, 1, \dots, 1) = u_1 + b(0, 1, 1, \dots, 1) = u_1 + c_1,$$

where $c_1 = b(0, 1, 1, \dots, 1)$. Analogous expressions can be found for the terms $b(1, 1, \dots, 1, u_i, 1, \dots, 1)$. Thus, if b is multiplicative we get

$$\begin{aligned} b(u_1, \dots, u_n) &= b(u_1, 1, 1, \dots, 1) \cdot b(1, u_2, 1, \dots, 1) \cdots b(1, 1, \dots, 1, u_n) \\ &= (c_1 + u_1)(c_2 + u_2) \cdots (c_n + u_n), \end{aligned}$$

which proves the result. □

Given a Boolean function b, the *negative Boolean function* b^* is defined by

$$b^*(u_1, \dots, u_n) = \overline{b}(\overline{u}_1, \dots, \overline{u}_n). \tag{2.21}$$

It is evident that

$$b^{**} = b.$$

A Boolean function is said to be *self-dual* if $b^* = b$.

2.47 Proposition. *Let b be a Boolean function.*
(a) *b is increasing if and only if* b^* *is increasing.*
(b) *b is multiplicative if and only if* b^* *is additive.*

The proof, which is easy, is left to the reader. More generally, one can show that computation of the negative Boolean polynomial amounts to changing $\cdot$ into $+$, and vice versa. For instance, if $b(u_1, u_2, u_3) = u_1\overline{u_2} + u_3$, then $b^*(u_1, u_2, u_3) = (u_1 + \overline{u_2})u_3$.

The remainder of this section is devoted to an important class of Boolean functions, the so-called *threshold functions*; refer to Muroga (1971) for a detailed exposition.

2.48 Definition. Given the scalars $w_1, w_2, \ldots, w_n, s \in \mathbb{R}$, the Boolean function given by

$$b(u_1, \ldots, u_n) = \begin{cases} 1, & \text{if } \Sigma_{i=1}^n w_i u_i \geq s, \\ 0, & \text{if } \Sigma_{i=1}^n w_i u_i < s, \end{cases}$$

is called a *threshold function*. The entries w_i are called the *weights* and s is called the *threshold*. The vector $(w_1, \ldots, w_n \mid s)$ is called the *realization vector*.

Using the convention that for a statement S we put $[S] = 1$ if S is true and 0 if S is false, we can also write

$$b(u_1, \ldots, u_n) = \left[\Sigma_{i=1}^n w_i u_i \geq s\right].$$

Since the rational numbers lie dense on the real line, it is possible to choose rational weights and thresholds. Multiplying the expression by their least common denominator, one obtains a threshold function whose realization vector contains only integers. From now on we work exclusively with integer weights and thresholds. One has the following alternative interpretation of this threshold function for the case where all weights as well as the threshold are nonnegative: order the values u_i counted w_i times in decreasing order. This results in a sequence consisting of 1's followed by 0's. Then $b(u_1, \ldots, u_n)$ is the value located at the sth position of this sequence.

Consider the threshold function

$$b(u_1, u_2, u_3) = \left[3u_1 - u_2 + 2u_3 \geq 3\right].$$

Replacing u_2 by $1 - \overline{u_2}$ we get

$$b(u_1, u_2, u_3) = \left[3u_1 + \overline{u_2} + 2u_3 \geq 4\right].$$

This means that negative weights correspond to complemented variables. It is easy to show that threshold functions which are increasing can be realized with realization vectors with nonnegative entries (and without complemented variables). If $n \geq 4$, there exist increasing Boolean functions which are not of threshold type. As a simple example, consider the function

$$b(u_1, u_2, u_3, u_4) = u_1 u_2 + u_3 u_4.$$

Assume that $b(u_1, u_2, u_3, u_4) = \left[w_1 u_1 + \cdots + w_4 u_4 \geq s\right]$. Then $w_1 + w_2 \geq s$ because $b(1, 1, 0, 0) = 1$; analogously, $w_3 + w_4 \geq s$. Assume $w_1 \geq w_2$ and $w_3 \geq w_4$; then $w_1 + w_3 \geq s$, which implies that $b(1, 0, 1, 0) = 1$, a contradiction.

If b is a threshold function with realization vector $(w_1, \ldots, w_n \mid s)$, then b^* is also a threshold function with realization vector $(w_1, \ldots, w_n \mid \Sigma w_i - s + 1)$. Consequently, b is self-dual ($b^* = b$) if and only if $s = \frac{1}{2}\Sigma w_i + \frac{1}{2}$. A rather special, but also important, example of a threshold function is the *rank function* r_s, where $s \leq n$. This function is realized by the vector $(1, 1, \ldots, 1 \mid s)$; in other words,

$$r_s(u_1, \ldots, u_n) = \left[\Sigma_{i=1}^{n} u_i \geq s\right].$$

It is easy to check that $r_s(u_1, \ldots, u_n)$ is the sth largest value of the u_i. In other words, the outcome is 1 if at least s of the u_i are 1, and 0 otherwise. One sees immediately that

$$r_1(u_1, \ldots, u_n) = u_1 + u_2 + \cdots + u_n,$$

and that

$$r_n(u_1, \ldots, u_n) = u_1 u_2 \cdots u_n.$$

Furthermore, we have the duality relation

$$r_s^* = r_{n-s+1}.$$

If n is odd and $s = \frac{1}{2}(n+1)$, then r_s is self-dual; this function is called the *median function*.

2.5. Bibliographical notes

Our standard reference on lattice theory is the treatise by Birkhoff (1967). A general discussion of abstract set theory can be found in Kuratowski and Mostowski (1976). The treatise by Halmos (1963) comprises a stimulating discussion on Boolean lattices; it also treats the regular (open) sets in considerable detail. Blumenthal and Menger (1970, pp. 42) present an interesting discussion of the Duality Principle. We point out that some authors use the terminology *co-prime* (resp. *prime*) instead of semi-atom (resp. dual semi-atom); see Gierz et al. (1980).

There is a considerable literature on Boolean functions and their applications to electric engineering; see, e.g., Biswas (1975), Harrison (1965), and Muroga (1971). This last reference is largely devoted to threshold logic. The function r_s is also often called the sth *order statistic*; consult David (1970) for background information.

CHAPTER THREE

Operators on complete lattices

Henk J.A.M. Heijmans
Formerly Centre for Mathematics and Computer Science, Amsterdam, Netherlands

Contents

The title of this chapter is somewhat misleading. It suggests that it deals with arbitrary operators between complete lattices. To a large extent, however, this chapter restricts attention to operators which have additional properties, such as being increasing or distributivity over infima, to mention a few. It is not the intention to undertake here a general study of these operators, but rather to set the stage for subsequent chapters where some special classes, relevant in the context of mathematical morphology, will be investigated in greater detail. Refer in particular to Chapter 5, where erosions and dilations are studied, to Chapter 6, where openings and closings are the central theme, and to Chapter 12, which is entirely devoted to morphological filters.

Section 3.1 presents some terminology, and Section 3.2 introduces the notion of adjunction, the main subject of Chapter 5. Openings and closings are introduced in Section 3.3; in Chapter 6 these operators are treated in more detail. Section 3.4 discusses conditional operators, and gives an abstract formulation of the local knowledge principle. Finally, Section 3.5 introduces a somewhat peculiar partial ordering on the lattice of operators, called activity ordering. This concept plays an important role in Chapter 13.

Advances in Imaging and Electron Physics, Volume 216
ISSN 1076-5670
https://doi.org/10.1016/bs.aiep.2020.07.003

3.1. Lattice operators

This chapter is concerned with functions mapping one complete lattice into another; henceforth the terminology *operator* will be used rather than function. As much as possible, operators will be denoted by Greek letters such as ϕ, ψ, θ, etc. Given two complete lattices $\mathcal{L}$ and $\mathcal{M}$, the set of all operators from $\mathcal{L}$ into $\mathcal{M}$ is denoted by $\mathcal{O}(\mathcal{L}, \mathcal{M})$. The set $\mathcal{O}(\mathcal{L}, \mathcal{M})$ inherits the partial ordering structure of $\mathcal{M}$: for operators ϕ, ψ from $\mathcal{L}$ into $\mathcal{M}$ we define $\phi \leq \psi$ if $\phi(X) \leq \psi(X)$ for every $X \in \mathcal{L}$. The set $\mathcal{O}(\mathcal{L}, \mathcal{M})$ becomes a complete lattice under this partial ordering; the infimum and supremum are given by

$$(\bigwedge_{i\in I} \psi_i)(X) = \bigwedge_{i\in I} \psi_i(X) \qquad \text{and} \qquad (\bigvee_{i\in I} \psi_i)(X) = \bigvee_{i\in I} \psi_i(X),$$

respectively, for every family of operators $\{\psi_i \mid i \in I\}$ in $\mathcal{O}(\mathcal{L}, \mathcal{M})$.

In fact, $\mathcal{O}(\mathcal{L}, \mathcal{M}) = \mathcal{M}^{\mathcal{L}}$, a lattice which has been considered in Example 2.10. It is important, however, to make a clear distinction between lattices of grey-scale functions, in this book denoted by $\mathsf{Fun}(E, \mathcal{T})$, and operators between two complete lattices. This distinction is expressed optimally by a different notation.

In Example 2.10 it was pointed out that additional properties of $\mathcal{M}$, such as modularity, distributivity, and completeness, carry over immediately to the lattice $\mathcal{O}(\mathcal{L}, \mathcal{M})$. The least element of $\mathcal{O}(\mathcal{L}, \mathcal{M})$ is the operator which maps every element of $\mathcal{L}$ onto the zero element of $\mathcal{M}$; we denote this operator by o. Similarly, the greatest element of $\mathcal{O}(\mathcal{L}, \mathcal{M})$, denoted by ι, is the operator which maps every element of $\mathcal{L}$ onto the unit element of $\mathcal{M}$.

Given an operator ψ from $\mathcal{L}$ into $\mathcal{M}$ and an operator ϕ from $\mathcal{M}$ into $\mathcal{N}$, application of ψ followed by application of ϕ yields an operator from $\mathcal{L}$ into $\mathcal{N}$; this operator is denoted by $\phi \circ \psi$, or briefly $\phi\psi$, and is called the *composition* of ψ and ϕ. Composition is an associative operation, that is, if ξ maps $\mathcal{N}$ into $\mathcal{P}$, then $\xi(\phi\psi) = (\xi\phi)\psi$. In this book the following fact will be used several times: if $\phi, \psi : \mathcal{M} \to \mathcal{N}$ are operators with $\phi \leq \psi$ and $\xi : \mathcal{L} \to \mathcal{M}$, then $\phi\xi \leq \psi\xi$.

3.1 Definition. The lattice operator $\psi : \mathcal{L} \to \mathcal{M}$ is called
(a) *increasing* if $X \leq Y$ implies $\psi(X) \leq \psi(Y)$ for all $X, Y \in \mathcal{L}$;
(b) *decreasing* if $X \leq Y$ implies $\psi(X) \geq \psi(Y)$ for all $X, Y \in \mathcal{L}$.

For instance the operator $\mathsf{card} : \mathcal{P}(\mathbb{Z}) \to [0, \infty]$ mapping a set of integers to the number of its elements (the cardinality of the set) is an increasing operator. Recall from Definition 2.6 that a lattice isomorphism is a bijective operator which distributes over infima and suprema. As a consequence, every isomorphism is increasing. On the other hand, every dual isomorphism is decreasing. In particular, if $\mathcal{L}$ is a Boolean lattice, then the operator mapping the element X to its complement X^* is a decreasing operator on $\mathcal{L}$. We denote the space of all increasing operators from $\mathcal{L}$ into $\mathcal{M}$ by $\mathcal{O}^+(\mathcal{L}, \mathcal{M})$. It is obvious that this space is a sublattice of $\mathcal{O}(\mathcal{L}, \mathcal{M})$. If $\mathcal{L} = \mathcal{M}$, then we write $\mathcal{O}(\mathcal{L})$ and $\mathcal{O}^+(\mathcal{L})$ instead of $\mathcal{O}(\mathcal{L}, \mathcal{L})$ and $\mathcal{O}^+(\mathcal{L}, \mathcal{L})$, respectively. We introduce the notation $\mathrm{id}_{\mathcal{L}}$, or briefly id, when no confusion is possible, for the operator on $\mathcal{L}$ mapping every element onto itself; this operator is called the *identity operator*.

It is evident that composition of two increasing operators gives an increasing operator. Furthermore, if $\phi, \psi : \mathcal{L} \to \mathcal{M}$ are operators such that $\phi \le \psi$, and if $\xi : \mathcal{M} \to \mathcal{N}$ is an increasing operator, then $\xi\phi \le \xi\psi$.

Assume that both lattices $\mathcal{L}$ and $\mathcal{M}$ have a negation denoted by $\nu_{\mathcal{L}}$ and $\nu_{\mathcal{M}}$, respectively. With every operator $\psi \in \mathcal{O}(\mathcal{L}, \mathcal{M})$ one can associate the *negative operator* $\psi^* \in \mathcal{O}(\mathcal{L}, \mathcal{M})$ as follows:

$$\psi^* = \nu_{\mathcal{M}} \circ \psi \circ \nu_{\mathcal{L}}. \tag{3.1}$$

When no confusion about the respective negations seems possible, we also put

$$\psi^*(X) = \left[\psi(X^*)\right]^*; \tag{3.2}$$

see also the discussion following Definition 2.29. If $\psi^* = \psi$, then ψ is called *self-dual* (with respect to $\nu_{\mathcal{L}}$ and $\nu_{\mathcal{M}}$). One can easily establish the following properties.

3.2 Proposition. *Let $\mathcal{L}$, $\mathcal{M}$ be complete lattices with a negation, and let ϕ, ψ, ψ_i ($i \in I$) be operators from $\mathcal{L}$ into $\mathcal{M}$.*

(a) *ψ is increasing if and only if ψ^* is increasing.*

(b) *$\phi \le \psi$ if and only if $\psi^* \le \phi^*$.*

(c) *$(\bigwedge_{i\in I} \psi_i)^* = \bigvee_{i\in I} \psi_i^*$ and $(\bigvee_{i\in I} \psi_i)^* = \bigwedge_{i\in I} \psi_i^*$.*

(d) *If $\mathcal{N}$ is a lattice with a negation and $\xi : \mathcal{M} \to \mathcal{N}$, then $(\xi\psi)^* = \xi^*\psi^*$.*

Furthermore, one can easily show

$$\psi^{**} = \psi. \tag{3.3}$$

Let ψ be an operator on $\mathcal{L}$, and let X be an element of $\mathcal{L}$ for which $\psi(X) = X$; then X is called *invariant under* ψ or, alternatively, a *fixpoint* of ψ. The set of all elements invariant under ψ is denoted by $\mathrm{Inv}(\psi)$ and called the *invariance domain* of ψ.

3.3 Tarski Fixpoint Theorem (weak version).
The invariance domain $\mathrm{Inv}(\psi)$ *of an increasing operator* ψ *on the complete lattice* $\mathcal{L}$ *is nonempty.*

Proof. Define $\mathcal{K} = \{X \in \mathcal{L} \mid \psi(X) \geq X\}$; then $O \in \mathcal{K}$. Define $A = \bigvee \mathcal{K}$. Since $A \geq X$ for every $X \in \mathcal{K}$ and ψ is increasing, we get $\psi(A) \geq \psi(X) \geq X$ for every $X \in \mathcal{K}$. Since A is the least upper bound of $\mathcal{K}$, this implies $\psi(A) \geq A$, and therefore $A \in \mathcal{K}$. Since $\psi(\psi(A)) \geq \psi(A)$, it follows that $\psi(A) \in \mathcal{K}$. But this gives $A \geq \psi(A)$, and so $A = \psi(A)$. We conclude that $A \in \mathrm{Inv}(\psi)$. □

Theorem 12.27 contains a stronger version of this result.

3.4 Definition. An operator ψ on $\mathcal{L}$ is called
(a) *extensive* if $\psi(X) \geq X$ for every $X \in \mathcal{L}$;
(a′) *anti-extensive* if $\psi(X) \leq X$ for every $X \in \mathcal{L}$.

Extensivity of an operator can also be expressed by means of the notation $\psi \geq \mathrm{id}$. Analogously, $\psi \leq \mathrm{id}$ denotes that ψ is anti-extensive.

3.5 Example. The operator on $\mathcal{P}(\mathbb{R}^d)$ given by

$$\psi(X) = X^{\circ -}$$

is increasing. It is not anti-extensive, for it maps every open set onto its closure. Neither is it extensive; if X is a nonempty set with empty interior, then $\psi(X) = \varnothing$. It is obvious that the invariance domain of ψ is $\mathcal{R}(\mathbb{R}^d)$, the regular closed sets of $\mathbb{R}^d$; refer to Section 2.3 for a discussion. The operator ψ restricted to the complete lattice $\mathcal{F}(\mathbb{R}^d)$, the closed sets, is anti-extensive; it has the same invariance domain as before, i.e., $\mathcal{R}(\mathbb{R}^d)$.

At several instances in this book, continuity requirements on operators turn up naturally. It is useful to distinguish two types of continuity conditions, namely, continuity in the lattice-theoretical sense and continuity in the topological sense. Chapter 7 examines the topological aspects of mathematical morphology including (semi-) continuity of morphological operators. In Chapter 13, (semi-) continuity of operators in the lattice-theoretical sense plays a crucial role. Below we give some preliminary definitions for operators which are increasing.

3.6 Definition. Let $\mathcal{L}, \mathcal{M}$ be complete lattices, and let $\psi \in \mathcal{O}^+(\mathcal{L}, \mathcal{M})$. We say that ψ is *lattice upper semi-continuous* (or *l.u.s.c.*) if for every chain $\mathcal{C}$ in $\mathcal{L}$ we have $\psi(\bigwedge \mathcal{C}) = \bigwedge_{X \in \mathcal{C}} \psi(X)$. Dually, we say that ψ is *lattice lower semi-continuous* (or *l.l.s.c.*) if for every chain $\mathcal{C}$ in $\mathcal{L}$ we have $\psi(\bigvee \mathcal{C}) = \bigvee_{X \in \mathcal{C}} \psi(X)$. We say that ψ is *lattice continuous* if it is both l.u.s.c and l.l.s.c.

Since lattice isomorphisms are distributive over infima and suprema, such operators are automatically lattice continuous. Often, it suffices to consider only countable chains. Given a sequence X_n in the complete lattice $\mathcal{L}$, we write $X_n \downarrow X$ if $X_1 \geq X_2 \geq X_3 \geq \cdots$ and $\bigwedge_{n \geq 1} X_n = X$. Dually, we write $X_n \uparrow X$ if $X_1 \leq X_2 \leq X_3 \leq \cdots$ and $\bigvee_{n \geq 1} X_n = X$.

3.7 Definition. Let $\mathcal{L}, \mathcal{M}$ be complete lattices and $\psi \in \mathcal{O}^+(\mathcal{L}, \mathcal{M})$. We say that ψ is $\downarrow$*-continuous* if $X_n \downarrow X$ in $\mathcal{L}$ implies $\psi(X_n) \downarrow \psi(X)$ in $\mathcal{M}$. Dually, we say that ψ is $\uparrow$*-continuous* if $X_n \uparrow X$ in $\mathcal{L}$ implies $\psi(X_n) \uparrow \psi(X)$ in $\mathcal{M}$. We say that ψ is $\updownarrow$*-continuous* if ψ is both $\downarrow$-continuous and $\uparrow$-continuous.

It is evident that, e.g., $\downarrow$-continuity is a weaker requirement than lattice upper semi-continuity. In Chapter 13 the reader can find a comprehensive treatment of lattice operators which are (semi-) continuous in the sense of Definition 3.7.

3.8 Lemma. *Let $\mathcal{L}, \mathcal{M}$ be complete lattices.*

(a) *Assume that every chain $\mathcal{C}$ in $\mathcal{L}$ contains a decreasing sequence C_n such that $\bigwedge \mathcal{C} = \bigwedge_{n \geq 1} C_n$. An operator $\psi : \mathcal{L} \to \mathcal{M}$ is l.u.s.c. if and only if it is $\downarrow$-continuous.*

(a′) *Assume that every chain $\mathcal{C}$ in $\mathcal{L}$ contains an increasing sequence C_n such that $\bigvee \mathcal{C} = \bigvee_{n \geq 1} C_n$. An operator $\psi : \mathcal{L} \to \mathcal{M}$ is l.l.s.c. if and only if it is $\uparrow$-continuous.*

Proof. The only if-part is obvious. To prove the if-part, suppose that $\mathcal{C}$ is a chain in $\mathcal{L}$. There is a nonincreasing sequence C_n in $\mathcal{C}$ such that $\bigwedge \mathcal{C} = \bigwedge_{n \geq 1} C_n$. Therefore, by the $\downarrow$-continuity of ψ,

$$\psi(\bigwedge \mathcal{C}) = \psi(\bigwedge_{n \geq 1} C_n) = \bigwedge_{n \geq 1} \psi(C_n) \geq \bigwedge_{X \in \mathcal{C}} \psi(X).$$

But the inequality $\psi(\bigwedge \mathcal{C}) \leq \bigwedge_{X \in \mathcal{C}} \psi(X)$ is trivially satisfied, and the result follows. □

3.9 Example. The condition in Lemma 3.8(a) as well as in (a′) holds if $\mathcal{L} = \mathcal{P}(E)$, with E a countable set. To prove this, assume that $\mathcal{C}$ is a chain in $\mathcal{P}(E)$ and put $X = \bigcap \mathcal{C}$. Assume that $X^c = \{h_1, h_2, \ldots\}$. We construct a decreasing sequence $C_n \in \mathcal{C}$ such that $h_1, h_2, \ldots, h_n \notin C_n$. First choose $C_1 \in \mathcal{C}$ such that $h_1 \notin C_1$. Such a C_1 must exist, since otherwise $h_1 \in \bigcap \mathcal{C} = X$. Define $C_2 = C_1$ if $h_2 \notin C_1$; if $h_2 \in C_1$, then we can repeat the foregoing argument to get a $C_2 \subseteq C_1$ such that $h_2 \notin C_2$. It is obvious that $X \subseteq C_n$ hence $X \subseteq \bigcap_{n\geq 1} C_n$. On the other hand, our construction is such that $h \in X^c$ implies that $h \in C_n^c$ from some n onward. This means that $X^c \subseteq \bigcup_{n\geq 1} C_n^c$ or, equivalently, $X \supseteq \bigcap_{n\geq 1} C_n$. Thus, the assertion follows.

3.2. Adjunctions

Adjunctions play an important role in this book and show up naturally on several occasions. They constitute one of the main ingredients of the theory of morphological operators, the central theme of this book. This section gives several basic results concerning adjunctions; furthermore, it discusses some simple examples. Refer to Chapter 5 for a comprehensive study of adjunctions which are invariant under a given automorphism group on $\mathcal{L}$.

3.10 Definition. Let $\mathcal{L}, \mathcal{M}$ be complete lattices, ε an operator from $\mathcal{L}$ into $\mathcal{M}$, and δ an operator from $\mathcal{M}$ into $\mathcal{L}$. The pair (ε, δ) is called an *adjunction* between $\mathcal{L}$ and $\mathcal{M}$ if

$$\delta(Y) \leq X \iff Y \leq \varepsilon(X), \tag{3.4}$$

for all $X \in \mathcal{L}$, $Y \in \mathcal{M}$. If $\mathcal{L} = \mathcal{M}$, then (ε, δ) is called an adjunction on $\mathcal{L}$.

From the definition it follows that $\delta(O) = O$ and $\varepsilon(I) = I$. To see the first relation, take $Y = O$ in (3.4). Since $O \leq \varepsilon(X)$ holds for every $X \in \mathcal{L}$, it follows that $\delta(O) \leq X$ also holds for every $X \in \mathcal{L}$; therefore, $\delta(O) = O$. It is clear from the definition that the operators ε and δ in an adjunction are dual in the sense of the Duality Principle: if (ε, δ) is an adjunction between $\mathcal{L}$ and $\mathcal{M}$, then (δ, ε) is an adjunction between the dual lattices $\mathcal{M}'$ and $\mathcal{L}'$.

3.11 Examples.

(a) Given a lattice isomorphism ψ between the complete lattices $\mathcal{L}$ and $\mathcal{M}$ with inverse ψ^{-1}; the pair (ψ, ψ^{-1}) is an adjunction between $\mathcal{L}$ and $\mathcal{M}$, and the pair (ψ^{-1}, ψ) is an adjunction between $\mathcal{M}$ and $\mathcal{L}$.

(b) Recall the definition of $\overline{\mathbb{R}}, \overline{\mathbb{Z}}$ from Example 2.7(a). Define $\varepsilon : \overline{\mathbb{R}} \to \overline{\mathbb{Z}}$ as follows: $\varepsilon(t)$ is the largest integer that is less than or equal to t and $\varepsilon(\infty) =$

∞. Furthermore, define $\delta : \overline{\mathbb{Z}} \to \overline{\mathbb{R}}$ by $\delta(n) = n$. It is easy to check that (ε, δ) is an adjunction between $\overline{\mathbb{R}}$ and $\overline{\mathbb{Z}}$.

(c) Let $\mathcal{T}$ be a complete lattice with least element O; let E be an arbitrary nonempty set, and let $\mathsf{Fun}(E, \mathcal{T})$ be the complete lattice of functions mapping E into $\mathcal{T}$; see Example 2.10. Define, for a given $t \in \mathcal{T}$, the operator $\varepsilon : \mathsf{Fun}(E, \mathcal{T}) \to \mathcal{P}(E)$ by

$$\varepsilon(F) = \{x \in E \mid F(x) \geq t\}.$$

Furthermore, define the operator $\delta : \mathcal{P}(E) \to \mathsf{Fun}(E, \mathcal{T})$ by

$$\delta(X)(x) = \begin{cases} t, & \text{if } x \in X, \\ O, & \text{otherwise.} \end{cases}$$

Then (ε, δ) is an adjunction between $\mathsf{Fun}(E, \mathcal{T})$ and $\mathcal{P}(E)$. In Chapters 10 and 11, which discuss the extension of set operators to function operators, these operators play an important role.

(d) Let D, E be arbitrary sets, $A \subseteq D$, $B \subseteq E$, $\mathcal{L} = \mathcal{P}(D)$, and $\mathcal{M} = \mathcal{P}(E)$. Define $\varepsilon : \mathcal{L} \to \mathcal{M}$ and $\delta : \mathcal{M} \to \mathcal{L}$ by

$$\varepsilon(X) = \begin{cases} E, & \text{if } A \subseteq X, \\ B, & \text{otherwise,} \end{cases}$$

$$\delta(Y) = \begin{cases} \varnothing, & \text{if } Y \subseteq B, \\ A, & \text{otherwise.} \end{cases}$$

The pair (ε, δ) constitutes an adjunction between $\mathcal{L}$ and $\mathcal{M}$.

3.12 Definition. Let $\mathcal{L}, \mathcal{M}$ be complete lattices. An operator $\varepsilon : \mathcal{L} \to \mathcal{M}$ satisfying $\varepsilon(\bigwedge_{i \in I} X_i) = \bigwedge_{i \in I} \varepsilon(X_i)$ for every collection $X_i \in \mathcal{L}$, $i \in I$, is called an *erosion*. An operator $\delta : \mathcal{M} \to \mathcal{L}$ satisfying $\delta(\bigvee_{i \in I} Y_i) = \bigvee_{i \in I} \delta(Y_i)$ for every collection $Y_i \in \mathcal{M}$, $i \in I$, is called a *dilation*.

Erosions and dilations are dual notions in the sense of the Duality Principle.

3.13 Theorem. *Let (ε, δ) be an adjunction between $\mathcal{L}$ and $\mathcal{M}$. Then ε is an erosion and δ is a dilation.*

Proof. Suppose that (ε, δ) is an adjunction between $\mathcal{L}$ and $\mathcal{M}$; we show that ε is an erosion. Suppose $X_i \in \mathcal{L}$ for $i \in I$; given $Y \in \mathcal{M}$, it holds that $\delta(Y) \leq$

$\bigwedge_{i \in I} X_i$ if and only if $\delta(Y) \leq X_i$ for every $i \in I$. This, however, is equivalent to $Y \leq \varepsilon(X_i)$ for every $i \in I$; that is, $Y \leq \bigwedge_{i \in I} \varepsilon(X_i)$. On the other hand, by the adjunction relation, $\delta(Y) \leq \bigwedge_{i \in I} X_i$ if and only if $Y \leq \varepsilon(\bigwedge_{i \in I} X_i)$. But this implies $\varepsilon(\bigwedge_{i \in I} X_i) = \bigwedge_{i \in I} \varepsilon(X_i)$. □

Theorem 3.13 shows that the left operator in an adjunction is an erosion and that the right operator is a dilation.

We state some elementary properties of adjunctions; in particular, we show that to every erosion ε there corresponds a unique dilation δ such that (ε, δ) forms an adjunction. Dually, to every dilation δ there corresponds a unique erosion ε such that (ε, δ) forms an adjunction.

3.14 Proposition. *Let (ε, δ) be an adjunction between $\mathcal{L}$ and $\mathcal{M}$; the following relations hold:*

$$\varepsilon\delta \geq \mathrm{id}_{\mathcal{M}}; \tag{3.5}$$

$$\delta\varepsilon \leq \mathrm{id}_{\mathcal{L}}; \tag{3.6}$$

$$\varepsilon\delta\varepsilon = \varepsilon; \tag{3.7}$$

$$\delta\varepsilon\delta = \delta; \tag{3.8}$$

$$\varepsilon(X) = \bigvee \{Y \in \mathcal{M} \mid \delta(Y) \leq X\}; \tag{3.9}$$

$$\delta(Y) = \bigwedge \{X \in \mathcal{L} \mid Y \leq \varepsilon(X)\}. \tag{3.10}$$

Proof. If (ε, δ) is an adjunction between $\mathcal{L}$ and $\mathcal{M}$, then

$$\delta(Y) \leq X \iff Y \leq \varepsilon(X),$$

for $X \in \mathcal{L}$, $Y \in \mathcal{M}$. Now (3.5) follows if we substitute $X = \delta(Y)$; similarly (3.6) follows if we substitute $Y = \varepsilon(X)$. Since ε, δ are increasing and (3.5) and (3.6) hold, it follows that $\varepsilon\delta\varepsilon \geq \varepsilon$ and $\varepsilon\delta\varepsilon \leq \varepsilon$, respectively; therefore, $\varepsilon\delta\varepsilon = \varepsilon$. The identity $\delta\varepsilon\delta = \delta$ is proved analogously. Finally, (3.9) follows from the observation

$$\varepsilon(X) = \bigvee \{Y \mid Y \leq \varepsilon(X)\} = \bigvee \{Y \mid \delta(Y) \leq X\}.$$

Eq. (3.10) follows by a similar argument. □

Theorem 3.13 states that the left operator in an adjunction is an erosion and that the right operator is a dilation. In fact, one can show that dilations and erosions always occur in pairs. Suppose that ε is an erosion; define the operator δ as in (3.10), that is, $\delta(Y) = \bigwedge \{X' \mid X' \in \mathcal{L} \text{ and } Y \leq \varepsilon(X')\}$. If

$\delta(Y) \leq X$, then, by applying ε on both sides and using that it distributes over infima, it follows that

$$\varepsilon\delta(Y) = \bigwedge\{\varepsilon(X') \mid X' \in \mathcal{L} \text{ and } Y \leq \varepsilon(X')\} \leq \varepsilon(X);$$

in particular, $Y \leq \varepsilon(X)$. Conversely, if $Y \leq \varepsilon(X)$, then by definition $\delta(Y) \leq X$; therefore, (ε, δ) is an adjunction. Dually, one shows that to every dilation $\delta : \mathcal{M} \to \mathcal{L}$ there corresponds a unique erosion $\varepsilon : \mathcal{L} \to \mathcal{M}$ such that (ε, δ) forms an adjunction. These facts are summarized in the following result.

3.15 Proposition. *To every erosion ε there corresponds a unique dilation δ such that (ε, δ) constitutes an adjunction. Dually, to every dilation δ there corresponds a unique erosion ε such that (ε, δ) constitutes an adjunction.*

If (ε, δ) is an adjunction, then ε is called the *left adjoint* of δ, and δ is called the *right adjoint* of ε.

3.16 Proposition.

(a) *Let (ε, δ) and (ε', δ') be adjunctions between $\mathcal{L}$ and $\mathcal{M}$. Then $\varepsilon' \leq \varepsilon$ if and only if $\delta' \geq \delta$.*

(b) *Let $(\varepsilon_i, \delta_i)$ be an adjunction between $\mathcal{L}$ and $\mathcal{M}$ for every i in the index set I. Then $(\bigwedge_{i \in I} \varepsilon_i, \bigvee_{i \in I} \delta_i)$ is an adjunction between $\mathcal{L}$ and $\mathcal{M}$ as well.*

(c) *Let (ε, δ) be an adjunction between $\mathcal{L}$ and $\mathcal{M}$, and let (ε', δ') be an adjunction between $\mathcal{M}$ and $\mathcal{N}$. Then $(\varepsilon'\varepsilon, \delta\delta')$ is an adjunction between $\mathcal{L}$ and $\mathcal{N}$.*

Proof. The proof is rather straightforward if one uses the adjunction relation (3.4). For the sake of exposition, the proof of (a) will be given. Assume $\varepsilon' \leq \varepsilon$ and take $Y \in \mathcal{M}$. From (3.5) it follows that $Y \leq \varepsilon'\delta'(Y)$ and hence that $Y \leq \varepsilon\delta'(Y)$. Using that (ε, δ) is an adjunction, it follows that $\delta(Y) \leq \delta'(Y)$, which was to be shown. The reverse implication follows by duality. □

We can give a more abstract formulation of the results stated in (a) and (b). Let $\mathcal{E}(\mathcal{L}, \mathcal{M})$ be the family of all erosions from $\mathcal{L}$ to $\mathcal{M}$, and let $\mathcal{D}(\mathcal{M}, \mathcal{L})$ be the family of all dilations from $\mathcal{M}$ to $\mathcal{L}$. Then $\mathcal{E}(\mathcal{L}, \mathcal{M})$ is an underlattice of $\mathcal{O}^+(\mathcal{L}, \mathcal{M})$, but *not* a sublattice: it has the same infimum as $\mathcal{O}^+(\mathcal{L}, \mathcal{M})$, but not the same supremum because the pointwise supremum of a family of erosions is not an erosion in general; similar remarks apply to $\mathcal{D}(\mathcal{M}, \mathcal{L})$. Define $H : \mathcal{E}(\mathcal{L}, \mathcal{M}) \to \mathcal{D}(\mathcal{M}, \mathcal{L})$ as the operator which maps an erosion ε onto its right adjoint. Then H is a bijection and by Proposition 3.16(a) it reverses the ordering: if $\varepsilon \le \varepsilon'$, then $H(\varepsilon) \ge H(\varepsilon')$. So H is a dual automorphism between $\mathcal{E}(\mathcal{L}, \mathcal{M})$ and $\mathcal{D}(\mathcal{M}, \mathcal{L})$. In particular, $H(\bigwedge_{i\in I} \varepsilon_i) = \bigvee_{i\in I} H(\varepsilon_i)$, which is a reformulation of Proposition 3.16(b).

In the chapters to follow the reader will get acquainted with various adjunctions relevant to the field of mathematical morphology. We consider two examples; the first one is an application from number theory and has nothing to do with morphology.

3.17 Examples.

(a) Let $\mathbb{N}$ be the complete lattice of nonnegative integers ordered by "$m \le n$ if m is a divisor of n"; cf. Example 2.7(b). Let P be the subset of all prime numbers. Given an integer n and a prime number p, define $\kappa(p, n)$ as the power of p occurring in the prime decomposition of n; in other words, $\kappa(p, n)$ is the largest integer k such that p^k divides n. For example, $\kappa(2, 24) = 3$. Define the operator $\varepsilon : \mathbb{N} \to \mathbb{N}$ by $\varepsilon(0) = 0$, $\varepsilon(1) = 1$, and $\varepsilon(n) = \Pi\{p \in P \mid \kappa(p, n) \ge 1\}$ (where Π denotes product) if $n > 1$. For example, $\varepsilon(24) = 2 \times 3 = 6$. Furthermore, define the operator $\delta : \mathbb{N} \to \mathbb{N}$ by $\delta(n) = n$ if $\kappa(p, n) = 0$ or 1 for every $p \in P$, and $\delta(n) = 0$ otherwise. We now show that

$$\delta(m) \le n \iff m \le \varepsilon(n).$$

The only nontrivial case is where $\delta(m) = m$. In this case $\kappa(p, m) = 0$ or 1 for every $p \in P$. Since $\delta(m) = m \le n$, we have $\kappa(p, n) \ge \kappa(p, m)$. But then $\kappa(p, \varepsilon(n)) \ge \kappa(p, m)$ as well; this implies $m \le \varepsilon(n)$. This proves that (ε, δ) defines an adjunction on $\mathbb{N}$. Note that in this particular example $\delta\varepsilon = \varepsilon^2 = \varepsilon \le \mathsf{id}$ and $\varepsilon\delta = \delta^2 = \delta \ge \mathsf{id}$.

(b) Recall the definition of a Boolean function from Section 2.4. Every multiplicative Boolean function $e\{0, 1\}^n \to \{0, 1\}$ given by

$$e(u_1, \ldots, u_n) = (c_1 + u_1)(c_2 + u_2) \cdots (c_n + u_n),$$

where $c_i \in \{0, 1\}$, is an erosion. We leave as an exercise to the reader to show that the adjoint dilation $d : \{0, 1\} \to \{0, 1\}^n$ is given by $d(0) = (0, 0, \ldots, 0)$ and $d(1) = (\bar{c}_1, \ldots, \bar{c}_n)$.

The remainder of this section is devoted to the case where both $\mathcal{L}$ and $\mathcal{M}$ have a negation.

3.18 Proposition. *Consider the complete lattices $\mathcal{L}$ and $\mathcal{M}$ and the operators $\varepsilon : \mathcal{L} \to \mathcal{M}$ and $\delta : \mathcal{M} \to \mathcal{L}$. If both lattices have a negation, then the pair (ε, δ)*

forms an adjunction between $\mathcal{L}$ and $\mathcal{M}$ if and only if the pair $(\delta^*, \varepsilon^*)$ forms an adjunction between $\mathcal{M}$ and $\mathcal{L}$.

Proof. We prove only the only if-statement. Assume that (ε, δ) defines an adjunction between $\mathcal{L}$ and $\mathcal{M}$. To show that $(\delta^*, \varepsilon^*)$ is an adjunction between $\mathcal{M}$ and $\mathcal{L}$ one must verify that

$$\varepsilon^*(X) \le Y \iff X \le \delta^*(Y),$$

for $X \in \mathcal{L}$ and $Y \in \mathcal{M}$. The inequality $\varepsilon^*(X) \le Y$ is equivalent to $(\varepsilon(X^*))^* \le Y$, that is, $Y^* \le \varepsilon(X^*)$. As (ε, δ) is an adjunction, this gives $\delta(Y^*) \le X^*$, that is, $\delta^*(Y) = (\delta(Y^*))^* \ge X$. This proves the implication $\Rightarrow$. The other implication is proved analogously. □

We give an alternative characterization of the negative adjunction $(\delta^*, \varepsilon^*)$ in the case where $\mathcal{L} = \mathcal{P}(D)$ and $\mathcal{M} = \mathcal{P}(E)$ for some arbitrary sets D, E. Define, for $x \in D$,

$$\check{\delta}(\{x\}) = \{y \in E \mid x \in \delta(\{y\})\}.$$

The operator $\check{\delta} : \mathcal{P}(D) \to \mathcal{P}(E)$ given by

$$\check{\delta}(X) = \bigcup_{x \in X} \check{\delta}(\{x\}) \tag{3.11}$$

defines a dilation. We call $\check{\delta}$ the *reflection* of δ. Denote its left adjoint erosion by $\check{\varepsilon}$.

3.19 Proposition. *If (ε, δ) is an adjunction between $\mathcal{P}(D)$ and $\mathcal{P}(E)$, then*

$$\check{\delta} = \varepsilon^* \qquad \text{and} \qquad \check{\varepsilon} = \delta^*. \tag{3.12}$$

Proof. It suffices to prove that $(\delta^*, \check{\delta})$ is an adjunction between $\mathcal{P}(E)$ and $\mathcal{P}(D)$. In fact, by Proposition 3.18, $(\delta^*, \varepsilon^*)$ also defines an adjunction; by uniqueness of adjoints this means $\check{\delta} = \varepsilon^*$. Then the second relation follows by adjunction. To prove that $(\delta^*, \check{\delta})$ is an adjunction we must show that for $X \subseteq D$ and $Y \subseteq E$,

$$\check{\delta}(X) \subseteq Y \iff X \subseteq \delta^*(Y).$$

$\Rightarrow$: Assume that $\check{\delta}(X) \subseteq Y$ and take $x \in X$; we must show that $x \in \delta^*(Y) = [\delta(Y^c)]^c$. Suppose $x \in \delta(Y^c)$; this gives that $\check{\delta}(\{x\}) \cap Y^c \ne \varnothing$ and hence that $\check{\delta}(X) \cap Y^c \ne \varnothing$. This contradicts the assumption that $\check{\delta}(X) \subseteq Y$.

$\Leftarrow$: Assume $X \subseteq \delta^*(Y)$, that is, $\delta(Y^c) \subseteq X^c$. Take $y \in \check{\delta}(X)$; it follows that $\delta(\{y\}) \cap X \neq \emptyset$. Since $\delta(Y^c) \subseteq X^c$, this implies $y \notin Y^c$, that is, $y \in Y$. □

3.3. Openings and closings

Openings and closings are operations which occur in several branches of mathematics, for instance, in topology; readers not acquainted with topology may refer to Chapter 7. Let E be a topological space; with every subset X of E one can associate its interior X°, the largest open set contained in X, and its closure X^-, the smallest closed set containing X. The following properties hold:

(i) $X \subseteq Y$ implies $X^\circ \subseteq Y^\circ$ and $X^- \subseteq Y^-$;
(ii) $X^\circ \subseteq X \subseteq X^-$;
(iii) $(X^\circ)^\circ = X^\circ$ and $(X^-)^- = X^-$.

For obvious reasons, the mapping $X \mapsto X^\circ$ on $\mathcal{P}(E)$ is called an opening, whereas the mapping $X \mapsto X^-$ is called a closing (or closure; Example 7.3 discusses the relation between closing and closure in more detail).

3.20 Definition. An operator ψ on the complete lattice $\mathcal{L}$ is called
(a) *idempotent* if $\psi^2 = \psi$ (here $\psi^2 = \psi \circ \psi$);
(b) a *filter* if ψ is increasing and idempotent;
(c) an *opening* if ψ is increasing, anti-extensive, and idempotent;
(c′) a *closing* if ψ is increasing, extensive, and idempotent.

Note that openings and closings are dual notions in the sense of the Duality Principle: ψ is an opening on $\mathcal{L}$ if and only if it is a closing on $\mathcal{L}'$. In this book, openings are denoted by α, α', etc., and closings by β, β', etc. Recall that the invariance domain $\mathrm{Inv}(\psi)$ of an operator ψ is defined as the set of all fixpoints of ψ. Furthermore, we define the *range* of ψ, $\mathrm{Ran}(\psi)$, by

$$\mathrm{Ran}(\psi) = \{\psi(X) \mid X \in \mathcal{L}\}.$$

It is obvious that $\mathrm{Inv}(\psi) \subseteq \mathrm{Ran}(\psi)$ for every operator ψ.

3.21 Lemma. *An operator ψ on a complete lattice is idempotent if and only if* $\mathrm{Inv}(\psi) = \mathrm{Ran}(\psi)$.

Proof. "only if": if $X \in \mathrm{Ran}(\psi)$, then $X = \psi(Y)$ for some $Y \in \mathcal{L}$. Now, if ψ is idempotent, then $\psi(X) = \psi^2(Y) = \psi(Y) = X$, and so $X \in \mathrm{Inv}(\psi)$.

"if": $\mathrm{Ran}(\psi) \subseteq \mathrm{Inv}(\psi)$ implies $\psi^2(X) = \psi(X)$ for $X \in \mathcal{L}$; thus ψ is idempotent. □

3.22 Lemma. *Let ψ be an increasing operator on a complete lattice $\mathcal{L}$.*
(a) *If ψ is anti-extensive, then* $\mathrm{Inv}(\psi)$ *is sup-closed.*
(a′) *If ψ is extensive, then* $\mathrm{Inv}(\psi)$ *is inf-closed.*

Proof. Assume that ψ is increasing and anti-extensive and that $X_i \in \mathrm{Inv}(\psi)$ for $i \in I$. Obviously, $\psi(\bigvee_{i\in I} X_i) \le \bigvee_{i\in I} X_i$ because $\psi \le \mathrm{id}$. Since ψ is increasing, it follows that $\psi(\bigvee_{i\in I} X_i) \ge \psi(X_i) = X_i$; therefore, $\psi(\bigvee_{i\in I} X_i) \ge \bigvee_{i\in I} X_i$. This proves the assertion. □

In Section 6.1 we will discuss the relation between sup-closed subsets of $\mathcal{L}$ and increasing, anti-extensive operators (in particular openings) in more detail. Here we confine ourselves to the following result; a more general statement can be found in Theorem 6.9.

3.23 Theorem. *Let $\mathcal{L}$ be a complete lattice.*
(a) *If α is an opening on $\mathcal{L}$, then*

$$\alpha(X) = \bigvee\{Y \in \mathrm{Inv}(\alpha) \mid Y \le X\}. \tag{3.13}$$

(a′) *If β is a closing on $\mathcal{L}$, then*

$$\beta(X) = \bigwedge\{Y \in \mathrm{Inv}(\beta) \mid Y \ge X\}. \tag{3.14}$$

Proof. Let α be an opening. If $Y \in \mathrm{Inv}(\alpha)$ with $Y \le X$, then $\alpha(X) \ge \alpha(Y) = Y$ and $\ge$ in (3.13) follows. On the other hand, $\alpha(X) \in \mathrm{Inv}(\alpha)$ and $\alpha(X) \le X$; now $\le$ follows. □

Let us apply this result to the opening $X \mapsto X^\circ$ on $\mathcal{P}(E)$, where E is topological space. It says that X° is the union of all open sets contained in X; note that this is precisely the definition of X°. Proposition 6.5 will show that not the entire set $\mathrm{Inv}(\alpha)$ is needed to build α; it suffices to have a subset $\mathcal{M}$ which sup-generates $\mathrm{Inv}(\alpha)$, that is, $\langle \mathcal{M} \mid \vee \rangle = \mathrm{Inv}(\alpha)$. For instance, X° can be obtained as the union of all open balls contained within X.

3.24 Theorem. *Let $\mathcal{L}$ be a complete lattice.*
(a) *Let α_1, α_2 be openings on $\mathcal{L}$. The following assertions are equivalent:*
  (i) $\alpha_1 \le \alpha_2$;
  (ii) $\alpha_1\alpha_2 = \alpha_2\alpha_1 = \alpha_1$;
  (iii) $\mathrm{Inv}(\alpha_1) \subseteq \mathrm{Inv}(\alpha_2)$.
In particular, $\alpha_1 = \alpha_2$ if and only if $\mathrm{Inv}(\alpha_1) = \mathrm{Inv}(\alpha_2)$.
(a′) *Let β_1, β_2 be closings on $\mathcal{L}$. The following assertions are equivalent:*
  (i) $\beta_1 \ge \beta_2$;

(ii) $\beta_1\beta_2 = \beta_2\beta_1 = \beta_1$;
(iii) $\mathrm{Inv}(\beta_1) \subseteq \mathrm{Inv}(\beta_2)$.
In particular, $\beta_1 = \beta_2$ *if and only if* $\mathrm{Inv}(\beta_1) = \mathrm{Inv}(\beta_2)$.

Proof. Let α_1, α_2 be openings.

(i) $\Rightarrow$ (ii): If $\alpha_1 \leq \alpha_2$, then $\alpha_1\alpha_2 \geq \alpha_1\alpha_1 = \alpha_1$. As the reverse inequality holds trivially, one gets $\alpha_1\alpha_2 = \alpha_1$. The identity $\alpha_2\alpha_1 = \alpha_1$ is proved analogously.

(ii) $\Rightarrow$ (iii): Let $X \in \mathrm{Inv}(\alpha_1)$, that is, $\alpha_1(X) = X$. Then $\alpha_2(X) = \alpha_2\alpha_1(X) = \alpha_1(X) = X$, and so $X \in \mathrm{Inv}(\alpha_2)$.

(iii) $\Rightarrow$ (i): As $\alpha_1(X) \in \mathrm{Inv}(\alpha_1) \subseteq \mathrm{Inv}(\alpha_2)$, one gets $\alpha_1(X) = \alpha_2\alpha_1(X) \leq \alpha_2(X)$. □

In the previous section we have already met an important example of openings and closings. In fact, let (ε, δ) be an adjunction between the complete lattices $\mathcal{L}$ and $\mathcal{M}$; then $\delta\varepsilon \leq \mathrm{id}_{\mathcal{L}}$ and $\varepsilon\delta\varepsilon = \varepsilon$, and hence $\delta\varepsilon\delta\varepsilon = \delta\varepsilon$. This implies that $\delta\varepsilon$ is an opening on $\mathcal{L}$; dually, $\varepsilon\delta$ is a closing on $\mathcal{M}$.

3.25 Theorem. *If (ε, δ) is an adjunction between the complete lattices $\mathcal{L}$ and $\mathcal{M}$, then $\delta\varepsilon$ is an opening on $\mathcal{L}$ and $\varepsilon\delta$ is a closing on $\mathcal{M}$. Furthermore,*

$$\mathrm{Inv}(\delta\varepsilon) = \mathrm{Ran}(\delta) \qquad \textit{and} \qquad \mathrm{Inv}(\varepsilon\delta) = \mathrm{Ran}(\varepsilon). \tag{3.15}$$

The equalities in (3.15) are straightforward. In fact, one can show that every opening and closing can be obtained by composing the dilation and the erosion of some adjunction. Refer to Example 6.19 for a precise formulation.

3.26 Definition. If (ε, δ) is an adjunction on $\mathcal{L}$, then the opening $\delta\varepsilon$ is called an *adjunctional opening*, and the closing $\varepsilon\delta$ is called an *adjunctional closing*.

Note that in this definition it is assumed explicitly that (ε, δ) is an adjunction acting on one lattice and not between two different ones.

We present some very simple (and somewhat artificial) examples; in forthcoming chapters the reader will become acquainted with some real-world examples.

3.27 Examples.

(a) The mapping $\alpha : \overline{\mathbb{R}} \to \overline{\mathbb{R}}$ given by "$\alpha(t)$ is the largest integer that is less than or equal to t, and $\alpha(\infty) = \infty$" defines an opening. Note that α coincides with the mapping ε defined in Example 3.11(b); the only difference here is that ε maps into $\overline{\mathbb{Z}}$.

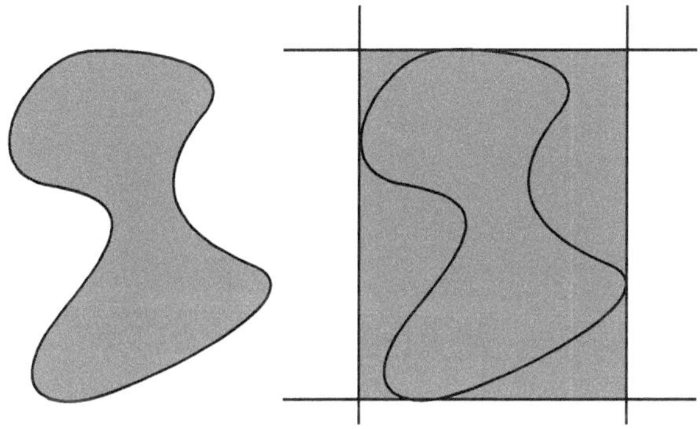

Figure 3.1 The closing $\beta(X)$ of Example 3.27(c) yields the smallest closed rectangle with horizontal and vertical edges which contains the original set X.

(b) The mapping ε defined in Example 3.17(a) is both an erosion and an opening.

(c) For a subset $X \subseteq \mathbb{R}^2$ we define $\beta(X)$ as the intersection of all closed half-planes with horizontal or vertical boundaries which contain X; see Fig. 3.1.

Note that $\beta(X)$ is a closed rectangle if X is bounded. Analogously, the closed convex hull operation, introduced in Example 2.16(b), and to be discussed in detail in Chapter 9, defines a closing.

(d) Consider the complete lattice $\mathbb{B}_n$ of Boolean functions of n variables. Let a be a fixed element of $\mathbb{B}_n$. Then the operator $b \mapsto ab$ on $\mathbb{B}_n$ defines an opening (as well as an erosion). Dually the operator $b \mapsto a + b$ defines a closing (and also a dilation).

We conclude this section with the following important result.

3.28 Theorem. *Let $\mathcal{L}$ be a complete lattice.*

(a) *Let α_i be openings for every i in some index set I; then $\bigvee_{i\in I} \alpha_i$ is an opening as well.*

(a′) *Let β_i be closings for every i in some index set I; then $\bigwedge_{i\in I} \beta_i$ is a closing as well.*

Proof. It is obvious that $\alpha = \bigvee_{i\in I} \alpha_i$ is anti-extensive and increasing; in particular, $\alpha^2 \leq \alpha$. On the other hand, $\alpha^2 \geq \alpha_i^2 = \alpha_i$, and hence $\alpha^2 \geq \bigvee_{i\in I} \alpha_i = \alpha$. This means that α is an opening. □

3.29 Example. We present a simple example which shows that neither the infimum nor the composition of two openings is an opening in general. Let

E be a nonempty set and $\mathcal{L} = \mathcal{P}(E)$. Define, for $A \subseteq E$, the opening α_A on $\mathcal{P}(E)$ by

$$\alpha_A(X) = \begin{cases} A, & \text{if } A \subseteq X, \\ \varnothing, & \text{otherwise.} \end{cases}$$

If A, B are two different subsets of E such that $\varnothing \neq A \cap B$, then

$$(\alpha_A \wedge \alpha_B)(X) = \begin{cases} A \cap B, & \text{if } A \cup B \subseteq X, \\ \varnothing, & \text{otherwise.} \end{cases}$$

This implies that $(\alpha_A \wedge \alpha_B)^2 = o$; hence $(\alpha_A \wedge \alpha_B)^2 \neq \alpha_A \wedge \alpha_B$. In particular, $\alpha_A \wedge \alpha_B$ is not an opening.

If B is a strict subset of A, then

$$\alpha_B \alpha_A(X) = \begin{cases} B, & \text{if } A \subseteq X, \\ \varnothing, & \text{otherwise.} \end{cases}$$

This gives $\alpha_A \alpha_B \alpha_A = o$, and so $(\alpha_B \alpha_A)^2 = o \neq \alpha_B \alpha_A$. Therefore, $\alpha_B \alpha_A$ is not an opening. In Example 13.34 it is shown that, under certain conditions, the iterates of $\alpha_1 \wedge \alpha_2$ and $\alpha_2 \alpha_1$, where α_1, α_2 are openings, converge to an opening.

Assume that α is an opening on $\mathcal{L}$; we say that $X \in \mathcal{L}$ is *open* with respect to α if $\alpha(X) = X$. Analogously, an element X is called *closed* with respect to the closing β if $\beta(X) = X$. It is obvious that $\alpha(X)$ is the largest element $\leq X$ that is open with respect to α, and that $\beta(X)$ is the smallest element $\geq X$ that is closed with respect to β.

3.30 Proposition. *Let (ε, δ) be an adjunction between $\mathcal{L}$ and $\mathcal{M}$.*

(a) *$X \in \mathcal{L}$ is open with respect to the opening $\delta\varepsilon$ if and only if $X = \delta(Y)$ for some $Y \in \mathcal{M}$.*

(a′) *$Y \in \mathcal{M}$ is closed with respect to the closing $\varepsilon\delta$ if and only if $Y = \varepsilon(X)$ for some $X \in \mathcal{L}$.*

Proof. If X is open with respect to $\delta\varepsilon$, then $\delta\varepsilon(X) = X$; that is, $X = \delta(Y)$, where $Y = \varepsilon(X)$. Conversely, if $X = \delta(Y)$, then $\delta\varepsilon(X) = \delta\varepsilon\delta(Y) = \delta(Y) = X$ by (3.8). □

3.31 Example. (Extension of openings)

Let $\mathcal{M}$ be a subset of a complete lattice $\mathcal{L}$, and let $\alpha : \mathcal{M} \to \mathcal{M}$ be a mapping with the properties of an opening; that is, α is increasing, anti-extensive, and idempotent. Define the extension $\overline{\alpha}$ to $\mathcal{L}$ in the following way:

$$\overline{\alpha}(X) = \bigvee\{\alpha(Y) \mid Y \in \mathcal{M} \text{ and } Y \leq X\}.$$

We show that $\overline{\alpha}$ is an opening on $\mathcal{L}$ whose restriction to $\mathcal{M}$ is α. It is easy to show that $\overline{\alpha}$ is increasing and anti-extensive and that $\overline{\alpha} = \alpha$ on $\mathcal{M}$. Thus it remains to demonstrate that $\overline{\alpha}$ is idempotent. From the anti-extensivity it is obvious that $\overline{\alpha}^2 \leq \overline{\alpha}$. We prove the reverse inequality. If $Y \in \mathcal{M}$ and $Y \leq X$, then $\alpha(Y) \leq \overline{\alpha}(X)$ by definition. Applying $\overline{\alpha}$ on both sides, we get $\overline{\alpha}\alpha(Y) = \alpha^2(Y) = \alpha(Y) \leq \overline{\alpha}^2(X)$. But then $\overline{\alpha}(X) \leq \overline{\alpha}^2(X)$.

We call $\overline{\alpha}$ the extension of α to $\mathcal{L}$. Dually, if $\beta : \mathcal{M} \to \mathcal{M}$ is a mapping with the properties of a closing, one can define an extension on $\mathcal{L}$ as follows

$$\overline{\beta}(X) = \bigwedge\{\beta(Y) \mid Y \in \mathcal{M} \text{ and } Y \geq X\}.$$

Then $\overline{\beta}$ is a closing whose restriction to $\mathcal{M}$ is β.

3.4. Conditional operators

Assume that $\mathcal{L}$ is a complete Boolean lattice and that $M \in \mathcal{L}$. Define $\mathcal{L}_{(\leq M)}$ as the set of all $X \in \mathcal{L}$ such that $X \leq M$. We call M the *mask element*. With the partial ordering inherited from $\mathcal{L}$ the set $\mathcal{L}_{(\leq M)}$ becomes a complete lattice with largest element M. Moreover, $\mathcal{L}_{(\leq M)}$ is Boolean, the negative operator being given by $\nu(X \mid\leq M) = X^* \wedge M$.

Suppose that (ε, δ) is an adjunction on $\mathcal{L}$; define the operators $\delta(\cdot \mid\leq M)$ and $\varepsilon(\cdot \mid\leq M)$ on $\mathcal{L}_{(\leq M)}$ by

$$\delta(X \mid\leq M) = \delta(X) \wedge M, \tag{3.16}$$

$$\varepsilon(X \mid\leq M) = \varepsilon(X \vee M^*) \wedge M. \tag{3.17}$$

3.32 Proposition. *The pair $(\varepsilon(\cdot \mid\leq M), \delta(\cdot \mid\leq M))$ defines an adjunction on $\mathcal{L}_{(\leq M)}$.*

Proof. We must show that for $X, Y \in \mathcal{L}_{(\leq M)}$,

$$\delta(X \mid\leq M) \leq Y \iff X \leq \varepsilon(Y \mid\leq M).$$

$\Rightarrow$: Assume that $\delta(X \mid\leq M) \leq Y$, that is, $\delta(X) \wedge M \leq Y$. Then $(\delta(X) \wedge M) \vee M^* = \delta(X) \vee M^* \leq Y \vee M^*$. This implies $\delta(X) \leq Y \vee M^*$. Since (ε, δ) is an adjunction, we get $X \leq \varepsilon(Y \vee M^*)$, and since $X \leq M$ we also get $X \leq \varepsilon(Y \vee M^*) \wedge M = \varepsilon(Y \mid\leq M)$.

$\Leftarrow$: If $X \le \varepsilon(Y \mid\le M)$, then $X \le \varepsilon(Y \vee M^*) \wedge M \le \varepsilon(Y \vee M^*)$. This means that $\delta(X) \le Y \vee M^*$, and therefore $\delta(X) \wedge M \le (Y \vee M^*) \wedge M = Y \wedge M = Y$. This proves the result. □

The operators $\varepsilon(\cdot \mid\le M)$ and $\delta(\cdot \mid\le M)$ are called the *conditional erosion* and *conditional dilation*, respectively. The compositions $\delta(\cdot \mid\le M)\varepsilon(\cdot \mid\le M)$ and $\varepsilon(\cdot \mid\le M)\delta(\cdot \mid\le M)$ are called the *conditional opening* and *conditional closing*, respectively. For the conditional closing one can easily derive the following expression:

$$\begin{aligned}\varepsilon\big(\delta(X \mid\le M) \mid\le M\big) &= \varepsilon\big((\delta(X) \wedge M) \vee M^*\big) \wedge M \\ &= \varepsilon\big(\delta(X) \vee M^*\big) \wedge M.\end{aligned}$$

Dually, define $\mathcal{L}_{(\ge M)}$ as the set of all $X \in \mathcal{L}$ for which $X \ge M$. This constitutes a Boolean lattice with complement operator $\nu(X \mid\ge M) = X^* \vee M$. Given the adjunction (ε, δ) on $\mathcal{L}$, we define

$$\delta(X \mid\ge M) = \delta(X \wedge M^*) \vee M, \tag{3.18}$$

$$\varepsilon(X \mid\ge M) = \varepsilon(X) \vee M. \tag{3.19}$$

The following analogue of Proposition 3.32 holds.

3.33 Proposition. *The pair* $(\varepsilon(\cdot \mid\ge M), \delta(\cdot \mid\ge M))$ *defines an adjunction on* $\mathcal{L}_{(\ge M)}$.

We call $\delta(\cdot \mid\ge M)$ and $\varepsilon(\cdot \mid\ge M)$ the *lower conditional dilation* and *erosion*, respectively. Note that we should, strictly speaking, refer to $\delta(\cdot \mid\le M)$ and $\varepsilon(\cdot \mid\le M)$ as the *upper conditional dilation* and *erosion*. Since the nomenclature "conditional dilation and erosion" has become standard in the literature, and since the upper conditional operators are used far more often in practice, we usually omit the adjective upper. Section 9.5 presents a comprehensive discussion of conditional (and geodesic) operators for binary images; furthermore, refer to Section 11.10 for applications of these operators to grey-scale images.

This section will be concluded with a formulation of the *local knowledge principle*. Here one should think of M as a window through which one envisages X, the object of interest. In other words, only $X \wedge M$ is known. To what extent can this "local knowledge" be used to compute $\delta(X)$?

3.34 Proposition. (Local Knowledge Principle)
Consider a complete Boolean lattice $\mathcal{L}$, an element $M \in \mathcal{L}$, and a dilation δ on $\mathcal{L}$. The largest element $W \in \mathcal{L}$ satisfying

$$\delta(X) \wedge W = \delta(X \wedge M) \wedge W, \qquad \textit{for all } X \in \mathcal{L}, \tag{3.20}$$

is $W = \delta^*(M)$.

Proof. Assume that W obeys (3.20). Substituting $X = M^*$ gives $\delta(M^*) \wedge W = \delta(O) \wedge W = O \wedge W = O$. Then $W \leq (\delta(M^*))^* = \delta^*(M)$. It remains to show that $W = \delta^*(M)$ solves (3.20). The inequality $\geq$ is obvious; to prove $\leq$ note that

$$\begin{aligned}\delta(X) &= \delta(X \wedge (M \vee M^*)) = \delta((X \wedge M) \vee (X \wedge M^*)) \\ &= \delta(X \wedge M) \vee \delta(X \wedge M^*).\end{aligned}$$

Then

$$\begin{aligned}\delta(X) \wedge \delta^*(M) &= \left[\delta(X \wedge M) \wedge \delta^*(M)\right] \vee \left[\delta(X \wedge M^*) \wedge \delta^*(M)\right] \\ &\leq \left[\delta(X \wedge M) \wedge \delta^*(M)\right] \vee \left[\delta(M^*) \wedge \delta^*(M)\right] \\ &= \delta(X \wedge M) \wedge \delta^*(M).\end{aligned}$$

Here we have used that

$$\delta(M^*) \wedge \delta^*(M) = \delta(M^*) \wedge \left(\delta(M^*)\right)^* = O.$$

Therefore, $\delta^*(M)$ is a solution, and the proof is finished. □

We leave it to the reader to verify that this result can be extended to the more general case where δ is an operator between two different lattices.

3.5. Activity ordering

Throughout this section it will be assumed that $\mathcal{L}$ is a complete Boolean lattice. Our goal is to define a partial ordering on $\mathcal{O}(\mathcal{L})$, called *activity ordering*, which enables us to compare the impact of two operators on an object $X \in \mathcal{L}$. Before we present a formal definition we try to capture the underlying idea.

Let E be a nonempty space, e.g., $\mathbb{R}^2$. Recall that the symmetric difference $X \,\Delta\, Y$ of two sets $X, Y \subseteq E$ comprises all points that lie either in X

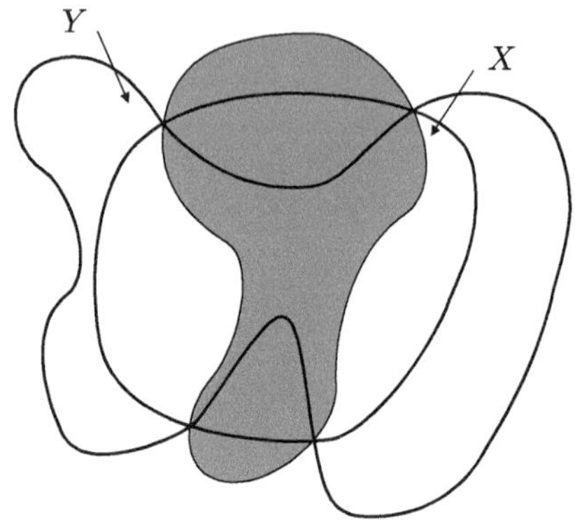

Figure 3.2 Let A be the shaded region; $X \preccurlyeq Y$ denotes that X resembles A more than Y.

or in Y but not in both; in other words,

$$X \triangle Y = (X \cap Y^c) \cup (X^c \cap Y).$$

Let A be a fixed subset of E; define a partial ordering $\preccurlyeq$ on $\mathcal{P}(E)$ as follows:

$$X \preccurlyeq Y \iff X \triangle A \subseteq Y \triangle A.$$

So $X \preccurlyeq Y$ expresses that X is more like A than Y; see Fig. 3.2. In fact,

$$X \preccurlyeq Y \iff \begin{cases} A \cap X \supseteq A \cap Y \\ A \cup X \subseteq A \cup Y. \end{cases}$$

It can be shown fairly easy that $(\mathcal{P}(E), \preccurlyeq)$ is a complete lattice (see what follows). It is obvious that the least element in this lattice is A and that the greatest element is A^c.

We use this example to define a partial ordering on $\mathcal{O}(\mathcal{L})$; here the role of A is being taken by the identity operator id. But note that one can choose any other operator here.

3.35 Definition. Given two operators ϕ, ψ on $\mathcal{L}$, we say that ψ *is more active than* ϕ, written $\phi \preccurlyeq \psi$, if

$$\mathrm{id} \wedge \psi \leq \mathrm{id} \wedge \phi, \tag{3.21}$$

$$\mathrm{id} \vee \psi \geq \mathrm{id} \vee \phi. \tag{3.22}$$

For example, if α_1, α_2 are openings on $\mathcal{L}$, then α_1 is more active than α_2 iff $\alpha_1 \leq \alpha_2$; cf. Theorem 3.24.

Let ν be the complement operator on $\mathcal{L}$. From the distributivity of $\mathcal{L}$ it follows that $\nu \vee (\mathrm{id} \wedge \psi) = (\nu \vee \mathrm{id}) \wedge (\nu \vee \psi) = \nu \vee \psi$. Using this observation (and its dual), it is easy to check that (3.21)–(3.22) are equivalent to

$$\nu \vee \psi \leq \nu \vee \phi, \tag{3.23}$$

$$\nu \wedge \psi \geq \nu \wedge \phi, \tag{3.24}$$

respectively.

3.36 Proposition. *Assume that $\mathcal{L}$ is a complete Boolean lattice. The relation $\preccurlyeq$ defines a partial ordering on $\mathcal{O}(\mathcal{L})$.*

Proof. Reflexivity and transitivity are easy. It remains to show that $\preccurlyeq$ is anti-symmetric. If $\psi \preccurlyeq \phi$ and $\phi \preccurlyeq \psi$, then

$$\begin{aligned} \psi &= \psi \wedge (\mathrm{id} \vee \nu) = (\psi \wedge \mathrm{id}) \vee (\psi \wedge \nu) \\ &= (\phi \wedge \mathrm{id}) \vee (\phi \wedge \nu) = \phi. \end{aligned}$$

This shows the result. □

Observe that the assumption that $\mathcal{L}$ is Boolean is essential. From now on we refer to $\preccurlyeq$ as the *activity ordering*. We list some basic properties of this ordering; the demonstration of these properties is easy and is therefore omitted.

Define, for a given operator ψ on $\mathcal{L}$, the operator ψ^{c} by $\psi^{\mathrm{c}}(X) = [\psi(X)]^*$.

3.37 Proposition. *Let $\mathcal{L}$ be an arbitrary complete Boolean lattice, and let ϕ, ψ, θ be arbitrary operators on $\mathcal{L}$. The following hold:*

(a) $\mathrm{id} \preccurlyeq \psi \preccurlyeq \nu$;

(b) $\psi \leq \phi \leq \mathrm{id} \Rightarrow \phi \preccurlyeq \psi$;

(b′) $\mathrm{id} \leq \phi \leq \psi \Rightarrow \phi \preccurlyeq \psi$;

(c) $\phi \preccurlyeq \psi$ *implies* $\theta \wedge \phi \preccurlyeq \theta \wedge \psi$ *and* $\theta \vee \phi \preccurlyeq \theta \vee \psi$;

(d) $\phi \preccurlyeq \psi \iff \phi^* \preccurlyeq \psi^* \iff \psi^{\mathrm{c}} \preccurlyeq \phi^{\mathrm{c}}$.

The following lemma provides a basis for subsequent results concerning the activity ordering.

3.38 Lemma. *Let $\mathcal{L}$ be a complete Boolean lattice, and let ϕ, ψ be operators on $\mathcal{L}$ such that $\phi \leq \psi$. Define*

$$\gamma = (\mathrm{id} \wedge \psi) \vee \phi = (\mathrm{id} \vee \phi) \wedge \psi, \tag{3.25}$$

$$\kappa = (\nu \wedge \psi) \vee \phi = (\nu \vee \phi) \wedge \psi. \tag{3.26}$$

Then:

(a) $\mathrm{id} \wedge \gamma = \mathrm{id} \wedge \psi$, $\mathrm{id} \vee \gamma = \mathrm{id} \vee \phi$, $\mathrm{id} \wedge \kappa = \mathrm{id} \wedge \phi$, $\mathrm{id} \vee \kappa = \mathrm{id} \vee \psi$.

(b) $\gamma \preccurlyeq \phi, \psi$; *and if* γ' *is any other operator that satisfies* $\gamma' \preccurlyeq \phi, \psi$, *then* $\gamma' \preccurlyeq \gamma$.

(c) $\phi, \psi \preccurlyeq \kappa$; *and if* κ' *is any other operator that satisfies* $\phi, \psi \preccurlyeq \kappa'$, *then* $\kappa \preccurlyeq \kappa'$.

(d) *For every operator* ξ *on* $\mathcal{L}$, *the equivalence* $\phi \leq \xi \leq \psi \iff \gamma \preccurlyeq \xi \preccurlyeq \kappa$ *holds.*

Proof. (a): Straightforward.

(b): Since $\mathrm{id} \wedge \gamma = \mathrm{id} \wedge \psi \geq \mathrm{id} \wedge \phi$ and $\mathrm{id} \vee \gamma = \mathrm{id} \vee \phi \leq \mathrm{id} \vee \psi$, it follows that $\gamma \preccurlyeq \phi, \psi$. If $\gamma' \preccurlyeq \phi, \psi$, then $\mathrm{id} \wedge \gamma' \geq \mathrm{id} \wedge \psi = \mathrm{id} \wedge \gamma$ and $\mathrm{id} \vee \gamma' \leq \mathrm{id} \vee \phi = \mathrm{id} \vee \gamma$, leading to $\gamma' \preccurlyeq \gamma$.

(c): Similar to (b).

(d): We prove $\Leftarrow$. The proof of $\Rightarrow$ proceeds along the same lines. Let $\gamma \preccurlyeq \xi \preccurlyeq \kappa$. Then

$$\begin{aligned} \phi &= (\mathrm{id} \vee \nu) \wedge \phi = (\mathrm{id} \wedge \phi) \vee (\nu \wedge \phi) \\ &= (\mathrm{id} \wedge \kappa) \vee (\nu \wedge \gamma) \leq (\mathrm{id} \wedge \xi) \vee (\nu \wedge \xi) \\ &= (\mathrm{id} \vee \nu) \wedge \xi = \xi. \end{aligned}$$

The inequality $\xi \leq \psi$ is proved analogously. □

We use Lemma 3.38 to define an activity infimum and supremum on $\mathcal{O}(\mathcal{L})$. In fact, let ψ_i, $i \in I$, be an arbitrary family of operators on $\mathcal{L}$. The operator γ' satisfies $\gamma' \preccurlyeq \psi_i$ for $i \in I$ iff

$$\mathrm{id} \wedge \gamma' \geq \mathrm{id} \wedge \psi_i \qquad \text{and} \qquad \mathrm{id} \vee \gamma' \leq \mathrm{id} \vee \psi_i,$$

for every $i \in I$. Putting $\phi = \bigwedge_{i \in I} \psi_i$ and $\psi = \bigvee_{i \in I} \psi_i$, this is equivalent to

$$\mathrm{id} \wedge \gamma' \geq \mathrm{id} \wedge \psi \qquad \text{and} \qquad \mathrm{id} \vee \gamma' \leq \mathrm{id} \vee \phi.$$

Here we have used that in a complete Boolean lattice the infinite distributivity laws hold; cf. Theorem 2.32. It follows with Lemma 3.38 that $\gamma = (\mathrm{id} \wedge \psi) \vee \phi$ satisfies these conditions and, moreover, that $\gamma' \preccurlyeq \gamma$ for every other such operator. But this means that γ is the greatest lower bound (i.e., infimum) of ψ_i with respect to the activity ordering; we denote this by

$$\gamma = \curlywedge_{i \in I} \psi_i.$$

Dually, $\kappa = (\nu \wedge \psi) \vee \phi$ is the least upper bound (i.e., supremum) with respect to the activity ordering, which we denote by

$$\kappa = \curlyvee_{i\in I} \psi_i.$$

3.39 Theorem. *Given a complete Boolean lattice $\mathcal{L}$, the poset $(\mathcal{O}(\mathcal{L}), \preccurlyeq)$ is a complete lattice. For an arbitrary family ψ_i, $i \in I$, of operators on $\mathcal{L}$, the activity infimum and supremum are given respectively by*

$$\curlywedge_{i\in I} \psi_i = \big[\mathrm{id} \wedge (\bigvee_{i\in I} \psi_i)\big] \vee (\bigwedge_{i\in I} \psi_i), \tag{3.27}$$

$$\curlyvee_{i\in I} \psi_i = \big[\nu \wedge (\bigvee_{i\in I} \psi_i)\big] \vee (\bigwedge_{i\in I} \psi_i). \tag{3.28}$$

3.40 Remark. Let ψ be an arbitrary operator on $\mathcal{L}$. The reader can easily verify the following identities:

$$o \curlywedge \psi = \mathrm{id} \wedge \psi, \quad o \curlyvee \psi = \nu \wedge \psi,$$
$$\iota \curlywedge \psi = \mathrm{id} \vee \psi, \quad \iota \curlyvee \psi = \nu \vee \psi.$$

These identities express the duality between the complete lattices $(\mathcal{O}(\mathcal{L}), \leq)$ and $(\mathcal{O}(\mathcal{L}), \preccurlyeq)$.

It is obvious (see also Proposition 3.37(a)) that id and ν are the least and greatest element of $(\mathcal{O}(\mathcal{L}), \preccurlyeq)$, respectively. Furthermore, every operator ψ has a unique complement in $(\mathcal{O}(\mathcal{L}), \preccurlyeq)$, namely, ψ^c.

3.41 Definition. Let ψ_i, $i \in I$, be arbitrary operators on the complete Boolean lattice $\mathcal{L}$. The operator $\gamma = \curlywedge_{i\in I} \psi_i$ is called the *centre* of the operators ψ_i, whereas the operator $\kappa = \curlyvee_{i\in I} \psi_i$ is called the *anti-centre*.

Observe that the centre is an increasing operator if every ψ_i is increasing; obviously this does not hold for the anti-centre.

We examine how the centre and anti-centre behave under negation. Using the explicit formulas (3.27) and (3.28) for the activity infimum and supremum, it follows immediately that

$$\curlywedge_{i\in I} \psi_i^* = \big(\curlywedge_{i\in I} \psi_i\big)^*, \qquad \curlyvee_{i\in I} \psi_i^* = \big(\curlyvee_{i\in I} \psi_i\big)^*, \tag{3.29}$$

$$\curlywedge_{i\in I} \psi_i^c = \big(\curlyvee_{i\in I} \psi_i\big)^c, \qquad \curlyvee_{i\in I} \psi_i^c = \big(\curlywedge_{i\in I} \psi_i\big)^c. \tag{3.30}$$

From these identities it is immediately clear how to obtain centres and anti-centres which are self-dual; see also Proposition 13.49.

3.42 Proposition. *If ψ_i, $i \in I$, is a family of operators such that with every ψ_i the negative operator ψ_i^* is also a member of the family, then both the centre $\gamma = \curlywedge_{i \in I} \psi_i$ and the anti-centre $\kappa = \curlyvee_{i \in I} \psi_i$ are self-dual operators.*

3.6. The centre on non-Boolean lattices

The previous section has discussed the centre of a family of operators on a complete Boolean lattice. Its definition was based on a somewhat peculiar partial ordering on the lattice of operators, namely, the activity ordering. The assumption that the underlying lattice $\mathcal{L}$ is Boolean is essential, for it guarantees that the binary relation $\preccurlyeq$ on $\mathcal{O}(\mathcal{L})$ is indeed a partial ordering. If this assumption is dropped, then $\preccurlyeq$ is no longer anti-symmetric; however, it is still transitive and reflexive.

In this section the results of the previous section will be extended in two ways. First, it is shown that many of the results can be generalized to the non-Boolean lattice $\mathrm{Fun}(E, \mathcal{T})$, where $\mathcal{T}$ is a complete chain. In fact, the activity ordering is still a partial ordering in this case. Furthermore, we can give an interesting geometric characterization of the centre operator in this case. Second, even if the lattice $\mathcal{L}$ is "only" modular, the definition of the centre operator still makes sense, although the activity ordering is not a partial ordering any longer.

3.43 Proposition. *Let $\mathcal{T}$ be a complete chain; the activity ordering $\preccurlyeq$ given by Definition 3.35 defines a partial ordering on* $\mathrm{Fun}(E, \mathcal{T})$.

To prove this, one has to establish the anti-symmetry property of $\preccurlyeq$, i.e., $\phi \preccurlyeq \psi$ and $\psi \preccurlyeq \phi$ implies $\phi = \psi$. This becomes easy if one makes the following observation: $\phi \preccurlyeq \psi$ iff $\phi(F)(x) \geq F(x)$ implies $\psi(F)(x) \geq \phi(F)(x) \geq F(x)$, and dually, if $\phi(F)(x) \leq F(x)$ implies $\psi(F)(x) \leq \phi(F)(x) \leq F(x)$, for every F and x.

Now we can define the centre γ of the operators $\phi \leq \psi$ in the usual way, i.e., $\gamma = (\mathrm{id} \wedge \psi) \vee \phi$.

3.44 Example. (Geometric characterization of the centre)
There exists a nice geometric characterization of the centre on the complete lattice $\mathrm{Fun}(E, \mathcal{T})$ when $\mathcal{T}$ is a complete chain. Let ψ_1, ψ_2 be operators on $\mathrm{Fun}(E, \mathcal{T})$; put $\phi = \psi_1 \wedge \psi_2$ and $\psi = \psi_1 \vee \psi_2$; let γ be the centre of ϕ and ψ. Define the median $m : \mathcal{T}^3 \to \mathcal{T}$ as follows: given t_1, t_2, t_3, let $m(t_1, t_2, t_3)$ be the value t_i which lies between the other two. In fact, m is

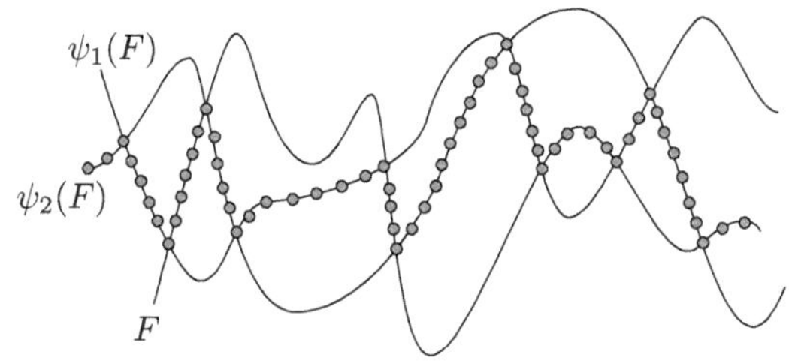

Figure 3.3 Geometrical characterization of the centre operator. The value $\gamma(F)(x)$ is the middle value of $F(x)$, $\psi_1(F)(x)$, $\psi_2(F)(x)$.

given by the formula

$$m(t_1, t_2, t_3) = (t_1 \wedge t_2) \vee (t_1 \wedge t_3) \vee (t_2 \wedge t_3).$$

We show that

$$\gamma(F)(x) = m(F(x), \psi_1(F)(x), \psi_2(F)(x)). \tag{3.31}$$

A geometrical illustration of this identity can be found in Fig. 3.3.

To prove (3.31), put $t = F(x)$ and $t_i = \psi_i(F)(x)$ for $i = 1, 2$. By definition,

$$\begin{aligned} \gamma(F)(x) &= (t \wedge (t_1 \vee t_2)) \vee (t_1 \wedge t_2) \\ &= (t \wedge t_1) \vee (t \wedge t_2) \vee (t_1 \wedge t_2) \\ &= m(t, t_1, t_2); \end{aligned}$$

this proves the assertion.

Assume that $\mathcal{L}$ is a modular complete lattice. Given two increasing operators ϕ and ψ on $\mathcal{L}$ such that $\phi \le \psi$; define the centre $\gamma = \gamma(\phi, \psi)$ by

$$\gamma = (\mathrm{id} \wedge \psi) \vee \phi; \tag{3.32}$$

cf. Definition 3.41. Since $\mathcal{L}$ is modular, it follows that $\gamma = (\mathrm{id} \vee \phi) \wedge \psi$.

3.45 Lemma. *Consider a modular complete lattice $\mathcal{L}$ and two increasing operators $\phi \le \psi$. Their centre γ satisfies*

(a) $\phi \le \gamma \le \psi$;

(b) $\mathrm{id} \wedge \gamma = \mathrm{id} \wedge \psi$, $\mathrm{id} \vee \gamma = \mathrm{id} \vee \phi$;

(c) $\gamma \preccurlyeq \phi, \psi$.

The proof is easy; see also Lemma 3.38.

3.7. Bibliographical notes

Most of the terminology encountered in this chapter is standard in the morphological literature. Some authors, however, speak of *algebraic openings* and *algebraic closings* instead of openings and closings. Furthermore, they use the terminology *morphological openings* and *morphological closings* instead of *adjunctional openings and closings*; refer to Section 6.9 for some additional remarks.

The Tarski Fixpoint Theorem 3.3 can be found, e.g., in Birkhoff (1967).

In most algebra books, e.g., in Birkhoff (1967), one encounters the terminology *join morphism* and *meet morphism* for dilation and erosion, respectively.

As explained in Section 1.4 adjunctions are a well-known concept in many fields of mathematics. The terminology "adjunction" is due to Gierz et al. (1980). Our concept of dilation is identical to the *residuated mapping* studied by Blyth and Janowitz (1972). We quote the following definition from their book.

3.46 Definition. An operator ψ between the posets $\mathcal{L}, \mathcal{M}$ is called *residuated* if and only if it is increasing, and there exists an increasing operator $\psi^+ : \mathcal{M} \to \mathcal{L}$, called the *residual* of ψ, such that

$$\psi^+\psi \geq \mathsf{id}_{\mathcal{L}} \qquad \text{and} \qquad \psi\psi^+ \leq \mathsf{id}_{\mathcal{M}}.$$

The residual mapping ψ^+, the analogue of our erosion, is unique. That ψ is a residuated mapping implies that (ψ^+, ψ) is an adjunction between $\mathcal{M}$ and $\mathcal{L}$. Blyth and Janowitz (1972) also introduce a *closure*; this notion coincides with our closing.

In a recent paper Banon and Barrera (1993) generalize the notion of an adjunction by dropping the assumption of increasingness.

Reflected dilations were introduced by Serra (1988, pp. 42,59).

The local knowledge principle plays a crucial role in Serra's book (Serra, 1982) where it is presented as one of the basic principles of mathematical morphology; the formulation in Proposition 3.34 is taken from Baddeley and Heijmans (1992).

The activity ordering, the centre and the anti-centre are inventions of Serra (1988); see also Meyer and Serra (1989a). Serra claims that the lattice $(\mathcal{O}(\mathcal{L}), \preccurlyeq)$ is distributive, assuming that $\mathcal{L}$ is Boolean; however, he does not give an explicit proof of this fact. Note that $(\mathcal{O}(\mathcal{L}), \preccurlyeq)$ is a Boolean lattice if Serra's claim is correct.

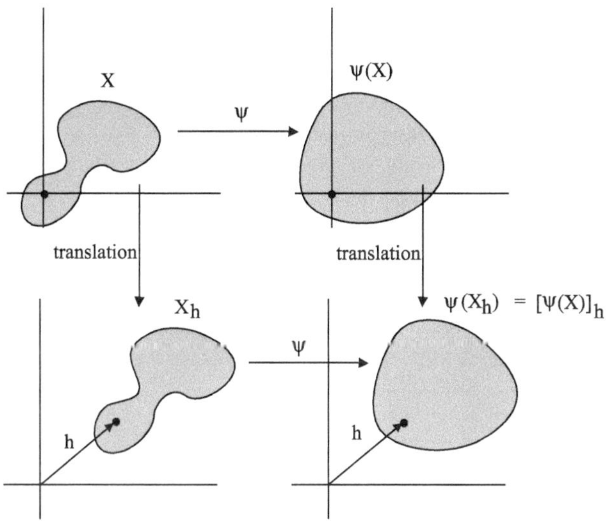

Figure 4.1 Translation invariance of an operator.

operator, briefly called an operator, we mean a mapping $\psi : \mathcal{P}(\mathbb{E}^d) \to \mathcal{P}(\mathbb{E}^d)$. This chapter deals exclusively with operators which are invariant under translations. Given $X \subseteq \mathbb{E}^d$ and $h \in \mathbb{E}^d$, the translate X_h is defined by

$$X_h = \{x + h \mid x \in X\}. \tag{4.1}$$

4.2 Definition. An operator ψ on $\mathcal{P}(\mathbb{E}^d)$ is called *translation invariant* if

$$\psi(X_h) = [\psi(X)]_h, \tag{4.2}$$

for every $X \subseteq \mathbb{E}^d$ and $h \in \mathbb{E}^d$; see also Fig. 4.1.

Apart from the identity operator, the simplest example of an operator which is translation invariant is the translation operator $X \mapsto X_a$, where a is a given vector in $\mathbb{E}^d$. As an instance of an operator which is not translation invariant, we mention the mapping $X \mapsto X \cup A$, where A is a given nonempty subset of $\mathbb{E}^d$. It is evident that composition, union, and intersection of translation invariant operators gives again a translation invariant operator.

4.3 Definition. An operator ψ is called *increasing* if $X \subseteq Y$ implies $\psi(X) \subseteq \psi(Y)$; it is called *decreasing* if $X \subseteq Y$ implies $\psi(X) \supseteq \psi(Y)$ (see also Definition 3.1).

For example, the operators $X \mapsto X_a$ and $X \mapsto X \cup A$ are increasing while the complement operator $X \mapsto X^c$ is decreasing. The *invariance domain* of an operator ψ, denoted by $\text{Inv}(\psi)$, is defined to be the set of all $X \subseteq \mathbb{E}^d$ that are invariant under ψ, i.e., such that $\psi(X) = X$; cf. Section 3.1.

The *scaling*, or *multiplication*, of a set $X \subseteq \mathbb{E}^d$ by a scalar $r \in \mathbb{E}$ is defined by

$$rX = \{rx \mid x \in X\}. \tag{4.3}$$

We use the convention that $0X = \{0\}$ if X is nonempty.

Given an operator ψ on $\mathcal{P}(\mathbb{E}^d)$, one can define the *negative operator* ψ^* by applying ψ to the background instead of to the foreground:

$$\psi^*(X) = [\psi(X^c)]^c; \tag{4.4}$$

cf. (3.2). It can readily be shown that ψ^* is increasing if and only if ψ is (cf. Proposition 3.2(a)). A similar statement holds for translation invariance.

4.4 Definition. An operator ψ is called *self-dual* if $\psi^* = \psi$.

The following sections discuss a number of translation invariant operators on $\mathcal{P}(\mathbb{E}^d)$ which are relevant in the context of mathematical morphology.

4.2. Hit-or-miss operator

The principal idea underlying many morphological operators is to probe a fixed pattern, called a *structuring element*, with the image and its background. The hit-or-miss operator formalizes this idea. Let $A, B \subseteq \mathbb{E}^d$; the *hit-or-miss operator* is defined by

$$X \circledast (A, B) = \{h \in \mathbb{E}^d \mid A_h \subseteq X \text{ and } B_h \subseteq X^c\}. \tag{4.5}$$

Note that the result is empty if $A \cap B \neq \varnothing$. The name hit-or-miss operator can be explained as follows: a point h lies in the hit-or-miss-transformed set $X \circledast (A, B)$ iff A_h does not hit X^c ("hit" in the sense of "intersect with") and B_h does not hit X.

The hit-or-miss operator is well suited to the task of locating points inside an object with certain (local) geometric properties, e.g., isolated points, edge points, corner points, and T-junctions. In the example depicted in Fig. 4.2, A and B have been chosen in such a way that $X \circledast (A, B)$ comprises the lower left corner points of the original image.

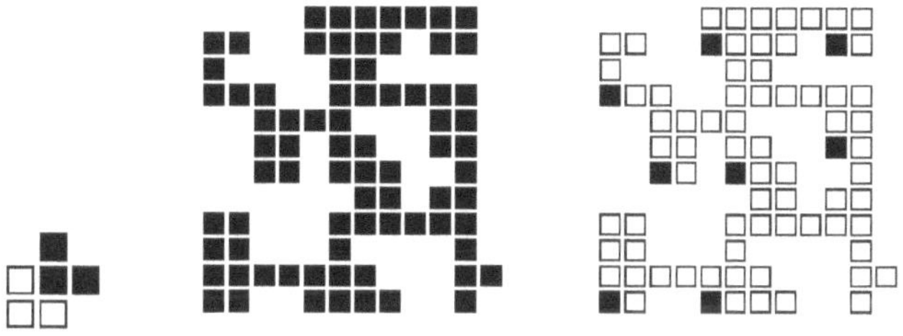

Figure 4.2 Localization of lower-left corner points in a discrete image by a hit-or-miss operator. From left to right: the structuring element (A, B) where A contains the black pixels and B the white pixels; the set X; the transformed set $X \circledast (A, B)$ (black pixels).

We list a number of elementary properties.

4.5 Proposition. *Let* $A, B, X \subseteq \mathbb{E}^d$, $h \in \mathbb{E}^d$ *and* $r \in \mathbb{E}$, $r > 0$. *Then*

$$X_h \circledast (A, B) = [X \circledast (A, B)]_h, \tag{4.6}$$

$$X^c \circledast (A, B) = X \circledast (B, A), \tag{4.7}$$

$$X \circledast (A_h, B_h) = [X \circledast (A, B)]_{-h}, \tag{4.8}$$

$$rX \circledast (rA, rB) = r[X \circledast (A, B)]. \tag{4.9}$$

Putting $T = (A, B)$, we also write $X \circledast T$ for $X \circledast (A, B)$. Note that (4.6) states that the hit-or-miss operator is translation invariant. Furthermore, the hit-or-miss operator $X \mapsto X \circledast (A, B)$ is increasing iff $B = \varnothing$ and decreasing iff $A = \varnothing$. The hit-or-miss operator is the basic ingredient for two other translation invariant operators which are not increasing, namely, the thickening operator and the thinning operator. For a structuring element $T = (A, B)$, the *thickening* and *thinning* of the set X are defined, respectively, by

$$X \odot T = X \cup (X \circledast T), \tag{4.10}$$

$$X \circledcirc T = X \setminus (X \circledast T). \tag{4.11}$$

Since $X \circledast T \subseteq X$ if $0 \in A$ and $X \circledast T = \varnothing$ if $A \cap B \neq \varnothing$, thickening leads to a result which is nontrivial if $A \cap B = \varnothing$ and $0 \notin A$. Similarly, thinning gives a nontrivial result if $A \cap B = \varnothing$ and $0 \notin B$. It is easy to show that thickening and thinning are complementary operators in the sense that

$$X^c \odot (A, B) = (X \circledcirc (B, A))^c. \tag{4.12}$$

Finally, one can show that property (4.9) also holds for thickenings and thinnings.

The hit-or-miss operator and the thinning and thickening operators derived from it play an important role in many morphological image processing algorithms. Several of them can be found in Serra (1988). Here we consider two illustrative examples. The first example describes an algorithm for the computation of the pseudo-convex hull of a set. The second example concerns a skeletonization algorithm. We emphasize that these algorithms are by no means optimal with respect to speed; they are merely presented as an illustration of the use of the hit-or-miss operator.

4.6 Example. (A pseudo-convex hull algorithm)
Convexity of an object in continuous Euclidean space $\mathbb{R}^d$ is a global property. To construct a convex hull of a discrete object using thickenings by finite structuring elements, one has to modify the definition of convexity. We introduce the following definition of convexity for subsets of the discrete space $\mathbb{Z}^2$. For a set $X \subseteq \mathbb{Z}^2$ we define the convex hull $\mathrm{co}_{45}(X)$ as the intersection of all discrete half-planes $\{(x, y) \mid ax + by \leq c\}$ for which $a, b = -1, 0, 1$ and $c \in \mathbb{Z}$. This means that we restrict ourselves to half-planes bounded by lines whose angle with the x-axis is an integer multiple of 45°.

A set $X \subseteq \mathbb{Z}^2$ is called *45°-convex* if it coincides with its convex hull, i.e., $X = \mathrm{co}_{45}(X)$. There is a simple algorithm, using thickenings, to construct this convex hull. Let $T_1, T_2, \ldots, T_8$ be as depicted in Fig. 4.3 and define

$$\psi(X) = (\cdots((X \odot T_1) \odot T_2) \odot \cdots \odot T_8).$$

Then iteration of ψ until convergence is reached yields the 45°-convex hull. The algorithm is illustrated in Fig. 4.3.

4.7 Example. (Skeleton)
In Euclidean space the *skeleton* of a set X is usually defined as the set of centres of maximal balls contained in X. Here a ball is said to be *maximal* if it is not contained in another ball which lies entirely inside X. This set is also called the *medial axis*; refer to Section 9.8 for further details. One may think of the skeleton as a set which is a union of arcs with the same homotopy as the original set X and which lies "in the middle of" X. Algorithms for the computation of a discrete version of the skeleton, sometimes called *homotopic thinning*, can be found at several places in the literature. As an example we mention the sequential thinning by the structuring elements $T_1, T_2, \ldots, T_8$ depicted in Fig. 4.4.

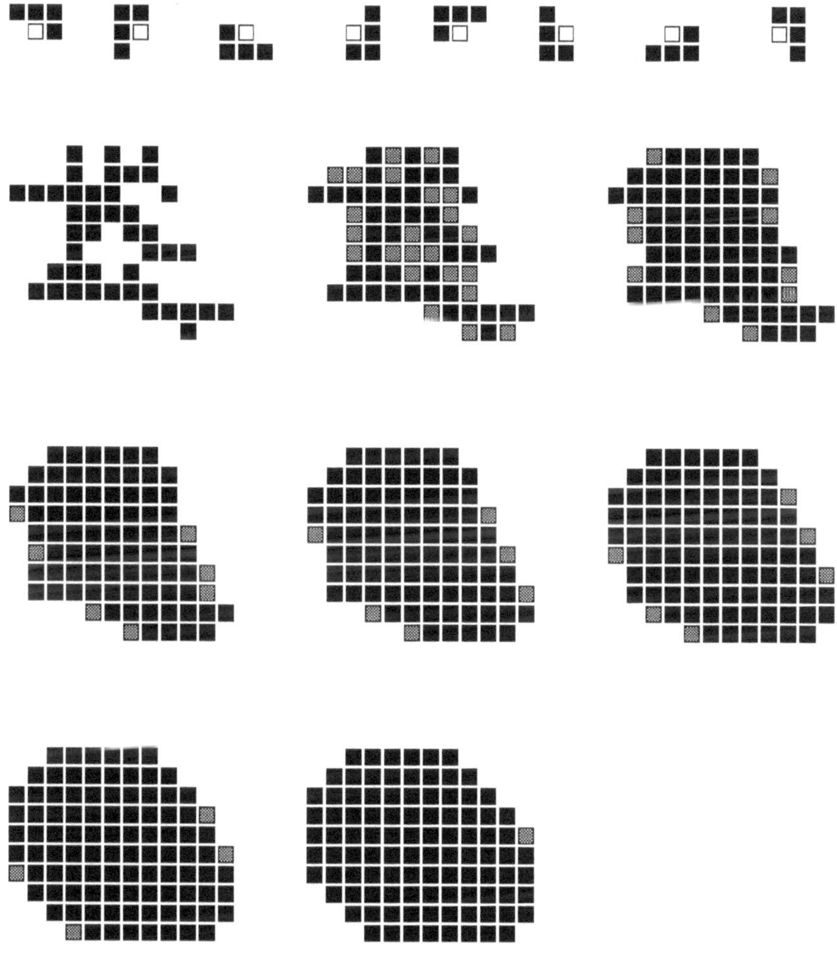

Figure 4.3 Top: the structuring elements $T_1, T_2, \ldots, T_8$. The 45°-convex hull of a set X (first object at the second row) is obtained by iteration of $\psi(X) = (\cdots((X \circledcirc T_1) \circledcirc T_2) \circledcirc \cdots \circledcirc T_8)$. The first object is the original set X; the second object is $\psi(X)$, etc. The final object (reached after seven iterations) shows $X = \mathrm{co}_{45}(X)$. The grey pixels represent those points which are added by subsequent iterations.

This is the set which results after iterated application of the operator ψ given by

$$\psi(X) = (\cdots((X \circledcirc T_1) \circledcirc T_2) \circledcirc \cdots \circledcirc T_8),$$

to the initial set X.

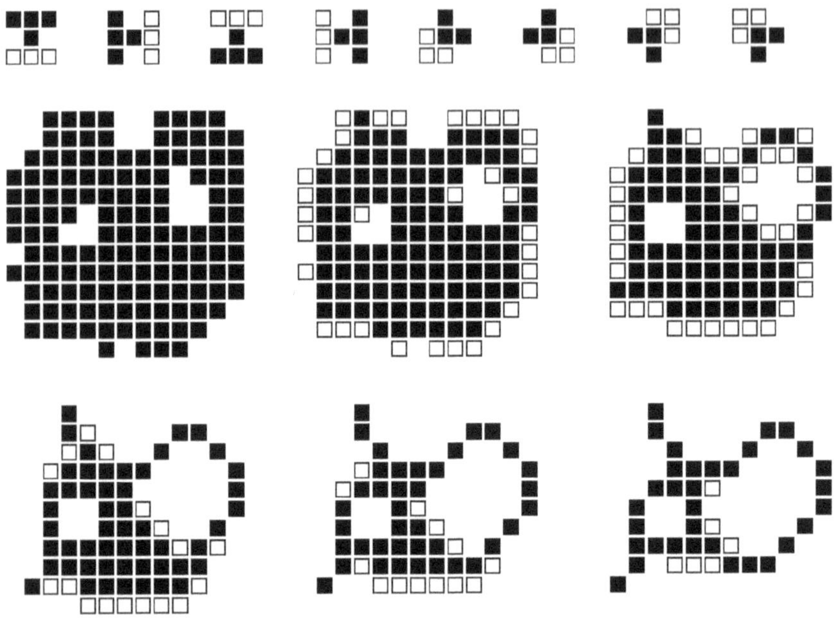

Figure 4.4 Top: the structuring elements $T_1, T_2, \ldots, T_8$. The homotopic thinning is found by iteration of $\psi(X) = (\cdots((X \circledcirc T_1) \circledcirc T_2) \circledcirc \cdots \circledcirc T_8)$. The first object is the original set X; the second object is $\psi(X)$, etc. The final object shows the homotopic thinning of X and is reached after five iterations. The white pixels represent those points which are deleted by subsequent iterations.

Both examples describe the construction of an idempotent operator by iteration of an operator which is not idempotent. Chapter 13 presents a comprehensive discussion of this method. In particular, this chapter gives (sufficient) conditions under which a sequence of iterates of some morphological operator converges. As we shall see, the two examples just given satisfy these conditions (because the operators involved use only finite structuring elements).

The hit-or-miss operator is just one particular kind of translation invariant operator on $\mathcal{P}(\mathbb{E}^d)$. There are several translation invariant operators which are not of hit-or-miss-type; however, every translation invariant operator on $\mathcal{P}(\mathbb{E}^d)$ can be represented as a union of hit-or-miss operators. Before we give a formal statement, we need some further definitions. First, we present another manifestation of the hit-or-miss operator, called the *wedge operator*:

$$X \circledwedge (A, B) = \{h \in \mathbb{E}^d \mid A_h \subseteq X \subseteq B_h\}. \tag{4.13}$$

Note that the outcome is the empty set if $A \not\subseteq B$. It is obvious that

$$X \circledcirc (A, B) = X \circledast (A, B^c). \tag{4.14}$$

For $A, B \subseteq \mathbb{E}^d$ the *interval* $[A, B]$ is the set of all $X \subseteq \mathbb{E}^d$ with $A \subseteq X \subseteq B$. Evidently, $[A, B]$ is nonempty if and only if $A \subseteq B$.

4.8 Definition. The *kernel* $\mathcal{V}(\psi)$ of an operator ψ on $\mathcal{P}(\mathbb{E}^d)$ is defined by

$$\mathcal{V}(\psi) = \{A \subseteq \mathbb{E}^d \mid 0 \in \psi(A)\}; \tag{4.15}$$

its *bi-kernel* $\mathcal{W}(\psi)$ is defined by

$$\mathcal{W}(\psi) = \{(A, B) \in \mathcal{P}(\mathbb{E}^d) \times \mathcal{P}(\mathbb{E}^d) \mid [A, B] \subseteq \mathcal{V}(\psi)\}. \tag{4.16}$$

Note that the kernel of a translation invariant operator is empty if and only if ψ maps every set onto the empty set. For, if $\psi(X) \neq \varnothing$ for some $X \subseteq \mathbb{E}^d$ and $h \in \psi(X)$, then, by translation invariance, $0 \in \psi(X_{-h})$; hence $X_{-h} \in \mathcal{V}(\psi)$. Furthermore, the bi-kernel is nonempty if and only if the kernel is nonempty. For, if $A \in \mathcal{V}(\psi)$, then $(A, A) \in \mathcal{W}(\psi)$.

4.9 Theorem. *Let ψ be a translation invariant operator on $\mathcal{P}(\mathbb{E}^d)$; then*

$$\psi(X) = \bigcup_{(A,B) \in \mathcal{W}(\psi)} X \circledcirc (A, B). \tag{4.17}$$

Proof. $\supseteq$: Let $h \in X \circledcirc (A, B)$ for some $(A, B) \in \mathcal{W}(\psi)$. Then $A_h \subseteq X \subseteq B_h$, and hence $X_{-h} \in [A, B] \subseteq \mathcal{V}(\psi)$. Therefore, $0 \in \psi(X_{-h})$, and with the translation invariance of ψ this gives $h \in \psi(X)$.

$\subseteq$: Let $h \in \psi(X)$, that is, $0 \in \psi(X_{-h})$. Then $(X_{-h}, X_{-h}) \in \mathcal{W}(\psi)$. Evidently, $h \in X \circledcirc (X_{-h}, X_{-h})$, whence $h \in \bigcup_{(A,B) \in \mathcal{W}(\psi)} X \circledcirc (A, B)$. □

Unfortunately, this result does not say anything about the family of structuring elements required for such a representation. This family is in general quite large.

4.3. Dilation and erosion

Section 3.2 has presented a theoretical treatment of adjunctions on complete lattices. Recall that an adjunction is a pair of operators, a dilation and an erosion, satisfying a certain relation. As the title indicates, this section also deals with dilations and erosions, but of a rather specific kind: the mappings considered here act on $\mathcal{P}(\mathbb{E}^d)$ and are translation invariant. Yet they fit inside the complete lattice framework of Chapters 2 and 3. Even though this framework is only of secondary importance here, we spend a few lines to point out the connection between the concrete operators discussed here and the adjunctions of Section 3.2.

Let A be a subset of $\mathbb{E}^d$; henceforth we refer to A as the structuring element. The *dilation* of a set X by A is

$$X \oplus A = \bigcup_{a \in A} X_a, \tag{4.18}$$

or alternatively,

$$X \oplus A = \{x + a \mid x \in X,\ a \in A\} = \bigcup_{x \in X} A_x. \tag{4.19}$$

In the mathematical literature $X \oplus A$ is usually called the *Minkowski sum* of the sets X and A.

From the latter characterization it is clear that

$$X \oplus A = A \oplus X;$$

in other words, Minkowski addition is a commutative operation. We show that

$$X \oplus A = \{h \in \mathbb{E}^d \mid \check{A}_h \cap X \neq \varnothing\}. \tag{4.20}$$

Here $\check{A}$ is the reflected set of A with respect to the origin, that is,

$$\check{A} = \{-a \mid a \in A\}.$$

To prove (4.20), observe that $h \in X \oplus A$ if and only if $h \in X_a$ for some $a \in A$. But this is equivalent to the assertion that $\{-a + h \mid a \in A\} \cap X \neq \varnothing$.

The *erosion* of a set X by the structuring element A, also known as *Minkowski subtraction*, is

$$X \ominus A = \bigcap_{a \in A} X_{-a}. \tag{4.21}$$

One can easily check that the erosion can also be expressed as

$$X \ominus A = \{h \in E \mid A_h \subseteq X\}. \tag{4.22}$$

Figure 4.5 From left to right: the original set, its dilation, and its erosion. (a) The continuous case; the structuring element is a disk. (b) The discrete case; the structuring element is a 3 × 3 square. Grey pixels are added, white pixels are deleted.

This formula means that the erosion of X by A comprises all points h such that the structuring element A located at h fits entirely inside X. Fig. 4.5 illustrates the dilation and the erosion for both continuous and discrete sets.

Combining (4.5) and (4.22), one finds that the hit-or-miss operator can be expressed in terms of erosions as

$$X \circledast (A, B) = (X \ominus A) \cap (X^c \ominus B). \tag{4.23}$$

The next two propositions summarize some elementary properties of dilation and erosion. The proofs are straightforward.

4.10 Proposition. (Properties of dilation)
For $A, B, X, Y \subseteq \mathbb{E}^d$, $h \in \mathbb{E}^d$, and $r \in \mathbb{E}$,

$$X \oplus \{h\} = X_h, \tag{4.24}$$

$$X_h \oplus A = X \oplus A_h = (X \oplus A)_h, \tag{4.25}$$

$$X \oplus A \subseteq X \oplus B \text{ if } A \subseteq B, \tag{4.26}$$

$$X \oplus (A \cup B) = (X \oplus A) \cup (X \oplus B), \tag{4.27}$$
$$X \oplus (A \cap B) \subseteq (X \oplus A) \cap (X \oplus B), \tag{4.28}$$
$$(X \oplus A) \oplus B = X \oplus (A \oplus B), \tag{4.29}$$
$$X \subseteq Y \Rightarrow X \oplus A \subseteq Y \oplus A, \tag{4.30}$$
$$rX \oplus rA = r(X \oplus A). \tag{4.31}$$

4.11 Proposition. (Properties of erosion)
For $A, B, X, Y \subseteq \mathbb{E}^d$, $h \in \mathbb{E}^d$, and $r \in \mathbb{E}$,

$$X \ominus \{h\} = X_{-h}, \tag{4.32}$$
$$X_h \ominus A = X \ominus A_{-h} = (X \ominus A)_h, \tag{4.33}$$
$$X \ominus A \supseteq X \ominus B \text{ if } A \subseteq B, \tag{4.34}$$
$$X \ominus (A \cup B) = (X \ominus A) \cap (X \ominus B), \tag{4.35}$$
$$X \ominus (A \cap B) \supseteq (X \ominus A) \cup (X \ominus B), \tag{4.36}$$
$$(X \ominus A) \ominus B = X \ominus (A \oplus B), \tag{4.37}$$
$$X \subseteq Y \Rightarrow X \ominus A \subseteq Y \ominus A, \tag{4.38}$$
$$rX \oplus rA = r(X \oplus A). \tag{4.39}$$

Discrete structuring elements which are used quite often in practice include

$$3 \times 3 \text{ square} = \begin{array}{ccc} \bullet & \bullet & \bullet \\ \bullet & \bullet & \bullet \\ \bullet & \bullet & \bullet \end{array}, \quad \text{rhombus} = \begin{array}{ccc} & \bullet & \\ \bullet & \bullet & \bullet \\ & \bullet & \end{array}$$

Properties (4.27), (4.29), (4.35), and (4.37) provide a method for decomposing erosions and dilations with large structuring elements. It is evident that such decomposition procedures are useful with regard to implementations of dilations and erosions. Some example of such decompositions in the 2-dimensional discrete case are

$$\begin{array}{cc} \bullet & \bullet \end{array} \oplus \begin{array}{c} \bullet \\ \bullet \end{array} = \begin{array}{cc} \bullet & \bullet \\ \bullet & \bullet \end{array}$$

$$\begin{array}{cc} \bullet & \bullet \\ \bullet & \bullet \end{array} \oplus \begin{array}{cc} \bullet & \bullet \\ \bullet & \bullet \end{array} = \begin{array}{ccc} \bullet & \bullet & \bullet \\ \bullet & \bullet & \bullet \\ \bullet & \bullet & \bullet \end{array}$$

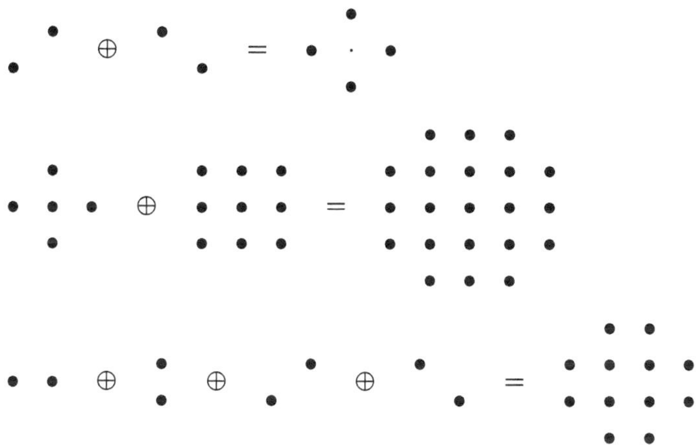

where the •'s represent points in the structuring element. The fourth identity decomposes the octagon as the Minkowski sum of the rhombus and the 3×3 square.

Dilation and erosion are dual operators in three different ways. First, they are dual in the sense of the Duality Principle 2.4. This fact has already been noted in Section 3.2; it also plays a role in Chapter 5 where we present a comprehensive treatment of erosions and dilations in the framework of complete lattices. Second,

$$(X \oplus A)^c = X^c \ominus \check{A} \qquad \text{and} \qquad (X \ominus A)^c = X^c \oplus \check{A}. \tag{4.40}$$

These relations state that dilating a set X by A gives the same result as eroding its complement X^c by the reflected structuring element $\check{A}$. Recalling Proposition 3.19, this means that the dilation $X \mapsto X \oplus \check{A}$ is the reflection of the dilation $X \mapsto X \oplus A$. To prove it, we use (4.20); this relation implies that $h \in (X \oplus A)^c$ iff $\check{A}_h \cap X = \varnothing$, that is, $\check{A}_h \subseteq X^c$. But this is equivalent to $h \in X^c \ominus \check{A}$.

The third duality between dilation and erosion is the adjunction relation (3.4), which, in our perspective, is the most important one:

$$Y \oplus A \subseteq X \iff Y \subseteq X \ominus A. \tag{4.41}$$

We prove the implication $\Rightarrow$. The inclusion $X \oplus A \subseteq Y$ gives that $X_a \subseteq Y$ for every $a \in A$. Then $X \subseteq Y_{-a}$ for every $a \in A$, whence $X \subseteq \bigcap_{a \in A} X_{-a} = X \ominus A$.

4.12 Remark. There is some confusion in the literature about the original definition of Minkowski subtraction and its relation to dilation and erosion. There is general agreement about $X \oplus A$ being the Minkowski addition. Some authors (including Serra and Matheron), however, call the operator $X \to X \oplus \check{A}$ the dilation by A. Our definition of Minkowski subtraction coincides with Hadwiger's original definition (Hadwiger, 1957). Serra, Matheron, and co-workers define Minkowski subtraction by $X \ominus A = \bigcap_{a \in A} X_a$; they call $X \ominus \check{A}$ the erosion by A. Note that this latter definition coincides with our nomenclature, though our notation is different. The motivation for our conventions lies in the fact that the duality relation between dilations and erosions comprised by the adjunction relation (4.41) is by far the most important one, and this fact is expressed in our notation.

Dilation and erosion by A are often denoted by δ_A and ε_A, respectively; in other words,

$$\delta_A(X) = X \oplus A, \tag{4.42}$$

$$\varepsilon_A(X) = X \ominus A. \tag{4.43}$$

Now the relations expressed by (4.41) can be restated in the following way.

4.13 Proposition. *Given a structuring element $A \subseteq \mathbb{E}^d$, the pair $(\varepsilon_A, \delta_A)$ defines an adjunction on the complete lattice $\mathcal{P}(\mathbb{E}^d)$.*

This result states that δ_A and ε_A are a dilation and an erosion, respectively, in the sense of Section 3.2; as such, these operators are distributive over union and intersection, respectively:

$$\Big(\bigcup_{i \in I} X_i\Big) \oplus A = \bigcup_{i \in I} (X_i \oplus A), \tag{4.44}$$

$$\Big(\bigcap_{i \in I} X_i\Big) \ominus A = \bigcap_{i \in I} (X_i \ominus A), \tag{4.45}$$

for an arbitrary collection $X_i \subseteq \mathbb{E}^d$. These identities can also be derived easily without reference to the framework of adjunctions.

4.14 Example. (Neighbourhood functions)
The dilation δ_A and erosion ε_A can be written as

$$\delta_A(X) = X \oplus A = \bigcup_{h \in X} A_h,$$

$$\varepsilon_A(X) = X \ominus A = \{h \in \mathbb{E}^d \mid A_h \subseteq X\},$$

respectively. These operators belong to the class of translation invariant operators. Dropping the translation invariance condition, one ends up with a class of dilations which is considerably larger. Let E be an arbitrary set, and let A be a mapping from E into $\mathcal{P}(E)$. Such a mapping is called a *neighbourhood function*, since $A(h)$ can be interpreted as a neighbourhood of h (though it is not required that $h \in A(h)$). Define

$$\delta_A(X) = \bigcup_{h \in X} A(h),$$

$$\varepsilon_A(X) = \{h \in \mathbb{E}^d \mid A(h) \subseteq X\};$$

then $(\varepsilon_A, \delta_A)$ is an adjunction on $\mathcal{P}(E)$. To see this we must show that

$$\delta_A(Y) \subseteq X \iff Y \subseteq \varepsilon_A(X),$$

for $X, Y \subseteq E$. We prove $\Rightarrow$; the other implication is proved analogously. The inclusion $\delta_A(Y) \subseteq X$ implies that $A(y) \subseteq X$ for every $y \in Y$. Then $A(y) \subseteq X$, and therefore $y \in \varepsilon_A(X)$. This means that $Y \subseteq \varepsilon_A(X)$, which was to be shown. The translation invariant dilations and erosions are obtained by choosing special neighbourhood functions, namely, those functions for which every $A(h)$ is the translate of a fixed set A along h, that is, $A(h) = A_h$. Neighbourhood functions which are not translation invariant come naturally into play if the universal set E is a bounded subset of the Euclidean plane $\mathbb{R}^d$ or of the discrete grid $\mathbb{Z}^d$. A computer screen is a well-known example. In these cases, a natural choice for the neighbourhood function is $A(h) = A_h \cap E$.

It is clear that dilation and erosion are increasing operators in the sense of Definition 4.3: if $X \subseteq Y$, then $X \oplus A \subseteq Y \oplus A$ and $X \ominus A \subseteq Y \ominus A$. The following result is known as *Matheron's representation theorem*.

4.15 Theorem. *Let ψ be an increasing translation invariant operator on $\mathcal{P}(\mathbb{E}^d)$. Then*

$$\psi(X) = \bigcup_{A \in \mathcal{V}(\psi)} X \ominus A = \bigcap_{A \in \mathcal{V}(\psi^*)} X \oplus \check{A}.$$

Note that this result can be regarded as a specialization of Theorem 4.9 to increasing operators. Here one uses essentially the fact that $X \circledS (A, \mathbb{E}^d) = X \ominus A$. These observations, in combination with Theorem 4.9, lead to a proof of the first identity. In fact, it is not difficult to prove this result directly. To prove the second identity, we use that $\psi^*(X) = \bigcup_{A \in \mathcal{V}(\psi^*)} X \ominus A$.

Substituting X^c for X, we get $[\psi(X)]^c = \bigcup_{A \in \mathcal{V}(\psi^*)} X^c \ominus A$. Taking complements on both sides, and using de Morgan's laws and (4.40) we arrive at the second identity.

At this point we devote a few extra words to the notion of a kernel. As

$$h \in \psi(X) \iff 0 \in \psi(X_{-h}) \iff X_{-h} \in \mathcal{V}(\psi),$$

if ψ is translation invariant, we get

$$\psi(X) = \{h \in \mathbb{E}^d \mid X_{-h} \in \mathcal{V}(\psi)\}; \tag{4.46}$$

this provides a reconstruction of the operator ψ from its kernel $\mathcal{V}(\psi)$. One easily establishes that for an arbitrary family of operators ψ_i:

$$\begin{aligned} \mathcal{V}(\bigvee_{i \in I} \psi_i) &= \bigcup_{i \in I} \mathcal{V}(\psi_i), \\ \mathcal{V}(\bigwedge_{i \in I} \psi_i) &= \bigcap_{i \in I} \mathcal{V}(\psi_i). \end{aligned} \tag{4.47}$$

In particular,

$$\psi \le \psi' \iff \mathcal{V}(\psi) \subseteq \mathcal{V}(\psi'). \tag{4.48}$$

The representation of an increasing translation invariant operator ψ as a union of erosions does not require the entire kernel $\mathcal{V}(\psi)$. For instance, if $A \in \mathcal{V}(\psi)$ and $A \subseteq B$, then $B \in \mathcal{V}(\psi)$ too; in other words, $\mathcal{V}(\psi)$ is an upper set. However, $X \ominus B$ is contained in $X \ominus A$, and hence B can be omitted without changing the result. To remove this redundancy we look for a subset of kernel elements still sufficiently large to recover the original operator ψ.

4.16 Definition. Given an increasing, translation invariant operator ψ on $\mathcal{P}(\mathbb{E}^d)$, the set $A \in \mathcal{V}(\psi)$ is called a *minimal kernel element* if $B \subseteq A$ and $B \neq A$ implies that $B \notin \mathcal{V}(\psi)$. The collection of all minimal kernel elements is called the *basis* of ψ and denoted by $\mathcal{V}_b(\psi)$.

Example 4.17(c) shows that there exist increasing, translation invariant operators without minimal kernel elements.

4.17 Examples.

(a) Let ε_A be the erosion given by (4.43); then $X \in \mathcal{V}(\varepsilon_A)$ iff $0 \in X \ominus A$, that is, $A \subseteq X$. In particular, A is the smallest kernel element, and so $\mathcal{V}_b(\psi) = \{A\}$.

(b) Let δ_A be the dilation given by (4.42). Then $X \in \mathcal{V}(\delta_A)$ iff $0 \in X \oplus A$, which by (4.20) is equivalent to $\check{A} \cap X \neq \varnothing$. The basis comprises all singletons $\{-a\}$ where $a \in A$.

(c) Let ψ be the operator on $\mathcal{P}(\mathbb{R})$ given by

$$\psi(X) = \begin{cases} \mathbb{R}, & \text{if } (-\infty, c) \subseteq X \text{ for some } c \in \mathbb{R}, \\ \varnothing, & \text{otherwise.} \end{cases}$$

The kernel of ψ comprises all half lines $(-\infty, c)$, $c \in \mathbb{R}$. Obviously, this family contains no minimal element. Putting $L = (-\infty, 0)$ we have $\mathcal{V}(\psi) = \{L_c \mid c \in \mathbb{R}\}$, and thus

$$\psi(X) = \bigcup_{c \in \mathbb{R}} X \ominus L_c = \bigcup_{c \in \mathbb{R}} (X \ominus L)_{-c} = (X \ominus L) \oplus \mathbb{R}.$$

Section 7.6 gives conditions which guarantee that the basis of an operator is nonempty and, moreover, that ψ can be recovered using only the minimal kernel elements in the erosion expansion of ψ.

4.4. Opening and closing

Dilation and erosion are not inverse operators. If a set X is eroded by A and subsequently dilated by A, one does not end up with the original set X but with a set which is smaller; this set is called the *opening* of X by A and denoted by $X \circ A$. In symbols,

$$X \circ A = (X \ominus A) \oplus A. \tag{4.49}$$

That $X \circ A \subseteq X$ follows easily from (4.41) by substituting $Y = X \ominus A$. Dually, dilation followed by erosion, both by the same structuring element A, returns a set which is larger than X; this set is called the *closing* of X by A, denoted by $X \bullet A$. So

$$X \bullet A = (X \oplus A) \ominus A. \tag{4.50}$$

From the fact that erosion is an increasing operator we get $\big((X \ominus A) \oplus A\big) \ominus A \subseteq X \ominus A$. On the other hand, by the extensivity of the closing operator, $\big((X \ominus A) \oplus A\big) \ominus A = (X \ominus A) \bullet A \supseteq X \ominus A$. This gives

$$\big((X \ominus A) \oplus A\big) \ominus A = X \ominus A. \tag{4.51}$$

Dually, one gets

$$\big((X \oplus A) \ominus A\big) \oplus A = X \oplus A. \tag{4.52}$$

A straightforward consequence of these relations is that opening and closing are idempotent operators, i.e.,

$$(X \circ A) \circ A = X \circ A, \tag{4.53}$$

$$(X \bullet A) \bullet A = X \bullet A. \tag{4.54}$$

We point out that these results also follow immediately from the general results about adjunctions stated in Proposition 3.14 and Theorem 3.25.

Applying (4.40) twice, we get

$$(X \circ A)^{c} = X^{c} \bullet \check{A} \qquad \text{and} \qquad (X \bullet A)^{c} = X^{c} \circ \check{A}. \tag{4.55}$$

In other words, opening an image has the same effect as closing the background by the reflected structuring element.

We give a geometrical interpretation of the opening and the closing. Since $X \ominus A = \{h \in \mathbb{E}^d \mid A_h \subseteq X\}$, it follows that

$$X \circ A = \bigcup \{A_h \mid h \in \mathbb{E}^d \text{ and } A_h \subseteq X\}; \tag{4.56}$$

that is, $X \circ A$ comprises all translates of the structuring element A that are contained inside X. The closing $X \bullet A$ is given by

$$\begin{aligned} X \bullet A &= (X \oplus A) \ominus A \\ &= \{k \in \mathbb{E}^d \mid A_k \subseteq \{h \in \mathbb{E}^d \mid \check{A}_h \cap X \neq \varnothing\}\} \\ &= \{k \in \mathbb{E}^d \mid h \in A_k \Rightarrow \check{A}_h \cap X \neq \varnothing\}. \end{aligned}$$

This yields that

$$X \bullet A = \{k \in \mathbb{E}^d \mid k \in \check{A}_h \Rightarrow \check{A}_h \cap X \neq \varnothing\}. \tag{4.57}$$

So a point k belongs to $X \bullet A$ if all translates $\check{A}_h$ that contain k intersect X; see Fig. 4.6 for an illustration.

If one takes the expression on the right-hand side of (4.56) and replaces union by intersection and the inclusion by the reverse inclusion, one obtains another closing, given by the expression $\bigcap\{A_h \mid h \in \mathbb{E}^d \text{ and } X \subseteq A_h\}$. We introduce the notation

$$\alpha_A(X) = \bigcup \{A_h \mid h \in \mathbb{E}^d \text{ and } A_h \subseteq X\}, \tag{4.58}$$

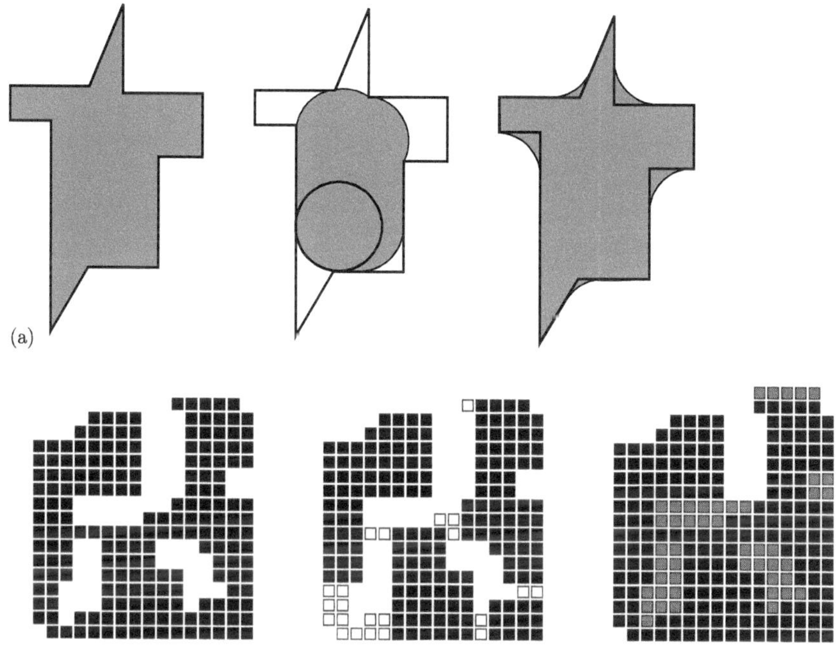

Figure 4.6 Left to right: the original set, its opening, and its closing. (a) The continuous case; the structuring element is a disk. (b) The discrete case; the structuring element is a 3×3 square.

$$\beta_A(X) = \bigcap \{A_h \mid h \in \mathbb{E}^d \text{ and } X \subseteq A_h\}. \tag{4.59}$$

Note that $\alpha_A(X) = X \circ A$ but $\beta_A(X) \neq X \bullet A$. Instead, one verifies readily that $(X^c \circ A^c)^c = \beta_A(X)$. In combination with (4.55) this leads to

$$\beta_A(X) = X \bullet (\check{A})^c. \tag{4.60}$$

We refer to α_A and β_A as the *structural opening* and *structural closing* since they involve only one structuring element A. Later, the reader will encounter openings and closings which are not of structural type.

4.18 Remarks.

(a) The opening α_A is an adjunctional opening (see Section 3.3) since $\alpha_A = \delta_A \varepsilon_A$, and the pair $(\varepsilon_A, \delta_A)$ constitutes an adjunction. Moreover, β_A is an adjunctional closing since, by (4.60), $\beta_A = \varepsilon_B \delta_B$ where $B = (\check{A})^c$. Actually, for translation invariant openings on $\mathcal{P}(\mathbb{E}^d)$ the class of adjunctional openings coincides with the class of structural openings; the same remark holds for closings. A systematic discussion of this issue can be found in Section 6.3.

(b) Generally speaking, structuring elements used in morphology are small sets. The structural closing β_A yields meaningful results only if A is large, however. For example, the closing $X \bullet A$, where A is the unit disk in $\mathbb{R}^2$, corresponds to the structural closing by the complement of the unit disk.

We sum up some properties of openings and closings. Those properties which have not yet been demonstrated can readily be verified by the reader.

4.19 Proposition. (Properties of opening)
For $A, X, Y \subseteq \mathbb{E}^d$, $h \in \mathbb{E}^d$ and $r \in \mathbb{E}$,

$$X \circ \{h\} = X, \tag{4.61}$$
$$X_h \circ A = (X \circ A)_h, \tag{4.62}$$
$$X \circ A_h = X \circ A, \tag{4.63}$$
$$X \subseteq Y \Rightarrow X \circ A \subseteq Y \circ A, \qquad \text{(increasing)} \tag{4.64}$$
$$rX \circ rA = r(X \circ A), \tag{4.65}$$
$$X \circ A \subseteq X, \qquad \text{(anti-extensive)} \tag{4.66}$$
$$(X \circ A) \circ A = X \circ A. \qquad \text{(idempotent)} \tag{4.67}$$

4.20 Proposition. (Properties of closing)
For $A, X, Y \subseteq \mathbb{E}^d$, $h \in \mathbb{E}^d$ and $r \in \mathbb{E}$,

$$X \bullet \{h\} = X, \tag{4.68}$$
$$X_h \bullet A = (X \bullet A)_h, \tag{4.69}$$
$$X \bullet A_h = X \bullet A, \tag{4.70}$$
$$X \subseteq Y \Rightarrow X \bullet A \subseteq Y \bullet A, \qquad \text{(increasing)} \tag{4.71}$$
$$rX \bullet rA = r(X \bullet A), \tag{4.72}$$
$$X \subseteq X \bullet A, \qquad \text{(extensive)} \tag{4.73}$$
$$(X \bullet A) \bullet A = X \bullet A. \qquad \text{(idempotent)} \tag{4.74}$$

It is not true in general that $A \subseteq B$ implies $X \circ A \subseteq X \circ B$ or $X \circ A \supseteq X \circ B$; see Fig. 4.7 for a counterexample.

One can obtain such inclusion properties under certain conditions on the structuring elements, however. A set $X \subseteq \mathbb{E}^d$ is called *A-open* if $X \circ A = X$. In other words, a set X is A-open if it is contained in the invariance

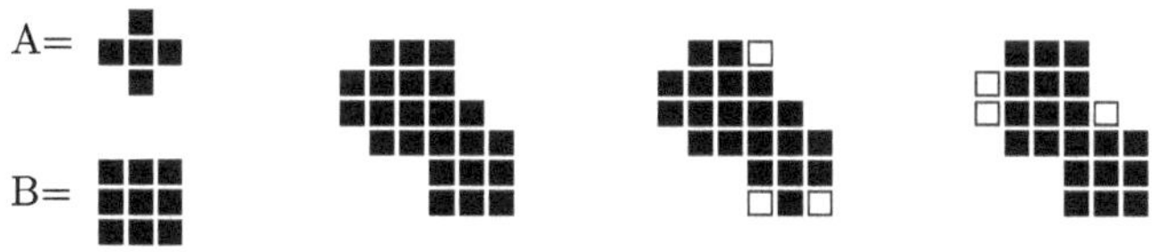

Figure 4.7 $A \subseteq B$ does not imply $X \circ A \subseteq X \circ B$ or $X \circ A \supseteq X \circ B$. From left to right: the structuring elements A and B; the set X; the opening $X \circ A$; the opening $X \circ B$.

domain of the operator α_A, that is, if $X \in \mathrm{Inv}(\alpha_A)$. Dually, X is said to be *A-closed* if $X \bullet A = X$.

4.21 Proposition. *X is A-open if and only if $X = Y \oplus A$ for some $Y \subseteq \mathbb{E}^d$.*

Proof. If X is A-open, then $X = (X \ominus A) \oplus A = Y \oplus A$, if $Y = X \oplus A$. On the other hand, if $X = Y \oplus A$, then $X \circ A = ((Y \oplus A) \ominus A) \oplus A = Y \oplus A = X$, by (4.52). □

4.22 Proposition. *Let $A, B \subseteq \mathbb{E}^d$ be such that A is B-open. For every $X \subseteq \mathbb{E}^d$,*

$$X \circ A \subseteq X \circ B, \tag{4.75}$$

$$(X \circ A) \circ B = (X \circ B) \circ A = X \circ A, \tag{4.76}$$

$$X \bullet A \supseteq X \bullet B, \tag{4.77}$$

$$(X \bullet A) \bullet B = (X \bullet B) \bullet A = X \bullet A. \tag{4.78}$$

Moreover, for these identities to hold the condition that A is B-open is necessary.

Proof. We prove (4.75) and (4.76). The other two relations follow by similar arguments. If A is B-open, then $A = B \oplus C$ for some $C \subseteq \mathbb{E}^d$. This gives

$$(X \circ A) \circ B = \big(((X \ominus A) \oplus C) \oplus B\big) \circ B = (X \ominus A) \oplus C \oplus B = X \circ A.$$

Since $X \circ A \subseteq X$, this means

$$X \circ A = (X \circ A) \circ B \subseteq X \circ B.$$

It remains to be shown that $(X \circ B) \circ A = X \circ A$. It is obvious that $(X \circ B) \circ A \subseteq X \circ A$. On the other hand, $(X \circ B) \circ A \supseteq (X \circ A) \circ A = X \circ A$.

That the given identities can only hold when A is B-open follows easily by substitution of $X = A$ in (4.75). In fact, this means that $A = A \circ A \subseteq A \circ B$. But the inclusion $A \circ B \subseteq A$ follows from the anti-extensivity of the opening; hence the assertion follows. □

The inclusion $X \circ A \subseteq X \circ B$ can also be expressed by saying that the opening by A is *more active* than the opening by B; cf. Section 3.5.

Proposition 4.22 forms the basis for the definition of a *size distribution*. Restrict, for the moment, to the Euclidean space $\mathbb{R}^d$. Let $A(r),\ r > 0$, be a collection of structuring elements with the property that $A(s)$ is $A(r)$-open for $s \geq r$. Suppose that an image X consists of several grains which differ in size and shape. Application of the opening $\alpha_r = \alpha_{A(r)}$ to X has the effect that all grains smaller than $A(r)$ (that is, grains which do not contain a translate of $A(r)$) are deleted. If, subsequently, we apply α_s with $s > r$ to the remainder, grains smaller than $A(s)$ are deleted. This procedure can be envisaged as a sieving process where one conceives of the α_r as a stack of sieves, r being a measure for the mesh width. Intuitively, as the mesh width of the sieve is increased, more of the image grains are falling through the sieve and the residual area of the filtered (sieved) image is decreasing monotonically. These residual areas form a size distribution that is indicative of the image structure.

4.23 Proposition. *If $A(r),\ r > 0$, is a collection of structuring elements such that $A(s)$ is $A(r)$-open for $s \geq r$, then the openings $\alpha_r = \alpha_{A(r)}$ satisfy the semigroup property*

$$\alpha_r \alpha_s = \alpha_s \alpha_r = \alpha_s, \qquad s \geq r. \tag{4.79}$$

Relation (4.79) is an immediate consequence of Proposition 4.22. A collection of openings $\alpha_r,\ r > 0$, which satisfies (4.79) is called a *granulometry*.

Taking the Lebesgue measure of the opened sets $\alpha_r(X)$ gives a nonincreasing function in the variable $r > 0$ called the *size distribution*. Note that a rigorous definition of size distributions requires that one restricts attention to sets X for which the openings $\alpha_r(X)$ are Lebesgue measurable. If one considers only closed subsets of $\mathbb{R}^d$ and uses compact structuring elements, then the opened sets are closed, too; see Section 7.6.

The family of openings $\alpha_r(X) = X \circ rB$, where B is the closed unit ball, is a prototype of a *Minkowski granulometry*. This is a granulometry $\alpha_r,\ r > 0$, which satisfies the following:

(i) every opening α_r is translation invariant;

(ii) the family $\alpha_r,\ r > 0$, is compatible under scalings, i.e.,

$$\alpha_r(rX) = r\alpha_1(X), \qquad r > 0,$$

for $X \subseteq \mathbb{R}^d$.

Section 9.6 presents a general theory of granulometries on $\mathcal{P}(\mathbb{R}^d)$. The main conclusion reached there is that for condition (4.79) to hold, convex structuring elements are a prerequisite. Section 6.7 presents a formal treatment of granulometries within the complete lattice framework, and Section 11.11 discusses granulometries for grey-scale functions.

Section 3.3 contains a formal treatment on openings and closings; it has been shown that every supremum of openings is an opening, and dually, that every infimum of closings is a closing; cf. Theorem 3.28. The next result states that every translation invariant opening on $\mathcal{P}(\mathbb{E}^d)$ can be decomposed as a supremum (read "union") of structural openings.

4.24 Theorem.

(a) *Let α be a translation invariant opening on $\mathcal{P}(\mathbb{E}^d)$; then*

$$\alpha = \bigvee_{A \in \mathrm{Inv}(\alpha)} \alpha_A. \tag{4.80}$$

(a′) *Let β be a translation invariant closing on $\mathcal{P}(\mathbb{E}^d)$; then*

$$\beta = \bigwedge_{A \in \mathrm{Inv}(\beta)} \beta_A. \tag{4.81}$$

Proof. Let $A \in \mathrm{Inv}(\alpha)$ and $X \subseteq \mathbb{E}^d$. If $A_h \subseteq X$ for some $h \in \mathbb{E}^d$, then $A_h = \alpha(A_h) \subseteq \alpha(X)$. This gives $\alpha_A(X) \subseteq \alpha(X)$, whence $\geq$ in (4.80) follows. To prove $\leq$ observe that $\alpha(X) \in \mathrm{Inv}(\alpha)$ for every $X \subseteq \mathbb{E}^d$. However, $\alpha(X) \subseteq \alpha_{\alpha(X)}(X)$, since $\alpha(X) \subseteq X$. Hence, the equality $\leq$ follows. □

Theorem 6.10 generalizes this result to arbitrary complete lattices. In general, not all structuring elements in $\mathrm{Inv}(\alpha)$ are needed to recover the opening α. Namely, if $A, B \in \mathrm{Inv}(\alpha)$ and B is A-open, then $\alpha_B \leq \alpha_A$, and one may delete the term α_B from the expression in (4.80).

The following results provide estimates for arbitrary increasing translation invariant operators in terms of openings and closings.

4.25 Proposition. *Let ψ be an increasing translation invariant operator on $\mathcal{P}(\mathbb{E}^d)$.*

(a) *If $A \subseteq \psi(A)$, then $\alpha_A \leq \psi$.*

(a′) *If $\psi(A) \subseteq A$, then $\psi \leq \beta_A$.*

Proof. If $A_h \subseteq X$, then $A_h \subseteq \psi(A_h) \subseteq \psi(X)$, since ψ is increasing and translation invariant. This implies the result. □

Note that $\alpha_A \leq \psi$ implies that $\alpha_A \leq \psi^n$ for every $n \geq 1$.

4.26 Corollary. *If ψ is an increasing translation invariant operator on $\mathcal{P}(\mathbb{E}^d)$ which is self-dual and if $\psi(A) = A$, then*

$$X \circ A \subseteq \psi(X) \subseteq X \bullet \check{A}.$$

Proof. That $X \circ A \subseteq \psi(X)$ follows from Proposition 4.25(a′). Furthermore, $\psi(A) = A$ in combination with the self-duality of ψ gives $\psi(A^c) = A^c$; hence $\psi \leq \beta_{A^c}$ by Proposition 4.25(a′). By (4.60), however, $\beta_{A^c}(X) = X \bullet \check{A}$, and the assertion follows. □

In Chapter 6 we discuss several methods of constructing openings and closings. Here we mention only one kind of opening (and closing), which is not a structural opening, namely, the annular opening. We say that the structuring element A is *symmetric* if $\check{A} = A$.

4.27 Proposition. *Let A be a symmetric structuring element. Then the mapping $X \mapsto (X \oplus A) \cap X$ defines an opening and the mapping $X \mapsto (X \ominus A) \cup X$ defines a closing.*

Proof. We show that $\psi(X) = (X \oplus A) \cap X$ is an opening. It is evident that ψ is increasing and anti-extensive. In particular, $\psi^2 \leq \psi$. If we can show that $\psi(X) \subseteq \psi(X) \oplus A$, then $\psi^2(X) = (\psi(X) \oplus A) \cap \psi(X) = \psi(X)$, and the result is proved. Assume that $h \in \psi(X) = (X \oplus A) \cap X$. There exist $x \in X$ and $a \in A$ such that $h = x + a$. Thus, $x = h - a = h + (-a)$. Since A is symmetric, $-a \in A$, whence $x \in (X \oplus A) \cap X = \psi(X)$. But then $h = x + a \in \psi(X) \oplus A$. □

If $0 \in A$, then $(X \oplus A) \cap X = X$ for every set X, and the resulting opening is trivial, namely, the identity operator. If $0 \notin A$, then the opening $X \mapsto (X \oplus A) \cap X$ is called the *annular opening*; dually, the closing $X \mapsto (X \ominus A) \cup X$ is called the *annular closing*. To understand this nomenclature refer to Fig. 4.8, where A is ring-shaped. A point x belongs to the opened set $(X \oplus A) \cap X$ if $x \in X$ and $x + a \in X$ for some $a \in A$.

Thus the central pixels disappear, whereas the pixels at the periphery are preserved. A set is invariant under this opening if it has the shape of a ring (annulus). Refer to Section 6.5 for a more general discussion.

Figure 4.8 Annular opening. From left to right: the structuring element A; the set X; the annular opening $X \cap (X \oplus A)$ (the white pixels are deleted).

4.5. Boolean functions

With every subset $X \subseteq \mathbb{E}^d$ one can associate its *characteristic function*; this has the value 1 at the points in X and the value 0 outside X. This characteristic function is denoted by the same symbol; thus $X(h) = 1$ if $h \in X$ and 0 if $h \in X^c$.

The computation of $X \oplus \check{A}$ and $X \ominus A$ at the point h requires only knowledge about $X(\cdot)$ inside the window A_h. In fact, if X takes the value 1 at all points of A_h, then $h \in X \ominus A$. If X takes the value 1 at at least one point of A_h, then $h \in X \oplus \check{A}$. Assume that A is a structuring element containing n points $a_1, a_2, \ldots, a_n$ and that b is a Boolean function of n variables; see Section 2.4. Define the operator ψ_b by

$$\psi_b(X) = \{h \in \mathbb{E}^d \mid b(X(a_1 + h), X(a_2 + h), \ldots, X(a_n + h)) = 1\}; \tag{4.82}$$

note that ψ_b depends on A, too. To get $X \ominus A$ one takes $b(u_1, \ldots, u_n) = u_1 u_2 \cdots u_n$; dually, to get $X \oplus \check{A}$ one chooses $b(u_1, \ldots, u_n) = u_1 + u_2 + \cdots + u_n$. It is plain that the operators ψ_b given by (4.82) are translation invariant. We mention some other elementary properties. First assume that b_i is a Boolean function in n variables for every i in some index set I. Then

$$\psi_{\bigvee_{i\in I} b_i} = \bigvee_{i\in I} \psi_{b_i}, \qquad \psi_{\bigwedge_{i\in I} b_i} = \bigwedge_{i\in I} \psi_{b_i}. \tag{4.83}$$

This implies in particular that

$$\psi_b \le \psi_{b'} \qquad \text{if } b \le b'. \tag{4.84}$$

Furthermore, one obtains the following relation for the negative operator:

$$(\psi_b)^* = \psi_{b^*}, \tag{4.85}$$

where b^* is given by (2.21).

4.28 Proposition. *Let b be a Boolean function in n variables.*

(a) ψ_b *is increasing if and only if b is increasing.*

(b) ψ_b *is a dilation if and only if b is additive.*

(b′) ψ_b *is an erosion if and only if b is multiplicative.*

The proofs of the if-statements are straightforward, and we leave them as an exercise to the reader. The only if-statements are an easy consequence of Theorem 4.30 following which shows how to compute b when ψ_b is given.

The operators ψ_b given by (4.82) use only a finite window to decide whether a point h lies in the transformed set $\psi_b(X)$, namely, $X \cap A_h$. This motivates the following definition.

4.29 Definition. Given a translation invariant operator ψ on $\mathcal{P}(\mathbb{E}^d)$, we say that ψ is a *finite window operator* if there exists a finite set $A \subseteq \mathbb{E}^d$ such that

$$h \in \psi(X) \iff h \in \psi(X \cap A'_h),$$

for every $h \in \mathbb{E}^d$, $X \subseteq \mathbb{E}^d$, and $A' \supseteq A$.

If, in addition, ψ is increasing, then it is sufficient to consider $A' = A$. In Definition 13.20 this definition will be extended to operators which are not translation invariant. Recall from Section 2.4 the convention that the notation $\big[S\big]$ means the Boolean variable which has the value 1 if the statement S is true and 0 if S is false.

4.30 Theorem. *Let $\psi : \mathcal{P}(\mathbb{E}^d) \to \mathcal{P}(\mathbb{E}^d)$ be a translation invariant finite window operator with window $A = \{a_1, a_2, \ldots, a_n\}$. Define the Boolean function b by*

$$b(u_1, \ldots, u_n) = \big[0 \in \psi(\{a_i \mid u_i = 1\})\big]; \tag{4.86}$$

then $\psi = \psi_b$.

Proof. Using the translation invariance and the assumption that ψ is finite with window A, we get

$$\begin{aligned} h \in \psi(X) &\iff h \in \psi(X \cap A_h) \\ &\iff 0 \in \psi(X_{-h} \cap A) \\ &\iff 0 \in \psi(\{a_i \mid a_i + h \in X\}) \\ &\iff 0 \in \psi(\{a_i \mid X(a_i + h) = 1\}) \end{aligned}$$

$$\iff b(X(a_1+h),\ldots,X(a_n+h))=1$$
$$\iff h\in\psi_b(X).$$

This proves the assertion. □

Using the expression for b given in (4.86), it is easy to prove the only if-statements in Proposition 4.28.

We point out that for finite window operators Matheron's representation theorem 4.15 is an immediate consequence of Theorem 2.44; the latter states that an increasing Boolean function can be represented as a sum-of-products, or alternatively, as a product-of-sums.

If b is the rank function r_s (see Section 2.4), then the resulting operator ψ_{r_s} is called *rank operator* and is denoted by $\rho_{A,s}$. It is easy to check that $h\in\rho_{A,s}(X)$ if and only if $X\cap A_h$ contains at least s points. Note that

$$\rho_{A,1}(X)=X\oplus\check{A} \qquad \text{and} \qquad \rho_{A,n}(X)=X\ominus A, \tag{4.87}$$

where $n=\text{card}(A)$. It is evident that

$$\rho_{A,n}\le\rho_{A,n-1}\le\cdots\le\rho_{A,1}. \tag{4.88}$$

Furthermore,

$$\rho^*_{A,s}=\rho_{A,n-s+1}. \tag{4.89}$$

In particular, this implies that the rank operator $\rho_{A,s}$ is self-dual if n is odd and $s=\frac{1}{2}(n+1)$. The corresponding operator $\rho_{A,(n+1)/2}$ is called *median operator* and is denoted by μ_A. In Fig. 4.9 we have illustrated all nine rank operators for the case where A is the 3×3 square.

The rank function is a special example of a more general class of Boolean functions, namely, the class of threshold functions; these functions have been discussed in Section 2.4. Let b be the Boolean threshold function with realization vector $(w_1,\ldots,w_n\mid s)$; assume that all entries are integer valued. Thus

$$b(u_1,\ldots,u_n)=\Big[\sum_{i=1}^n w_iu_i\ge s\Big].$$

Furthermore, let $A=\{a_1,a_2,\ldots,a_n\}$ be a finite structuring element. The corresponding operator ψ_b, called *weighted rank operator*, is given by

$$\psi_b(X)=\{x\in\mathbb{E}^d\mid\sum_{i=1}^n w_iX(x+a_i)\ge s\}.$$

Figure 4.9 From left to right and top to bottom: the original image X and the transformed images $\rho_{A,n}(X)$, where A is the 3×3 square and $n = 1, 2, \ldots, 9$. Note that the foreground X is white and that the background X^c is black.

We present an alternative expression for weighted rank operators which emphasizes its convolution-like nature. Consider w as a function from $\mathbb{E}^d$ into $\mathbb{Z}$ with finite domain A; let $w(a_i) = w_i$. The operator ψ_b can also be written as $\psi_b(X) = X\star_s w$, where

$$X\star_s w = \{x \in \mathbb{E}^d \mid \sum_{h \in \mathbb{E}^d} w(h)X(x+h) \geq s\}. \tag{4.90}$$

Denoting by Σw the sum over all weights, we have the duality relation

$$(X^c \star_s w)^c = X \star_{\Sigma w - s + 1} w. \tag{4.91}$$

If $w = A$, where A is the characteristic function of a finite structuring element, then

$$\rho_{A,s}(X) = X\star_s A. \tag{4.92}$$

This also explains the nomenclature "weighted rank operator". If the weights as well as the threshold are positive integers one has the following interpretation: a point x belongs to $X\star_s w$ if the sth value of the sequence obtained by putting the values $X(x+h)$ (each counted $w(h)$ times) in decreasing order equals 1.

An important field of application of weighted rank operators is the detection of geometric structures (corner points, boundaries, T-junctions) in noisy images. The next example illustrates this point.

4.31 Example. (Hit-or-miss operator)
Consider the hit-or-miss operator $X \mapsto X \circledast (B, C)$ where $B \cap C = \varnothing$. This operator can be written in terms of a Boolean function as follows. Let $B = \{a_1, a_2, \ldots, a_m\}$ and $C = \{a_{m+1}, a_{m+2}, \ldots, a_n\}$; define $A = B \cup C$ and

$$b(u_1, \ldots, u_n) = u_1 u_2 \cdots u_m \overline{u_{m+1} u_{m+2}} \cdots \overline{u_n}.$$

From (4.82) we get that $h \in \psi_b(X)$ if and only if $a_1 + h, a_2 + h, \ldots, a_m + h \in X$ and $a_{m+1} + h, a_{m+2} + h, \ldots, a_n + h \notin X$, that is, $B_h \subseteq X$ and $C_h \subseteq X^c$. This means that $\psi_b(X) = X \circledast (B, C)$. This operator can also be expressed as a weighted rank operator. If w is the weight function $B(\cdot) - C(\cdot)$, i.e.,

$$w(h) = \begin{cases} 1, & \text{if } h \in B, \\ -1, & \text{if } h \in C, \\ 0, & \text{otherwise,} \end{cases}$$

then

$$X \circledast (B, C) = X\star_m w. \tag{4.93}$$

The verification of this identity is left to the reader.

There are essentially two different ways to modify the performance of the hit-or-miss operator using the weighted rank representation (4.90). The simplest is to take a threshold which is smaller than m, leaving the weights unchanged. But one can also change the weights to values different from ± 1. Both approaches are illustrated by a concrete example. Suppose one wants to extract horizontal edges in a binary image by using the structuring element

$$(B, C) = \begin{array}{ccccc} \circ & \circ & \circ & \circ & \circ \\ \bullet & \bullet & \underline{\bullet} & \bullet & \bullet \\ \circ & \circ & \circ & \circ & \circ \end{array}$$

where the black dots correspond to points in B and the white dots to points in C. If the original image is corrupted by noise, then the performance of the hit-or-miss operator may be rather poor, in the sense that many edge pixels are not detected. If one allows that at most i pixels in the 5×3 neighbourhood of a point are affected by the noise process, then one has to replace the threshold $m = 5$ (which corresponds to the hit-or-miss operator) by $m = 5 - i$. Note that m may become negative. For instance, if $i = 2$, then the weighted rank operator classifies the underscored pixel in

$$\begin{array}{ccccc} \circ & \circ & \circ & \bullet & \circ \\ \bullet & \bullet & \underline{\circ} & \bullet & \bullet \\ \circ & \circ & \circ & \circ & \circ \end{array}$$

as horizontal edge pixel. In this approach it is irrelevant whether a pixel close to the centre or one which is far off has been distorted by noise. If one wants to take the distance to the centre point into account, one can use weights whose absolute values are different from 1. For example, one can take

$$w = \begin{pmatrix} -1 & -2 & -3 & -2 & -1 \\ 2 & 3 & \underline{7} & 3 & 2 \\ -1 & -2 & -3 & -2 & -1 \end{pmatrix}$$

and threshold 14. (Note that our threshold s must satisfy $s \leq 17$.) This operator classifies the centres of

$$\begin{array}{ccccc} \circ & \circ & \circ & \circ & \circ \\ \bullet & \circ & \underline{\bullet} & \bullet & \bullet \\ \circ & \circ & \circ & \circ & \circ \end{array} \qquad \text{and} \qquad \begin{array}{ccccc} \bullet & \circ & \circ & \circ & \circ \\ \bullet & \bullet & \underline{\bullet} & \bullet & \circ \\ \circ & \circ & \circ & \circ & \circ \end{array}$$

as edge points. If $s = 11$, then more points are classified as edge points, e.g., the centres of

$$\begin{array}{ccccc} \bullet & \bullet & \circ & \circ & \circ \\ \bullet & \bullet & \underline{\bullet} & \bullet & \circ \\ \circ & \circ & \circ & \circ & \bullet \end{array} \qquad \text{and} \qquad \begin{array}{ccccc} \circ & \circ & \bullet & \circ & \circ \\ \bullet & \bullet & \underline{\bullet} & \bullet & \bullet \\ \circ & \circ & \bullet & \circ & \circ \end{array}$$

4.6. Grey-scale morphology

The morphological operators discussed so far are based on set-theoretical operations such as union, intersection, complementation as well as translation. To extend these operators to grey-scale images there are essentially two alternative approaches. Either one has to give a set-based representation of grey-scale images, or one should construct grey-scale morphological operators from scratch, e.g., by looking for a lattice representation of grey-scale images. In this volume both approaches will get ample attention. The set-based approach will be treated in detail in Chapter 11; the direct approach in terms of complete lattices will be discussed at several places, e.g., in Section 5.7. Chapter 10 presents an axiomatic approach to grey-scale morphology combining both viewpoints.

The *umbra* approach introduced by Sternberg (1981), is based on the simple observation that the points on and below the graph of a function yield a set to which morphological set operations can be applied. We devote a few words to this approach below but we postpone a more thorough discussion until Section 11.6.

A second approach starts with the representation of a function as a family of threshold sets. Such a set comprises all points at which the function exceeds a given threshold. One can apply the morphological set operators, discussed in the previous sections, to these threshold sets; the resulting sets correspond to another function, the transformed grey-scale image. This method will be described in detail in Chapter 11.

Throughout this book grey-scale images are represented mathematically by functions $F : E \to \mathcal{T}$; here E is the domain space (usually $\mathbb{R}^d$ or $\mathbb{Z}^d$) and $\mathcal{T}$ is the grey-value set. We consider many different grey-value sets in this book, e.g., the infinite sets $\overline{\mathbb{R}}, \overline{\mathbb{Z}}, \overline{\mathbb{R}}_+, \overline{\mathbb{Z}}_+$, and the finite set $\{0, 1, 2, \ldots, N\}$. An essential requirement on $\mathcal{T}$ is that it defines a complete lattice.

This section does not pursue a general treatment of grey-scale morphology; its intention, rather, is to give the reader a flavour of the basic ideas. Throughout this section we restrict attention to the grey-value set $\overline{\mathbb{R}}$.

Let $\mathrm{Fun}(\mathbb{E}^d, \overline{\mathbb{R}})$, or briefly $\mathrm{Fun}(\mathbb{E}^d)$, denote the collection of all functions $F : \mathbb{E}^d \to \overline{\mathbb{R}}$. The function analogues of union and intersection are supremum and infimum, respectively. Let F_i, $i \in I$, be a collection of functions; the supremum and infimum are given by

$$(\bigvee_{i\in I} F_i)(x) = \bigvee\{F_i(x) \mid i \in I\}, \tag{4.94}$$

$$(\bigwedge_{i\in I} F_i)(x) = \bigwedge\{F_i(x) \mid i \in I\}, \tag{4.95}$$

respectively; cf. Example 2.10. Instead of set inclusion one has to consider the partial ordering given by

$$F \le G \qquad \text{iff} \qquad F(x) \le G(x), \text{ for every } x \in \mathbb{E}^d.$$

The counterpart of the set complement for functions is the *function negative*, defined by

$$F^*(x) = -F(x). \tag{4.96}$$

Although the mapping $F \mapsto F^*$ does not have the nice properties of a lattice complement (and $\mathrm{Fun}(\mathbb{E}^d)$ is *not* a Boolean lattice; cf. Example 2.30(b)) it defines a negation on $\mathrm{Fun}(\mathbb{E}^d)$; see also Section 2.2. As much as possible we denote operators on $\mathrm{Fun}(\mathbb{E}^d)$ by uppercase Greek letters. The operator Ψ is called *increasing* if $F \le F'$ implies that $\Psi(F) \le \Psi(F')$. Furthermore, define the *negative operator* Ψ^* of the operator Ψ by

$$\Psi^*(F) = -\Psi(-F). \tag{4.97}$$

A straightforward yet useful method to transform a grey-scale image is by changing the relative scale of grey-values. Such transforms can be used, for instance, to achieve higher contrast, to suppress certain ranges of grey-values, etc. Following Serra (1982) we call such grey-scale transforms anamorphoses.

4.32 Definition. An *anamorphosis* is a function $a : \overline{\mathbb{R}} \to \overline{\mathbb{R}}$ which is continuous and increasing. If, in addition, the function a is strictly increasing, then it is called a *strong anamorphosis*.

The function a given by $a(t) = 0$ if $t \le 0$ and $a(t) = t^2$ if $t > 0$ is an anamorphosis; the function $a(t) = e^t$ is a strong anamorphosis. Note that both mappings have range $[0, \infty]$.

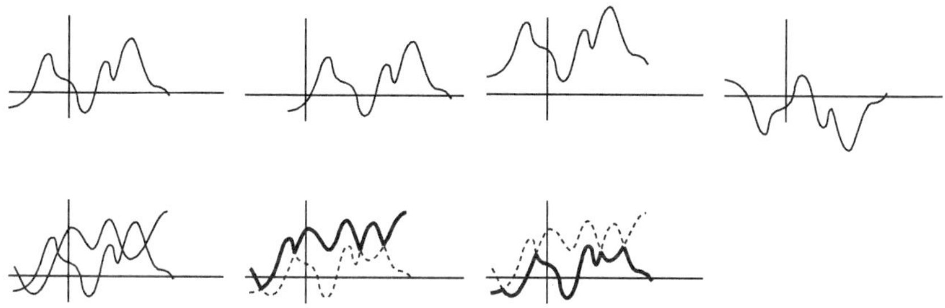

Figure 4.10 Top, from left to right: F, F_h, $F+v$, $F^* = -F$. Bottom, from left to right: F and G, $F \vee G$, $F \wedge G$.

If F is a function and $h \in \mathbb{E}^d$, then the *horizontal* (or *spatial*) translate F_h of F is defined by

$$F_h(x) = F(x - h). \tag{4.98}$$

The *vertical* translate $F + v$, where $v \in \mathbb{R}$, is defined by

$$(F + v)(x) = F(x) + v. \tag{4.99}$$

Some of these definitions are illustrated in Fig. 4.10.

There are essentially two alternative ways to define scalar multiplications (or scalings) of functions. Define the *T-scaling* of a function F by the factor $r \in \mathbb{E}$ by

$$(r \cdot F)(x) = \begin{cases} rF(x/r), & \text{if } x/r \in \mathbb{E}^d, \\ -\infty, & \text{otherwise.} \end{cases} \tag{4.100}$$

Here the prefix "T" refers to the fact that this multiplication acts both in the horizontal (spatial) and vertical (grey-scale) direction. By allowing multiplication in horizontal direction only, we obtain the *H-scaling* given by

$$(r \,\mathbf{.}\, F)(x) = \begin{cases} F(x/r), & \text{if } x/r \in \mathbb{E}^d, \\ -\infty, & \text{otherwise.} \end{cases} \tag{4.101}$$

Both scalings are illustrated in Fig. 4.11.

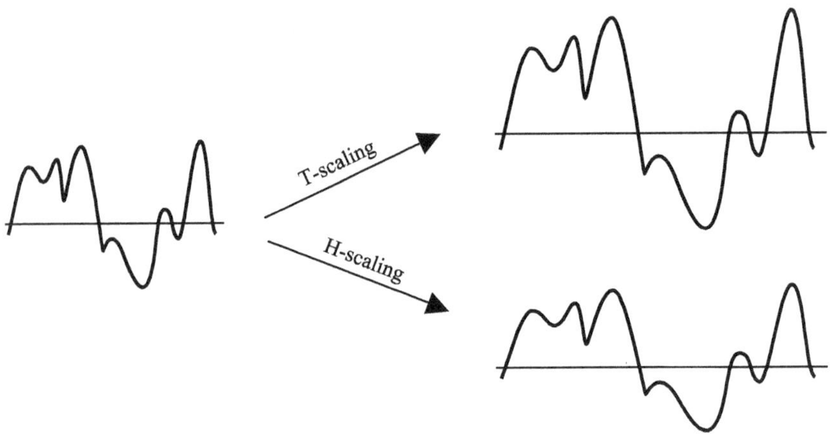

Figure 4.11 T-scaling and H-scaling.

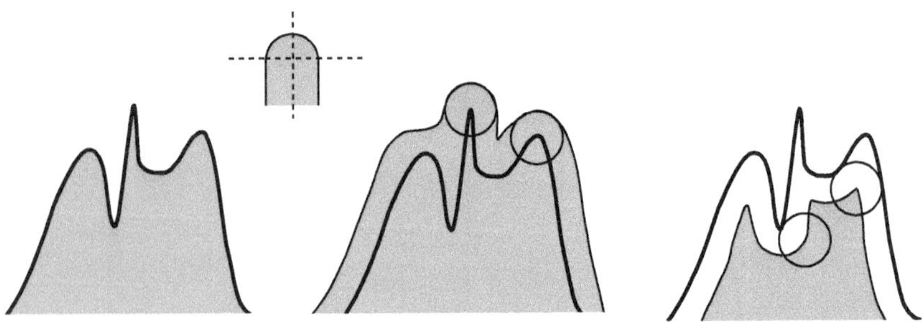

Figure 4.12 Minkowski sum and difference of two functions F and G. The structuring function is $G(x) = 1 - x^2$ for $|x| \leq 1$ and $G(x) = -\infty$ elsewhere. In this case Minkowski addition and subtraction can be visualized by dilating the umbra of the function F (see the following) by the unit disk.

The Minkowski sum and difference of two functions F and G are defined by

$$(F \oplus G)(x) = \bigvee_{h \in \mathbb{E}^d} [F(x-h) + G(h)], \tag{4.102}$$

$$(F \ominus G)(x) = \bigwedge_{h \in \mathbb{E}^d} [F(x+h) - G(h)]; \tag{4.103}$$

see Fig. 4.12 for an illustration. In the case of ambiguous expressions we use the convention that $s + t = -\infty$ if $s = -\infty$ or $t = -\infty$, and $s - t = +\infty$ if $s = +\infty$ or $t = -\infty$.

One can easily show that

$$F \oplus G = G \oplus F. \tag{4.104}$$

Define the operators $\Delta_G, \mathcal{E}_G$ on $\mathrm{Fun}(\mathbb{E}^d)$ by

$$\Delta_G(F) = F \oplus G, \qquad \mathcal{E}_G(F) = F \ominus G. \tag{4.105}$$

The function G is called *additive structuring function.*

4.33 Proposition. *For every* $G \in \mathrm{Fun}(\mathbb{E}^d)$ *the pair* $(\mathcal{E}_G, \Delta_G)$ *defines an adjunction on* $\mathrm{Fun}(\mathbb{E}^d)$*. In other words,*

$$F \oplus G \le F' \iff F \le F' \ominus G,$$

for $F, F' \in \mathrm{Fun}(\mathbb{E}^d)$.

Proof. We prove the implication $\Rightarrow$. The inequality $F \oplus G \le F'$ means that

$$\bigvee_{h \in \mathbb{E}^d} (F(x-h) + G(h)) \le F'(x),$$

for every $x \in \mathbb{E}^d$. This means that $F(x-h) + G(h) \le F'(x)$ for every $h, x \in \mathbb{E}^d$. But this is equivalent to $F(y) \le F'(y+h) - G(h)$ for $h, y \in \mathbb{E}^d$; here we have substituted $y = x - h$. Therefore,

$$F(y) \le \bigwedge_{h \in \mathbb{E}^d} (F'(y+h) - G(h)) = (F' \ominus G)(y).$$

This implies the result. □

This result implies in particular that $\mathcal{E}_G$ is an erosion and that Δ_G is a dilation. So for a given collection of functions F_i, $i \in I$,

$$(\bigvee_{i \in I} F_i) \oplus G = \bigvee_{i \in I} (F_i \oplus G), \tag{4.106}$$

$$(\bigwedge_{i \in I} F_i) \ominus G = \bigwedge_{i \in I} (F_i \ominus G). \tag{4.107}$$

The following proposition summarizes some further properties; these can be verified by the reader rather easily.

4.34 Proposition. (Properties of dilation)
Let $F, F', G, H \in \mathrm{Fun}(\mathbb{E}^d)$, $h \in \mathbb{E}^d$, $v \in \mathbb{R}$, *and* $r \in \mathbb{E}$;

$$F_h \oplus G = F \oplus G_h = (F \oplus G)_h, \tag{4.108}$$

$$(F + v) \oplus G = F \oplus (G + v) = (F \oplus G) + v, \tag{4.109}$$

$$F \le F' \Rightarrow F \oplus G \le F' \oplus G, \tag{4.110}$$

$$G \le H \Rightarrow F \oplus G \le F \oplus H, \tag{4.111}$$

$$F \oplus (G \vee H) = (F \oplus G) \vee (F \oplus H), \tag{4.112}$$

$$F \oplus (G \wedge H) \le (F \oplus G) \wedge (F \oplus H), \tag{4.113}$$

$$(F \oplus G) \oplus H = F \oplus (G \oplus H), \tag{4.114}$$

$$r \cdot (F \oplus G) = (r \cdot F) \oplus (r \cdot G), \tag{4.115}$$

$$r \,\bullet\, (F \oplus G) = (r \,\bullet\, F) \oplus (r \,\bullet\, G). \tag{4.116}$$

In fact, properties (4.110)–(4.113) follow immediately from (4.104) and (4.106). They have been included for the sake of completeness.

4.35 Proposition. (Properties of erosion)
Let $F, F', G, H \in \mathrm{Fun}(\mathbb{E}^d)$, $h \in \mathbb{E}^d$, $v \in \mathbb{R}$, *and* $r \in \mathbb{E}$;

$$F_h \ominus G = F \ominus G_{-h} = (F \ominus G)_h, \tag{4.117}$$

$$(F + v) \ominus G = F \ominus (G - v) = (F \ominus G) + v, \tag{4.118}$$

$$F \le F' \Rightarrow F \ominus G \le F' \ominus G, \tag{4.119}$$

$$G \le H \Rightarrow F \ominus G \ge F \ominus H, \tag{4.120}$$

$$F \ominus (G \vee H) = (F \ominus G) \wedge (F \ominus H), \tag{4.121}$$

$$F \ominus (G \wedge H) \ge (F \ominus G) \vee (F \ominus H), \tag{4.122}$$

$$(F \ominus G) \ominus H = F \ominus (G \oplus H), \tag{4.123}$$

$$r \cdot (F \ominus G) = (r \cdot F) \ominus (r \cdot G), \tag{4.124}$$

$$r \,\bullet\, (F \ominus G) = (r \,\bullet\, F) \ominus (r \,\bullet\, G). \tag{4.125}$$

There exist alternative expressions for $F \oplus G$ and $F \ominus G$ which have a straightforward geometric interpretation. Define

$$\check{G}(x) = G(-x). \tag{4.126}$$

For $F, G \in \mathrm{Fun}(\mathbb{E}^d)$ we have

$$(F \oplus G)(x) = \inf\{v \in \mathbb{R} \mid -(\check{G})_x + v \ge F\}, \tag{4.127}$$

$$(F \ominus G)(x) = \sup\{v \in \mathbb{R} \mid G_x + v \leq F\}. \tag{4.128}$$

Here we derive the second identity. The first one can be proved analogously.

$$\begin{aligned}(F \ominus G)(x) &= \bigwedge_{h \in \mathbb{E}^d} [F(x+h) - G(h)] \\ &= \bigvee \{v \in \mathbb{R} \mid \forall h \in \mathbb{E}^d :\ v \leq F(x+h) - G(h)\} \\ &= \bigvee \{v \in \mathbb{R} \mid \forall y \in \mathbb{E}^d :\ v \leq F(y) - G(y-x)\} \\ &= \bigvee \{v \in \mathbb{R} \mid G_x + v \leq F\}.\end{aligned}$$

Besides the adjunction relationship expressed by Proposition 4.33, there exist yet two other dualities between grey-scale dilations and erosions. The lattice $\mathrm{Fun}(\mathbb{E}^d)$ is self-dual in the sense of the Duality Principle; dilations and erosions are dual notions in this sense. And finally, the negation $F \mapsto F^* = -F$ defined by (4.96) leads to the duality relation

$$(F \oplus G)^* = F^* \ominus \check{G} \qquad \text{and} \qquad (F \ominus G)^* = F^* \oplus \check{G}. \tag{4.129}$$

An operator Ψ on $\mathrm{Fun}(\mathbb{E}^d)$ is called *translation invariant* if for $h \in \mathbb{E}^d$ and $v \in \mathbb{R}$ we have

$$\Psi(F_h + v) = [\Psi(F)]_h + v. \tag{4.130}$$

Note that this definition includes invariance under horizontal as well as vertical translations. A function operator Ψ satisfying (4.130) is called a *T-operator.* A dilation with this property is called a T-dilation; T-erosions, T-openings, etc., are defined similarly. In Chapter 11 we also introduce H-operators; such operators are only required to be invariant under horizontal translations. Finally, the *kernel* $\mathcal{V}(\Psi)$ of a function operator Ψ is defined by

$$\mathcal{V}(\Psi) = \{G \in \mathrm{Fun}(\mathbb{E}^d) \mid \Psi(G)(0) \geq 0\}. \tag{4.131}$$

4.36 Theorem. (Representation of T-operators)
Given an increasing T-operator Ψ on $\mathrm{Fun}(\mathbb{E}^d)$*; then*

$$\Psi(F) = \bigvee_{G \in \mathcal{V}(\Psi)} F \ominus G, \tag{4.132}$$

and dually,

$$\Psi(F) = \bigwedge_{G \in \mathcal{V}(\Psi^*)} F \oplus \check{G}. \tag{4.133}$$

Proof. Let F' denote the right-hand side of (4.132). Assume further that $\Psi(F)(x) \geq t$; we show that $F'(x) \geq t$. Since $\Psi(F_{-x} - t)(0) \geq 0$, we get $F_{-x} - t \in \mathcal{V}(\psi)$. Furthermore, by (4.128),

$$\begin{aligned}(F \ominus (F_{-x} - t))(x) &= \sup\{v \in \mathbb{R} \mid (F_{-x} - t)_x + v \leq F\} \\ &= \sup\{v \in \mathbb{R} \mid F + v \leq F + t\} \\ &\geq t.\end{aligned}$$

This gives $F'(x) \geq t$.

Assume conversely that $F'(x) > t$; we show that $\Psi(F)(x) \geq t$. Apparently, $(F \ominus G)(x) > t$ for some $G \in \mathcal{V}(\psi)$. From (4.128) we deduce that $G_x + t \leq F$. But then

$$\Psi(F)(x) = \Psi(F_{-x})(0) \geq \Psi(G + t)(0) \geq t.$$

This concludes the proof of (4.132).

To prove (4.133), one uses that

$$\Psi^*(F) = \bigvee_{G \in \mathcal{V}(\Psi^*)} F \ominus G.$$

Substituting $-F$, taking the negation on both sides, and using (4.129) gives the identity. □

The *domain* $\mathrm{dom}(G)$ of a function $G \in \mathrm{Fun}(\mathbb{E}^d)$ is defined by

$$\mathrm{dom}(G) = \{x \in \mathbb{E}^d \mid G(x) > -\infty\}. \tag{4.134}$$

In practice, one uses structuring functions with bounded domain. If the structuring function G takes only the value 0 on its domain, then it can be represented by a set, namely, $\mathrm{dom}(G)$; in that case the dilation $F \oplus G$ and erosion $F \ominus G$ become

$$(F \oplus G)(x) = \bigvee_{h \in \mathrm{dom}(G)} F(x - h),$$

$$(F \ominus G)(x) = \bigwedge_{h \in \mathrm{dom}(G)} F(x + h).$$

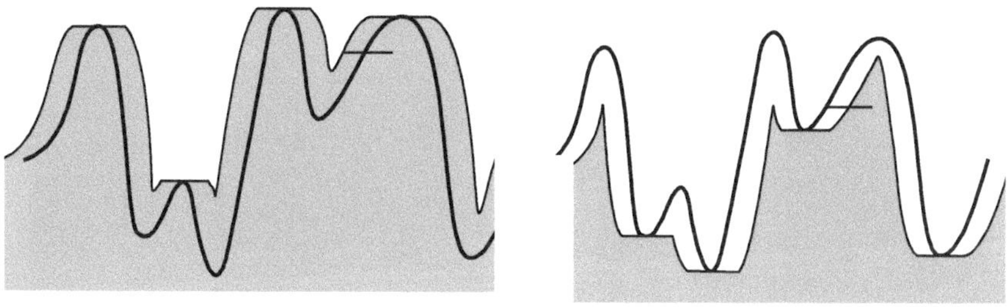

Figure 4.13 Dilation and erosion by a flat structuring function.

We use the convention that for a function F and a set A,

$$(F \oplus A)(x) = \bigvee_{h \in A} F(x-h), \tag{4.135}$$

$$(F \ominus A)(x) = \bigwedge_{h \in A} F(x+h); \tag{4.136}$$

see Figs. 4.13 and 4.18 for an illustration.

In this context A is called *flat structuring function*. Dilation and erosion by a flat structuring function are examples of *flat operators*. Though a formal definition of flat operators is postponed until Chapter 11, we try to give the reader some intuition for this concept. Every increasing set operator ψ can be extended to a function operator Ψ in the following way: represent a function F by its threshold sets $\{x \mid F(x) \geq t\}$, where $t \in \overline{\mathbb{R}}$. Apply ψ to these sets, and reconstruct a function $\Psi(F)$ from the transformed sets $\psi(\{x \mid F(x) \geq t\})$. This procedure is worked out in detail in Chapter 11. Flat operators have several interesting properties. Characteristic for such operators is that they commute with anamorphoses; see Theorem 11.12 for a precise statement.

Flat structuring functions are used, e.g., to define the *morphological gradient operator*. Recall that for a continuously differentiable function F on $\mathbb{R}^d$ the gradient ∇F is the d-vector $(\partial F/\partial x_1, \partial F/\partial x_2, \ldots, \partial F/\partial x_d)$. The morphological gradient is defined by

$$\operatorname{grad}(F) = \lim_{r \downarrow 0} \frac{1}{2r}\big[(F \oplus rB) - (F \ominus rB)\big], \tag{4.137}$$

where B is the d-dimensional ball with radius 1. It is easy to show that

$$\operatorname{grad}(F) = \|\nabla F\|,$$

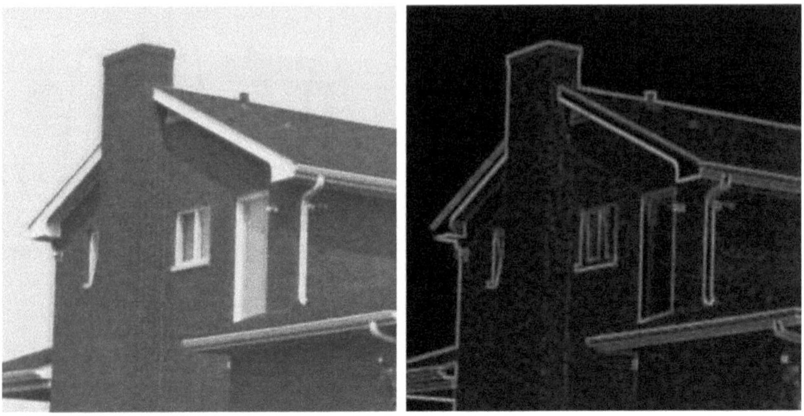

Figure 4.14 Morphological gradient $\frac{1}{2}\left[(F \oplus B) - (F \ominus B)\right]$ of an image; B is the 3×3 square.

if F is continuously differentiable; here $\| \cdot \|$ denotes the Euclidean norm. For a discrete image one replaces (4.137) by

$$\mathrm{grad}(F) = \frac{1}{2}\left[(F \oplus B) - (F \ominus B)\right],$$

where B is some discretization of the unit ball; see Section 9.9. In the 2-dimensional case one usually takes for B the 3×3 square containing nine points. Fig. 4.14 illustrates the discrete gradient for a specific image.

Just as in the binary case, erosion of a function F by the structuring function G followed by a dilation by G gives the opening of F by G:

$$F \circ G = (F \ominus G) \oplus G. \tag{4.138}$$

The properties of this opening are the same as in the binary case: it is increasing, anti-extensive, and idempotent. Moreover, this opening is invariant under horizontal and vertical translations; following our convention, we call this opening *T-opening*. Dually, dilation followed by erosion gives a T-closing:

$$F \bullet G = (F \oplus G) \ominus G. \tag{4.139}$$

This operator is increasing, extensive, and idempotent. The grey-scale opening and closing are negative operators in the following sense:

$$(F^* \circ G)^* = F \bullet \check{G}. \tag{4.140}$$

Both operators are illustrated in Fig. 4.15.

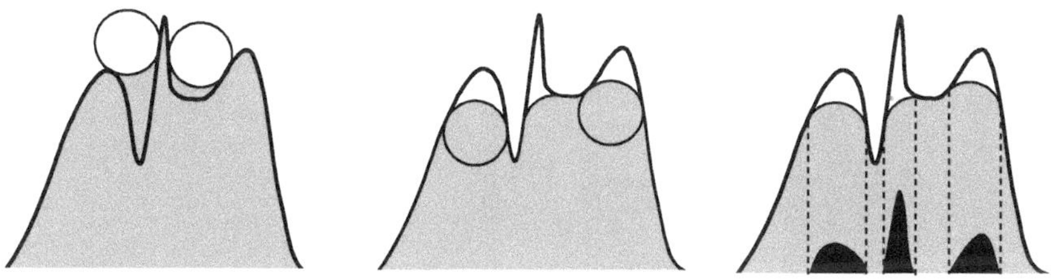

Figure 4.15 Grey-scale opening and closing. Let G be the structuring function associated with the unit ball, i.e., $G(x) = 1 - |x|^2$. From left to right: the closing $F \bullet G$, the opening $F \circ G$, and the top-hat operator $F - F \circ G$ (see the following).

Most of the properties of set openings and closings stated in Section 4.4, such as Propositions 4.21 and 4.22, carry over to the function case almost verbatim.

4.37 Proposition. *F is G-open if and only if $F = F' \oplus G$ for some function F'.*

4.38 Proposition. *Let $G, H \subseteq \mathbb{E}^d$ be such that G is H-open. For every function F,*

$$F \circ G \subseteq F \circ H, \tag{4.141}$$

$$(F \circ G) \circ H = (F \circ H) \circ G = F \circ G, \tag{4.142}$$

$$F \bullet G \supseteq F \bullet H, \tag{4.143}$$

$$(F \bullet G) \bullet H = (F \bullet H) \bullet G = F \bullet G. \tag{4.144}$$

Moreover, for these identities to hold the condition G is H-open is necessary.

Furthermore, the analogue of Theorem 4.24 holds; the proof is nearly identical and will be omitted.

4.39 Theorem.

(a) *Let α be a T-opening on* $\mathrm{Fun}(\mathbb{E}^d)$*; then*

$$\alpha(F) = \bigvee_{G \in \mathrm{Inv}(\alpha)} F \circ G. \tag{4.145}$$

(a′) *Let β be a T-closing on* $\mathrm{Fun}(\mathbb{E}^d)$*; then*

$$\beta(F) = \bigwedge_{G \in \mathrm{Inv}(\beta)} F \bullet G. \tag{4.146}$$

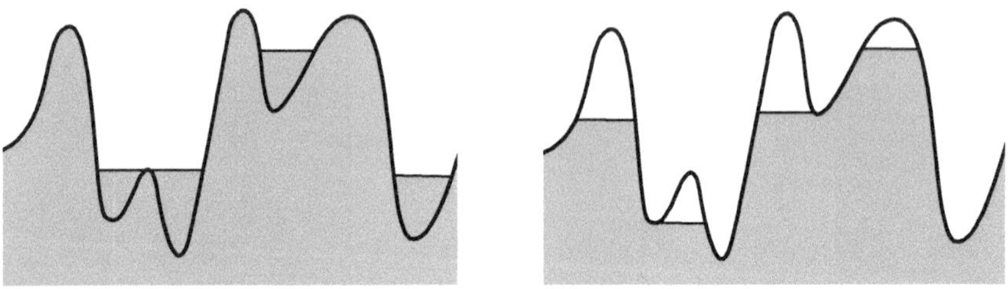

Figure 4.16 Flat closing and opening.

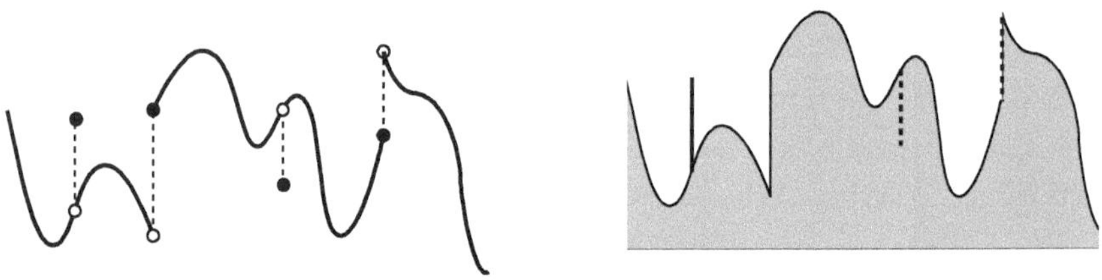

Figure 4.17 The umbra $\mathcal{U}(F)$ (right) of a function F (left).

Here the invariance domain of an operator Ψ on Fun($\mathbb{E}^d$) is defined as usual, i.e., $\text{Inv}(\Psi) = \{F \in \text{Fun}(\mathbb{E}^d) \mid \Psi(F) = F\}$. If A is a flat structuring function, then $F \circ A$ denotes the opening $(F \ominus A) \oplus A$; this opening is called a *flat opening*. Analogously, one defines the *flat closing* $F \bullet A$. An illustration of flat openings and closings can be found in Fig. 4.16; see also Fig. 4.18.

This seems to be an appropriate place to devote a few words to the concept of an umbra. The *umbra* $\mathcal{U}(F)$ of a function F is defined by

$$\mathcal{U}(F) = \{(x, t) \in \mathbb{E}^d \times \overline{\mathbb{R}} \mid t \leq F(x)\}; \tag{4.147}$$

see Fig. 4.17.

It has become very popular among researchers in morphology to extend binary morphological operators to grey-scale images by means of the *umbra transform*. The basic idea is to apply a set operator $\psi : \mathcal{P}(\mathbb{E}^{d+1}) \to \mathcal{P}(\mathbb{E}^{d+1})$ to the umbra of a function defined on $\mathbb{E}^d$. A technical complication is that, contrary to what is often asserted in the morphological literature, the transform $\psi(U)$ of an umbra need not be an umbra. Refer to Fig. 11.4 for a counterexample. Section 11.6 presents a comprehensive discussion of the umbra transform. Until then we use umbras only as a visualization tool; in fact we have given applications in Figs. 4.12, 4.13, 4.15, and 4.16.

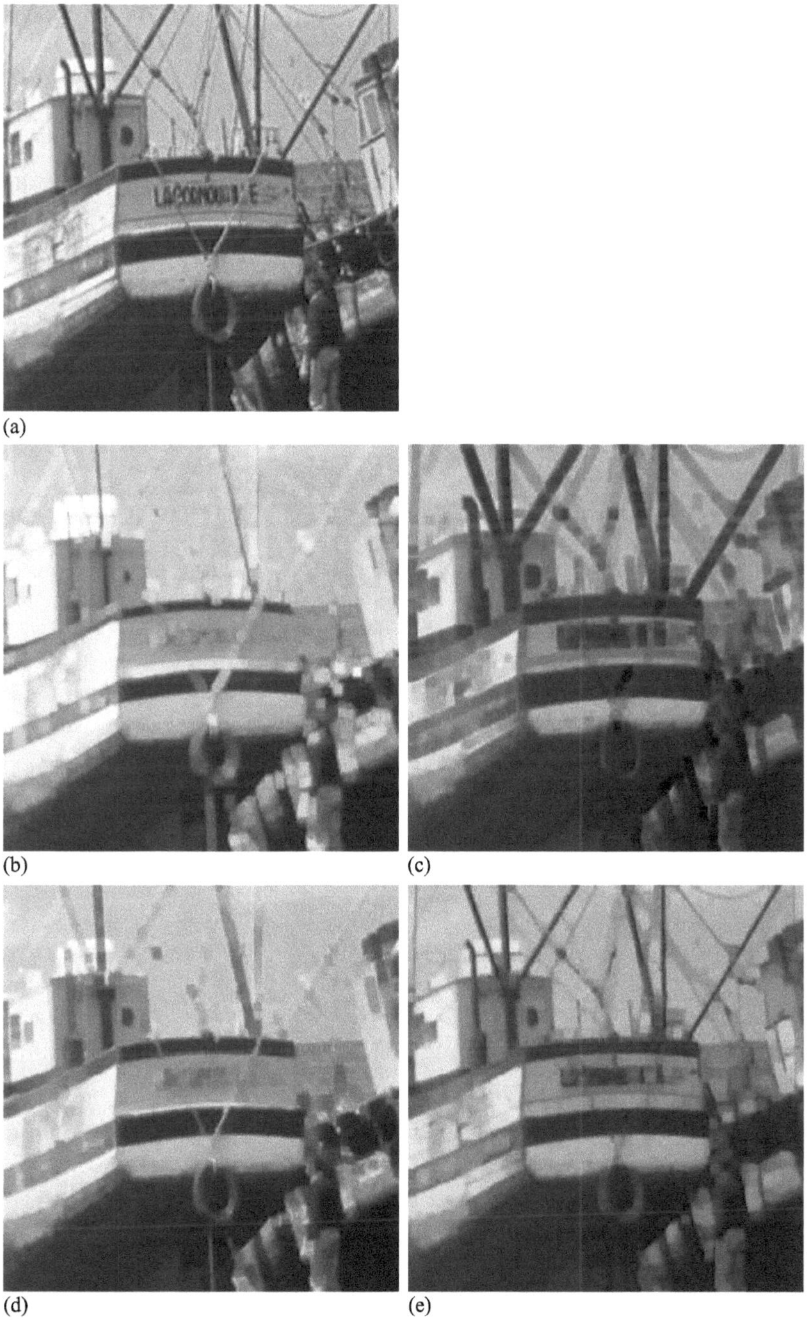

Figure 4.18 (a) Original image; (b) flat dilation; (c) flat erosion; (d) flat opening; (e) flat closing with 5×5 square.

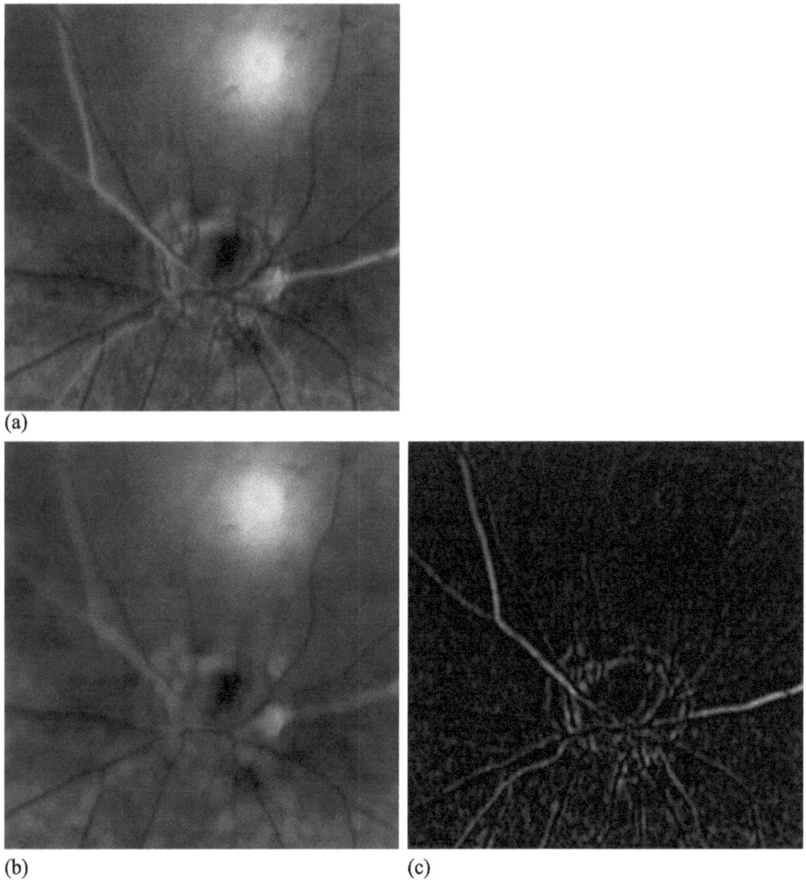

Figure 4.19 Rolling ball opening and top-hat operator. (a) Original image, (b) rolling ball opening, and (c) top-hat operator (multiplied).

Fig. 4.19 shows that the opening of an umbra by a ball gives again an umbra. Note that the structuring function G corresponding to this opening is given by $G(x) = (1 - \|x\|^2)^{1/2}$. This opening is called the *rolling ball opening*, since it has the effect of rolling a ball along the undersurface of the umbra and removing all parts that this ball cannot enter. Dually, one can define the *rolling ball closing* as the transform obtained by rolling the ball over the surface of the umbra.

Every opening $\alpha(F)$ lies below the function F itself, and so the difference $\Psi(F) = F - \alpha(F)$ is a function which is nonnegative on the domain of F. If, for instance, α is the rolling ball opening, then $\Psi(F)$ contains the narrow peaks of F; see Fig. 4.19. The operator Ψ is called the *top-hat operator*. In Example 11.19 we show that $\alpha(F - \alpha(F)) = 0$ everywhere.

We briefly discuss the class of flat function operators associated with the finite window operators on $\mathcal{P}(\mathbb{E}^d)$. A comprehensive discussion is postponed until Section 11.4. Let b be an increasing Boolean function. We can extend b to a function $b^\sim : \overline{\mathbb{R}}^n \to \overline{\mathbb{R}}$ as follows: replace a product by an infimum, a sum by a supremum, 0 by $-\infty$, and 1 by ∞. For example, if $b(x_1, x_2, x_3) = x_1 x_2 + x_3$, then $b^\sim(t_1, t_2, t_3) = (t_1 \wedge t_2) \vee t_3$. Henceforth we denote an increasing Boolean function and its extension to $\overline{\mathbb{R}}^n$ by the same symbol.

Now, if $A = \{a_1, a_2, \ldots, a_n\}$ is a subset of $\mathbb{E}^d$ and b an increasing Boolean function of n variables, then we can define the operator Ψ_b on $\mathrm{Fun}(\mathbb{E}^d)$ by

$$\Psi_b(F)(x) = b(F(x + a_1), \ldots, F(x + a_n)).$$

It is easy to show that Ψ_b is an increasing T-operator. In fact, it will be demonstrated in Section 11.4 that the operator Ψ_b is the flat extension of the set operator ψ_b given by (4.82). In particular, if Ψ_b is applied to the characteristic function of a set X, then the result is the characteristic function of $\psi_b(X)$. It follows with little effort that properties (4.83)–(4.85) and Proposition 4.28 can be extended to the grey-scale case. If b is a threshold function, the resulting function operator Ψ_b is called a *weighted rank operator*. Let $w : \mathbb{E}^d \to \mathbb{N}$ be a weight function with finite domain, and let $s \in \mathbb{N}$ be a threshold. An extension of the convolution-like notation of (4.91) to grey-scale functions is possible but results in untransparent expressions like

$$(F \star_s w)(x) = \sup\{t \in \overline{\mathbb{R}} \mid \sum_{h \in \mathbb{E}^d} w(h)\big[F(x + h) \geq t\big] \geq s\}; \tag{4.148}$$

see Section 11.4 for more details. Instead we can also interpret $(F \star_s w)(x)$ as the sth value in the series obtained by arranging the values $F(x + h)$ occurring $w(h)$ times in decreasing order. In the special case where all nonzero weights $w(h)$ are equal to 1, we speak of a *rank operator*. Analogously to (4.92) this operator is denoted by

$$\rho_{A,s}(F) = F \star_s A. \tag{4.149}$$

Note that $(F \star_s A)(x)$ is the sth largest member of the sequence $F(x + a_1)$, $F(x + a_2), \ldots, F(x + a_n)$. Analogous to (4.88),

$$\rho_{A,n} \leq \rho_{A,n-1} \leq \cdots \leq \rho_{A,1}. \tag{4.150}$$

If n is odd and $s = \frac{1}{2}(n+1)$, then $\rho_{A,s}$ is a self-dual operator, called the *median operator*; this operator is denoted by μ_A.

4.7. Bibliographical notes

The two standard references on mathematical morphology are the treatises by Matheron (1975) and Serra (1982). The first book comprises a deep mathematical exposition on the theory of random sets and integral geometry; in this context mathematical morphology comes into play as a natural tool. In fact, many of the theoretical results in morphology (in particular those in Sections 4.3 and 4.4) can be found in Matheron (1975). The book by Serra (1982) is obligatory reading to everybody who wants to learn something about morphology. It includes most of the material covered by this chapter, albeit that Serra puts the emphasis on the practical aspects rather than on the underlying theory. A third major reference is the recent treatise by Schmitt and Vincent (2010); this book is eminently suited for those readers interested in algorithmic aspects. It discusses all basic morphological tools and explains in considerable detail how to exploit them for specific tasks. The book by Coster and Chermant (1985) contains a general exposition on morphology-based methods of image analysis; this book is addressed mainly to an audience of applied researchers in material science and cell biology. The book by Preston and Duff (1984) contains a general discussion on *cellular logic* and *cellular automata* and their applications to picture processing; they consider algorithmic aspects as well as hardware implementations. Finally, a very elementary exposition on mathematical morphology is the book by Giardina and Dougherty (1988).

There exist also a number of books in the area of image processing and computer vision containing some basic material on morphology; we mention the second edition of Pratt's treatise on image analysis (Pratt, 1991) and the books by Dougherty and Giardina (1987b, Chapter 3), Haralick and Shapiro (1992, Chapter 5), Pitas and Venetsanopoulos (1990, Chapter 6), and Russ (1992, Chapter 6). Furthermore, the recent volume edited by Dougherty (1993) gives a good impression of some recent developments. Finally, we mention the tutorial papers by Maragos (1987), Haralick et al. (1987), and Heijmans (1992b).

There is an extensive literature on set theory. For our goals it suffices to mention the books by Hausdorff (1962), Kuratowski (1972), and Kuratowski and Mostowski (1976).

The hit-or-miss operator, discussed in detail in Serra (1982), has the effect of a *template matching* as described by Crimmins and Brown (1985); see also Haralick and Shapiro (1992, §5.2.3). The skeleton algorithm in Example 4.7 is due to Levialdi (1971); see also Serra (1982, Exercise XI.I.4). The modification of the hit-or-miss operator discussed in Example 4.31 may give a better performance in the presence of noise. Bloomberg and Maragos (1990) discuss a generalization of this operator based on rank operators. In fact, for a pair of structuring elements (A, B) they define the (s, t)th *rank hit-or-miss operator* by (A, B) as follows:

$$X \circledast_{s,t} (A, B) = (X \star_s A) \cap (X^{c} \star_t B),$$

where $s = 1, 2, \ldots, \mathrm{card}(A)$, $t = 1, 2, \ldots, \mathrm{card}(B)$, and where $\star_s$ is given by (4.92). The bi-kernel as well as the wedge operator were introduced by Banon and Barrera (1991); Theorem 4.9 results from their work. Actually, the results they achieved apply to arbitrary complete lattices.

Vincent (1991) presents an efficient algorithm for the Minkowski addition of two arbitrary shapes. Ghosh (1994) introduces the notion of *negative shape* to make Minkowski addition into a group operation on the family of convex polygons. All basic properties of Minkowski addition and subtraction can be found in Hadwiger (1957).

As noted, decomposition of structuring elements into smaller parts is of great practical value, since this technique can be used to obtain fast implementations of morphological operators. Exhaustive treatments can be found in the papers by Xu (1991) and Zhuang and Haralick (1986). We do not give any details here but point out only that discrete convexity plays an important role in these theories. To understand this, one should observe that a compact set $A \subseteq \mathbb{R}^d$ is convex if and only if

$$(r + s)A = rA \oplus sA,$$

for $r, s > 0$. This relation is also valid in the 2-dimensional discrete case, but fails in higher dimensions. The 5×5 square can be decomposed as the Minkowski addition of two 3×3 squares. But one has also

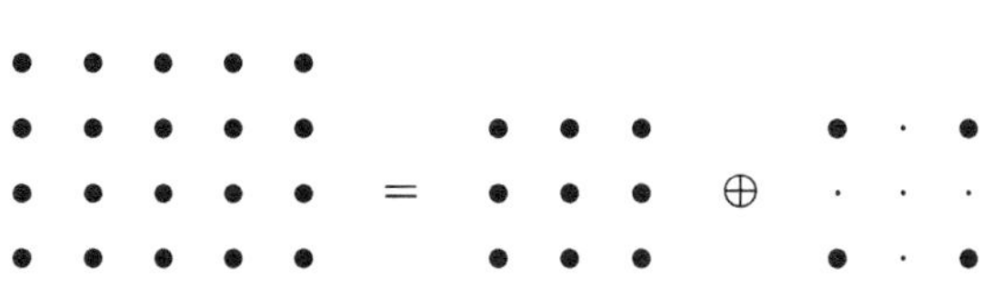

Note that in this decomposition one of the 3×3 squares is replaced by its extreme points. This observation may lead to a substantial reduction of the number of computations needed to perform dilations and erosions with large structuring elements. The larger the structuring elements, the more significant such reductions become. We refer to Engbers et al. (2001), where this property is used to achieve *logarithmic decompositions.*

The representation theorem 4.15 is due to Matheron (1975). The notions *minimal kernel element* and *basis of an operator* stem from Maragos (1985); see also Maragos (1989b). Structural openings and closings have been introduced for the first time by Ronse and Heijmans (1991); for some comments about the relation between this notion, the adjunctional opening and the morphological opening defined by Serra and his co-workers (Serra, 1988), we refer to Chapter 6.

Granulometries are treated in great detail in Matheron (1975); its practical applications have been discussed by Serra (1988, Chapter X). Consult Section 9.10 for additional references. Note that Matheron and Serra use the terminology Euclidean granulometry instead of Minkowski granulometry.

The annular opening is due to Serra (1988, Section 5.4).

Median statistics was invented by Tukey (1977), who used it for nonlinear smoothing of data. The utilization of this operator in image processing has been described by many authors; we mention in particular the study by Justusson (1981). There is a great deal of literature about the construction of morphological operators (such as rank operators) from Boolean functions. Such studies are often motivated by the need for modifications of the median operator. It is impossible to mention all literature on this subject. The interested reader is referred to the papers by Nodes and Gallagher (1982), Bovik et al. (1983), and Wendt et al. (1986) and the references given there. The relations between the operators examined in these papers (actually, they are usually called "filters" there) and mathematical morphology is pointed out by Maragos and Schafer (1987b); Maragos and Schafer (1987a) discuss the embedding of linear translation invariant operators into the morphological framework. The paper by van den Boomgaard (1990) presents some interesting applications of threshold logic in morphology. Wilson (1993) uses threshold logic to construct a method for training structuring elements to be used for the localization of given patterns in an image. Finally, we point out that the operators resulting from Boolean logic are known under many different names. The rank operator, e.g., is also called *order statistic filter* (Maragos & Schafer, 1987b), *percentile filter* (van den Boomgaard, 1990), or Ξ-*filter* (Preston, 1983). Furthermore, in this context, the terminology *voting logic* is often encountered.

It is difficult to trace the history of grey-scale morphology, and we shall resist this temptation. It suffices to mention some of the early work in this field. The Minkowski addition and subtraction of two functions as well as the umbra transform are originally due to Sternberg (1981). Goetcherian (1980) uses concepts from *fuzzy logic* to define a number of morphological operations (e.g., skeletonization) on grey-scale images. Without doubt, however, the main developments are due to the Fontainebleau school. Yet, since most of their work appeared in internal reports of the Ecole des Mines de Paris, their methods remained obscure until the appearance in 1982 of the book by Serra (1982). We refer to Chapter XII of that volume and a recent paper by Serra (1993) for a historical discussion.

The representation theorem 4.36 was proved independently by Maragos (1989b) for upper semi-continuous functions and Giardina and Dougherty (1988). The rolling ball opening occurs for the first time in the work of Sternberg (1982).

CHAPTER FIVE

Adjunctions, dilations, and erosions

Henk J.A.M. Heijmans
Formerly Centre for Mathematics and Computer Science, Amsterdam, Netherlands

Contents

In Chapter 3 we have given formal definitions of dilations and erosions between two complete lattices. We have seen two important applications of this definition: Minkowski addition and subtraction for sets and for functions. In both cases the resulting operators are invariant under translations. In the function case, one can distinguish two kind of translations, spatial translations and grey-value translations. The present chapter unifies all these different cases into one consistent algebraic framework. As a matter of fact, it develops a theory of dilations and erosions between arbitrary complete

Advances in Imaging and Electron Physics, Volume 216
ISSN 1076-5670
https://doi.org/10.1016/bs.aiep.2020.07.005

lattices which are invariant under a given automorphism group. In the context of morphology on the Euclidean space discussed previously this means that one can replace translations by any other transformation group such as rotations and multiplications. To a certain extent, the algebraic aspects of morphology are independent of the underlying transformation groups. But it is evident that the particular choice of this group has an enormous impact on the geometrical nature of the resulting operators. This is illustrated by several examples.

5.1. General properties of adjunctions

Throughout this chapter, $\mathcal{L}$ and $\mathcal{M}$ are complete lattices. We recall the definition of an adjunction from Section 3.2 along with some basic properties. Let ε be an operator from $\mathcal{L}$ into $\mathcal{M}$ and δ an operator from $\mathcal{M}$ into $\mathcal{L}$; the pair (ε, δ) is called an adjunction between $\mathcal{L}$ and $\mathcal{M}$ if

$$\delta(Y) \leq X \iff Y \leq \varepsilon(X), \tag{5.1}$$

for all $X \in \mathcal{L},\ Y \in \mathcal{M}$. In particular,

$$\varepsilon(I_{\mathcal{L}}) = I_{\mathcal{M}} \quad \text{and} \quad \delta(O_{\mathcal{M}}) = O_{\mathcal{L}}. \tag{5.2}$$

If (ε, δ) is an adjunction, it follows automatically that ε is an erosion and δ a dilation. In particular, both operators are increasing. If $\varepsilon : \mathcal{L} \to \mathcal{M}$ is an erosion, then the mapping $\delta : \mathcal{M} \to \mathcal{L}$ given by

$$\delta(Y) = \bigwedge \{X \in \mathcal{L} \mid Y \leq \varepsilon(X)\} \tag{5.3}$$

defines a dilation, and the pair (ε, δ) is an adjunction between $\mathcal{L}$ and $\mathcal{M}$. Conversely, if $\delta : \mathcal{M} \to \mathcal{L}$ is a dilation, then the mapping

$$\varepsilon(X) = \bigvee \{Y \in \mathcal{M} \mid \delta(Y) \leq X\} \tag{5.4}$$

defines an erosion, and the pair (ε, δ) is an adjunction between $\mathcal{L}$ and $\mathcal{M}$. The operator ε is called the *left adjoint* of δ, whereas δ is called the *right adjoint* of ε. Furthermore,

$$\varepsilon\delta \geq \mathsf{id}_{\mathcal{M}} \quad \text{and} \quad \delta\varepsilon \leq \mathsf{id}_{\mathcal{L}}; \tag{5.5}$$

also,

$$\varepsilon\delta\varepsilon = \varepsilon \quad \text{and} \quad \delta\varepsilon\delta = \delta. \tag{5.6}$$

The class of dilations is closed under suprema, whereas the class of erosions is closed under infima; cf. Proposition 3.16(b). The operators ε and δ are not inverses of each other in general. If one of the operators is an isomorphism, however, then both operators are isomorphisms, and they are each other's inverses.

5.1 Proposition. *Assume that ψ is a bijection from $\mathcal{L}$ to $\mathcal{M}$. The operator ψ is an isomorphism if and only if ψ is both an erosion and a dilation; in that case (ψ, ψ^{-1}) and (ψ^{-1}, ψ) are adjunctions.*

To prove this, note that an isomorphism distributes over infima as well as over suprema. Furthermore, $\psi(X) \le Y$ iff $X \le \psi^{-1}(Y)$, and $\psi^{-1}(Y) \le X$ iff $Y \le \psi(X)$. With these observations, the proof becomes trivial.

We have the following abstract version of Matheron's representation theorem (Theorem 4.15).

5.2 Theorem.

(a) *Every increasing operator $\psi : \mathcal{L} \to \mathcal{M}$ with $\psi(I) = I$ can be decomposed as a supremum of erosions from $\mathcal{L}$ into $\mathcal{M}$.*

(a′) *Every increasing operator $\psi : \mathcal{L} \to \mathcal{M}$ with $\psi(O) = O$ can be decomposed as an infimum of dilations from $\mathcal{L}$ into $\mathcal{M}$.*

Proof. Let $A \in \mathcal{L}$ and define ε_A by

$$\varepsilon_A(X) = \begin{cases} I, & \text{if } X = I, \\ \psi(A), & \text{if } I \neq X \ge A, \\ O, & \text{otherwise.} \end{cases}$$

One verifies easily that ε_A is an erosion; we prove that $\psi = \bigvee_{A \in \mathcal{L}} \varepsilon_A$. The inequality $\ge$ follows immediately from the observation that $\psi \ge \varepsilon_A$. On the other hand, since $\varepsilon_A(A) = \psi(A)$, we also get the inequality $\le$. □

In this chapter we are particularly interested in representation theorems for erosions and dilations. First, we consider the case where both $\mathcal{L}$ and $\mathcal{M}$ are lattices of functions; we explain in detail how to decompose dilations and erosions in such cases. These decompositions are used in Section 5.7, which is entirely devoted to grey-scale morphology. Furthermore, they are used to give an alternative proof of Theorem 2.44 concerning the representation of increasing Boolean functions.

Let $\mathcal{T}$ be some complete lattice, and let D, E be arbitrary sets. We consider dilations $\delta : \mathsf{Fun}(D, \mathcal{T}) \to \mathsf{Fun}(E, \mathcal{T})$. For $h \in D$ and $t \in \mathcal{T}$, the *pulse function* $f_{h,t}$ is defined by

$$f_{h,t}(x) = \begin{cases} t, & \text{if } x = h, \\ O, & \text{if } x \neq h, \end{cases} \tag{5.7}$$

for $x \in D$. Here O is the least element of $\mathcal{T}$. Every element $F \in \mathrm{Fun}(D, \mathcal{T})$ can be written as

$$F = \bigvee_{x \in D} f_{x,F(x)}. \tag{5.8}$$

In other words, pulse functions comprise a sup-generating family in the complete lattice $\mathrm{Fun}(D, \mathcal{T})$; see also Example 2.38(a). Define, for $x \in D$ and $y \in E$, the mapping $\delta_{x,y} : \mathcal{T} \to \mathcal{T}$ by

$$\delta_{x,y}(t) = \delta(f_{x,t})(y), \qquad t \in \mathcal{T}.$$

It is obvious that $\delta_{x,y}$ is a dilation on $\mathcal{T}$. If $F \in \mathrm{Fun}(D, \mathcal{T})$ and $y \in E$, then using (5.8) one derives

$$\delta(F)(y) = \delta(\bigvee_{x \in D} f_{x,F(x)})(y) = \bigvee_{x \in D} \delta(f_{x,F(x)})(y) = \bigvee_{x \in D} \delta_{x,y}(F(x)).$$

On the other hand, since a supremum of dilations is again a dilation, it is clear that every operator $\delta : \mathrm{Fun}(D, \mathcal{T}) \to \mathrm{Fun}(E, \mathcal{T})$ that is of the form above defines a dilation. Let $\varepsilon_{y,x}$ be the left adjoint erosion of $\delta_{x,y}$; define $\varepsilon : \mathrm{Fun}(E, \mathcal{T}) \to \mathrm{Fun}(D, \mathcal{T})$ by

$$\varepsilon(G)(x) = \bigwedge_{y \in E} \varepsilon_{y,x}(G(y)).$$

We show that the pair (ε, δ) defines an adjunction between $\mathrm{Fun}(E, \mathcal{T})$ and $\mathrm{Fun}(D, \mathcal{T})$. Thereto we must prove that for $F \in \mathrm{Fun}(D, \mathcal{T})$ and $G \in \mathrm{Fun}(E, \mathcal{T})$,

$$\delta(F) \leq G \iff F \leq \varepsilon(G).$$

Now

$$\begin{aligned} \delta(F) \leq G &\iff \forall y \in E : \quad \delta(F)(y) \leq G(y) \\ &\iff \forall y \in E : \quad \bigvee_{x \in D} \delta_{x,y}(F(x)) \leq G(y) \end{aligned}$$

$$\begin{aligned}
&\iff \forall y \in E\ \forall x \in D: \quad \delta_{x,y}(F(x)) \le G(y) \\
&\iff \forall y \in E\ \forall x \in D: \quad F(x) \le \varepsilon_{y,x}(G(y)) \\
&\iff \forall x \in D: \quad F(x) \le \bigwedge_{y \in E} \varepsilon_{y,x}(G(y)) \\
&\iff \forall x \in D: \quad F(x) \le \varepsilon(G)(x) \\
&\iff F \le \varepsilon(G).
\end{aligned}$$

This proves our claim. Thus we have established the only if-part of the following result. The if-part is straightforward.

5.3 Proposition. *Let $\mathcal{T}$ be a complete lattice and D, E arbitrary sets. The pair (ε, δ) is an adjunction between* $\mathrm{Fun}(E, \mathcal{T})$ *and* $\mathrm{Fun}(D, \mathcal{T})$ *if and only if for every $x \in D$, $y \in E$ there exists an adjunction $(\varepsilon_{y,x}, \delta_{x,y})$ on $\mathcal{T}$ such that*

$$\delta(F)(y) = \bigvee_{x \in D} \delta_{x,y}(F(x)), \tag{5.9}$$

$$\varepsilon(G)(x) = \bigwedge_{y \in E} \varepsilon_{y,x}(G(y)), \tag{5.10}$$

for $x \in D$, $y \in E$ and $F \in$ $\mathrm{Fun}(D, \mathcal{T})$, *$G \in$* $\mathrm{Fun}(E, \mathcal{T})$.

5.4 Example. (Boolean functions)
The lattice $\mathbb{B}_n$ of all Boolean functions of n variables (see Section 2.4) can be identified with the lattice of all operators $b : \{0, 1\}^D \to \{0, 1\}^E$ where D is a set containing n elements and E is a singleton. Instead we write $b : \{0, 1\}^n \to \{0, 1\}$. According to the previous decomposition result, a dilation $d : \{0, 1\}^n \to \{0, 1\}$ can be written as

$$d(x_1, \ldots, x_n) = \bigvee_{i=1}^{n} d_i(x_i),$$

where $d_i : \{0, 1\} \to \{0, 1\}$ is a dilation for every $i = 1, \ldots, n$. But a dilation on $\{0, 1\}$ is either constantly 0 or the identity map id. Using the convention of Section 2.4, we get that d is a sum of x_i's, or equivalently, an additive Boolean function. Dually, an erosion $e : \{0, 1\}^n \to \{0, 1\}$ is a product of x_i's, or equivalently, a multiplicative Boolean function. Let b be an increasing Boolean function. If $b(0, \ldots, 0) \neq 0$, then b is identically 1, and so it can be written as an empty infimum. If $b(0, \ldots, 0) = 0$, then we may apply Theorem 5.2, which says that b can be written as an infimum of dilations, or in other words, as a product-of-sums. Dually, every increasing Boolean function can be decomposed as sum-of-products. This gives an alternative proof of Theorem 2.44.

5.2. $\mathbb{T}$-invariance: the abelian case

In Chapter 4 we have discussed morphological operators which are translation invariant. The main goal of this chapter is to generalize the property of translation invariance to the complete lattice framework. To do so, we note that the family of translations constitutes a (abelian) group of automorphisms on the space $\mathcal{P}(\mathbb{R}^d)$. This observation forms the basis for the theory developed in this section. For simplicity, we deal exclusively with adjunctions on a lattice $\mathcal{L}$; however, most of the results can be generalized to adjunctions acting between two different lattices.

Let $\mathcal{L}$ be a complete lattice. By $\mathrm{Aut}(\mathcal{L})$ we denote the set of all automorphisms on $\mathcal{L}$; it is obvious that $\mathrm{Aut}(\mathcal{L})$ is a group. Note that $\mathrm{Aut}(\mathcal{L})$ coincides with $\mathrm{Aut}(\mathcal{L}')$, where $\mathcal{L}'$ is the opposite lattice of $\mathcal{L}$. If ψ is an operator on $\mathcal{L}$, $\tau \in \mathrm{Aut}(\mathcal{L})$, and ψ commutes with τ, i.e., $\tau\psi = \psi\tau$, then ψ also commutes with τ^{-1}, the inverse of τ. This follows easily by applying τ^{-1} on both sides of the identity $\tau\psi = \psi\tau$. Furthermore, if ψ commutes with τ_1 and τ_2, then ψ also commutes with $\tau_2\tau_1$ and $\tau_1\tau_2$. This implies that the subset of $\mathrm{Aut}(\mathcal{L})$ which commutes with a given operator ψ (or with every element of a family of operators) is a subgroup of $\mathrm{Aut}(\mathcal{L})$.

Let $\mathbb{T}$ be a subgroup of $\mathrm{Aut}(\mathcal{L})$; an operator ψ on $\mathcal{L}$ is called $\mathbb{T}$-*invariant* if

$$\psi\tau = \tau\psi, \qquad \tau \in \mathbb{T}. \tag{5.11}$$

An operator which is $\mathbb{T}$-invariant is called a $\mathbb{T}$-*operator*. Similarly, a dilation on $\mathcal{L}$ which is $\mathbb{T}$-invariant is called a $\mathbb{T}$-*dilation*, etc. Note that in the extreme case where $\mathbb{T} = \{\mathrm{id}\}$ every operator is automatically a $\mathbb{T}$-operator. Therefore, the case where no invariance is presupposed can be considered as a special case of $\mathbb{T}$-invariance. If ψ_i ($i \in I$) are $\mathbb{T}$-operators, then, since $\tau(\bigvee_{i\in I}\psi_i) = \bigvee_{i\in I}\tau\psi_i = \bigvee_{i\in I}\psi_i\tau = (\bigvee_{i\in I}\psi_i)\tau$, we get that $\bigvee_{i\in I}\psi_i$ is a $\mathbb{T}$-operator as well. The same remark applies to infima. In fact the following result holds.

5.5 Proposition. *Let $\mathbb{T}$ be a group of automorphisms on $\mathcal{L}$. The family of $\mathbb{T}$-operators (resp. increasing $\mathbb{T}$-operators) is a complete sublattice of $\mathcal{O}(\mathcal{L})$ (resp. $\mathcal{O}^+(\mathcal{L})$) which contains* id *and is closed under composition.*

We denote the lattice of $\mathbb{T}$-operators on $\mathcal{L}$ by $\mathcal{O}_{\mathbb{T}}(\mathcal{L})$ and the lattice of increasing $\mathbb{T}$-operators by $\mathcal{O}^+_{\mathbb{T}}(\mathcal{L})$.

5.6 Proposition. *Let ψ be an operator on $\mathcal{L}$. Then $\bigvee_{\tau \in \mathbb{T}} \tau\psi\tau^{-1}$ is the smallest $\mathbb{T}$-operator $\geq \psi$ and $\bigwedge_{\tau \in \mathbb{T}} \tau\psi\tau^{-1}$ is the largest $\mathbb{T}$-operator $\leq \psi$.*

Proof. We prove the first statement. It is easy to show that $\bigvee_{\tau \in \mathbb{T}} \tau\psi\tau^{-1}$ is a $\mathbb{T}$-operator. Let ϕ be a $\mathbb{T}$-operator $\geq \psi$; then $\tau\phi\tau^{-1} \geq \tau\psi\tau^{-1}$ since τ is increasing. However, $\tau\phi\tau^{-1} = \phi$, and we get $\phi \geq \tau\psi\tau^{-1}$. As this holds for every $\tau \in \mathbb{T}$, we get $\phi \geq \bigvee_{\tau \in \mathbb{T}} \tau\psi\tau^{-1}$; this shows the result. □

If (ε, δ) is an adjunction on $\mathcal{L}$, we can show that when either of these two operators is $\mathbb{T}$-invariant, then both are.

5.7 Proposition. *Let (ε, δ) be an adjunction on $\mathcal{L}$; then ε is a $\mathbb{T}$-operator if and only if δ is a $\mathbb{T}$-operator.*

Proof. Assume that ε is a $\mathbb{T}$-operator; take $\tau \in \mathbb{T}$ and $X, Y \in \mathcal{L}$. Then

$$\begin{aligned} \delta(\tau(X)) \leq Y &\iff \tau(X) \leq \varepsilon(Y) \\ &\iff X \leq \tau^{-1}\varepsilon(Y) = \varepsilon\tau^{-1}(Y) \\ &\iff \delta(X) \leq \tau^{-1}(Y) \\ &\iff \tau\delta(X) \leq Y. \end{aligned}$$

As this holds for arbitrary $X, Y \in \mathcal{L}$, we get $\delta\tau = \tau\delta$. □

Adjunctions in which both operators are $\mathbb{T}$-invariant are called *$\mathbb{T}$-adjunctions*. Many results stated for adjunctions remain valid for $\mathbb{T}$-adjunctions. For instance, if $(\varepsilon_i, \delta_i)$ is a $\mathbb{T}$-adjunction for every i in some index set I, then $(\bigwedge_{i \in I} \varepsilon_i, \bigvee_{i \in I} \delta_i)$ is again a $\mathbb{T}$-adjunction (cf. Proposition 3.16(b)).

Proposition 5.1 says that (τ^{-1}, τ) is an adjunction for every $\tau \in \mathbb{T}$. It is *not* a $\mathbb{T}$-adjunction in general, however. In order that every (τ^{-1}, τ) is a $\mathbb{T}$-adjunction, one has to assume that $\mathbb{T}$ is abelian. In that case $(\bigwedge_{i \in I} \tau_i, \bigvee_{i \in I} \tau_i^{-1})$ is a $\mathbb{T}$-adjunction for an arbitrary family $\tau_i \subset \mathbb{T}$, $i \in I$. Proposition 5.14 will give conditions under which every $\mathbb{T}$-adjunction has this form.

If $\mathbb{T}$ is abelian, then every of its elements is an increasing $\mathbb{T}$-operator, and, by Proposition 5.5, suprema and infima of elements in $\mathbb{T}$ also are $\mathbb{T}$-operators. More generally, the class of operators which is closed under suprema, infima and compositions and which contains $\mathbb{T}$ is a subset of $\mathcal{O}^{+}_{\mathbb{T}}(\mathcal{L})$. In a sense, Theorem 5.22 will provide conditions under which this class coincides with $\mathcal{O}^{+}_{\mathbb{T}}(\mathcal{L})$. To gain some further insight, consider the following example.

5.8 Example. Let $\mathcal{L} = \mathcal{P}(\mathbb{R}^d)$, and let $\mathbb{T}$ be the abelian group of all translations on $\mathbb{R}^d$. For $h \in \mathbb{R}^d$ we define the translation operator τ_h on $\mathcal{P}(\mathbb{R}^d)$ by

$$\tau_h(X) = X_h.$$

In Section 4.3 we have seen that the pair of operators

$$\delta_A(X) = X \oplus A, \qquad \varepsilon_A(X) = X \ominus A$$

is an adjunction on $\mathcal{P}(\mathbb{R}^d)$. Since both operators are $\mathbb{T}$-invariant, it is a $\mathbb{T}$-adjunction. Eqs. (4.18) and (4.21) show that δ_A and ε_A can be written as

$$\delta_A = \bigvee_{a \in A} \tau_a \qquad \text{and} \qquad \varepsilon_A = \bigwedge_{a \in A} \tau_a^{-1}.$$

Here we have used that $\tau_a^{-1} = \tau_{-a}$. Assume, on the other hand, that (ε, δ) is a $\mathbb{T}$-adjunction on $\mathcal{P}(\mathbb{R}^d)$. Define $A = \delta(\{0\})$; then

$$\begin{aligned}\delta(X) &= \delta\Big(\bigcup_{x \in X} \{x\}\Big) = \delta\Big(\bigcup_{x \in X} \tau_x\{0\}\Big)\\ &= \bigcup_{x \in X} \tau_x \delta(\{0\}) = \bigcup_{x \in X} A_x\\ &= X \oplus A.\end{aligned}$$

This proves that every translation invariant dilation on $\mathcal{P}(\mathbb{R}^d)$ is a Minkowski addition, and dually, that every translation invariant erosion is a Minkowski subtraction.

One may wonder if in analogy with this example and under the assumption that $\mathbb{T}$ is abelian, every $\mathbb{T}$-adjunction on a complete lattice $\mathcal{L}$ is of the form $(\bigwedge_{i \in I} \tau_i^{-1}, \bigvee_{i \in I} \tau_i)$, where $\{\tau_i \mid i \in I\}$ is a subset of $\mathbb{T}$. A moment of reflection shows that the answer is negative. E.g., in the extreme case where $\mathbb{T}$ contains only the identity operator id, the only adjunctions of this form are (ι, o) and (id, id). But there are less trivial counterexamples.

5.9 Example. Let $\mathcal{L} = \mathcal{P}(\mathbb{R}^2)$, and let $\mathbb{T}$ be the group of translations along the x-axis. Let L be a line parallel to the x-axis, and define

$$\delta(X) = X \cap L.$$

Since $(\bigcup X_i) \cap L = \bigcup (X_i \cap L)$, it follows that δ is a dilation. Moreover, since L is invariant under translations in the x-direction, δ is a $\mathbb{T}$-dilation. However, $\delta(X)$ cannot be written as a union of horizontal translates of X.

In Example 5.8 we have used essentially two properties of the Euclidean space $\mathbb{R}^d$ to show that every $\mathbb{T}$-dilation can be written as a union of translates. First, every element of $\mathcal{P}(\mathbb{R}^2)$ is a union of points (singletons), and second, the group of translations is transitive on these points. The latter means that for every two points there exists a translation which maps one point to the other. In the assumption below, which we make throughout the remainder of this section, we generalize this property for arbitrary complete lattices.

5.10 Basic Assumption. $\mathcal{L}$ is a complete lattice which possesses a sup-generating family ℓ and $\mathbb{T}$ is an abelian automorphism group on $\mathcal{L}$ such that

- **(i)** ℓ is invariant under $\mathbb{T}$; that is, if $x \in \ell$ and $\tau \in \mathbb{T}$, then $\tau(x) \in \ell$;
- **(ii)** $\mathbb{T}$ is transitive on ℓ; that is, for every $x, y \in \ell$ there exists a $\tau \in \mathbb{T}$ such that $\tau(x) = y$.

Elements of ℓ are denoted by lowercase letters x, y, z, etc. Note that in Example 5.8 the Basic Assumption is satisfied if ℓ is the set of singletons (i.e., the atoms of $\mathcal{P}(\mathbb{R}^d)$). Other examples can be found in what follows; see in particular Section 5.7, where we consider the complete lattice of grey-scale functions.

We make the following important observation. Suppose that $\mathcal{L}$ is a complete lattice with a sup-generating family ℓ and that $\mathbb{T}$ is an automorphism group on $\mathcal{L}$ which acts transitively on ℓ. If the lattice $\mathcal{L}$ is atomic, then ℓ must coincide with the set of atoms, because (i) an atom cannot be decomposed as a supremum of other lattice elements, and so the atoms are contained in ℓ; and (ii) every element of Aut(L) maps an atom onto an atom, and so, by transitivity, ℓ cannot contain elements other than atoms. For this reason there is no danger in using the notation ℓ for the sup-generating family, since in the case that $\mathcal{L}$ is atomic, ℓ is precisely the collection of all atoms.

5.11 Remark. Note that the Basic Assumption is not self-dual. This means the following: if a complete lattice $\mathcal{L}$ with the automorphism group $\mathbb{T}$ satisfies this assumption, there is no guarantee that the dual lattice $\mathcal{L}'$ satisfies this assumption as well. In fact, the Basic Assumption on $\mathcal{L}$ induces a *dual Basic Assumption* on $\mathcal{L}'$, which reads as follows:

$\mathcal{L}'$ is a complete lattice which possesses an inf-generating family ℓ' and $\mathbb{T}$ is an abelian automorphism group on $\mathcal{L}'$ such that

- **(i)** *ℓ' is invariant under $\mathbb{T}$;*
- **(ii)** *$\mathbb{T}$ is transitive on ℓ'.*

For example, the singletons constitute a sup-generating family in the lattice $\mathcal{F}(\mathbb{R}^d)$. These induce the inf-generating family comprising the open sets $\mathbb{R}^d \setminus \{x\}$ in the dual lattice of open sets $\mathcal{G}(\mathbb{R}^d)$. The translation group acts transitively on both families.

If $\mathcal{L}$ is Boolean, then the Basic Assumption holds if and only if the dual Basic Assumption holds. Section 5.3 explains that this observation is also true for lattices with a $\mathbb{T}$-compatible negation.

Just like in the Euclidean case, the Basic Assumption implies that $\mathbb{T}$ acts *simply transitively*, or *regularly*, on ℓ. This means that for every $x, y \in \ell$, there is a unique $\tau \in \mathbb{T}$ such that $\tau(x) = y$. Indeed, suppose that $\tau_1(x) = \tau_2(x) = y$; then $\tau_1^{-1}\tau_2(x) = x$. Now, for every $z \in \ell$, there is some $\tau_3 \in \ell$ such that $\tau_3(x) = z$, and so $\tau_1^{-1}\tau_2(z) = (\tau_1^{-1}\tau_2)\tau_3(x) = \tau_3(\tau_1^{-1}\tau_2)(x) = \tau_3(x) = z$. Thus $\tau_1^{-1}\tau_2$ fixes every element of ℓ, and, as $\tau_1^{-1}\tau_2$ commutes with the supremum and ℓ is sup-generating, this means that $\tau_1^{-1}\tau_2$ fixes every element of $\mathcal{L}$; in other words, $\tau_1 = \tau_2$. The distinction between the simply transitive and multi-transitive case will become important in Section 5.8, which deals with the nonabelian case.

One can endow ℓ with a group structure isomorphic to $\mathbb{T}$. Fix an element $o \in \ell$ and call this the *origin* of ℓ. For any $x \in \ell$, let τ_x be the unique element of $\mathbb{T}$ which satisfies $\tau_x(o) = x$. Define a binary addition $+$ on ℓ by

$$x + y = \tau_x\tau_y(o) = \tau_x(y) = \tau_y(x).$$

Then $\tau_{x+y} = \tau_x\tau_y$, expressing the isomorphy with $\mathbb{T}$. Thus $(\ell, +)$ is an abelian group with neutral element o. The inverse of x is denoted by $-x$, i.e., $\tau_{-x} = \tau_x^{-1}$, and the subtraction $-$ is defined by

$$x - y = x + (-y) = \tau_x\tau_y^{-1}(o) = \tau_x(-y) = \tau_y^{-1}(x).$$

5.12 Remark. If $\mathcal{L} = \mathcal{P}(E)$ and E has an abelian group structure $+$ we can define an automorphism group on $\mathcal{L}$ by putting

$$\tau_h(X) = X_h = \{x + h \mid x \in X\},$$

for $h \in E$ and $X \in \mathcal{P}(E)$. In fact, it does not matter whether we start with this automorphism group or with the group structure on E. If $\mathcal{L}$ is not isomorphic to some $\mathcal{P}(E)$, however, it may be inadequate to start with a group structure on ℓ. For example, let $\mathcal{L}$ be the set consisting of $\mathbb{R}^2$, the lines passing through the origin, and the singletons $\{x\}$. It is obvious that

$\mathcal{L}$ ordered by inclusion is a complete lattice. The set ℓ of singletons is sup-generating and has the translation group structure $+$ of $\mathbb{R}^2$. The extension of the mapping $\tau_h : \{x\} \mapsto \{x+h\}$ on ℓ, where $h \in \mathbb{R}^2$, to $\mathcal{L}$ is given by

$$\tau_h(X) = \bigvee_{x \in X} \tau_h\{x\}.$$

It is obvious that this supremum equals $\mathbb{R}^2$ if the translation of X along h is not a line passing through the origin. But then τ_h is not an automorphism.

The group structure on ℓ can be used to define Minkowski operations on $\mathcal{L}$. Let, for $X \in \mathcal{L}$, $h \subset \ell$,

$$\ell(X) = \{x \in \ell \mid x \leq X\} \tag{5.12}$$

and

$$X_h = \tau_h(X). \tag{5.13}$$

Define the Minkowski operations $\oplus$ and $\ominus$ on $\mathcal{L}$ as follows:

$$X \oplus Y = \bigvee_{y \in \ell(Y)} X_y, \tag{5.14}$$

$$X \ominus Y = \bigwedge_{y \in \ell(Y)} X_{-y}. \tag{5.15}$$

If $x, y \in \ell$, then

$$x \oplus y = x + y, \qquad x \ominus y = x - y. \tag{5.16}$$

We prove the first identity; the second follows by duality. By definition, $x \oplus y = \bigvee_{y' \in \ell(y)} (x + y')$, which is $\geq x + y$ since $y \in \ell(y)$. On the other hand, since $y' \leq y$ if $y' \in \ell(y)$, we get $x + y' = y' + x = \tau_x(y') \leq \tau_x(y) \leq x + y$; from this we derive that $x \oplus y \leq x + y$.

Let $X \in \mathcal{L}$ and $y \in \ell$; by virtue of (5.16),

$$\begin{aligned} X \oplus y &= \bigvee_{y' \in \ell(y)} X_{y'} = \bigvee_{y' \in \ell(y)} \tau_{y'}(X) \\ &= \bigvee_{y' \in \ell(y)} \tau_{y'}\Big(\bigvee_{x \in \ell(X)} x\Big) = \bigvee_{y' \in \ell(y)} \bigvee_{x \in \ell(X)} \tau_{y'}(x) \\ &= \bigvee_{x \in \ell(X)} \bigvee_{y' \in \ell(y)} (x + y') = \bigvee_{x \in \ell(X)} x \oplus y \end{aligned}$$

$$= \bigvee_{x \in \ell(X)} (x + y) = \tau_y(X).$$

A similar expression can be derived for $X \ominus y$. We have

$$X \oplus y = X_y, \qquad X \ominus y = X_{-y}. \tag{5.17}$$

5.13 Proposition. *For every $X, Y \in \mathcal{L}$ the following relations hold:*

$$X \oplus Y = Y \oplus X = \bigvee\{x + y \mid x \in \ell(X), y \in \ell(Y)\}, \tag{5.18}$$

$$X \ominus Y = \bigvee\{h \in \ell \mid Y_h \le X\}. \tag{5.19}$$

Proof. The first statement is easy; we prove the second. Set $W = \bigvee\{h \in \ell \mid Y_h \le X\}$, and let $h \in \ell$ be such that $Y_h \le X$. Then, for every $y \in \ell(Y)$, $h_y = y + h = y_h \le Y_h \le X$, and so $h = (h_y)_{-y} \le X_{-y}$. In view of (5.15), this means that $h \le X \ominus Y$. This implies $W \le X \ominus Y$. Now let $h \in \ell(X \ominus Y)$; then $h \le X_{-y}$ for every $y \in \ell(Y)$, and so $y_h = y + h = h_y \le (X_{-y})_y = X$. Since $Y_h = \bigvee_{y \in \ell(Y)} y_h$, we get $Y_h \le X$, and so $h \le W$. This shows that $X \ominus Y \le W$; thus the equality follows. □

Note that (5.18) and (5.19) generalize properties (4.19) and (4.22), respectively.

Given $A \in \mathcal{L}$, define the operators δ_A and ε_A by

$$\delta_A(X) = X \oplus A, \qquad \varepsilon_A(X) = X \ominus A. \tag{5.20}$$

Note that we can also write

$$\delta_A = \bigvee_{a \in \ell(A)} \tau_a, \qquad \varepsilon_A = \bigwedge_{a \in \ell(A)} \tau_a^{-1}, \tag{5.21}$$

from which it follows immediately that $(\varepsilon_A, \delta_A)$ is a $\mathbb{T}$-adjunction.

5.14 Proposition. *For every $A \in \mathcal{L}$, the pair $(\varepsilon_A, \delta_A)$ is a $\mathbb{T}$-adjunction. Conversely, every $\mathbb{T}$-adjunction is of this form.*

Proof. It remains to show that every $\mathbb{T}$-adjunction on $\mathcal{L}$ is of the form $(\varepsilon_A, \delta_A)$. Let (ε, δ) be a $\mathbb{T}$-adjunction on $\mathcal{L}$; define $A = \delta(o)$. We show that $\delta = \delta_A$. For $x \in \ell$,

$$\delta(x) = \delta\big(\tau_x(o)\big) = \tau_x\big(\delta(o)\big) = \tau_x(A).$$

Applying (5.14) and Proposition 5.13, we find

$$\begin{aligned}\delta(X) &= \delta\big(\bigvee \ell(X)\big) = \bigvee_{x\in\ell(X)} \delta(x)\\ &= \bigvee_{x\in\ell(X)} \tau_x(A) = A\oplus X = X\oplus A\\ &= \delta_A(X),\end{aligned}$$

for every $X\in\mathcal{L}$. By the uniqueness of the left adjoint of a dilation, this means that $\varepsilon=\varepsilon_A$. □

Let $A, B\in\mathcal{L}$, and consider the expression $(X\oplus A)\oplus B$. Since $(X\oplus A)\oplus B=\delta_B\delta_A(X)$ and the composition of two $\mathbb{T}$-dilations is again a $\mathbb{T}$-dilation, we may conclude that $\delta_B\delta_A=\delta_C$, where $C=\delta_B\delta_A(o)=A\oplus B$. This gives

$$(X\oplus A)\oplus B = X\oplus(A\oplus B). \tag{5.22}$$

The left adjoint of $\delta_B\delta_A$ is $\varepsilon_A\varepsilon_B$, or alternatively, ε_C; this means that

$$(X\ominus B)\ominus A = X\ominus(A\oplus B). \tag{5.23}$$

We have already pointed out that the family of $\mathbb{T}$-dilations is closed under suprema and dually, that the family of $\mathbb{T}$-erosions is closed under infima. Let $A_i\in\mathcal{L}$ for all i in some index set I. Then $\bigvee_{i\in I}\delta_{A_i}$ is a $\mathbb{T}$-dilation, and hence this expression equals δ_B for some $B\in\mathcal{L}$. Application to the element o gives $B=\bigvee_{i\in I}A_i$. The left adjoint erosion is ε_B, or alternatively, $\bigwedge_{i\in I}\varepsilon_{A_i}$. We have thus derived the following result.

5.15 Theorem.

(a) *The mapping $A\mapsto\delta_A$ defines an isomorphism between $\mathcal{L}$ and the complete lattice of $\mathbb{T}$-dilations. In particular, given $A_i\in\mathcal{L}$, $i\in I$, one has*

$$\bigvee_{i\in I}\delta_{A_i} = \delta_{\bigvee_{i\in I}A_i}\,. \tag{5.24}$$

(a′) *The mapping $A\mapsto\varepsilon_A$ defines a dual isomorphism between $\mathcal{L}$ and the complete lattice of $\mathbb{T}$-erosions. In particular, given $A_i\in\mathcal{L}$, $i\in I$, one has*

$$\bigwedge_{i\in I}\varepsilon_{A_i} = \varepsilon_{\bigvee_{i\in I}A_i}\,. \tag{5.25}$$

To conclude this section, we consider the following problem. Are there automorphisms of $\mathcal{L}$ outside $\mathbb{T}$ with which all $\mathbb{T}$-adjunctions commute? Suppose that σ is such an automorphism; then

$$\sigma\tau = \tau\sigma \qquad \text{for every } \tau \in \mathbb{T}.$$

Conversely, if σ is an automorphism commuting with every $\tau \in \mathbb{T}$, then σ commutes with all $\mathbb{T}$-adjunctions.

5.16 Proposition. *Suppose that $\ell \cup \{O, I\}$ is inf-closed. If $\sigma \in \mathrm{Aut}(\mathcal{L})$ and σ commutes with every element of $\mathbb{T}$, then $\sigma \in \mathbb{T}$.*

Proof. Since σ is a $\mathbb{T}$-erosion, $\sigma = \varepsilon_A$ for some $A \in \mathcal{L}$. Then $\sigma(o) = o \ominus A = \bigwedge\{-a \mid a \in \ell(A)\}$. As $\ell \cup \{O, I\}$ is inf-closed, it follows that $\sigma(o) \in \ell \cup \{O, I\}$. From the fact that σ is an automorphism one concludes immediately that $\sigma(o) \in \ell$, say $\sigma(o) = x$. But σ is also a $\mathbb{T}$-dilation; therefore, $\sigma = \delta_B$ for some $B \in \mathcal{L}$. One gets $B = \sigma(o) = x$, and so $\sigma = \delta_x = \tau_x \in \mathbb{T}$. □

This result applies in particular to the case where ℓ is the set of atoms of $\mathcal{L}$, e.g., $\mathcal{L} = \mathcal{P}(E)$.

Note that the problem just addressed should not be confused with the situation where a given adjunction has some additional symmetry. In the Euclidean case $\mathcal{L} = \mathcal{P}(\mathbb{R}^2)$, for example, it is easy to construct dilations and erosions which are not only translation invariant, but are also invariant with respect to rotations about the origin. In fact, Section 5.9 shows that all dilations and erosions using a structuring element which is rotation symmetric have this property.

In the case where the dual Basic Assumption holds, we may derive results which are very similar to the foregoing. Again one fixes an origin o' in ℓ', the inf-generating family, and defines $\tau_{x'}$ to be the unique element of $\mathbb{T}$ which maps o' to x' for every $x' \in \ell'$. Defining $\ell'(X) = \{x' \in \ell' \mid X \le x'\}$, one shows that every erosion can be decomposed as

$$\begin{aligned}\varepsilon(X) &= \varepsilon\Big(\bigwedge_{x' \in \ell'(X)} x'\Big) = \bigwedge_{x' \in \ell'(X)} \varepsilon(x') \\ &= \bigwedge_{x' \in \ell'(X)} \varepsilon\tau_{x'}(o') = \bigwedge_{x' \in \ell'(X)} \tau_{x'}\varepsilon(o').\end{aligned}$$

Putting $A = \varepsilon(o')$, one arrives at

$$\varepsilon(X) = \bigwedge_{x' \in \ell'(X)} \tau_{x'}(A) = \bigwedge_{a' \in \ell'(A)} \tau_{a'}(X).$$

The right adjoint dilation is given by

$$\delta(X) = \bigvee_{a' \in \ell'(A)} \tau_{a'}^{-1}(X).$$

We leave all further details as an exercise to the reader.

5.3. Self-dual and Boolean lattices

If the lattice $\mathcal{L}$ has a negation which is compatible with the automorphism group $\mathbb{T}$ (see the definition that follows), then one can establish an interesting duality relation between dilations and erosions. The Boolean lattices form an important example. Another example are the grey-scale functions, treated in Section 5.7.

Let $\mathcal{L}$ be a complete lattice for which the Basic Assumption 5.10 is satisfied. Recall from Definition 2.29 that a negation is a dual automorphism ν with $\nu^2 = \mathrm{id}$.

5.17 Definition. Let $\mathcal{L}$ be a complete lattice and $\mathbb{T}$ an automorphism group on $\mathcal{L}$. A negation ν on $\mathcal{L}$ is said to be $\mathbb{T}$-*compatible* if

$$\nu\tau\nu \in \mathbb{T},$$

for every $\tau \in \mathbb{T}$.

As before, we use the notation $X^* = \nu(X)$. For every $h \in \ell$, we define $\check{h} \in \ell$ by

$$\nu\tau_h\nu = \tau_{\check{h}}. \tag{5.26}$$

Here τ_h is the unique element of $\mathbb{T}$ which maps the origin o to h. Define the *reflection* of an element $A \in \mathcal{L}$ by

$$\check{A} = \bigvee\{-\check{a} \mid a \in \ell(A)\}. \tag{5.27}$$

Whenever convenient, we write $h^{\check{}}$ instead of $\check{h}$ and $A^{\check{}}$ instead of $\check{A}$. Taking the inverse in (5.26) of both sides, we get

$$(-h)^{\check{}} = -\check{h}. \tag{5.28}$$

Also, (5.26) implies $\tau_h = \nu\tau_{\check{h}}\nu$; hence

$$h^{\check{}\check{}} = h. \tag{5.29}$$

From (5.24), we get

$$\delta_{\check{A}} = \bigvee_{a\in\ell(A)} \delta_{-\check{a}} = \bigvee_{a\in\ell(A)} \tau_{-\check{a}}. \tag{5.30}$$

Furthermore, (5.25) gives

$$\varepsilon_{\check{A}} = \bigwedge_{a\in\ell(A)} \varepsilon_{-\check{a}} = \bigwedge_{a\in\ell(A)} \tau_{\check{a}}. \tag{5.31}$$

Here we have also used (5.17), which says that $\delta_h = \tau_h$ and $\varepsilon_h = \tau_{-h}$ for every $h \in \ell$. Relation (5.26) can be reformulated as

$$(X^*)_h = (X_{\check{h}})^*.$$

By (5.31), we get

$$\begin{aligned} X^* \oplus A &= \bigvee_{a\in\ell(A)} (X^*)_a = \bigvee_{a\in\ell(A)} (X_{\check{a}})^* \\ &= (\bigwedge_{a\in\ell(A)} X_{\check{a}})^* = (X \ominus \check{A})^*. \end{aligned}$$

A similar expression can be derived for $X^* \ominus A$, and we arrive at the following result.

5.18 Proposition. *If $\mathcal{L}$ has a $\mathbb{T}$-compatible negation $X \mapsto X^*$, then*

$$(X^* \oplus A)^* = X \ominus \check{A} \qquad \textit{and} \qquad (X^* \ominus A)^* = X \oplus \check{A}, \tag{5.32}$$

for every $X, A \in \mathcal{L}$.

Twofold application of these identities leads to $X \oplus A^{\vee\vee} = X \oplus A$; this gives

$$A^{\vee\vee} = A. \tag{5.33}$$

As we pointed out in Remark 5.11, the Basic Assumption is not self-dual. If $\mathcal{L}$ satisfies the Basic Assumption and has a $\mathbb{T}$-compatible negation ν, however, then the dual Basic Assumption is satisfied automatically. For one easily shows that the family $\{\nu(x) \mid x \in \ell\}$ defines an inf-generating family on which $\mathbb{T}$ acts transitively.

Assume that $\mathcal{L}$ is a Boolean lattice, and put $\nu(X) = X^*$. If τ is an automorphism on $\mathcal{L}$, then

$$\tau(X^*) = [\tau(X)]^*,$$

and thus $\nu\tau\nu = \tau$. This means in particular that $\check{h} = h$ and that

$$\check{A} = \bigvee\{-a \mid a \in \ell(A)\}. \tag{5.34}$$

5.19 Corollary. *Let $\mathcal{L}$ be a Boolean lattice, and let the reflection $\check{A}$ of an element $A \in \mathcal{L}$ be given by* (5.34). *Then, for $A, X \in \mathcal{L}$,*

$$(X^* \oplus A)^* = X \ominus \check{A} \qquad \textit{and} \qquad (X^* \ominus A)^* = X \oplus \check{A}. \tag{5.35}$$

5.4. Representation theorems

Consider the situation of Section 5.2, that is, $\mathcal{L}$ is a complete lattice, $\mathbb{T}$ is an abelian group of automorphisms on $\mathcal{L}$, and ℓ is a sup-generating family in $\mathcal{L}$ such that the Basic Assumption is satisfied. We can extend the notion of a kernel given in Definition 4.8 as follows. Let ψ be a $\mathbb{T}$-operator on $\mathcal{L}$; the *kernel* $\mathcal{V}(\psi)$ of ψ is defined by

$$\mathcal{V}(\psi) = \{A \in \mathcal{L} \mid o \leq \psi(A)\}. \tag{5.36}$$

Here o is the origin in ℓ. Note that this definition depends on the underlying automorphism group $\mathbb{T}$ and that a more appropriate nomenclature would have been $\mathbb{T}$-*kernel of* ψ. As in most situations only one automorphism group plays a role, however, we omit the prefix "$\mathbb{T}$-". It is evident that the kernel of a $\mathbb{T}$-operator ψ is empty if and only if ψ maps every element of $\mathcal{L}$ onto O. For $h \in \ell$, $X \in \mathcal{L}$, we find

$$h \leq \psi(X) \iff o \leq \psi(X_{-h}) \iff X_{-h} \in \mathcal{V}(\psi),$$

and we obtain

$$\psi(X) = \bigvee\{h \in \ell \mid X_{-h} \in \mathcal{V}(\psi)\}; \tag{5.37}$$

cf. (4.46). This relation shows how to reconstruct a $\mathbb{T}$-operator from its kernel. Recall that a subset $\mathcal{H}$ of $\mathcal{L}$ is called an upper set if $A \in \mathcal{H}, X \in \mathcal{L}$ and $X \geq A$ implies that $X \in \mathcal{H}$.

5.20 Proposition. *A $\mathbb{T}$-operator ψ is increasing if and only if $\mathcal{V}(\psi)$ is an upper set.*

Proof. It is evident that the kernel of an increasing $\mathbb{T}$-operator is an upper set. Conversely, assume that $\mathcal{V}(\psi)$ is an upper set; we show that ψ is increasing. If $X \leq Y$ and $h \leq \psi(X)$ for some $h \in \ell$, then $X_{-h} \in \mathcal{V}(\psi)$. Since

$X_{-h} \le Y_{-h}$, also $Y_{-h} \in \mathcal{V}(\psi)$. This means, however, that $h \le \psi(Y)$. Therefore, $\psi(X) \le \psi(Y)$. □

It is obvious that $\psi \le \psi'$ implies that $\mathcal{V}(\psi) \subseteq \mathcal{V}(\psi')$. By (5.37), the converse also holds. In fact, one can establish the following stronger result. Its proof is straightforward.

5.21 Proposition. *Let ψ_i, $i \in I$, be $\mathbb{T}$-operators; then*

$$\mathcal{V}(\bigwedge_{i\in I} \psi_i) = \bigcap_{i\in I} \mathcal{V}(\psi_i), \tag{5.38}$$

$$\mathcal{V}(\bigvee_{i\in I} \psi_i) \supseteq \bigcup_{i\in I} \mathcal{V}(\psi_i). \tag{5.39}$$

If $\mathcal{L} = \mathcal{P}(E)$ (in which case ℓ comprises the singletons), then equality in (5.39) *holds.*

It is not difficult to think of an example where the opposite inclusion in (5.39) does not hold. Let $\mathcal{L}$ consist of the closed subsets of $\mathbb{R}^d$, and let $\mathbb{T}$ be the group of translations. We denote the translation along h by τ_h. Let h_n be a sequence in $\mathbb{R}^d$ converging to some $h \in \mathbb{R}^d$, and assume that $h_n \ne h$ for every n. It is evident that

$$\mathcal{V}(\tau_{h_n}) = \{X \subseteq \mathbb{R}^d \mid X \text{ closed and } -h_n \in X\}.$$

However, $\bigvee \tau_{h_n}(\{-h\}) = \overline{\{h_n - h \mid n \ge 1\}}$, which contains 0, and so $\{h\} \in \mathcal{V}(\bigvee_{n\ge 1} \tau_{h_n})$, whereas $\{h\}$ is not contained in any of the $\mathcal{V}(\tau_{h_n})$.

Let $A \in \mathcal{L}$, and let ε_A be the $\mathbb{T}$-erosion defined in (5.20); then $X \in \mathcal{V}(\varepsilon_A)$ if and only if $o \le X \ominus A$. According to the proof of Proposition 5.13, this is equivalent to $A \le X$, and so

$$\mathcal{V}(\varepsilon_A) = \{X \mid X \ge A\}. \tag{5.40}$$

5.22 Theorem. (Representation of $\mathbb{T}$-operators)
Suppose that the Basic Assumption 5.10 is satisfied. Every increasing $\mathbb{T}$-operator ψ can be represented as a supremum of erosions; more precisely

$$\psi = \bigvee_{A\in\mathcal{V}(\psi)} \varepsilon_A. \tag{5.41}$$

Proof. Since $\mathcal{V} = \mathcal{V}(\psi)$ is an upper set, $X \in \mathcal{V}$ if $X \ge A$ and $A \in \mathcal{V}$. So it follows from Proposition 5.21 and (5.40) that $\psi \ge \varepsilon_A$ if $A \in \mathcal{V}$; this implies

$\psi \geq \bigvee_{A\in\mathcal{V}} \varepsilon_A$. To prove the reverse inequality, we must show that $\psi(X) \leq \bigvee_{A\in\mathcal{V}} X \ominus A$ for each $X \in \mathcal{L}$. Take $h \in \ell(\psi(X))$; since $h \leq \psi(X)$, we have $o \leq \psi(X)_{-h} = \psi(X_{-h})$, and so $X_{-h} \in \mathcal{V}(\psi)$. As $h \oplus X_{-h} = X_{-h+h} = X$, we obtain by adjunction that $h \leq X \ominus X_{-h}$. As $X_{-h} \in \mathcal{V}(\psi)$, we get $h \leq X \ominus X_{-h} \leq \bigvee_{A\in\mathcal{V}(\psi)} \varepsilon_A(X)$; this implies $\psi(X) \leq \bigvee_{A\in\mathcal{V}(\psi)} \varepsilon_A(X)$. □

When comparing this result with Theorem 5.2(a), one notes that in the statement of the latter result it was assumed that $\psi(I) = I$. Here such a requirement is not necessary since, as a consequence of $\mathbb{T}$-invariance, $\psi(I) < I$ implies that ψ is identically O; therefore, $\mathcal{V}(\psi) = \varnothing$. In that case, (5.41) is still valid since an empty supremum is O. The Basic Assumption is not sufficient to derive a decomposition of ψ as an infimum of $\mathbb{T}$-dilations. In fact, in Section 5.5 we give an example of an increasing $\mathbb{T}$-operator on $\mathcal{F}(\mathbb{R}^d)$ which is not decomposable as an infimum of $\mathbb{T}$-dilations. The Duality Principle gives the opposite of Theorem 5.22, however: if the dual Basic Assumption (see Remark 5.11) is satisfied, every increasing $\mathbb{T}$-operator can be represented as an infimum of dilations. We do not give an exact formula like (5.41), however, for this would require further definitions such as the "dual kernel" of an operator.

In Section 5.3 we showed that the existence of a negation ν on $\mathcal{L}$ compatible with the group $\mathbb{T}$ implies the validity of the dual Basic Assumption. The negative of an operator ψ is defined as usual:

$$\psi^* = \nu\psi\nu. \tag{5.42}$$

It is evident that ψ^* is increasing if and only if ψ is increasing and that ψ^* is a $\mathbb{T}$-operator if and only if ψ is a $\mathbb{T}$-operator.

5.23 Theorem. *Let $\mathcal{L}$ be a complete lattice for which the Basic Assumption is satisfied; assume further that $\mathcal{L}$ possesses a $\mathbb{T}$-compatible negation. Every increasing $\mathbb{T}$-operator ψ can be written as*

$$\psi(X) = \bigvee_{A\in\mathcal{V}(\psi)} X \ominus A, \tag{5.43}$$

and dually, as

$$\psi(X) = \bigwedge_{A\in\mathcal{V}(\psi^*)} X \oplus \check{A}. \tag{5.44}$$

The proof of the identity in (5.44) is very much the same as the proof of the second identity in Theorem 4.15.

Observe that the latter result applies in particular to the case where $\mathcal{L}$ is a Boolean lattice.

The kernel of an increasing $\mathbb{T}$-operator ψ is an upper set, that is, $A \in \mathcal{V}(\psi)$ and $A \subseteq B$ means that $B \in \mathcal{V}(\psi)$. This implies in particular that the representation $\psi = \bigvee_{A \in \mathcal{V}(\psi)} \varepsilon_A$ is redundant in the following sense: if $A, B \in \mathcal{V}(\psi)$ and $A \leq B$, then $\varepsilon_A \geq \varepsilon_B$, and so omission of ε_B from this representation has no effect on the outcome. In Chapter 4 we attempted to remove the redundancy by introducing the concepts *minimal kernel element* and *basis* of an operator; cf. Definition 4.16. These definitions can be extended to the complete lattice framework.

5.24 Definition. Let ψ be an increasing $\mathbb{T}$-operator. An element $A \in \mathcal{V}(\psi)$ is called a *minimal kernel element* if $B < A$ implies that $B \notin \mathcal{V}(\psi)$. The *basis* of ψ, denoted $\mathcal{V}_b(\psi)$, is defined as the collection of all minimal kernel elements of ψ.

If $\mathcal{M}$ is a poset, then $A \in \mathcal{M}$ is called a *minimal element* if there does not exist an element $X \in \mathcal{M}$ with $X < A$. In this sense a minimal kernel element is nothing but a minimal element of the kernel (which is a poset). The following well-known result provides sufficient conditions for the existence of minimal elements.

5.25 Zorn's Lemma. *If each chain in a poset $\mathcal{M}$ has a lower bound, then $\mathcal{M}$ contains a minimal element.*

It is our goal to find conditions under which an increasing $\mathbb{T}$-operator has a minimal representation as a supremum of erosions. In fact, this amounts to the following problem: given A with $\psi(A) \geq o$, find a minimal element $A_0 \leq A$ such that $\psi(A_0) \geq o$. Recall the definition of a l.u.s.c. operator, Definition 3.6.

5.26 Lemma. *Consider an increasing $\mathbb{T}$-operator ψ on $\mathcal{L}$ which is l.u.s.c. For every $A \in \mathcal{V}(\psi)$ there exists a minimal kernel element $A_0 \leq A$.*

Proof. Let $A \in \mathcal{V}(\psi)$, and define the poset $\mathcal{M} = \{X \in \mathcal{V}(\psi) \mid X \leq A\}$. To show that $\mathcal{M}$ obeys the assumptions in Zorn's lemma, let $\mathcal{C}$ be a chain in $\mathcal{M}$. Define $C = \bigwedge \mathcal{C}$; since ψ is l.u.s.c. and $\psi(X) \geq o$ for every $X \in \mathcal{C}$, we have $\psi(C) = \bigwedge_{X \in \mathcal{C}} \psi(X) \geq o$. Therefore, $C \in \mathcal{M}$ and C is a lower bound of $\mathcal{C}$. Zorn's lemma gives that $\mathcal{M}$ has a minimal element A_0. □

5.27 Theorem. (Basis Representation of $\mathbb{T}$-operators)
Let $\mathcal{L}$ be a complete lattice for which the Basic Assumption is satisfied. Every increasing $\mathbb{T}$-operator on $\mathcal{L}$ which is l.u.s.c. has a nonempty basis $\mathcal{V}_b(\psi)$ and can be represented by

$$\psi(X) = \bigvee_{A \in \mathcal{V}_b(\psi)} X \ominus A. \tag{5.45}$$

If, moreover, there exists a $\mathbb{T}$-compatible negation on $\mathcal{L}$, then every increasing l.l.s.c. operator ψ satisfies

$$\psi(X) = \bigwedge_{A \in \mathcal{V}_b(\psi^*)} X \oplus \check{A}. \tag{5.46}$$

Proof. The first relation follows immediately from Theorem 5.22 and the previous result. To prove the dual result in (5.46), we note that ψ^* is a l.u.s.c. operator iff ψ is l.l.s.c. Hence ψ^* has an expansion as in (5.45), that is,

$$\psi^*(X) = \bigvee_{A \in \mathcal{V}_b(\psi^*)} X \ominus A.$$

Therefore, using (5.32), we get

$$\begin{aligned}\psi^*(X) &= (\psi(X^*))^* = \Big(\bigvee_{A \in \mathcal{V}_b(\psi^*)} X^* \ominus A\Big)^* \\ &= \bigwedge_{A \in \mathcal{V}_b(\psi^*)} (X^* \ominus A)^* = \bigwedge_{A \in \mathcal{V}_b(\psi^*)} X \oplus \check{A}.\end{aligned}$$

This completes the proof. □

5.5. Translation invariant morphology

In this section we consider four different complete lattices corresponding to the cases where an image is (i) a subset of $\mathbb{R}^d$ or $\mathbb{Z}^d$, (ii) a closed subset of $\mathbb{R}^d$, (iii) a convex subset of $\mathbb{R}^d$, and (iv) a matrix whose entries are subsets of $\mathbb{R}^d$ or $\mathbb{Z}^d$. In all these examples we choose as automorphism group $\mathbb{T}$ the group of translations.

5.5.1 The Boolean lattice $\mathcal{P}(\mathbb{E}^d)$

Recall that the notation $\mathbb{E}^d$ stands for either the Euclidean space $\mathbb{R}^d$ or the discrete space $\mathbb{Z}^d$. In Chapter 4 we have presented a comprehensive discussion of morphological operators on $\mathcal{P}(\mathbb{E}^d)$ which are invariant under translations. In the present section we show that this case fits into the general lattice framework. First recall that $\mathcal{P}(\mathbb{E}^d)$ is a complete Boolean lattice, the complement of an element being the ordinary set complement. The lattice $\mathcal{P}(\mathbb{E}^d)$ is atomic; the set of atoms ℓ comprises the singletons $\{x\}$, $x \in \mathbb{E}^d$. For convenience of notation we identify ℓ with $\mathbb{E}^d$. Let $h \in \mathbb{E}^d$; define the operator τ_h on $\mathcal{P}(\mathbb{E}^d)$ by

$$\tau_h(X) = X_h;$$

here X_h is the translate of X along the vector h. Thus τ_h defines an automorphism on $\mathcal{P}(\mathbb{E}^d)$ and the collection

$$\mathbb{T} = \{\tau_h \mid h \in \mathbb{E}^d\}$$

is an abelian group, called the *group of translations*. Since for any two points $x, y \in \mathbb{E}^d$, there is a unique translation mapping x onto y (namely, τ_{y-x}), the Basic Assumption holds. The corresponding group operation on $\mathbb{E}^d$ is vector addition. Then the Minkowski operations given by (5.14) and (5.15) coincide with the original Minkowski addition and subtraction as defined in Section 4.3. The properties stated in Propositions 4.10 and 4.11 (except (4.31) and (4.39)) follow immediately from the abstract results. Since $\mathcal{P}(\mathbb{E}^d)$ is a Boolean lattice, we can apply Corollary 5.19; we find

$$(X^c \oplus A)^c = X \ominus \check{A} \qquad \text{and} \qquad (X^c \ominus A)^c = X \oplus \check{A}, \tag{5.47}$$

where $\check{A} = -A$. (cf. (4.40)). Finally, the representation theorem 4.15 can be regarded as an application of Theorem 5.23. As a matter of fact, one may substitute an arbitrary abelian group for $\mathbb{E}^d$ here. In Section 5.6 we will consider the group of rotations and scalar multiplications in the plane.

In Example 3.9 it was shown that an operator ψ on $\mathcal{P}(E)$, with E a countable set, is l.u.s.c. if and only if it is $\downarrow$-continuous. This implies, e.g., that on $\mathcal{P}(\mathbb{Z}^d)$ every $\downarrow$-continuous translation invariant operator has a basis representation. Moreover, since the complement operator is compatible with translations, the dual basis representation in (5.46) holds also. Section 13.5 explains that every morphological operator on $\mathcal{P}(\mathbb{Z}^d)$ using only finite structuring elements is $\downarrow$-continuous as well as $\uparrow$-continuous.

5.5.2 The closed sets $\mathcal{F}(\mathbb{R}^d)$

In Example 2.9 we have seen that $\mathcal{F}(\mathbb{R}^d)$ is a complete lattice with supremum and infimum given, respectively, by

$$\bigvee_{i\in I} X_i = \overline{\bigcup_{i\in I} X_i},$$
$$\bigwedge_{i\in I} X_i = \bigcap_{i\in I} X_i.$$

Furthermore, the singletons are atoms and turn $\mathcal{F}(\mathbb{R}^d)$ into an atomic lattice. An important difference from $\mathcal{P}(\mathbb{R}^d)$ is that it is not Boolean; even worse, it is not self-dual. If, as in the previous example, $\mathbb{T}$ denotes the translation group on $\mathbb{R}^d$, then the Basic Assumption is satisfied. The group induced on $\mathbb{R}^d$ is vector addition, and the Minkowski operations resulting from (5.14) and (5.15), which we denote by $\overline{\oplus}$ and $\overline{\ominus}$, are given by

$$X \overline{\oplus} A = \bigvee_{a\in A} X_a = \overline{\bigcup_{a\in A} X_a}, \tag{5.48}$$
$$X \overline{\ominus} A = \bigwedge_{a\in A} X_{-a} = \bigcap_{a\in A} X_{-a}. \tag{5.49}$$

Note that $X \overline{\ominus} A = X \ominus A$, where $\ominus$ is the classical Minkowski subtraction. In accordance with our conventions, we define $\delta_A(X) = X \overline{\oplus} A$ and $\varepsilon_A(X) = X \overline{\ominus} A$. The representation theorem 5.22 states that every increasing $\mathbb{T}$-operator ψ on $\mathcal{F}(\mathbb{R}^d)$ can be written as $\psi = \bigvee_{A\subset\mathcal{V}(\psi)} \varepsilon_A$. In Section 7.7 we explain that under some additional continuity requirements on ψ we can prove the validity of a basis representation as in Theorem 5.27.

The lattice $\mathcal{F}(\mathbb{R}^d)$ provided with the translation group $\mathbb{T}$ does not satisfy the dual Basic Assumption, and so one cannot conclude that an increasing $\mathbb{T}$-operator has a decomposition as an infimum of dilations. Indeed, one can show by means of a counterexample that such a decomposition does not exist in general. Let $r > 0$; define the closed set C_r by

$$C_r = \{x \in \mathbb{R}^d \mid |x| \geq r\}.$$

For a closed set $X \subseteq \mathbb{R}^d$ we define the *outer radius* $R(X)$ as the radius of the smallest circumscribing circle and $R(X) = +\infty$ if X is unbounded; see Fig. 5.1.

5.28 Lemma. *For every closed set A,*

$$C_r \oplus A = \mathbb{R}^d \iff R(A) \geq r.$$

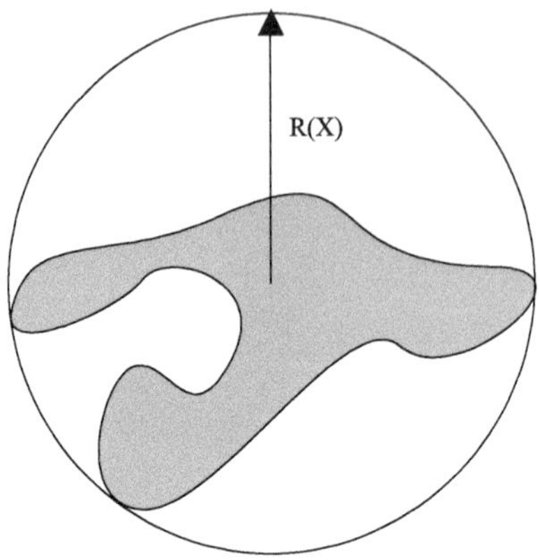

Figure 5.1 The outer radius $R(X)$ of a set X.

Proof. Let D_r be the open disk with radius r. Using (5.47) in $\mathcal{P}(\mathbb{R}^d)$, we find

$$(C_r \overline{\oplus} \check{A})^c = (\overline{C_r \oplus \check{A}})^c = \left[(C_r \oplus \check{A})^c\right]^\circ = (D_r \ominus A)^\circ.$$

By the symmetry of C_r we conclude that $C_r \overline{\oplus} A = \mathbb{R}^d$ if and only if $C_r \overline{\oplus} \check{A} = \mathbb{R}^d$; this holds if and only if $(D_r \ominus A)^\circ = \varnothing$. If $R(A) < r$, then, modulo a translation, we have $A \subseteq D_s$ for some $s < r$. This means $(D_r \ominus A)^\circ \subseteq (D_r \ominus D_s)^\circ = D_{r-s} \neq \varnothing$. Thus $R(A) < r$ implies that $C_r \oplus A \neq \mathbb{R}^d$. On the other hand, if $R(A) \geq r$, then $D_r \ominus A = \varnothing$ since no translate of A fits inside the open set D_r. For if it would, then since A is closed, it would also fit inside some smaller open disk D_s, $s < r$, yielding that $R(A) \leq s$, a contradiction. This completes the proof. □

Now let $C = C_1$, and define the operator $\psi : \mathcal{F}(\mathbb{R}^d) \to \mathcal{F}(\mathbb{R}^d)$ as follows:

$$\psi(X) = \bigcap \{\tau_h(C) \mid h \in \mathbb{R}^d \text{ and } X \subseteq \tau_h(C)\}. \tag{5.50}$$

Note that $\psi(X) = \mathbb{R}^d$ if $X \not\subseteq \tau_h(C)$ for every $h \in \mathbb{R}^d$. It is obvious that ψ is increasing and $\mathbb{T}$-invariant. In Example 6.17 we shall see that ψ is a so-called *structural closing*. Assume that δ_A is a translation invariant dilation on $\mathcal{F}(\mathbb{R}^d)$ such that $\delta_A \geq \psi$. Since $\psi(C_r) = \mathbb{R}^d$ if $r < 1$, we conclude that $\delta_A(C_r) = C_r \overline{\oplus} A = \mathbb{R}^d$ if $r < 1$. From Lemma 5.28 we conclude $R(A) \geq r$ if $r < 1$, and so $R(A) \geq 1$. But then, applying this lemma once again, we find $\delta_A(C) = \mathbb{R}^d$. Suppose that ψ can be written as an infimum of translation invariant dilations; it follows that $\psi(C) = \mathbb{R}^d$. It follows from (5.50), however, that $\psi(C) = C$, a contradiction.

5.5.3 Convex subsets of $\mathbb{R}^d$

The space $\mathcal{C}(\mathbb{R}^d)$ consisting of all convex subsets of $\mathbb{R}^d$ ordered by inclusion is a complete lattice with supremum and infimum given, respectively, by

$$\bigvee_{i \in I} X_i = \mathrm{co}(\bigcup_{i \in I} X_i),$$
$$\bigwedge_{i \in I} X_i = \bigcap_{i \in I} X_i.$$

Here co(X) denotes the convex hull of X; see also Example 2.16. The lattice $\mathcal{C}(\mathbb{R}^d)$ is atomic, the singletons being the atoms. The lattice is not distributive (and therefore not Boolean), however, and it does not allow a dual automorphism. As before, let $\mathbb{T}$ be the translation group. Evidently, the Basic Assumption is satisfied; this implies that $\mathbb{T}$-dilations and $\mathbb{T}$-erosions on $\mathcal{C}(\mathbb{R}^d)$ are of the form

$$\delta_A(X) = \mathrm{co}(\bigcup_{a \in A} X_a) = \mathrm{co}(X \oplus A), \tag{5.51}$$

$$\varepsilon_A(X) = \bigcap_{a \in A} X_{-a} = X \ominus A, \tag{5.52}$$

where $A \in \mathcal{C}(\mathbb{R}^d)$ and where $\oplus, \ominus$ denote the Minkowski addition and subtraction on $\mathcal{P}(\mathbb{R}^d)$. In Proposition 9.4 it will be demonstrated that $X \oplus A$ is convex if X, A are convex; so we may write

$$\delta_A(X) = X \oplus A.$$

Given $A, B \subseteq \mathbb{R}^d$, by the commutativity and distributivity properties of dilation it follows that

$$\bigvee_{a \in A} \bigvee_{b \in B} (a + b) = \mathrm{co}[(\bigvee_{a \in A} a) \oplus (\bigvee_{b \in B} b)];$$

this implies

$$\mathrm{co}(A \oplus B) = \mathrm{co}(A) \oplus \mathrm{co}(B). \tag{5.53}$$

For instance, let X, Y be two convex polyhedra with vertex sets A and B, respectively; then $X \oplus Y$ is the convex polyhedron spanned by the vertex set $A \oplus B$ comprising the points $a+b$, where $a \in A$ and $b \in B$.

5.5.4 Matrix morphology

Wilson (1992) discusses applications of mathematical morphology to character recognition. In his approach, Wilson does not restrict himself to one image but considers a matrix of images. In this subsection we present only the underlying theoretical concepts and show how they fit into the complete lattice framework; for further motivation and applications the reader should refer to Wilson (1992).

We restrict attention to the binary case here, but it is not difficult to extend the theory to the case of grey-scale functions. In matrix morphology, an object is defined as a matrix of sets: $X = \left[X_{ij}\right]$ where $i, j = 1, 2, \ldots, N$ (for notational convenience we consider only square matrices) and $X_{ij} \subseteq \mathbb{E}^d$. For $h \in \mathbb{E}^d$ we define the translation of X along h by $\tau_h(X) = \left[(X_{ij})_h\right]$. The transpose X^T of X is defined by $(X^T)_{ij} = X_{ji}$. Let $\mathcal{L}$ be the set of all objects X; define an ordering on $\mathcal{L}$ as follows:

$$X \leq Y \iff X_{ij} \subseteq Y_{ij} \text{ for every } i, j.$$

It is obvious that $\mathcal{L}$ is a complete Boolean lattice. The supremum of a family $X_k,\ k \in K$, is given by

$$\Big(\bigvee_{k\in K} X_k\Big)_{ij} = \Big(\bigcup_{k\in K} (X_k)_{ij}\Big).$$

The infimum is given by a similar expression. Define the binary operations $\boxplus$ and $\boxminus$ on $\mathcal{L}$ by

$$(X \boxplus A)_{ij} = \bigcup_{k=1}^{N} X_{ik} \oplus A_{kj},$$

$$(X \boxminus A)_{ij} = \bigcap_{k=1}^{N} X_{ik} \ominus A_{jk}.$$

Note the order of the indices in the last expression; instead one can also write

$$(X \boxminus A)_{ij} = \bigcap_{k=1}^{N} X_{ik} \ominus A^T_{kj}.$$

It is easy to show that the pair $(\varepsilon_A, \delta_A)$, where $\varepsilon_A(X) = X \boxminus A$ and $\delta_A(X) = X \boxplus A$, defines an adjunction on $\mathcal{L}$ for every $A \in \mathcal{L}$. Moreover, both operators are invariant under the translations $X \mapsto \tau_h(X)$. In fact, these simple observations yield a proof of many of the results shown by Wilson (1992).

5.6. Polar morphology

Translation symmetry need not be a useful assumption under all circumstances. In fact, there occur situations where *rotation symmetry* is more appropriate. In such cases one has to develop algorithms which take this symmetry into account. This may be done by replacing the translation group discussed previously by a group which includes all rotations. In this section we consider the abelian group of rotations and scalar multiplications on the plane. Section 5.8 deals with the case where both translations and rotations are considered. This latter case is considerably more difficult since the rotation–translation group is neither abelian nor simply transitive.

Consider the Boolean lattice $\mathcal{P}(E)$ where $E = \mathbb{R}^2 \setminus \{0\}$. The points in E are parametrized with polar coordinates (r, ϑ), where r is the radius $(r > 0)$ and ϑ the angle taken modulo 2π $(0 \le \vartheta < 2\pi)$. Define a group operation $\cdot$ on E by

$$(r, \vartheta) \cdot (r', \vartheta') = (rr', \vartheta + \vartheta'), \tag{5.54}$$

where $\vartheta + \vartheta'$ is taken modulo 2π. The neutral element is $(1, 0)$, and the inverse of (r, ϑ) is $(1/r, -\vartheta)$.

Note that the group operation $\cdot$ can be represented conveniently as a multiplication on the complex plane. Define an abelian group $\mathbb{T} = \{\tau_{r,\vartheta} \mid r > 0\ ,\ 0 \le \vartheta < 2\pi\}$ of automorphisms on $\mathcal{P}(E)$ by putting

$$\tau_{r,\vartheta}(X) = \{(r, \vartheta) \cdot (s, \varphi) \mid (s, \varphi) \in X\},$$

for $X \subseteq E$. The mapping $\tau_{r,\vartheta}$ is the composition of a scalar multiplication relative to the origin with a factor r and a rotation around the origin by an angle ϑ; see Fig. 5.2. In this context, a $\mathbb{T}$-operator is a mapping which is invariant under rotations and scalar multiplications. The group $\mathbb{T}$ yields its own class of Minkowski additions and subtractions (i.e., dilations and erosions). These operations are given by the algebraic formulas in (5.14)

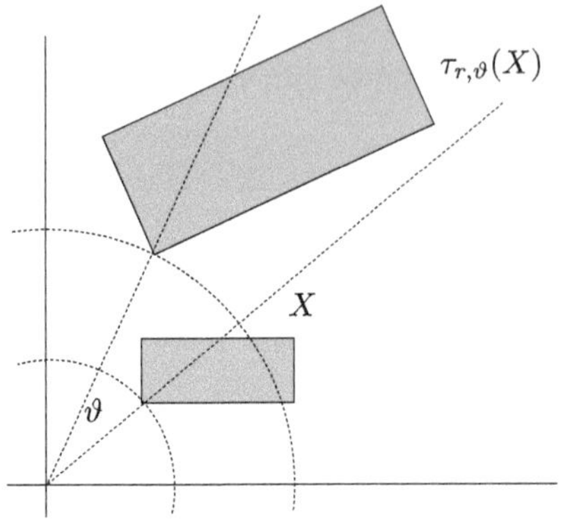

Figure 5.2 The automorphism $\tau_{r,\vartheta}$.

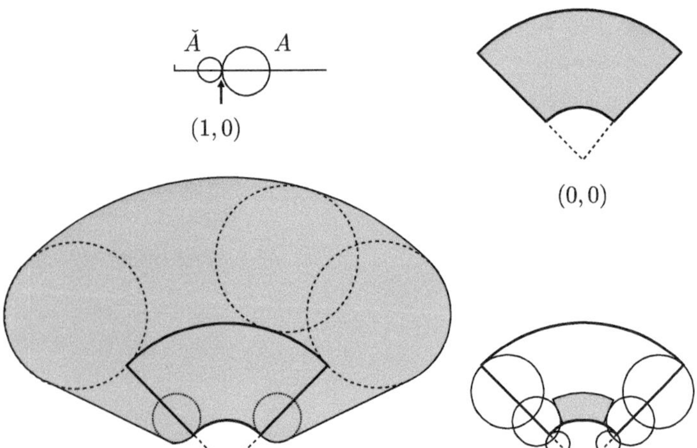

Figure 5.3 Polar dilation and erosion. Top: structuring elements A and $\breve{A}$ and set X. Bottom: polar dilation and erosion of X.

and (5.15), that is,

$$X \oplus A = \bigcup_{(p,\alpha)\in A} \tau_{p,\alpha}(X),$$

$$X \ominus A = \bigcap_{(p,\alpha)\in A} \tau_{p,\alpha}^{-1}(X).$$

The geometrical nature of these operations, however, differs substantially from their translation invariant counterparts; see Fig. 5.3.

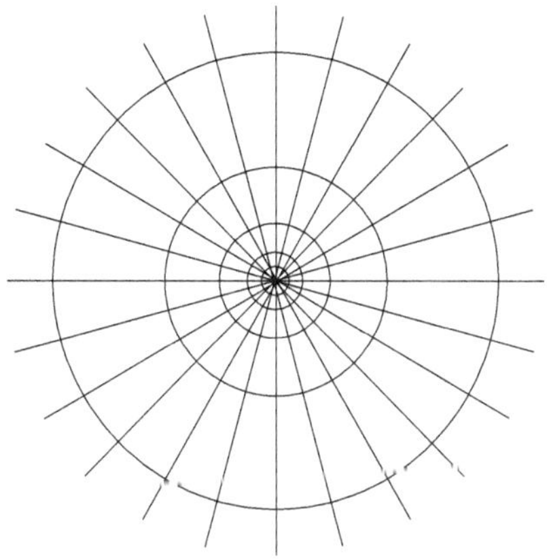

Figure 5.4 Polar grid.

We can unify the $\mathbb{T}$-operators resulting from this rotation–multiplication group under the heading *polar morphology* and refer to the operators given above as *polar dilation* and *polar erosion*. The algebraic properties of these operators are exactly the same as in the translation invariant case.

To implement polar morphological operators on a digital computer, one must work on a *polar grid*. This is obtained by discretizing the circle $[0, 2\pi)$ and the radial axis $r > 0$ in a regular fashion. Let $r > 1$, and let $n > 1$ be an integer. The set $\{(r^i, 2\pi j/n) \mid i \in \mathbb{Z}\ ,\ j = 0, 1, \ldots, n-1\}$ constitutes a polar grid; see Fig. 5.4 for an illustration.

5.7. Grey-scale functions

In Section 4.6 we have discussed morphological operators on the lattice of grey-scale functions which are invariant under horizontal and vertical translations. We briefly recall some notation. By $\mathrm{Fun}(E, \mathcal{T})$ we denote the complete lattice of all functions $F : E \to \mathcal{T}$. Whenever it is clear from the context which grey-value set $\mathcal{T}$ is meant, the argument $\mathcal{T}$ is suppressed. The horizontal translate F_h of a function F (where $h \in \mathbb{R}^d$) is defined by

$$F_h(x) = F(x - h).$$

5.7.1 Additive structuring functions

Throughout this subsection, $\mathcal{T} = \overline{\mathbb{R}}$. For $F \in \mathrm{Fun}(\mathbb{R}^d)$ and $v \in \mathbb{R}$, the vertical translate $F + v$ of F is defined by

$$(F + v)(x) = F(x) + v.$$

The *negative function* F^* of F is

$$F^*(x) = -F(x);$$

see (4.96). The family ℓ of *pulse functions* $\{f_{h,t} \mid h \in \mathbb{R}^d \text{ and } t \in \mathbb{R}\}$ (cf. Example 2.38(a) and (5.7)) defines a sup-generating family in $\mathrm{Fun}(\mathbb{R}^d)$. For $h \in \mathbb{R}^d$ and $v \in \mathbb{R}$, we define the automorphism $\tau_{h,v}$ on $\mathrm{Fun}(\mathbb{R}^d)$ by

$$\big(\tau_{h,v}(F)\big)(x) = F(x - h) + v.$$

The class $\mathbb{T} = \{\tau_{h,v} \mid h \in \mathbb{R}^d, v \in \mathbb{R}\}$ constitutes an abelian automorphism group on $\mathcal{L}$ which is transitive on ℓ. This follows directly from the observation that $\tau_{h,v} f_{x,t} = f_{x+h,t+v}$. If one chooses $f_{0,0}$ as the origin in ℓ, one gets a group operation $+$ on ℓ by putting

$$f_{x,t} + f_{x',t'} = f_{x+x',t+t'}.$$

The corresponding Minkowski addition and subtraction on $\mathrm{Fun}(\mathbb{R}^d)$ are

$$(F \oplus G)(x) = \bigvee_{h \in \mathbb{R}^d} [F(x - h) + G(h)], \tag{5.55}$$

$$(F \ominus G)(x) = \bigwedge_{h \in \mathbb{R}^d} [F(x + h) - G(h)]; \tag{5.56}$$

cf. (4.102) and (4.103). Let $\Delta_G, \mathcal{E}_G$ be the operators defined by $\Delta_G(F) = F \oplus G$ and $\mathcal{E}_G(F) = F \ominus G$, respectively. These operators, called *grey-scale dilation* and *grey-scale erosion*, are invariant under horizontal and vertical translations. From Proposition 5.14 the following result follows immediately.

5.29 Proposition. *For every* $G \in \mathrm{Fun}(\mathbb{R}^d)$*, the pair* $(\mathcal{E}_G, \Delta_G)$ *defines a* $\mathbb{T}$*-adjunction on* $\mathrm{Fun}(\mathbb{R}^d)$*; moreover, every* $\mathbb{T}$*-adjunction is of this form.*

From the identity

$$\big[\tau_{h,v}(F^*)\big]^* = \tau_{h,-v}(F),$$

it follows that the negation $F \mapsto F^*$ is compatible with the group $\mathbb{T}$. Following the definition in (5.26) we write $(f_{x,t})\check{} = f_{x,-t}$. According to (5.27) the reflection of the structuring function G is

$$\check{G}(x) = G(-x). \tag{5.57}$$

Application of Proposition 5.18 gives (4.129), i.e.,

$$(F^* \oplus G)^* = F \ominus \check{G} \qquad \text{and} \qquad (F^* \ominus G)^* = F \oplus \check{G}. \tag{5.58}$$

Furthermore, Theorem 5.23 provides an alternative proof of the representation Theorem 4.36.

5.30 Remark. Proposition 5.16 states that the only automorphisms on $\mathcal{L}$ that commute with every $\mathbb{T}$-adjunction are the elements of $\mathbb{T}$ if $\ell \cup \{O, I\}$ is inf-closed. To show that the latter condition is necessary, take $\ell = \{f_{x,t} \mid x \in \mathbb{R}^d, t \in \mathbb{Q}\}$ and $\mathbb{T} = \{\tau_{h,v} \mid h \in \mathbb{R}^d, v \in \mathbb{Q}\}$. Again, the Basic Assumption is satisfied, and the resulting class of $\mathbb{T}$-adjunctions coincides with that given before. But a translation $\tau_{h,v}$ with $v \notin \mathbb{Q}$, although not contained in $\mathbb{T}$, commutes with every adjunction in this class.

5.7.2 Multiplicative structuring functions

The grey-value set $\mathcal{T} = \overline{\mathbb{R}}_+$ differs from $\overline{\mathbb{R}}$ in that it is not a group under addition. In particular, this means that one cannot define grey-scale adjunctions as in (5.55)–(5.56). The observation that $(0, \infty)$ is an abelian group under scalar multiplication suggests the following alternative approach. Define

$$\ell = \{f_{x,t} \mid x \in \mathbb{R}^d \ , \ t > 0\};$$

this defines a sup-generating family in $\text{Fun}(\mathbb{R}^d, \overline{\mathbb{R}}_+)$. Consider the group

$$\dot{\mathbb{T}} = \{\dot{\tau}_{h,v} \mid h \in \mathbb{R}^d, \ v > 0\},$$

where $\dot{\tau}_{h,v}$ is the automorphism on $\text{Fun}(\mathbb{R}^d, \overline{\mathbb{R}}_+)$ given by

$$(\dot{\tau}_{h,v}F)(x) = vF(x - h).$$

Thus $\dot{\tau}_{h,v}$ is a composition of a horizontal translation and a grey-scale multiplication. Now $\dot{\mathbb{T}}$ is an abelian group with

$$\dot{\tau}_{h,v}\dot{\tau}_{h',v'} = \dot{\tau}_{h+h',vv'},$$

and, evidently, it is transitive on ℓ. Therefore, the Basic Assumption is satisfied. Choosing $f_{0,1}$ as the origin of ℓ (as before, the choice of the origin is strongly suggested by the parametrization of $\mathbb{T}$), one finds that $\dot{\mathbb{T}}$ induces a group structure $\dot{+}$ on ℓ given by

$$f_{x,t} \dot{+} f_{x',t'} = f_{x+x',tt'}.$$

The resulting Minkowski addition $\dot{\oplus}$ and subtraction $\dot{\ominus}$ are given by

$$(F \dot{\oplus} G)(x) = \bigvee_{h \in \mathbb{R}^d} [F(x-h)G(h)], \tag{5.59}$$

$$(F \dot{\ominus} G)(x) = \bigwedge_{h \in \mathbb{R}^d} [F(x+h)/G(h)], \tag{5.60}$$

with the convention that $ts = 0$ when $t = 0$ or $s = 0$ and $t/s = \infty$ when $t = \infty$ or $s = 0$. For obvious reasons we call G a *multiplicative structuring function*.

One can define a negation on $\text{Fun}(\mathbb{R}^d, \overline{\mathbb{R}}_+)$ by

$$F^*(x) = 1/F(x).$$

From the identity

$$\left[\tau_{h,v}(F^*)\right]^* = \tau_{h,1/v}(F),$$

it follows that this negation is compatible with $\mathbb{T}$. Just as in the additive case, the reflection $\check{G}$ of G defined in (5.27) is given by

$$\check{G}(x) = G(-x);$$

thus one obtains the duality relations

$$\frac{1}{F \dot{\oplus} G} = \frac{1}{F} \dot{\ominus} \check{G}, \tag{5.61}$$

$$\frac{1}{F \dot{\ominus} G} = \frac{1}{F} \dot{\oplus} \check{G}. \tag{5.62}$$

There exists a one–one relation between the multiplicative case and the additive case discussed in the previous subsection; we describe this relation in more detail. Let e be the *anamorphosis* from $\text{Fun}(\mathbb{R}^d, \overline{\mathbb{R}})$ to $\text{Fun}(\mathbb{R}^d, \overline{\mathbb{R}}_+)$ given by

$$[e(F)](x) = \exp(F(x)); \tag{5.63}$$

its inverse e^{-1} is

$$[e^{-1}(F)](x) = \log(F(x)).$$

Clearly,

$$e\tau_{h,v} = \dot{\tau}_{h,\exp(v)}e,$$

for $h \in \mathbb{R}^d$, $v \in \mathbb{R}$. Therefore, if ψ is a $\mathbb{T}$-operator on Fun$(\mathbb{R}^d, \overline{\mathbb{R}})$, then $\dot{\psi}$ given by

$$\dot{\psi} = e\psi e^{-1}$$

defines a $\dot{\mathbb{T}}$-operator on Fun$(\mathbb{R}^d, \overline{\mathbb{R}}_+)$. We have, in particular,

$$F \dot{\oplus} G = \exp(\log F \oplus \log G),$$
$$F \dot{\ominus} G = \exp(\log F \ominus \log G).$$

Conversely, if $\dot{\psi}$ is a $\dot{\mathbb{T}}$-operator on Fun$(\mathbb{R}^d, \overline{\mathbb{R}}_+)$, then

$$\psi = e^{-1}\dot{\psi}e$$

is a $\mathbb{T}$-operator on Fun$(\mathbb{R}^d, \overline{\mathbb{R}})$.

5.8. $\mathbb{T}$-invariance: the nonabelian case

Section 5.2 contains a systematic exposition on $\mathbb{T}$-adjunctions. It was assumed that the automorphism group $\mathbb{T}$ is abelian and acts transitively on some sup-generating family ℓ of $\mathcal{L}$. Under these assumptions, $\mathbb{T}$ is simply transitive on ℓ: for every pair $x, y \in \ell$ there is a unique $\tau \in \mathbb{T}$ carrying x to y. In this section the assumption that $\mathbb{T}$ is abelian is dropped. As a result one has to distinguish between two different cases, namely, the simply transitive and the multi-transitive case. It is hardly necessary to remark that the first case is easier than the second one. Throughout this section we restrict ourselves to the complete Boolean lattice $\mathcal{P}(E)$.

5.8.1 Homogeneous spaces

Suppose that $\mathbb{T}$ is an automorphism group on $\mathcal{P}(E)$. As a result, every element of $\mathbb{T}$ maps a singleton $\{x\}$ onto another singleton $\{y\}$; this means that every element of $\mathbb{T}$ can be considered as a member of the permutation group of E (i.e., the group of all bijections). Throughout this section we make the following assumption.

5.31 Assumption. The group $\mathbb{T}$ acts transitively on E.

In the literature, the pair $(\mathbb{T}, E)$, where E is a nonempty set and $\mathbb{T}$ a transitive permutation group on E, is sometimes called a *homogeneous space*. The reader may refer to Section 5.10 for more bibliographical information. We start with some examples.

5.32 Examples.

(a) $E = \mathbb{R}^d$, $\mathbb{T}$ = translation group; see Section 5.5.

(b) $E = \mathbb{R}^2 \setminus \{0\}$, $\mathbb{T}$ = group of rotations and multiplications. The corresponding $\mathbb{T}$-operators, called polar operators, were treated in Section 5.6.

(c) $E = S^2$, the sphere in $\mathbb{R}^3$ and $\mathbb{T} = \mathrm{SO}(3)$, containing all rotations in $\mathbb{R}^3$. This group is transitive but not abelian. A point x is a fixpoint of every rotation around the axis through the origin and x.

(d) $E = \mathbb{R}^d$ and $\mathbb{T} = \mathrm{E}^+(d)$, the group generated by translations and rotations. This group, sometimes called *Euclidean motion group* or *group of rigid motions*, is transitive but not abelian. It is not simply transitive, as every point is invariant under all rotations around that point. This example will be treated in Section 5.9.

We present another example which we use throughout this section to visualize the abstract concepts and results.

5.33 Example. (Motions on the hexagonal grid)
Let E be the hexagonal grid in 2-dimensional space, and let $\mathbb{T}$ be the group containing all translations and rotations leaving the grid invariant. Note that there are essentially six rotations, namely, the rotations about angles of $k \cdot 60°$, where $k = 0, 1, \ldots, 5$. We call this group the *hexagonal motion group*. Let $\vec{v}_k$ denote the result of the rotation of the vector $\vec{v}_0 = (1, 0)$ around the origin about an angle of $k \cdot 60°$.

Every pair $(x, \vec{v}_k)$, x being a point on the hexagonal grid E and $k \in \{0, 1, \ldots, 5\}$, corresponds to a unique motion $\tau \in \mathbb{T}$ which maps $(0, \vec{v}_0)$ onto $(x, \vec{v}_k)$. In fact, τ is a translation along x followed by a rotation around x about an angle $k \cdot 60°$, or equivalently, a rotation around 0 about an angle of $k \cdot 60°$ followed by a translation along x; we call $(x, \vec{v}_k)$ a *motion vector*. The neutral element of $\mathbb{T}$ is represented by $(0, \vec{v}_0)$. This example is illustrated in Fig. 5.5; (a) shows a subset of $\mathbb{T}$, and in (c) motion τ, consisting of a rotation followed by a translation, is applied to a set $X \subseteq E$.

Define the *stabilizer* Σ_x of an element $x \in E$ as the subset of $\mathbb{T}$ containing all actions τ that leave x invariant:

$$\Sigma_x = \{\tau \in \mathbb{T} \mid \tau(x) = x\}. \tag{5.64}$$

If $\mathbb{T}$ is the hexagonal motion group, then Σ_x comprises all rotations about the point x; it is evident that Σ_x is a subgroup of $\mathbb{T}$. As in Section 5.2, we

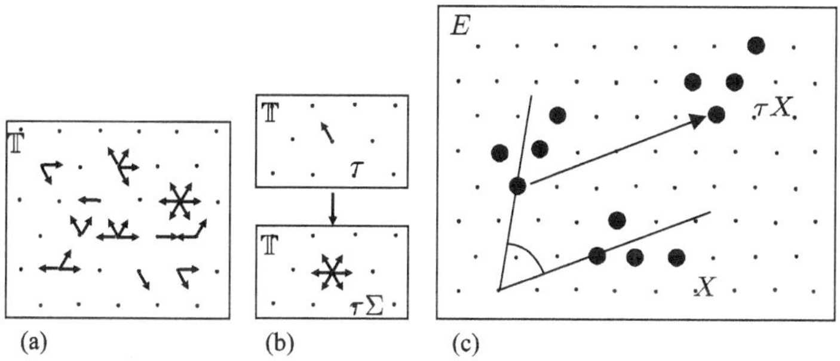

Figure 5.5 Motions on the hexagonal grid; (a) subset of $\mathbb{T}$, (b) the left coset $\tau\Sigma$ of τ, and (c) motion τ applied to a set $X \subseteq E$.

fix an origin o in E and define

$$\Sigma = \Sigma_o = \{\tau \in \mathbb{T} \mid \tau(o) = o\}; \tag{5.65}$$

or, in terms of motion vectors,

$$\Sigma = \{(o, \vec{v}_i) \mid i = 0, 1, \ldots, 5\}.$$

In the simply transitive case one has $\Sigma = \{\mathrm{id}\}$. Define an equivalence relation on $\mathbb{T}$ by

$$\tau \sim \tau' \iff \tau(o) = \tau'(o).$$

The collection of all elements equivalent to τ is the *left coset*

$$\tau\Sigma = \{\tau\sigma \mid \sigma \in \Sigma\}.$$

For the hexagonal motions the equivalence class corresponding to motion vector $(x, \vec{v}_k)$ is the set $\{(x, \vec{v}_i) \mid i = 0, 1, \ldots, 5\}$; see Fig. 5.5(b).

It is obvious that there exists a one–one correspondence between the equivalence classes $\mathbb{T}\backslash\Sigma$, called the *left coset space associated with* Σ, and the elements of E. Namely, the function which maps the left coset $\tau\Sigma$ onto $\tau(o)$ defines a bijection between $\mathbb{T}\backslash\Sigma$ and E. This observation provides the basis for the characterization of $\mathbb{T}$-adjunctions on the lattice $\mathcal{P}(E)$. Following the ideas of Section 5.2, one might attempt to define $\mathbb{T}$-adjunctions on $\mathcal{P}(E)$ as follows. Let $\mathbb{A} \subseteq \mathbb{T}$, and define $\Delta_{\mathbb{A}}$ and $\mathcal{E}_{\mathbb{A}}$ by

$$\delta_{\mathbb{A}}(X) = \bigcup_{\tau \in \mathbb{A}} \tau(X),$$

$$\varepsilon_{\mathbb{A}}(X) = \bigcap_{\tau \in \mathbb{A}} \tau^{-1}(X).$$

One easily checks that this pair indeed defines an adjunction on $\mathcal{P}(E)$. Unfortunately, $\delta_{\mathbb{A}}$ and $\varepsilon_{\mathbb{A}}$ are *not* $\mathbb{T}$-invariant unless $\mathbb{T}$ is abelian; therefore, we must look for a different approach. The approach taken in this section consists of two steps. The first step is based on the observation that $\mathbb{T}$ constitutes an automorphism group on the Boolean lattice $\mathcal{P}(\mathbb{T})$ as well. Consequently, one may consider $\mathbb{T}$-operators on this larger lattice. The reason why this is easier than working on $\mathcal{P}(E)$ directly is that $\mathbb{T}$ is simply transitive on $\mathbb{T}$. The second observation is that $\mathbb{T}$ yields two classes of transformations on $\mathcal{P}(\mathbb{T})$, the *left* and *right translations*. The mapping $\mathbb{X} \mapsto \tau\mathbb{X}$, where $\tau \in \mathbb{T}$ and $\tau\mathbb{X} = \{\tau\xi \mid \xi \in \mathbb{X}\}$, is a left translation, whereas the mapping $\mathbb{X} \mapsto \mathbb{X}\tau = \{\xi\tau \mid \xi \in \mathbb{X}\}$ is a right translation. Note that in the abelian case left and right translations coincide. Below, these observations are used to define $\mathbb{T}$-adjunctions on $\mathcal{P}(\mathbb{T})$. As a second step, we define a projection operator from $\mathcal{P}(\mathbb{T})$ to $\mathcal{P}(E)$ and a lift operator from $\mathcal{P}(E)$ to $\mathcal{P}(\mathbb{T})$ and use them to construct $\mathbb{T}$-adjunctions on $\mathcal{P}(E)$ from $\mathbb{T}$-adjunctions on $\mathcal{P}(\mathbb{T})$. To distinguish between both spaces we denote operators on $\mathcal{P}(\mathbb{T})$ by uppercase Greek letters and operators on $\mathcal{P}(E)$ by lowercase Greek letters.

5.8.2 $\mathbb{T}$-operators on $\mathcal{P}(\mathbb{T})$

Above we have observed that $\mathbb{T}$ defines an automorphism group on $\mathcal{P}(\mathbb{T})$ by interpreting each of its elements as a left translation:

$$\tau\mathbb{X} = \{\tau\xi \mid \xi \in \mathbb{X}\},$$

for $\mathbb{X} \subseteq \mathbb{T}$. So there are essentially three different ways to interpret an element $\tau \in \mathbb{T}$. First, by their very definition, they are automorphisms on $\mathcal{P}(E)$; second, they define permutations on E; and third, they can be used in their quality of left translations on $\mathcal{P}(\mathbb{T})$. It is usually clear from the context which interpretation is meant, and we do not distinguish between them in our notation. (Note that $\mathbb{T}$ has yet a fourth interpretation, namely, as right translations on $\mathcal{P}(\mathbb{T})$.)

An operator Ψ on $\mathcal{P}(\mathbb{T})$ is called a $\mathbb{T}$-*operator* if

$$\Psi\tau = \tau\Psi \qquad \text{for every } \tau \in \mathbb{T}.$$

In accordance with our convention, τ is interpreted as a left translation on $\mathcal{P}(\mathbb{T})$ here. It is evident that every right translation is a $\mathbb{T}$-operator. Define the Minkowski operations $\overline{\oplus}$ and $\overline{\ominus}$ on $\mathcal{P}(\mathbb{T})$ by

$$\mathbb{X} \overline{\oplus} \mathbb{A} = \bigcup_{\xi \in \mathbb{X}} \xi \mathbb{A} = \bigcup_{\alpha \in \mathbb{A}} \mathbb{X}\alpha, \tag{5.66}$$

$$\mathbb{X} \overline{\ominus} \mathbb{A} = \bigcap_{\alpha \in \mathbb{A}} \mathbb{X}\alpha^{-1} = \{\tau \in \mathbb{T} \mid \tau \mathbb{A} \subseteq \mathbb{X}\}. \tag{5.67}$$

The proof of the last equality in (5.67) is left as an exercise to the reader. We write

$$\Delta_{\mathbb{A}}(\mathbb{X}) = \mathbb{X} \overline{\oplus} \mathbb{A}, \qquad \mathcal{E}_{\mathbb{A}}(\mathbb{X}) = \mathbb{X} \overline{\ominus} \mathbb{A}. \tag{5.68}$$

The proof of the following result is analogous to that of Proposition 5.14.

5.34 Proposition. *For every $\mathbb{A} \subseteq \mathbb{T}$, the pair $(\mathcal{E}_{\mathbb{A}}, \Delta_{\mathbb{A}})$ defines a $\mathbb{T}$-adjunction on $\mathcal{P}(\mathbb{T})$. Conversely, every $\mathbb{T}$-adjunction is of this form.*

In fact, $\Delta_{\mathbb{A}}$ and $\mathcal{E}_{\mathbb{A}}$ are a union and an intersection of *right* translations, respectively. Since they commute with *left* translations, they are $\mathbb{T}$-operators. Such a combination of left and right translations is only possible on the larger space $\mathcal{P}(\mathbb{T})$. It follows easily that

$$\Delta_{\mathbb{A}} \Delta_{\mathbb{B}} = \Delta_{\mathbb{B} \overline{\oplus} \mathbb{A}}, \qquad \mathcal{E}_{\mathbb{A}} \mathcal{E}_{\mathbb{B}} = \mathcal{E}_{\mathbb{A} \oplus \mathbb{B}}. \tag{5.69}$$

Since $\mathcal{P}(\mathbb{T})$ is a Boolean lattice, one can define the complemented operator Ψ^* of the operator Ψ in the usual way. It is easy to check that Ψ^* is a $\mathbb{T}$-operator if and only if Ψ is. Analogous to Corollary 5.19 there exists the following duality relation between $\Delta_{\mathbb{A}}$ and $\mathcal{E}_{\mathbb{A}}$:

$$\Delta_{\mathbb{A}}^* = \mathcal{E}_{\check{\mathbb{A}}} \qquad \text{and} \qquad \mathcal{E}_{\mathbb{A}}^* = \Delta_{\check{\mathbb{A}}}, \tag{5.70}$$

where $\check{\mathbb{A}}$ is the reflection of $\mathbb{A}$, given by

$$\check{\mathbb{A}} = \{\alpha^{-1} \mid \alpha \in \mathbb{A}\}. \tag{5.71}$$

Define the kernel $\mathcal{V}(\Psi)$ by

$$\mathcal{V}(\Psi) = \{\mathbb{A} \in \mathcal{P}(\mathbb{T}) \mid \mathsf{id} \in \Psi(\mathbb{A})\}.$$

If one compares this definition of the kernel with that in (5.36), one notes that here $\{\mathsf{id}\}$ plays the role of the origin. The following analogue of Theorem 5.23 holds.

5.35 Theorem. (Representation of $\mathbb{T}$-operators)
Every increasing $\mathbb{T}$-operator Ψ on $\mathcal{P}(\mathbb{T})$ has the following representations:

$$\Psi(\mathbb{X}) = \bigcup_{\mathbb{A}\in\mathcal{V}(\Psi)} \mathbb{X}\,\overline{\ominus}\,\mathbb{A}, \tag{5.72}$$

$$\Psi(\mathbb{X}) = \bigcap_{\mathbb{A}\in\mathcal{V}(\Psi^*)} \mathbb{X}\,\overline{\oplus}\,\check{\mathbb{A}}. \tag{5.73}$$

5.8.3 Projection and lift operator

Let us return to our original problem, the characterization of $\mathbb{T}$-operators on $\mathcal{P}(E)$. We make use of the preceding results and two new notions, the projection operator and lift operator. Recall that o is the origin in E. The *projection* $\pi : \mathcal{P}(\mathbb{T}) \to \mathcal{P}(E)$ is defined by

$$\pi(\mathbb{X}) = \{\xi(o) \mid \xi \in \mathbb{X}\}; \tag{5.74}$$

the *lift* $\lambda : \mathcal{P}(E) \to \mathcal{P}(\mathbb{T})$ is the mapping given by

$$\lambda(X) = \{\tau \in \mathbb{T} \mid \tau(o) \in X\}. \tag{5.75}$$

It is obvious that λ is a dilation, since

$$\lambda(X) = \bigcup_{x\in X}\{\tau \in \mathbb{T} \mid \tau(o) = x\} = \bigcup_{x\in X}\lambda(\{x\}),$$

cf. Proposition 5.37. In the case of the hexagonal motions π maps a motion vector $(x, \vec{v}_k)$ to the point x (hence the name "projection"), whereas λ maps a point x to the set of all motion vectors $(x, \vec{v}_i)$, $i = 0, 1, \ldots, 5$. See Fig. 5.6 for an illustration.

Proposition 5.37 below lists some properties of π and λ and two other useful operators, namely, the erosion $\mathcal{E}_\Sigma$ and the dilation Δ_Σ on $\mathcal{P}(\mathbb{T})$. The dilation Δ_Σ augments a set $\mathbb{X} \subseteq \mathbb{T}$ by all transformations τ which are equivalent to some $\xi \in \mathbb{X}$ (i.e., $\tau(o) = \xi(o)$). Dually, the erosion $\mathcal{E}_\Sigma$ preserves only those elements of a set $\mathbb{X}$ for which the entire corresponding equivalent class is contained in $\mathbb{X}$.

5.36 Proposition.

(a) $\Delta_\Sigma = {\Delta_\Sigma}^2 = \mathcal{E}_\Sigma\Delta_\Sigma$; *in particular,* Δ_Σ *is a closing.*

(b) $\mathcal{E}_\Sigma = {\mathcal{E}_\Sigma}^2 = \Delta_\Sigma\mathcal{E}_\Sigma$; *in particular,* $\mathcal{E}_\Sigma$ *is an opening.*

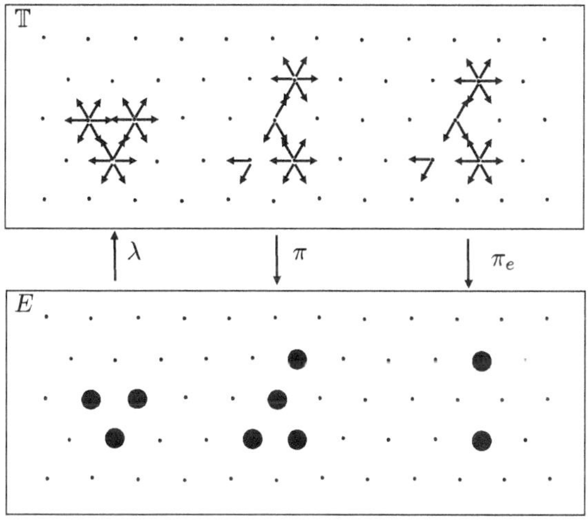

Figure 5.6 The operators λ, π, π_e.

Proof. From the fact that Σ is a subgroup of $\mathbb{T}$, it follows that $\Sigma \,\overline{\oplus}\, \Sigma = \Sigma$. In combination with (5.69) this implies the first equality in (a). To prove the second, use that

$$\mathbb{X} \oplus \Sigma \,\overline{\oplus}\, \Sigma \leq \mathbb{X} \,\overline{\oplus}\, \Sigma \iff \mathbb{X} \,\overline{\oplus}\, \Sigma \leq (\mathbb{X} \,\overline{\oplus}\, \Sigma) \,\overline{\ominus}\, \Sigma;$$

this follows from the fact that $(\mathcal{E}_\Sigma, \Delta_\Sigma)$ is an adjunction. Because the inequality on the left-hand side is (trivially) satisfied, the inequality on the right-hand side holds as well. Since id $\in \Sigma$, $\mathbb{X} \,\overline{\oplus}\, \Sigma \geq (\mathbb{X} \,\overline{\oplus}\, \Sigma) \,\overline{\ominus}\, \Sigma$, and the proof of (a) is completed. Now (b) follows by duality. □

The operators $\mathcal{E}_\Sigma, \Delta_\Sigma$ are illustrated in Fig. 5.7.

As a final preparation, define a second projection operator $\pi_e : \mathcal{P}(\mathbb{T}) \to \mathcal{P}(E)$ by

$$\pi_e = \pi \mathcal{E}_\Sigma. \tag{5.76}$$

It follows immediately that

$$\pi_e(\mathbb{X}) = \{\xi(o) \mid \xi \in \mathbb{X} \text{ and } \xi\Sigma \subseteq \mathbb{X}\}. \tag{5.77}$$

In terms of the hexagonal motion example, the difference between π and π_e consists hereof that π maps every motion vector $(x, \vec{v}_k) \in \mathbb{X}$ to the point x,

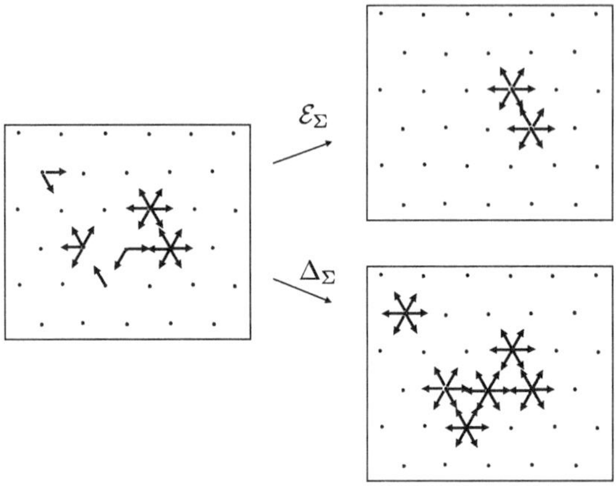

Figure 5.7 The adjunction $(\mathcal{E}_\Sigma, \Delta_\Sigma)$.

whereas π_e maps $(x, \vec{v}_k)$ to x if and only if $(x, \vec{v}_i) \in \mathbb{X}$ for all $i = 0, 1, \ldots, 5$. The next result lists several properties of $\pi, \pi_e, \lambda, \Delta_\Sigma$ and $\mathcal{E}_\Sigma$.

5.37 Proposition.

(a) *$\pi, \pi_e, \lambda, \Delta_\Sigma, \mathcal{E}_\Sigma$ are increasing $\mathbb{T}$-operators.*

(b) *(λ, π) is an adjunction between $\mathcal{P}(E)$ and $\mathcal{P}(\mathbb{T})$.*

(c) *(π_e, λ) is an adjunction between $\mathcal{P}(\mathbb{T})$ and $\mathcal{P}(E)$.*

(d) *The following relations hold:*

$$\pi\lambda = \pi_e\lambda = \mathrm{id}, \tag{5.78}$$

$$\lambda\pi = \Delta_\Sigma, \tag{5.79}$$

$$\lambda\pi_e = \mathcal{E}_\Sigma, \tag{5.80}$$

$$\Delta_\Sigma\lambda = \mathcal{E}_\Sigma\lambda = \lambda, \tag{5.81}$$

$$\pi\Delta_\Sigma = \pi. \tag{5.82}$$

(e) *For $X \subseteq E$, $\mathbb{X} \subseteq \mathbb{T}$,*

$$\lambda(X^c) = [\lambda(X)]^c, \tag{5.83}$$

$$\pi_e(\mathbb{X}^c) = [\pi(\mathbb{X})]^c. \tag{5.84}$$

Proof. (a): That Δ_Σ and $\mathcal{E}_\Sigma$ are increasing is obvious; for the other operators, this is a consequence of (b) and (c). Furthermore, it is evident that all operators are $\mathbb{T}$-invariant.

(b): We must show that

$$\pi(\mathbb{X}) \subseteq Y \iff \mathbb{X} \subseteq \lambda(Y),$$

for $\mathbb{X} \subseteq \mathbb{T}$, $Y \subseteq E$. Using the definitions of π and λ, we get

$$\begin{aligned}\pi(\mathbb{X}) \subseteq Y &\iff \forall \xi \in \mathbb{X} : \xi(o) \in Y \\ &\iff \forall \xi \in \mathbb{X} : \xi \in \lambda(Y) \\ &\iff \mathbb{X} \subseteq \lambda(Y).\end{aligned}$$

(c): We must show that

$$\lambda(X) \subseteq \mathbb{Y} \iff X \subseteq \pi_e(\mathbb{Y}),$$

for $X \subseteq E, \mathbb{Y} \subseteq \mathbb{T}$. To prove $\Rightarrow$, assume that $\lambda(X) \subseteq \mathbb{Y}$, and take $x \in X$. Use that $\pi_e(\mathbb{Y}) = \{\xi(o) \mid \xi\Sigma \subseteq \mathbb{Y}\}$ and that $\{\tau \in \mathbb{T} \mid \tau(o) = x\} \subseteq \mathbb{Y}$. There is at least one $\tau_x \in \mathbb{T}$ such that $\tau_x(o) = x$; hence $\tau_x\Sigma \subseteq \mathbb{Y}$. This means that $\tau_x(o) = x \in \pi_e(\mathbb{Y})$. The reverse implication is proved analogously.

(d): $\pi\lambda = \mathrm{id}$ follows immediately. To show that $\pi_e\lambda = \mathrm{id}$, observe first that $\pi_e\lambda$ is a closing, and hence $\pi_e\lambda \geq \mathrm{id}$. On the other hand, since $\mathcal{E}_\Sigma \leq \mathrm{id}$, one gets

$$\pi_e\lambda = \pi\mathcal{E}_\Sigma\lambda \leq \pi\lambda = \mathrm{id};$$

thus equality follows.

$\lambda\pi = \Delta_\Sigma$: For $\mathbb{X} \subseteq \mathbb{T}$ we have

$$\begin{aligned}\lambda\pi(\mathbb{X}) &= \{\tau \in \mathbb{T} \mid \tau(o) = \xi(o) \text{ and } \xi \in \mathbb{X}\} \\ &= \bigcup\{\xi\Sigma \mid \xi \in \mathbb{X}\} \\ &= \mathbb{X} \,\overline{\oplus}\, \Sigma.\end{aligned}$$

$\lambda\pi_e = \mathcal{E}_\Sigma$: Using Proposition 5.36(b), we get $\lambda\pi_e = \lambda\pi\mathcal{E}_\Sigma = \Delta_\Sigma\mathcal{E}_\Sigma = \mathcal{E}_\Sigma$.

$\Delta_\Sigma\lambda = \lambda$: We conclude from Proposition 3.16(c) that $\Delta_\Sigma\lambda$ is a dilation from $\mathcal{P}(E)$ to $\mathcal{P}(\mathbb{T})$ with adjoint erosion

$$\pi_e\mathcal{E}_\Sigma = \pi\mathcal{E}_\Sigma\mathcal{E}_\Sigma = \pi\mathcal{E}_\Sigma = \pi_e;$$

here we have used Proposition 5.36(b). But the adjoint dilation of π_e is λ, and by the uniqueness of adjoints the conclusion follows.

$$\begin{array}{ccc} \mathcal{P}(\mathbb{T}) & \overset{\Psi}{\longleftrightarrow} & \mathcal{P}(\mathbb{T}) \\ \lambda\big\uparrow\big\downarrow\pi & & \pi_e\big\downarrow\big\uparrow\lambda \\ \mathcal{P}(E) & \overset{\psi}{\longleftrightarrow} & \mathcal{P}(E) \end{array}$$

Figure 5.8 Schematic representation of the adjunction relations between $\mathcal{P}(E)$ and $\mathcal{P}(\mathbb{T})$. The vertical arrows on the left-hand side have to be interpreted as "(λ, π) is an adjunction between $\mathcal{P}(E)$ and $\mathcal{P}(\mathbb{T})$". Similarly, the vertical arrows on the right-hand side have to be interpreted as "(π_e, λ) is an adjunction between $\mathcal{P}(\mathbb{T})$ and $\mathcal{P}(E)$".

$\mathcal{E}_\Sigma\lambda = \lambda$: Since (λ, π) is an adjunction, $\lambda\pi\lambda = \lambda$. Using that $\lambda\pi = \Delta_\Sigma$, we get

$$\mathcal{E}_\Sigma\lambda = \mathcal{E}_\Sigma\lambda\pi\lambda = \mathcal{E}_\Sigma\Delta_\Sigma\lambda = \Delta_\Sigma\lambda = \lambda.$$

$\pi\Delta_\Sigma = \pi$: Again we use Proposition 3.16(c) and conclude that $\pi\Delta_\Sigma$ is a dilation with adjoint erosion $\mathcal{E}_\Sigma\lambda = \lambda$. But λ has the adjoint dilation π and, by uniqueness, $\pi\Delta_\Sigma = \pi$.

(e): The first assertion is obvious. To prove the second, we use $\pi\lambda = \mathrm{id}$ and $\lambda\pi = \Delta_\Sigma$. Therefore

$$\begin{aligned} \pi(\mathbb{X})^c &= \pi\lambda(\pi(\mathbb{X})^c) = \pi\left[(\lambda\pi(\mathbb{X}))^c\right] \\ &= \pi\left[\Delta_\Sigma(\mathbb{X})^c\right] = \pi\mathcal{E}_\Sigma(\mathbb{X}^c) \\ &= \pi_e(\mathbb{X}^c). \end{aligned}$$

At the penultimate step we have used (5.70) with $\mathbb{A} = \Sigma$ and $\check{\Sigma} = \Sigma$. □

The diagram in Fig. 5.8 shows a schematic representation of the various operators and their interrelations.

5.8.4 $\mathbb{T}$-operators on $\mathcal{P}(E)$

Let $\psi : \mathcal{P}(E) \to \mathcal{P}(E)$ be an arbitrary operator; the diagram in Fig. 5.8 shows that there are two alternative ways to define an operator on $\mathcal{P}(\mathbb{T})$. Namely,

$$\Psi_+ = \Psi_+(\psi) = \lambda\psi\pi, \tag{5.85}$$

$$\Psi_- = \Psi_-(\psi) = \lambda\psi\pi_e. \tag{5.86}$$

It is evident that $\Psi_- \le \Psi_+$ whenever ψ is increasing. Furthermore, using (5.78) it is easy to see that ψ can be recovered from Ψ_+ and Ψ_- as follows:

$$\psi = \pi\Psi_+\lambda = \pi\Psi_-\lambda. \tag{5.87}$$

Note that in these expressions π may be replaced by π_e. Furthermore, Ψ_+ and Ψ_- inherit several nice properties of ψ. The most important ones are listed in the next result.

5.38 Proposition. *Let ψ be an operator on $\mathcal{P}(E)$, and let Ψ_+ and Ψ_- on $\mathcal{P}(\mathbb{T})$ be given by* (5.85) *and* (5.86), *respectively.*

- **(a)** *If ψ is increasing, then Ψ_+, Ψ_- are increasing.*
- **(b)** *If ψ is a $\mathbb{T}$-operator, then Ψ_+, Ψ_- are $\mathbb{T}$-operators.*
- **(c)** *If ψ is increasing and extensive, then Ψ_+ is increasing and extensive.*
- **(d)** *If ψ is increasing and anti-extensive, then Ψ_- is increasing and anti-extensive.*
- **(e)** *If ψ is idempotent, then Ψ_+ and Ψ_- are idempotent.*
- **(f)** *If ψ is a filter, then Ψ_+ and Ψ_- are filters.*
- **(g)** *If ψ is a closing, then Ψ_+ is a closing.*
- **(h)** *If ψ is an opening, then Ψ_- is an opening.*

The proof of the result is straightforward. To prove, e.g., the statement in (c) assume that ψ is increasing and extensive; then $\Psi_+ = \lambda\psi\pi \geq \lambda\,\mathrm{id}\,\pi = \lambda\pi \geq \mathrm{id}$ since $\lambda\pi$ is a closing. Furthermore, from the duality results in Proposition 5.37(e) one obtains easily that

$$\Psi_+(\psi^*) = \left[\Psi_-(\psi)\right]^*. \tag{5.88}$$

Note that this relation implies

$$\Psi_-(\psi^*) = \left[\Psi_+(\psi)\right]^*.$$

On the other hand, with any operator Ψ on $\mathcal{P}(\mathbb{T})$ one can associate two operators ψ_+, ψ_- on $\mathcal{P}(E)$ given by

$$\psi_+ = \psi_+(\Psi) = \pi\Psi\lambda, \tag{5.89}$$

$$\psi_- = \psi_-(\Psi) = \pi_e\Psi\lambda. \tag{5.90}$$

It is *not* possible, however, to recover Ψ from ψ_+ or ψ_- (or both). This is due to the fact that $\mathcal{P}(\mathbb{T})$ has a richer structure than $\mathcal{P}(E)$. In terms of our prototype example: there is only one way to project a motion vector $(x, \vec{v}_k)$ to the point x, but there are many ways to lift x to a set of motion vectors which have first coordinate x. Again, many properties of Ψ carry over to ψ_+ and ψ_-.

5.39 Proposition. *Consider an operator Ψ on $\mathcal{P}(\mathbb{T})$, and let ψ_+, ψ_- on $\mathcal{P}(E)$ be given by* (5.89) *and* (5.90), *respectively.*

(a) *If Ψ is increasing, then ψ_+, ψ_- are increasing.*
(b) *If Ψ is a $\mathbb{T}$-operator, then ψ_+, ψ_- are $\mathbb{T}$-operators.*
(c) *If Ψ is increasing and extensive, then ψ_+, ψ_- are increasing and extensive.*
(d) *If Ψ is increasing and anti-extensive, then ψ_+, ψ_- are increasing and anti-extensive.*
(e) *If Ψ is a closing, then ψ_- is a closing.*
(f) *If Ψ is an opening, then ψ_+ is an opening.*

To prove, for example, (e) assume that Ψ is a closing. Then $\psi_- = \pi_e \Psi \lambda \geq \pi_e \lambda = \mathrm{id}$; hence ψ_- is extensive. Therefore, $\psi_-^2 \geq \psi_-$. On the other hand,

$$\psi_-^2 = \pi_e \Psi \lambda \pi_e \Psi \lambda = \pi_e \Psi \mathcal{E}_\Sigma \Psi \lambda \leq \pi_e \Psi^2 \lambda = \pi_e \Psi \lambda = \psi_-.$$

This gives the result.

Concerning duality one easily shows

$$\psi_+(\Psi^*) = \left[\psi_-(\Psi)\right]^*. \tag{5.91}$$

An important question is whether the intertwining relations (5.85)–(5.86) and (5.89)–(5.90) map adjunctions onto adjunctions. The next result answers this question affirmatively.

5.40 Proposition.
(a) *If (ε, δ) is an adjunction on $\mathcal{P}(E)$, then $(\Psi_-(\varepsilon), \Psi_+(\delta))$ is an adjunction on $\mathcal{P}(\mathbb{T})$.*
(b) *If $(\mathcal{E}, \Delta)$ is an adjunction on $\mathcal{P}(\mathbb{T})$, then $(\psi_-(\mathcal{E}), \psi_+(\Delta))$ is an adjunction on $\mathcal{P}(E)$.*
Moreover, if one of the operators is $\mathbb{T}$-invariant, then all of them are.

This result is an immediate consequence of the fact that both (π_e, λ) and (λ, π) are adjunctions. The statement concerning $\mathbb{T}$-invariance follows from Proposition 5.7.

5.41 Remark. One can interpret Ψ_+, Ψ_- as operators from $\mathcal{O}^+(\mathcal{P}(E))$ to $\mathcal{O}^+(\mathcal{P}(\mathbb{T}))$ and ψ_+, ψ_- as operators from $\mathcal{O}^+(\mathcal{P}(\mathbb{T}))$ to $\mathcal{O}^+(\mathcal{P}(E))$. As a matter of fact, (Ψ_+, ψ_+) constitutes an adjunction between $\mathcal{O}^+(\mathcal{P}(E))$ and $\mathcal{O}^+(\mathcal{P}(\mathbb{T}))$ and (ψ_-, Ψ_-) constitutes an adjunction between $\mathcal{O}^+(\mathcal{P}(\mathbb{T}))$ and $\mathcal{O}^+(\mathcal{P}(E))$. To prove, e.g., the first assertion one must show that

$$\pi \Psi \lambda \leq \psi \iff \Psi \leq \lambda \psi \pi,$$

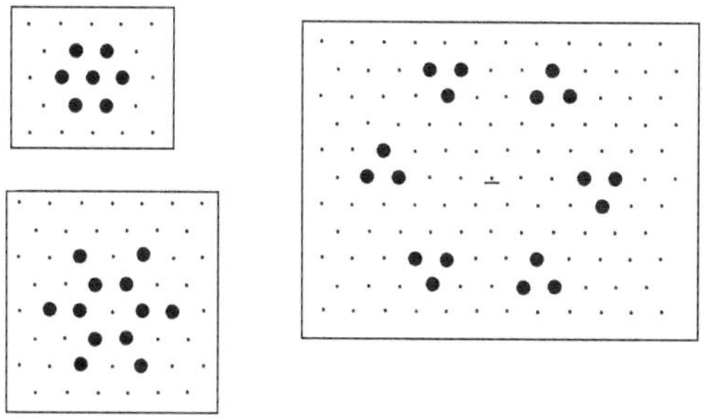

Figure 5.9 Σ-invariant subsets of E.

for all increasing operators ψ on $\mathcal{P}(E)$ and Ψ on $\mathcal{P}(\mathbb{T})$. To prove $\Rightarrow$ observe that $\pi\Psi\lambda \le \psi$ implies that $\lambda\pi\Psi\lambda\pi \le \lambda\psi\pi$. Now, since $\lambda\pi = \Delta_\Sigma$ and $\Delta_\Sigma \ge$ id, one finds

$$\Psi \le \Delta_\Sigma \Psi \Delta_\Sigma \le \lambda\psi\pi;$$

hence $\Rightarrow$ follows. The reverse inequality is proved similarly.

A set $A \subseteq E$ is called Σ-*invariant* if $A = \Sigma A$; here $\Sigma A = \bigcup_{\sigma\in\Sigma} \sigma(A)$. For the hexagonal motions, a set A is Σ-invariant if it is invariant under rotations around the origin. Some examples are depicted in Fig. 5.9.

5.42 Lemma.
(a) *If* $\mathbb{A} \subseteq \mathbb{T}$, *then*

$$\pi(\mathbb{A} \mathbin{\overline{\oplus}} \Sigma) = \pi(\mathbb{A}), \tag{5.92}$$

$$\pi(\Sigma \mathbin{\overline{\oplus}} \mathbb{A}) = \Sigma\pi(\mathbb{A}). \tag{5.93}$$

(b) *If* $A \subseteq E$, *then*

$$\lambda(A) \mathbin{\overline{\oplus}} \Sigma = \lambda(A), \tag{5.94}$$

$$\Sigma \mathbin{\overline{\oplus}} \lambda(A) = \lambda(\Sigma A). \tag{5.95}$$

Proof. (a): Relation (5.92) follows immediately from (5.82). To prove (5.93), we use that π is a $\mathbb{T}$-dilation:

$$\pi(\Sigma \mathbin{\overline{\oplus}} \mathbb{A}) = \pi(\bigcup_{\sigma\in\Sigma} \sigma\mathbb{A}) = \bigcup_{\sigma\in\Sigma} \sigma(\pi(\mathbb{A})) = \Sigma\pi(\mathbb{A}).$$

(b): Analogous proof. □

If $A \subseteq E$ is an arbitrary set, then $(\mathcal{E}_{\lambda(A)}, \Delta_{\lambda(A)})$ defines a $\mathbb{T}$-adjunction on $\mathcal{P}(\mathbb{T})$, and by Proposition 5.40(b), $(\psi_-(\mathcal{E}_{\lambda(A)}), \psi_+(\Delta_{\lambda(A)}))$ defines a $\mathbb{T}$-adjunction on $\mathcal{P}(E)$. Define

$$\delta_A = \psi_+(\Delta_{\lambda(A)}), \qquad \varepsilon_A = \psi_-(\mathcal{E}_{\lambda(A)}),$$

that is,

$$\delta_A(X) = \pi\big[\lambda(X) \mathbin{\overline{\oplus}} \lambda(A)\big], \tag{5.96}$$

$$\varepsilon_A(X) = \pi_e\big[\lambda(X) \mathbin{\overline{\ominus}} \lambda(A)\big]. \tag{5.97}$$

Using (5.94), (5.69) and Proposition 5.36(a), one can also write

$$\begin{aligned}\delta_A &= \pi\,\Delta_{\lambda(A)}\lambda = \pi\,\Delta_{\lambda(A)\overline{\oplus}\Sigma}\lambda = \pi\,\Delta_\Sigma\Delta_{\lambda(A)}\lambda \\ &= \pi\,\mathcal{E}_\Sigma\Delta_\Sigma\Delta_{\lambda(A)}\lambda = \pi_e\Delta_{\lambda(A)\overline{\oplus}\Sigma}\lambda = \pi_e\Delta_{\lambda(A)}\lambda;\end{aligned}$$

in other words,

$$\delta_A(X) = \pi_e\big[\lambda(X) \mathbin{\overline{\oplus}} \lambda(A)\big]. \tag{5.98}$$

The next result shows that every $\mathbb{T}$-adjunction on $\mathcal{P}(E)$ is of the form $(\varepsilon_A, \delta_A)$ and, moreover, that one can always restrict attention to structuring elements which are Σ-invariant.

5.43 Theorem. *Every $\mathbb{T}$-adjunction on $\mathcal{P}(E)$ is of the form $(\varepsilon_A, \delta_A)$, where A is a Σ-invariant subset of E and where δ_A, ε_A are given by* (5.96)–(5.97). *For ε_A one can also write*

$$\varepsilon_A(X) = \pi\big[\lambda(X) \mathbin{\overline{\ominus}} \lambda(A)\big]. \tag{5.99}$$

Proof. Assume that (ε, δ) is a $\mathbb{T}$-adjunction on $\mathcal{P}(E)$, and let $\mathcal{E} = \Psi_-(\varepsilon) = \lambda\varepsilon\pi_e$ and $\Delta = \Psi_+(\delta) = \lambda\delta\pi$. Then, by Proposition 5.40(a), $(\mathcal{E}, \Delta)$ is a $\mathbb{T}$-adjunction on $\mathcal{P}(\mathbb{T})$. From Proposition 5.34, we conclude that $(\mathcal{E}, \Delta) = (\mathcal{E}_{\mathbb{A}}, \Delta_{\mathbb{A}})$ for some $\mathbb{A} \subseteq \mathbb{T}$. Using Proposition 5.37(d), we get

$$\Delta_\Sigma\Delta_{\mathbb{A}} = \Delta_\Sigma\lambda\delta\pi = \lambda\delta\pi = \Delta_{\mathbb{A}},$$

and also

$$\Delta_{\mathbb{A}}\Delta_\Sigma = \lambda\delta\pi\,\Delta_\Sigma = \lambda\delta\pi = \Delta_{\mathbb{A}}.$$

With (5.69), this leads to

$$\mathbb{A}\,\overline{\oplus}\,\Sigma = \Sigma\,\overline{\oplus}\,\mathbb{A} = \mathbb{A}. \tag{5.100}$$

Furthermore,

$$\varepsilon = \pi_e \mathcal{E}_{\mathbb{A}} \lambda \quad \text{and} \quad \delta = \pi \Delta_{\mathbb{A}} \lambda.$$

We consider the expression for δ in more detail. Since $\pi = \pi \Delta_\Sigma$, by (5.82),

$$\delta = \pi \Delta_{\mathbb{A}} \lambda = \pi \Delta_\Sigma \Delta_{\mathbb{A}} \lambda = \pi \Delta_{\mathbb{A} \overline{\oplus} \Sigma} \lambda = \pi \Delta_{\lambda \pi(\mathbb{A})} \lambda.$$

Here we have used (5.79). Writing $A = \pi(\mathbb{A})$, we get $\delta = \pi \Delta_{\lambda(A)} \lambda$, or alternatively,

$$\delta(X) = \pi\big(\lambda(X)\,\overline{\oplus}\,\lambda(A)\big) = \delta_A(X).$$

From (5.93) and (5.100) we conclude

$$\Sigma A = \Sigma \pi(\mathbb{A}) = \pi(\Sigma\,\overline{\oplus}\,\mathbb{A}) = \pi(\mathbb{A}) = A;$$

therefore, A is Σ-invariant. That $\varepsilon = \varepsilon_A$ follows by the adjunction relation. To prove (5.99), use (5.69), (5.95), and the Σ-invariance of A:

$$\begin{aligned}
\varepsilon_A &= \pi_e \mathcal{E}_{\lambda(A)} \lambda = \pi \mathcal{E}_\Sigma \mathcal{E}_{\lambda(A)} \lambda \\
&= \pi \mathcal{E}_{\Sigma \overline{\oplus} \lambda(A)} \lambda = \pi \mathcal{E}_{\lambda(\Sigma A)} \lambda \\
&= \pi \mathcal{E}_{\lambda(A)} \lambda.
\end{aligned}$$

This concludes the proof. □

We point out that π_e in (5.97) may *not* be replaced by π in general; Theorem 5.43 shows that such a substitution is allowed if A is Σ-invariant.

Let $A \subseteq E$. Define the reflection of A by

$$\check{A} = \pi\big[\lambda(A)^{\check{}}\big], \tag{5.101}$$

where $\lambda(A)^{\check{}}$ is given by (5.71).

5.44 Lemma. *For every $A \subseteq E$ the reflection $\check{A}$ is Σ-invariant. If A is Σ-invariant, then $\lambda(\check{A}) = \lambda(A)^{\check{}}$ and $A^{\check{}\check{}} = A$.*

Proof. Let $h \in \check{A}$ and $\sigma \in \Sigma$; then $h = \tau^{-1}(o)$ for some $\tau \in \lambda(A)$. Since $\tau\sigma^{-1} \in \lambda(A)$ as well, $(\tau\sigma^{-1})^{-1}(o) = \sigma\tau^{-1}(o) = \sigma(h) \in \check{A}$. Hence $\check{A}$ is Σ-invariant. Now assume that A is Σ-invariant. Using (5.79) and (5.95), one gets

$$\begin{aligned}\lambda(\check{A}) &= \lambda\pi\left[\lambda(A)^{\check{}}\right] = \Delta_\Sigma\left[\lambda(A)^{\check{}}\right] = \{\alpha^{-1}\sigma \mid \alpha \in \lambda(A),\ \sigma \in \Sigma\} \\ &= \{(\sigma\alpha)^{-1} \mid \alpha \in \lambda(A),\ \sigma \in \Sigma\} = \left[\Sigma \,\overline{\oplus}\, \lambda(A)\right]^{\check{}} = \lambda(\Sigma A)^{\check{}} \\ &= \lambda(A)^{\check{}}.\end{aligned}$$

To prove the second relation note that

$$\lambda(A^{\check{}\check{}}) = \lambda(\check{A})^{\check{}} = [\lambda(A)]^{\check{}\check{}} = \lambda(A).$$

Application of π on both sides gives the desired equality. □

5.45 Proposition. *Let A be Σ-invariant and let $\check{A}$ be given by* (5.101)*; then*

$$(\delta_A)^* = \varepsilon_{\check{A}} \qquad \textit{and} \qquad (\varepsilon_A)^* = \delta_{\check{A}}.$$

Proof. Let $A, X \subseteq E$. Then

$$\begin{aligned}\delta_A(X^c) &= \pi\left[\lambda(X^c) \,\overline{\oplus}\, \lambda(A)\right] = \pi\left[\lambda(X)^c \,\overline{\oplus}\, \lambda(A)\right] \\ &= \pi\left[(\lambda(X) \,\overline{\ominus}\, \lambda(A)^{\check{}})^c\right] = \left(\pi_e\left[\lambda(X) \,\overline{\ominus}\, \lambda(\check{A})\right]\right)^c \\ &= \left[\varepsilon_{\check{A}}(X)\right]^c.\end{aligned}$$

Here we have subsequently used (5.83), (5.70), (5.84), and Lemma 5.44. The second relation follows by duality. □

As usual, the kernel of a $\mathbb{T}$-operator ψ on $\mathcal{P}(E)$ is defined by

$$\mathcal{V}(\psi) = \{A \subseteq E \mid o \in \psi(A)\}.$$

Let ψ be an increasing $\mathbb{T}$-operator on $\mathcal{P}(E)$ and define $\Psi = \Psi_+(\psi) = \lambda\psi\pi$. Then Ψ is an increasing $\mathbb{T}$-operator on $\mathcal{P}(\mathbb{T})$, and the representation theorem 5.35 gives $\Psi = \bigvee_{\mathbb{A}\in\mathcal{V}(\Psi)} \mathcal{E}_{\mathbb{A}}$. Since $\psi = \pi\Psi\lambda$ and π distributes over unions,

$$\psi = \bigvee_{\mathbb{A}\in\mathcal{V}(\Psi)} \pi\mathcal{E}_{\mathbb{A}}\lambda.$$

Define $A = \pi(\mathbb{A})$; from (5.81), (5.69), and (5.79), one gets

$$\begin{aligned}\pi\mathcal{E}_{\mathbb{A}}\lambda &= \pi\mathcal{E}_{\mathbb{A}}\mathcal{E}_{\Sigma}\lambda = \pi\mathcal{E}_{\mathbb{A}\overline{\oplus}\Sigma}\lambda \\ &= \pi\mathcal{E}_{\Delta_{\Sigma}(\mathbb{A})}\lambda = \pi\mathcal{E}_{\lambda\pi(\mathbb{A})}\lambda \\ &= \pi\mathcal{E}_{\lambda(A)}\lambda.\end{aligned}$$

Recall that the kernel of Ψ is given by $\mathcal{V}(\Psi) = \{\mathbb{A} \subseteq \mathbb{T} \mid \mathrm{id} \in \Psi(\mathbb{A})\}$. It is evident that $\mathrm{id} \in \Psi(\mathbb{A}) = \lambda\psi\pi(\mathbb{A})$ iff $0 \in \psi\pi(\mathbb{A})$. Combination of these facts leads to $\psi = \bigvee_{A\in\mathcal{V}(\psi)} \pi\mathcal{E}_{\lambda(A)}\lambda$. We summarize our findings in the following result.

5.46 Theorem. (Representation of $\mathbb{T}$-operators)
Let ψ be an increasing $\mathbb{T}$-operator on $\mathcal{P}(E)$. Then ψ can be written as

$$\psi(X) = \bigvee_{A\in\mathcal{V}(\psi)} \pi\big[\lambda(X)\,\overline{\ominus}\,\lambda(A)\big], \tag{5.102}$$

and dually as

$$\psi(X) = \bigwedge_{A\in\mathcal{V}(\psi^*)} \pi_e\big[\lambda(X)\,\overline{\oplus}\,\lambda(A)\check{}\,\big]. \tag{5.103}$$

Proof. The representation in (5.102) follows from the arguments given previously. To prove (5.103), apply (5.102) to ψ^*; this gives

$$\psi^*(X) = \bigvee_{A\in\mathcal{V}(\psi^*)} \pi\big[\lambda(X)\,\overline{\ominus}\,\lambda(A)\big].$$

Substituting X^c for X and taking complements on both sides leads to

$$\begin{aligned}\psi(X) &= \Big(\bigvee_{A\in\mathcal{V}(\psi^*)} \pi\big[\lambda(X^c)\,\overline{\ominus}\,\lambda(A)\big]\Big)^c \\ &= \bigwedge_{A\in\mathcal{V}(\psi^*)} \big(\pi[\lambda(X^c)\,\overline{\ominus}\,\lambda(A)]\big)^c \\ &= \bigwedge_{A\in\mathcal{V}(\psi^*)} \pi_e\big[(\lambda(X^c)\,\overline{\ominus}\,\lambda(A))^c\big] \\ &= \bigwedge_{A\in\mathcal{V}(\psi^*)} \pi_e\big[\lambda(X)\,\overline{\oplus}\,\lambda(A)\check{}\,\big];\end{aligned}$$

here we have used Proposition 5.37(e) and (5.70). □

We point out the following important difference between this result and the classical representation theorem 4.15 (see also Theorem 5.23); the latter states that a translation invariant operator can be decomposed as a union of erosions. The operator $\pi\mathcal{E}_{\lambda(A)}\lambda$ in (5.102), however, is *not* an erosion in general. More precisely, Theorem 5.43 states that this operator defines an erosion if A is Σ-invariant, and in that case one may replace π by π_e. In the forthcoming section this point will be illustrated by means of an example. Analogously, the operator $\pi_e\Delta_{\lambda(A)^{\check{}}}\lambda$ is a dilation if and only if A is Σ-invariant.

5.9. Translation–rotation morphology

In this section the abstract results of the previous section are applied to the case where $E = \mathbb{R}^2$ and $\mathbb{T}$ is the group of rotations and translations; cf. Example 5.32(d). Throughout this section $\oplus$ and $\ominus$ denote the classical Minkowski sum and difference; to denote the generalized Minkowski sum and difference on $\mathbb{T}$, we use the notation $\overline{\oplus}$ and $\overline{\ominus}$. The elements $\tau \in \mathbb{T}$ are called *motions* and can be decomposed as

$$\tau = T_h R_\varphi = R_\varphi T_{R_{-\varphi}h},$$

where R_φ is the rotation around the origin about an angle φ and T_h is the translation along the vector h. The action of R_φ on the point $x = (x_1, x_2)$ can be described by the matrix multiplication

$$R_\varphi x = \begin{pmatrix} \cos\varphi & -\sin\varphi \\ \sin\varphi & \cos\varphi \end{pmatrix} \begin{pmatrix} x_1 \\ x_2 \end{pmatrix}.$$

The stabilizer Σ contains all rotations about the origin, that is,

$$\Sigma = \{R_\varphi \mid 0 \le \varphi < 2\pi\}.$$

In Chapter 4 we have discussed a large class of morphological operators invariant under translations. The following result shows how these operators can be modified in such a manner that they become $\mathbb{T}$-operators, i.e., operators which are invariant both under translations and rotations. Note that this result is merely a variant of Proposition 5.6.

5.47 Proposition. *Let ψ be a translation invariant operator on $\mathcal{P}(\mathbb{R}^2)$. The operator*

$$\underline{\psi} = \bigcap_{0\le\varphi<2\pi} R_\varphi^{-1}\psi R_\varphi \tag{5.104}$$

is the largest $\mathbb{T}$-operator $\le \psi$; dually, the operator

$$\overline{\psi} = \bigcup_{0\le\varphi<2\pi} R_\varphi^{-1}\psi R_\varphi \tag{5.105}$$

is the smallest $\mathbb{T}$-operator $\ge \psi$.

Proof. That both $\underline{\psi}$ and $\overline{\psi}$ are $\mathbb{T}$-operators follows from the following two observations:

$$[R_\varphi^{-1}\psi R_\varphi]R_\vartheta = R_\vartheta[R_{\varphi+\vartheta}^{-1}\psi R_{\varphi+\vartheta}]$$

and

$$R_\varphi^{-1}\psi R_\varphi T_h = R_\varphi^{-1}\psi T_{R_\varphi h}R_\varphi = R_\varphi^{-1}T_{R_\varphi h}\psi R_\varphi = T_h R_\varphi^{-1}\psi R_\varphi.$$

The second relation gives that $R_\varphi^{-1}\psi R_\varphi$ is translation invariant if ψ is. With these facts the demonstration becomes analogous to the proof of Proposition 5.6. □

Now assume that $A \subseteq E$, and let $X \mapsto X \oplus A$ be the translation invariant dilation by A. From Proposition 3.16(b) it follows that

$$\delta(X) = \bigcup_{0\le\varphi<2\pi} R_\varphi^{-1}(R_\varphi X \oplus A) \tag{5.106}$$

defines a dilation. The previous result says that it is invariant both under rotations and translations. One can rewrite (5.106) as

$$\begin{aligned}
\delta(X) &= \bigcup_{0\le\varphi<2\pi} R_\varphi^{-1}\Big(\bigcup_{x\in X}\bigcup_{a\in A}(R_\varphi x + a)\Big) \\
&= \bigcup_{0\le\varphi<2\pi}\bigcup_{x\in X}\bigcup_{a\in A} R_\varphi^{-1}(R_\varphi x + a) \\
&= \bigcup_{x\in X}\bigcup_{0\le\varphi<2\pi}\bigcup_{a\in A}(x + R_\varphi^{-1}a) \\
&= \bigcup_{x\in X}\Big(x + \bigcup_{0\le\varphi<2\pi} R_\varphi^{-1}A\Big)
\end{aligned}$$

$$= \bigcup_{x \in X} (x + \Sigma A).$$

In other words,

$$\delta(X) = X \oplus \Sigma A. \tag{5.107}$$

Note that ΣA is a rotation invariant structuring element. Let λ be the lift operator defined in the previous section. It is obvious that $\lambda(X) = \{T_x R_\varphi \mid x \in X,\ 0 \le \varphi < 2\pi\}$. A straightforward calculation shows

$$\lambda(X) \mathbin{\overline{\oplus}} \lambda(A) = \lambda(X \oplus \Sigma A);$$

therefore

$$\pi\left[\lambda(X) \mathbin{\overline{\oplus}} \lambda(A)\right] = X \oplus \Sigma A.$$

This shows that the dilation δ given in (5.106), or equivalently (5.107), coincides with the expression given in (5.96). By adjunction, a similar statement holds for the erosion.

5.48 Theorem. *Every adjunction* (ε, δ) *on* $\mathcal{P}(\mathbb{R}^2)$ *which is invariant under translations and rotations is of the form*

$$\delta(X) = \delta_A(X) = X \oplus A,$$
$$\varepsilon(X) = \varepsilon_A(X) = X \ominus A,$$

where A is a rotation invariant structuring element.

A major implication of this result is that the class of dilations and erosions which are invariant under rotations and translations is rather meager; as a result there are many $\mathbb{T}$-operators that cannot be represented as a union of $\mathbb{T}$-erosions or as an intersection of $\mathbb{T}$-dilations. Two examples of such operators are the *modified dilation* and *erosion*, defined respectively by

$$\tilde{\delta}_A(X) = \bigcap_{0 \le \varphi < 2\pi} (X \oplus R_\varphi(A)), \tag{5.108}$$

$$\tilde{\varepsilon}_A(X) = \bigcup_{0 \le \varphi < 2\pi} (X \ominus R_\varphi(A)); \tag{5.109}$$

here $A \subseteq \mathbb{R}^2$ is an arbitrary set. Using

$$X \oplus R_\varphi(A) = R_\varphi^{-1}(R_\varphi(X) \oplus A), \tag{5.110}$$

$$X \ominus R_\varphi(A) = R_\varphi^{-1}(R_\varphi(X) \ominus A), \tag{5.111}$$

one easily derives that $\tilde{\delta}_A, \tilde{\varepsilon}_A$ are $\mathbb{T}$-operators. Observe that $\tilde{\delta}_A$ is the extension of the Minkowski addition $X \mapsto X \oplus A$ according to (5.104) and that $\tilde{\varepsilon}_A$ is the extension of the Minkowski subtraction $X \mapsto X \ominus A$ according to (5.105). Observe also that $\tilde{\delta}_A$ is not a dilation and that $\tilde{\varepsilon}_A$ is not an erosion. A straightforward computation shows

$$\lambda(X) \,\overline{\oplus}\, \lambda(A)^{\check{}} = \{T_{x+R_\varphi(-a)}R_\varphi \mid x \in X, a \in A, 0 \le \varphi < 2\pi\}.$$

This implies $h \in \pi_e(\lambda(X) \,\overline{\oplus}\, \lambda(A)^{\check{}})$ iff $h \in X \oplus R_\varphi(-A)$ for every $0 \le \varphi < 2\pi$; here $-A = \{-a \mid a \in A\}$. This shows

$$\pi_e\big[\lambda(X) \,\overline{\oplus}\, \lambda(A)^{\check{}}\big] = \tilde{\delta}_{-A}(X).$$

Note that $-A$ is different from $\check{A}$ given by (5.101) since, by Lemma 5.44, $\check{A}$ is rotation invariant. Similarly, it can be shown

$$\pi\big[\lambda(X) \,\overline{\ominus}\, \lambda(A)\big] = \tilde{\varepsilon}_A(X).$$

These identities in combination with Theorem 5.46 imply that every increasing operator on $\mathcal{P}(\mathbb{R}^2)$ which is invariant under rotations and translations can be represented as a union of modified erosions and, dually, as an intersection of modified dilations.

5.49 Theorem. (Translation–rotation invariant operators)
Let ψ be an increasing operator on $\mathcal{P}(\mathbb{R}^2)$ which is invariant under rotations and translations; then

$$\psi(X) = \bigcup_{A \in \mathcal{V}(\psi)} \bigcup_{0 \le \varphi < 2\pi} X \ominus R_\varphi(A); \tag{5.112}$$

dually,

$$\psi(X) = \bigcap_{A \in \mathcal{V}(\psi^*)} \bigcap_{0 \le \varphi < 2\pi} X \oplus R_\varphi(-A). \tag{5.113}$$

It is also possible to give a direct proof of this result by exploiting the classical representation theorem 4.15; we do this for (5.112). Every translation invariant operator ψ can be written as

$$\psi(X) = \bigcup_{A \in \mathcal{V}(\psi)} X \ominus A.$$

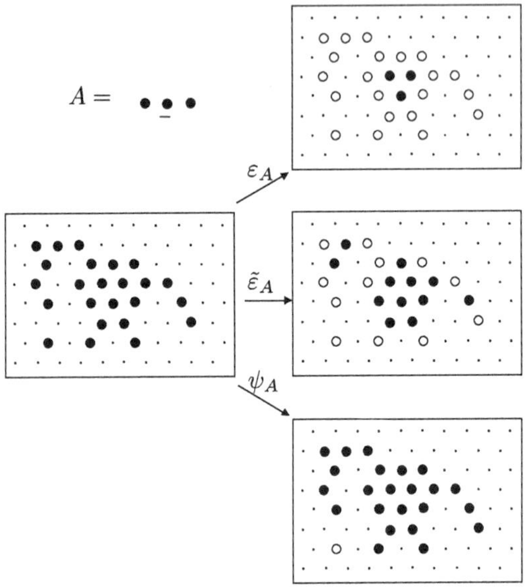

Figure 5.10 The erosion ε_A, the modified erosion $\tilde{\varepsilon}_A$, and the operator ψ_A.

If ψ is also rotation invariant, then $\psi = R_\varphi^{-1}\psi R_\varphi$ for every $\varphi \in [0, 2\pi)$; hence, by (5.111),

$$\psi(X) = R_\varphi^{-1}\Big(\bigcup_{A\in\mathcal{V}(\psi)} R_\varphi(X) \ominus A\Big)$$
$$= \bigcup_{A\in\mathcal{V}(\psi)} X \ominus R_\varphi(A).$$

Taking the union over all φ gives the result.

It is not difficult to find $\mathbb{T}$-operators which cannot be represented as a union of $\mathbb{T}$-erosions (and neither as an intersection of $\mathbb{T}$-dilations). For example, the modified erosion $\tilde{\varepsilon}_A$, where A is a structuring element which is not rotation symmetric, cannot be written as a union of $\mathbb{T}$-erosions ε_B. In Fig. 5.10 we have illustrated both types of erosions. Consider, as a second example, the operator

$$\psi_A(X) = \{h \in \mathbb{R}^2 \mid A \subseteq \Sigma(X_{-h})\}.$$

Evidently, ψ_A is a $\mathbb{T}$-operator. It is easy to check that $h \in \tilde{\varepsilon}_A(X)$ if and only if $A \subseteq R_\varphi(X_{-h})$ for some φ. On the other hand, $h \in \psi_A(X)$ if and only if $A \subseteq \Sigma(X_{-h})$. This implies that $\tilde{\varepsilon}_A \leq \psi_A$.

Assume that for some $B \subseteq \mathbb{R}^2$ we have $\varepsilon_B \leq \psi_A$, that is,

$$X \ominus \Sigma B \subseteq \psi_A(X).$$

Then $0 \in \Sigma B \ominus \Sigma B \subseteq \psi_A(\Sigma B)$, which means that $A \subseteq \Sigma B$. This implies that $\Sigma A \subseteq \Sigma B$, and hence that $\varepsilon_B \leq \varepsilon_A$. Therefore, if ψ_A were a union of $\mathbb{T}$-erosions, we would obtain $\psi_A = \varepsilon_A$; however, this identity is not true in general. A counterexample is provided by Fig. 5.10 (in terms of hexagonal motions). The reader can readily verify that for this particular structuring element the operator ψ_A deletes all isolated points.

5.10. Bibliographical notes

The extension of mathematical morphology to complete lattices was initiated by Matheron and Serra in (Serra, 1988); see also the bibliographical notes at the end of Chapter 3. Their work was extended by the author in (Heijmans, 1987); this paper is devoted to a study of $\mathbb{T}$-operators on a complete atomic lattice under the assumption that $\mathbb{T}$ is an abelian automorphism group. The polar operators treated in Section 5.6 occur for the first time in this paper. Roerdink and Heijmans (1988) discuss this example in more detail. A first comprehensive account of $\mathbb{T}$-adjunctions on complete lattices is given by Heijmans and Ronse (1990). In fact, Sections 5.1–5.7 are based on this paper. Some related ideas can be found in a paper by Maragos (1990).

In this chapter we used some elementary concepts, e.g., transitivity of groups and homogeneous spaces, from the theory of geometrical transformation groups. For background information on these concepts, the reader may refer to any textbook on geometry and group theory. For example, the books by Berger (1989) and Miller (1972) provide excellent introductions. Zorn's Lemma 5.25, which is equivalent to the Axiom of Choice, can be found in most textbooks on set theory; see, for example, Kuratowski (1972). Multiplicative structuring functions seem to occur for the first time in a paper by Herman (1988) who utilized them for edge enhancement in x-ray images.

The material in Sections 5.8 and 5.9 is due to Roerdink (1989, 1990a, 1990c, 1993). By the fact that an element $\tau \in \mathbb{T}$ can be regarded a left (choice made in Section 5.4) and right translation one can introduce left and right adjunctions on $\mathcal{P}(\mathbb{T})$. Since in our approach adjunctions on $\mathcal{P}(\mathbb{T})$ are only used as an intermediate step, namely, in the construction of adjunctions on $\mathcal{P}(E)$, it is immaterial which of them is used. Roerdink (1990b) examines $\mathbb{T}$-operators associated with the rotation group on the 3-sphere (cf. Example 5.32(c)) in considerable detail. Roerdink (1993) discusses the relevance of morphological $\mathbb{T}$-operators for shape description. In this paper one finds the following definition: two sets $X, Y \subseteq E$ have the same $\mathbb{T}$-shape if $Y = \tau(X)$ for some $\tau \in \mathbb{T}$. He points out the relation between this definition and Felix Klein's Erlanger Programm (Klein, 1872). Finally, Roerdink speculates about application of $\mathbb{T}$-operators to path planning problems. In yet another paper Roerdink (1994) extends morphology to arbitrary manifolds by requiring that operators are invariant under translations along geodesics.

CHAPTER SIX

Openings and closings

Henk J.A.M. Heijmans
Formerly Centre for Mathematics and Computer Science, Amsterdam, Netherlands

Contents

This chapter deals exclusively with openings and closings. The importance of these operators can hardly be overestimated. Definitions of openings and closings have already been given in Section 3.3; we recall these definitions here for the reader's convenience. An opening (resp. closing) is an increasing anti-extensive (resp. extensive) operator which is idempotent. It is obvious that openings and closings are dual notions in the sense of the Duality Principle: if ψ is an opening on the complete lattice $\mathcal{L}$, then it is a closing on the opposite lattice $\mathcal{L}'$, and vice versa. This chapter restricts attention to openings most of the time; the reader will have no difficulty in formulating the corresponding results for closings. Openings and closings on $\mathcal{P}(\mathbb{R}^d)$ which are translation invariant have been discussed in Section 4.4. Here the abstract approach initiated in the previous chapter will be extended; this leads naturally to two new concepts, namely, $\mathbb{T}$-openings and $\mathbb{T}$-closings; here $\mathbb{T}$ is an automorphism group on a complete lattice $\mathcal{L}$.

Sections 6.1–6.4 examine the algebraic properties of $\mathbb{T}$-openings. In particular, it discusses the one–one relation between an opening and its invariance domain. We introduce the notion of a structural opening and show that these constitute a basis for the class of all openings. Besides the structural openings, there is a second class which is important, the class of adjunctional openings. These are openings of the form $\delta\varepsilon$, where (ε, δ) is

Advances in Imaging and Electron Physics, Volume 216
ISSN 1076-5670
https://doi.org/10.1016/bs.aiep.2020.07.006

an adjunction on $\mathcal{L}$. In some cases, such as in the Euclidean translation invariant case, both classes coincide. More often, however, they are quite different. Sections 6.5–6.8 are devoted to annular openings, openings and inf-overfilters, and incidence structures, respectively.

6.1. Algebraic theory of $\mathbb{T}$-openings

The prototypical example of an opening is the operator α on $\mathcal{P}(E)$, E being some topological space, given by

$$\alpha(X) = X^\circ,$$

for $X \subseteq E$. Here X° denotes the interior of X. It is evident that this operator is increasing, anti-extensive, and idempotent; therefore, it is an opening. Recall from Section 3.1 that the *invariance domain* $\mathrm{Inv}(\alpha)$ of the operator α is defined as the set of fixpoints, i.e., $\mathrm{Inv}(\alpha) = \{X \mid \alpha(X) = X\}$. In the present example the invariance domain is the family $\mathcal{G}(E)$ of open subsets of E; note that this family is closed under union. For a given X, the interior X° is the union of all sets in $\mathcal{G}(E)$ that are subsets of X, that is,

$$\alpha(X) = \bigcup\{Y \in \mathcal{G}(E) \mid Y \subseteq X\}.$$

This section discusses in great detail the relation between an opening α on the one hand and its invariance domain $\mathrm{Inv}(\alpha)$ on the other. We start with a lemma which extends the results in Theorem 3.24(a).

6.1 Lemma. *Let α be an opening, and let ψ be an increasing anti-extensive operator. The following assertions are equivalent:*

(i) $\alpha \le \psi$;
(ii) $\alpha\psi = \alpha$;
(iii) $\psi\alpha = \alpha$;
(iv) $\mathrm{Inv}(\alpha) \subseteq \mathrm{Inv}(\psi)$.

Proof. (i) ⇒ (ii): As α is increasing and idempotent and $\alpha \le \psi \le \mathrm{id}$, it follows that $\alpha = \alpha\alpha \le \alpha\psi \le \alpha\,\mathrm{id} = \alpha$, that is, $\alpha = \alpha\psi$.

(ii) ⇒ (iii): As α and ψ are anti-extensive, one gets $\alpha\psi\alpha \le \psi\alpha \le \alpha$. Since $\alpha\psi = \alpha$, it follows that $\alpha\psi\alpha = \alpha\alpha = \alpha$. Therefore, $\alpha = \alpha\psi\alpha \le \psi\alpha \le \alpha$; in other words $\psi\alpha = \alpha$.

(iii) ⇒ (iv): If $X \in \mathrm{Inv}(\alpha)$, then $\alpha(X) = X$, and as $\psi\alpha = \alpha$ this gives $\psi(X) = \psi\alpha(X) = \alpha(X) = X$, that is, $X \in \mathrm{Inv}(\psi)$. Thus $\mathrm{Inv}(\alpha) \subseteq \mathrm{Inv}(\psi)$.

(iv) $\Rightarrow$ (i): For $X \in \mathcal{L}$, $\alpha(X) \in \mathrm{Inv}(\alpha)$, and as $\mathrm{Inv}(\alpha) \subseteq \mathrm{Inv}(\psi)$ it follows that $\psi\alpha(X) = \alpha(X)$. Thus $\psi\alpha = \alpha$; this means that $\alpha = \psi\alpha \leq \psi\,\mathrm{id} = \psi$. □

If one assumes that ψ is an opening as well, one obtains the following important result.

6.2 Corollary. *Let α_1, α_2 be openings on $\mathcal{L}$; then $\alpha_1 \leq \alpha_2$ if and only if* $\mathrm{Inv}(\alpha_1) \subseteq \mathrm{Inv}(\alpha_2)$. *In particular, $\alpha_1 = \alpha_2$ if and only if* $\mathrm{Inv}(\alpha_1) = \mathrm{Inv}(\alpha_2)$.

For completeness we recall the following result from Section 3.3; cf. Theorem 3.28.

6.3 Theorem.

(a) *If α_i is an opening for every $i \in I$, then $\bigvee_{i \in I} \alpha_i$ is an opening as well.*

(a′) *If β_i is a closing for every $i \in I$, then $\bigwedge_{i \in I} \beta_i$ is a closing as well.*

Let $\mathbb{T}$ be an arbitrary group of automorphisms on $\mathcal{L}$. In accordance with the convention introduced in Section 5.2, an opening (resp. closing) which is $\mathbb{T}$-invariant is called a $\mathbb{T}$-opening (resp. $\mathbb{T}$-closing). Define, for a set $\mathcal{M} \subseteq \mathcal{L}$,

$$\mathbb{T}\mathcal{M} = \{\tau(X) \mid X \in \mathcal{M}\ ,\ \tau \in \mathbb{T}\}.$$

The set $\mathcal{M}$ is said to be $\mathbb{T}$-*invariant* if $\mathbb{T}\mathcal{M} = \mathcal{M}$. Note that $\mathcal{M}$ is always a subset of $\mathbb{T}\mathcal{M}$ because $\mathrm{id} \in \mathbb{T}$. It is evident that $\mathbb{T}\mathcal{M}$ is $\mathbb{T}$-invariant; in fact, this is the smallest $\mathbb{T}$-invariant subset of $\mathcal{L}$ that contains $\mathcal{M}$. We recall the following notation from Section 2.1: $\langle \mathcal{M} \mid \vee \rangle$ denotes the smallest sup-closed subset of $\mathcal{L}$ containing $\mathcal{M}$; dually, $\langle \mathcal{M} \mid \wedge \rangle$ denotes the smallest inf-closed subset of $\mathcal{L}$ containing $\mathcal{M}$. It is easy to show that both $\langle \mathcal{M} \mid \vee \rangle$ and $\langle \mathcal{M} \mid \wedge \rangle$ are $\mathbb{T}$-invariant if $\mathcal{M}$ is. In other words,

$$\langle \mathbb{T}\mathcal{M} \mid \vee \rangle = \mathbb{T}\langle \mathbb{T}\mathcal{M} \mid \vee \rangle \quad \text{and} \quad \langle \mathbb{T}\mathcal{M} \mid \wedge \rangle = \mathbb{T}\langle \mathbb{T}\mathcal{M} \mid \wedge \rangle, \tag{6.1}$$

for every $\mathcal{M} \subseteq \mathcal{L}$. Note however that the set $\mathbb{T}\langle \mathcal{M} \mid \vee \rangle$ can be strictly smaller than $\langle \mathbb{T}\mathcal{M} \mid \vee \rangle$. For example, let $\mathcal{L} = \mathcal{P}(\mathbb{R}^d)$, $\mathbb{T}$ the group of translations, and $\mathcal{M}$ the open balls centred at the origin. Then $\langle \mathbb{T}\mathcal{M} \mid \vee \rangle$ contains all open subsets of $\mathbb{R}^d$, whereas $\mathbb{T}\langle \mathcal{M} \mid \vee \rangle$ contains only the open balls.

Define, for $\mathcal{M} \subseteq \mathcal{L}$, the operator $\alpha^{\mathbb{T}}(\mathcal{M})$ on $\mathcal{L}$ by

$$\alpha^{\mathbb{T}}(\mathcal{M})(X) = \bigvee \{\tau(A) \mid A \in \mathcal{M}\ ,\ \tau \in \mathbb{T}\ ,\ \tau(A) \leq X\}, \tag{6.2}$$

for $X \in \mathcal{L}$. For the sake of brevity we write $\alpha(\mathcal{M})$ instead of $\alpha^{\{\mathrm{id}\}}(\mathcal{M})$. If, for example, $\mathcal{L}$ is the Boolean lattice $\mathcal{P}(\mathbb{R}^d)$, $\mathbb{T}$ the translation group and

$\mathcal{M}$ the open balls centred at the origin, then $\alpha^{\mathbb{T}}(\mathcal{M})(X)$ is the union of all open balls that are contained in X; in other words, $\alpha^{\mathbb{T}}(\mathcal{M})(X) = X^\circ$.

6.4 Proposition. *For every* $\mathcal{M} \subseteq \mathcal{L}$,

$$\alpha^{\mathbb{T}}(\mathcal{M}) = \alpha(\mathbb{T}\mathcal{M}); \tag{6.3}$$

this operator defines a $\mathbb{T}$*-opening with invariance domain* $\langle \mathbb{T}\mathcal{M} \mid \vee \rangle$.

Proof. Put $\alpha = \alpha^{\mathbb{T}}(\mathcal{M})$. From (6.2) it follows immediately that

$$\alpha(X) = \bigvee \{A \mid A \in \mathbb{T}\mathcal{M},\ A \leq X\}; \tag{6.4}$$

hence (6.3) follows. It is easy to check that α defines an increasing anti-extensive $\mathbb{T}$-operator and that $\alpha(A) = A$ for $A \in \mathbb{T}\mathcal{M}$. Hence, if $A \leq X$ for some $A \in \mathbb{T}\mathcal{M}$, then, since α is increasing, $A = \alpha(A) \leq \alpha(X)$; this means $\alpha^2(X) \geq \alpha(X)$. Since α is anti-extensive, $\alpha^2 \leq \alpha$. Therefore, α is idempotent, and hence an opening. From (6.4) it follows immediately that $\mathrm{Inv}(\alpha) \subseteq \langle \mathbb{T}\mathcal{M} \mid \vee \rangle$. On the other hand, since $\mathbb{T}\mathcal{M} \subseteq \mathrm{Inv}(\alpha)$ and $\mathrm{Inv}(\alpha)$ is sup-closed, $\langle \mathbb{T}\mathcal{M} \mid \vee \rangle \subseteq \mathrm{Inv}(\alpha)$, and the theorem is proved. □

The opening $\alpha^{\mathbb{T}}(\mathcal{M})$ coincides with the identity operator on $\mathcal{M}$; therefore $\alpha^{\mathbb{T}}(\mathcal{M})$ is sometimes called the *least extension of the identity on* $\mathcal{M}$. The next result is an immediate consequence of the fact that an opening is completely determined by its invariance domain.

6.5 Proposition. *If* $\mathcal{M}, \mathcal{M}' \subseteq \mathcal{L}$ *and* $\langle \mathbb{T}\mathcal{M} \mid \vee \rangle = \langle \mathbb{T}\mathcal{M}' \mid \vee \rangle$, *then* $\alpha^{\mathbb{T}}(\mathcal{M}) = \alpha^{\mathbb{T}}(\mathcal{M}')$. *In particular,*

$$\alpha^{\mathbb{T}}(\mathcal{M}) = \alpha^{\mathbb{T}}(\tau\mathcal{M}) \qquad \text{for } \tau \in \mathbb{T}. \tag{6.5}$$

The *greatest extension of the identity on* $\mathcal{M}$ is denoted by $\beta^{\mathbb{T}}(\mathcal{M})$ and is given by

$$\beta^{\mathbb{T}}(\mathcal{M})(X) = \bigwedge \{\tau(A) \mid A \in \mathcal{M},\ \tau \in \mathbb{T},\ X \leq \tau(A)\}. \tag{6.6}$$

Analogous to Proposition 6.4 one can show that $\beta^{\mathbb{T}}(\mathcal{M})$ is a $\mathbb{T}$-closing with invariance domain $\langle \mathbb{T}\mathcal{M} \mid \wedge \rangle$.

For an operator ψ we define the *extensivity domain* $\mathrm{Inv}_+(\psi)$ and the *anti-extensivity domain* $\mathrm{Inv}_-(\psi)$ by

$$\mathrm{Inv}_+(\psi) = \{X \in \mathcal{L} \mid \psi(X) \geq X\}, \tag{6.7}$$

$$\mathrm{Inv}_{-}(\psi) = \{X \in \mathcal{L} \mid \psi(X) \le X\}. \tag{6.8}$$

Clearly, $\psi(X) \ge X$ is equivalent to $X \wedge \psi(X) = X$, and similarly, $\psi(X) \le X$ is equivalent to $X \vee \psi(X) = X$. Therefore

$$\mathrm{Inv}_{+}(\psi) = \mathrm{Inv}(\mathrm{id} \wedge \psi), \qquad \mathrm{Inv}_{-}(\psi) = \mathrm{Inv}(\mathrm{id} \vee \psi). \tag{6.9}$$

6.6 Proposition. *Let ψ be an increasing operator on $\mathcal{L}$.*
(a) $\mathrm{Inv}_{+}(\psi)$ *is sup-closed;*
(a′) $\mathrm{Inv}_{-}(\psi)$ *is inf-closed.*

Proof. Let $X_i \in \mathrm{Inv}_{+}(\psi)$ for $i \in I$, and define $X = \bigvee_{i\in I} X_i$. Then $\psi(X) \ge \psi(X_i) \ge X_i$ for $i \in I$, and hence $\psi(X) \ge \bigvee_{i\in I} X_i = X$. This proves the result. □

If ψ is anti-extensive, then $\psi = \mathrm{id} \wedge \psi$, and a combination of (6.9) and Proposition 6.6 gives the following result.

6.7 Proposition. *Let ψ be an increasing operator.*
(a) *If ψ is anti-extensive, then* $\mathrm{Inv}(\psi)$ *is sup-closed.*
(a′) *If ψ is extensive, then* $\mathrm{Inv}(\psi)$ *is inf-closed.*

Denote by $\mathcal{P}(\mathcal{L})$ the complete Boolean lattice of all subsets of $\mathcal{L}$ ordered by inclusion $\subseteq$, and let $\mathcal{P}'(\mathcal{L})$ be its dual, i.e., the subsets of $\mathcal{L}$ ordered by the opposite inclusion $\supseteq$. The next result describes the relationship between an opening and its invariance domain.

6.8 Proposition.
(a) *The pair* $(\mathrm{Inv}_{+}, \alpha^{\mathbb{T}})$ *defines an adjunction between $\mathcal{O}^{+}_{\mathbb{T}}(\mathcal{L})$ and $\mathcal{P}(\mathcal{L})$.*
(a′) *The pair* $(\beta^{\mathbb{T}}, \mathrm{Inv}_{-})$ *defines an adjunction between $\mathcal{P}'(\mathcal{L})$ and $\mathcal{O}^{+}_{\mathbb{T}}(\mathcal{L})$.*

Proof. We must show that

$$\alpha^{\mathbb{T}}(\mathcal{M}) \le \psi \iff \mathcal{M} \subseteq \mathrm{Inv}_{+}(\psi),$$

for every $\psi \in \mathcal{O}^{+}_{\mathbb{T}}(\mathcal{L})$ and $\mathcal{M} \subseteq \mathcal{L}$.

$\Rightarrow$: Assume that $\alpha^{\mathbb{T}}(\mathcal{M}) \le \psi$ and $X \in \mathcal{M}$. We must show that $\psi(X) \ge X$. From $X \in \mathcal{M}$ we conclude that $\alpha^{\mathbb{T}}(\mathcal{M})(X) = X$, whence the assertion follows.

$\Leftarrow$: Assume $\mathcal{M} \subseteq \mathrm{Inv}_{+}(\psi)$. Since the set $\mathrm{Inv}_{+}(\psi)$ is sup-closed and $\mathbb{T}$-invariant, it follows that $\langle \mathbb{T}\mathcal{M} \mid \vee \rangle \subseteq \mathrm{Inv}_{+}(\psi)$. Therefore, if $X \in \mathcal{L}$, then $\alpha^{\mathbb{T}}(\mathcal{M})(X) \in \mathrm{Inv}_{+}(\psi)$; hence $\psi(X) \ge \psi(\alpha^{\mathbb{T}}(\mathcal{M})(X)) = \alpha^{\mathbb{T}}(\mathcal{M})(X)$. This concludes the proof. □

This result leads to the following identities:

$$\alpha^{\mathbb{T}}(\bigcup_{i\in I}\mathcal{M}_i)=\bigvee_{i\in I}\alpha^{\mathbb{T}}(\mathcal{M}_i), \tag{6.10}$$

$$\beta^{\mathbb{T}}(\bigcup_{i\in I}\mathcal{M}_i)=\bigwedge_{i\in I}\beta^{\mathbb{T}}(\mathcal{M}_i); \tag{6.11}$$

here $\mathcal{M}_i \subseteq \mathcal{L}$ for $i \in I$. We mention some further implications of Proposition 6.8. An important consequence is the following theorem, which states that any opening (closing) can be retrieved from its invariance domain. This result can be interpreted as an extension of Corollary 6.2, which says that openings are uniquely determined by their invariance domain. Note that for $\mathbb{T}=\{\mathrm{id}\}$ this result coincides with Theorem 3.23.

6.9 Theorem.

(a) *If α is a $\mathbb{T}$-opening, then $\alpha=\alpha^{\mathbb{T}}(\mathrm{Inv}(\alpha))$.*

(a′) *If β is a $\mathbb{T}$-closing, then $\beta=\beta^{\mathbb{T}}(\mathrm{Inv}(\beta))$.*

Proof. Let α be an opening, and put $\alpha'=\alpha^{\mathbb{T}}(\mathrm{Inv}(\alpha))$. Applying Inv_+ on both sides and using that for anti-extensive operators the invariance domain and extensivity domain coincide, one arrives at the identity

$$\mathrm{Inv}(\alpha')=\mathrm{Inv}_+(\alpha^{\mathbb{T}}(\mathrm{Inv}_+(\alpha))).$$

Since the pair $(\mathrm{Inv}_+,\alpha^{\mathbb{T}})$ is an adjunction, the right-hand side equals $\mathrm{Inv}_+(\alpha)=\mathrm{Inv}(\alpha)$ (cf. (3.7)). Application of Corollary 6.2 gives $\alpha=\alpha'$. □

A trivial but important consequence of this result is that every opening α obeys the equation

$$\alpha(X)=\bigvee\{A\in\mathrm{Inv}(\alpha)\mid A\le X\}, \tag{6.12}$$

and dually, that every closing β obeys

$$\beta(X)=\bigwedge\{B\in\mathrm{Inv}(\beta)\mid X\le B\}; \tag{6.13}$$

see Theorem 3.23.

By Theorem 3.25 the operator $\alpha^{\mathbb{T}}\circ\mathrm{Inv}_+$ defines an opening on $\mathcal{O}^+_{\mathbb{T}}(\mathcal{L})$ which maps every increasing $\mathbb{T}$-operator ψ onto a $\mathbb{T}$-opening $\check{\psi}$, called the *lower envelope of ψ*. It follows immediately that

$$\mathrm{Inv}(\check{\psi})=\mathrm{Inv}_+(\psi)=\mathrm{Inv}(\mathrm{id}\wedge\psi). \tag{6.14}$$

The lower envelope plays an important role in Chapter 12. There it is shown among other things that $\check{\psi}$ is the largest opening $\leq \psi$. Note, however, that this result can also be established with the results mentioned previously.

Dually, the operator $\beta^{\mathbb{T}} \circ \mathrm{Inv}_-$ defines a closing on $\mathcal{O}^+_{\mathbb{T}}(\mathcal{L})$ which maps an increasing $\mathbb{T}$-operator ψ onto a $\mathbb{T}$-closing $\hat{\psi}$, called the *upper envelope of* ψ. We have

$$\mathrm{Inv}(\hat{\psi}) = \mathrm{Inv}_-(\psi) = \mathrm{Inv}(\mathrm{id} \vee \psi), \tag{6.15}$$

and in Section 12.2 it will be demonstrated that $\hat{\psi}$ is the smallest closing $\geq \psi$. Theorem 3.25 gives also that $\mathrm{Inv}_+ \circ \alpha^{\mathbb{T}}$ is a closing on $\mathcal{P}(\mathcal{L})$. It follows with Proposition 6.4 that it maps a set $\mathcal{M} \subseteq \mathcal{L}$ onto $\langle \mathbb{T}\mathcal{M} \mid \vee \rangle$. Dually, $\mathrm{Inv}_- \circ \beta^{\mathbb{T}}$ is an opening on $\mathcal{P}'(\mathcal{L})$, and therefore a closing on $\mathcal{P}(\mathcal{L})$; it maps a set $\mathcal{M} \subseteq \mathcal{L}$ onto $\langle \mathbb{T}\mathcal{M} \mid \wedge \rangle$.

An important implication of the identity (6.10) is that the opening $\alpha^{\mathbb{T}}(\mathcal{M})$ can be written as

$$\alpha^{\mathbb{T}}(\mathcal{M}) = \bigvee_{A \in \mathcal{M}} \alpha^{\mathbb{T}}(\{A\}).$$

The operator

$$\alpha^{\mathbb{T}}_A = \alpha^{\mathbb{T}}(\{A\}) \tag{6.16}$$

is called the *structural* $\mathbb{T}$-*opening* by A. The invariance domain of $\alpha^{\mathbb{T}}_A$ is $\langle \mathbb{T}A \mid \vee \rangle$, where $\mathbb{T}A = \{\tau(A) \mid \tau \in \mathbb{T}\}$. When $\mathbb{T} = \{\mathrm{id}\}$ we write α_A instead of $\alpha^{\{\mathrm{id}\}}_A$. Dually, the *structural* $\mathbb{T}$-*closing* by A is defined by

$$\beta^{\mathbb{T}}_A = \beta^{\mathbb{T}}(\{A\}). \tag{6.17}$$

Illustrations of the structural opening and closing in the translation invariant case can be found in Section 4.4. The following result is a direct consequence of Theorem 6.9 and relations (6.10)–(6.11); a particular case is given in Theorem 4.24.

6.10 Theorem.

(a) *If* α *is a* $\mathbb{T}$-*opening, then*

$$\alpha = \bigvee_{A \in \mathrm{Inv}(\alpha)} \alpha^{\mathbb{T}}_A. \tag{6.18}$$

(a′) *If β is a $\mathbb{T}$-closing, then*

$$\beta = \bigwedge_{A \in \mathrm{Inv}(\beta)} \beta_A^{\mathbb{T}}. \tag{6.19}$$

From (6.5) and its analogue for closings, one achieves that

$$\alpha_A^{\mathbb{T}} = \alpha_{\tau(A)}^{\mathbb{T}} \quad \text{and} \quad \beta_A^{\mathbb{T}} = \beta_{\tau(A)}^{\mathbb{T}}, \qquad \text{for } \tau \in \mathbb{T}. \tag{6.20}$$

This means in particular that the representations in Theorem 6.10 are redundant. For $A \in \mathrm{Inv}(\alpha)$ implies that $\tau(A) \in \mathrm{Inv}(\alpha)$ as well, but according to (6.20) the expansion (6.18) of α as a supremum of structural openings requires only one member of the family $\{\tau(A) \mid \tau \in \mathbb{T}\}$. Yet this is not the only redundancy in this expansion. Assume that $A \in \mathrm{Inv}(\alpha)$ and that $B = \bigvee_{i \in I} \tau_i(A)$ for some collection $\{\tau_i \mid i \in I\} \subseteq \mathbb{T}$. Then $B \in \mathrm{Inv}(\alpha)$ as well, since the invariance domain is $\mathbb{T}$-invariant and closed under suprema. It follows immediately, however, that $\alpha_B^{\mathbb{T}} \le \alpha_A^{\mathbb{T}}$. This implies that the structural opening $\alpha_B^{\mathbb{T}}$ is redundant in the expansion (6.18). In fact, B satisfies the identity

$$\alpha_A^{\mathbb{T}}(B) = B,$$

and we say that *B is A-open* (with respect to $\mathbb{T}$). Dually, B is said to be *A-closed* (with respect to $\mathbb{T}$) if

$$\beta_A^{\mathbb{T}}(B) = B.$$

The latter identity means that B can be written as an infimum of transforms $\tau(A)$ of A.

6.11 Proposition. *Let $A, B \in \mathcal{L}$.*
(a) *$\alpha_B^{\mathbb{T}} \le \alpha_A^{\mathbb{T}}$ if and only if B is A-open with respect to $\mathbb{T}$.*
(a′) *$\beta_B^{\mathbb{T}} \ge \beta_A^{\mathbb{T}}$ if and only if B is A-closed with respect to $\mathbb{T}$.*

Proof. (a): The if-part follows from the foregoing arguments. To prove the only if-part assume that $\alpha_B^{\mathbb{T}} \le \alpha_A^{\mathbb{T}}$. Then $B = \alpha_B^{\mathbb{T}}(B) \le \alpha_A^{\mathbb{T}}(B)$. By virtue of the anti-extensivity of $\alpha_A^{\mathbb{T}}$, $\alpha_A^{\mathbb{T}}(B) \le B$, and from this equality follows. This concludes the proof. □

6.12 Example. Let $p : \mathcal{L} \to \{0, 1\}$ be an increasing function. The operator α_p given by

$$\alpha_p(X) = \begin{cases} X, & \text{if } p(X) = 1, \\ \varnothing, & \text{if } p(X) = 0, \end{cases}$$

is an opening. On $\mathcal{P}(E)$, for example, one can take $p(X) = [X \cap A \neq \varnothing]$, where A is a given set.

On $\mathcal{F}(\mathbb{R}^d)$ one can choose $p(X) = [\text{area}(X) \geq a]$; here area$(X)$ denotes the Lebesgue measure of X and $a > 0$.

6.13 Example. (Generalized convex hull)
Edelsbrunner et al. (1983) introduce a generalization of the convex hull of a finite set of points in the plane. The present example shows how their definition fits into the framework of structural closings. First, however, we recall from Example 2.16(b) the following definition: the closed convex hull $\overline{\text{co}}(X)$ of a set $X \subseteq \mathbb{R}^2$ is the smallest closed convex set that contains X. In Proposition 9.7 it is shown that $\overline{\text{co}}(X)$ is the intersection of all closed half-planes that contain X. Let L be a closed half-plane in $\mathbb{R}^2$. The operator

$$\beta_L(X) = \bigcap\{L_h \mid h \in \mathbb{R}^2 \text{ and } X \subseteq L_h\}$$

is a structural translation invariant closing on $\mathcal{P}(\mathbb{R}^2)$ and

$$\overline{\text{co}}(X) = \bigcap\{\beta_L(X) \mid L \text{ is a closed half-plane}\}.$$

The idea of Edelsbrunner et al. (1983) is that the half-plane is replaced by some other geometrical shape. In fact they choose the interior and the exterior of a circle. Define, for $p \in \mathbb{R}$, the closings β_p on $\mathcal{P}(\mathbb{R}^2)$ as follows: if $p \neq 0$, then $\beta_p(X)$ is the intersection of all closed disks with radius $1/p$ which contain X; if $p = 0$, then $\beta_0(X)$ is the intersection of all closed half-planes which contain X. Here a closed disk with radius $r < 0$ is defined as the closed outer region of a circle with radius $-r$; see Fig. 6.1 for an illustration.

Furthermore, we use the convention that an empty intersection is $\mathbb{R}^2$. Fig. 6.1 illustrates the effect of these closings on a particular set X. It is obvious that $\beta_p(X)$ increases with p. This implies

$$\beta_p\beta_q = \beta_q\beta_p = \beta_p \qquad \text{if } p \geq q.$$

Edelsbrunner et al. (1983) call $\beta_p(X)$ the *p-convex hull* of X.

6.14 Example. (Polar openings)
In Section 5.6 we have studied $\mathbb{T}$-adjunctions, $\mathbb{T}$ being the group of rotations and scalar multiplications in the plane. In this example we explain by means of a pictorial example what the effect of this group structure is on the resulting openings; see Fig. 6.2.

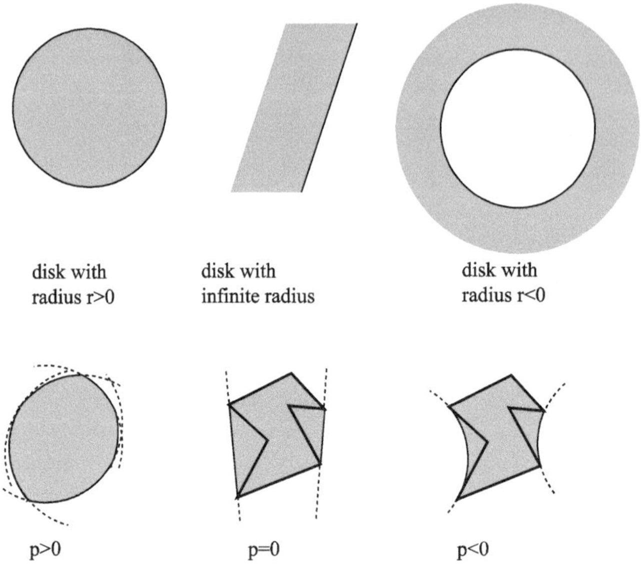

Figure 6.1 Top: generalized closed disks. Bottom: the closings $\beta_p(X)$ for $p > 0,\ p = 0$, and $p < 0$.

Let A be the disk of radius r $(0 < r < 1)$ and centre $(1, 0)$. Assume that the set $X \subseteq \mathbb{R}^2 \setminus \{0\}$ comprises disks with different radii. If $\mathbb{T}$ is the group of translations, then $\alpha_A^{\mathbb{T}}$ preserves all disks with radius $\geq r$. In the polar case, however, a disk is preserved if its radius is $\geq d \cdot r$, where d is the distance between the centre of the disk and the origin.

6.2. Self-dual and Boolean lattices

As already outlined, openings and closings are dual notions in the sense of the Duality Principle. If the lattice $\mathcal{L}$ is Boolean or, more generally, self-dual, then there exists a second duality relation between openings and closings. Suppose that there exists a negation ν on $\mathcal{L}$; as usual we put $X^* = \nu(X)$ and $\psi^* = \nu\psi\nu$. Define $\mathcal{M}^* = \{X^* \mid X \in \mathcal{M}\}$ for $\mathcal{M} \subseteq \mathcal{L}$. Let ψ be an arbitrary operator on $\mathcal{L}$ and $X \in \mathrm{Inv}_+(\psi)$; then $\psi^*(X^*) = [\psi(X)]^* \leq X^*$, and hence $X^* \in \mathrm{Inv}_-(\psi^*)$. Dually, if $X \in \mathrm{Inv}_-(\psi)$, then $X^* \in \mathrm{Inv}_+(\psi^*)$. Therefore, the following identities hold:

$$\left[\mathrm{Inv}_+(\psi)\right]^* = \mathrm{Inv}_-(\psi^*) \qquad \text{and} \qquad \left[\mathrm{Inv}_-(\psi)\right]^* = \mathrm{Inv}_+(\psi^*). \tag{6.21}$$

Recall the definition of a $\mathbb{T}$-compatible negation from Definition 5.17.

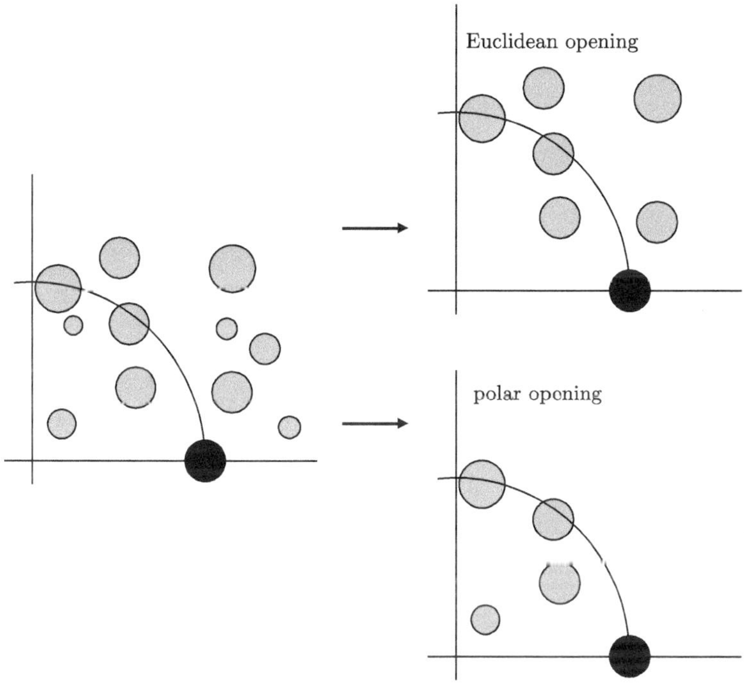

Figure 6.2 Euclidean opening versus polar opening. Left: the set X (grey) and the structuring element (black). Right: the Euclidean opening and the polar opening.

6.15 Theorem. *Assume there exists a* $\mathbb{T}$*-compatible negation on* $\mathcal{L}$*; then*

$$[\alpha^{\mathbb{T}}(\mathcal{M})]^* = \beta^{\mathbb{T}}(\mathcal{M}^*) \qquad \textit{and} \qquad [\beta^{\mathbb{T}}(\mathcal{M})]^* = \alpha^{\mathbb{T}}(\mathcal{M}^*), \tag{6.22}$$

for every $\mathcal{M} \subseteq \mathcal{L}$.

Proof. We prove only the first identity. Let $X \in \mathcal{L}$; then

$$\begin{aligned}
[\alpha^{\mathbb{T}}(\mathcal{M})(X^*)]^* &= \Big[\bigvee\{\tau(A) \mid A \in \mathcal{M}\,,\ \tau \in \mathbb{T}\,,\ \tau(A) \le X^*\}\Big]^* \\
&= \bigwedge\{(\tau(A))^* \mid A \in \mathcal{M}\,,\ \tau \in \mathbb{T}\,,\ (\tau(A))^* \ge X\} \\
&= \bigwedge\{\tau^*(A^*) \mid A \in \mathcal{M}\,,\ \tau \in \mathbb{T}\,,\ \tau^*(A^*) \ge X\} \\
&= \bigwedge\{\tau(B) \mid B \in \mathcal{M}^*\,,\ \tau \in \mathbb{T}\,,\ \tau(B) \ge X\} \\
&= \beta^{\mathbb{T}}(\mathcal{M}^*)(X);
\end{aligned}$$

this proves the assertion. Here we have used in the fourth identity that τ^* ranges over the whole group $\mathbb{T}$ if τ does. □

For structural openings and closings (6.22) specializes to

$$(\alpha_A^{\mathbb{T}})^* = \beta_{A^*}^{\mathbb{T}} \qquad \text{and} \qquad (\beta_A^{\mathbb{T}})^* = \alpha_{A^*}^{\mathbb{T}}. \tag{6.23}$$

Recall that $\check{\psi} = \alpha^{\mathbb{T}}(\mathrm{Inv}_+(\psi))$; combination of (6.21) and (6.22) gives

$$(\psi^*)^{\vee} = \alpha^{\mathbb{T}}(\mathrm{Inv}_+(\psi^*)) = \alpha^{\mathbb{T}}([\mathrm{Inv}_-(\psi)]^*) = \left[\beta^{\mathbb{T}}(\mathrm{Inv}_-(\psi))\right]^*.$$

Using that $\hat{\psi} = \beta^{\mathbb{T}}(\mathrm{Inv}_-(\psi))$, it follows that

$$(\psi^*)^{\vee} = (\hat{\psi})^* \qquad \text{and} \qquad (\psi^*)^{\wedge} = (\check{\psi})^*. \tag{6.24}$$

A completely different proof of these identities will be given in Section 12.2.

6.3. Adjunctional openings and closings

If (ε, δ) is an adjunction between the complete lattices $\mathcal{L}$ and $\mathcal{M}$, then $\delta\varepsilon$ is an opening on $\mathcal{L}$ with invariance domain $\mathrm{Ran}(\delta)$ and $\varepsilon\delta$ is a closing on $\mathcal{M}$ with invariance domain $\mathrm{Ran}(\varepsilon)$; see Theorem 3.25. In Example 6.19 it will be demonstrated that every opening α on $\mathcal{L}$ can be written as a composition of an erosion and a dilation which constitute an adjunction between $\mathcal{L}$ and some other complete lattice $\mathcal{M}$. This is no longer true if it is required that $\mathcal{L} = \mathcal{M}$. Recall the following concepts from Definition 3.26. Let (ε, δ) be an adjunction on $\mathcal{L}$; then $\delta\varepsilon$ is called an *adjunctional opening* and $\varepsilon\delta$ an *adjunctional closing*.

Two major problems considered in this section are:

- What is the relation between structural and adjunctional openings (closings)?
- Can every opening (closing) be represented as a supremum of adjunctional openings (infimum of adjunctional closings)?

Again we restrict attention to openings; analogous results for closings follow from the Duality Principle. In Section 4.4 (in particular, in Remark 4.18(a)) it has been argued that in the Euclidean case both classes of openings and closings coincide. In fact the following identities hold in this case:

$$\alpha_A^{\mathbb{T}}(X) = (X \ominus A) \oplus A \qquad \text{and} \qquad \beta_A^{\mathbb{T}}(X) = (X \oplus A) \ominus A;$$

here $\mathbb{T}$ denotes the translation group. Therefore, the questions we have posed have an affirmative answer in this special case.

In general, the problem is more complex; in fact, consider the case where $\mathcal{L}$ is the Boolean lattice $\mathcal{P}(E)$ and where $\mathbb{T} = \{\mathrm{id}\}$. Let $B \subseteq E$; the structural opening α_B is given by

$$\alpha_B(X) = \begin{cases} B, & \text{if } B \subseteq X, \\ \varnothing, & \text{otherwise.} \end{cases}$$

Define

$$\varepsilon_B(X) = \begin{cases} E, & \text{if } B \subseteq X, \\ \varnothing, & \text{otherwise,} \end{cases} \qquad \delta_B(X) = \begin{cases} \varnothing, & \text{if } X = \varnothing, \\ B, & \text{if } X \neq \varnothing; \end{cases}$$

one can easily check that $(\varepsilon_B, \delta_B)$ is an adjunction on $\mathcal{P}(E)$ and that $\delta_B\varepsilon_B = \alpha_B$. So in this case every structural opening is an adjunctional opening. In combination with Theorem 6.10(a) this implies in particular that every opening is a supremum of adjunctional openings. However, there are many adjunctional openings that are not structural. Take, for example, $E = \mathbb{R}^d$. It is obvious that the opening $X \mapsto X \circ A$, where $A \subseteq \mathbb{R}^d$, is an adjunctional opening. However, this opening is not structural with respect to the trivial automorphism group $\{\mathrm{id}\}$.

These examples indicate clearly the role of $\mathbb{T}$. To stress this role even further we anticipate on the results of Section 6.4, where it is shown that there are more structural openings than adjunctional openings if $\mathbb{T}$ is multi-transitive.

In Section 5.2 we have introduced the Basic Assumption 5.10, which requires that the lattice $\mathcal{L}$ contains a sup-generating family ℓ on which $\mathbb{T}$ acts transitively. Under this assumption one can define a Minkowski addition and subtraction on $\mathcal{L}$ and show that every $\mathbb{T}$-adjunction is of the form $(\varepsilon_A, \delta_A)$ with $\delta_A(X) = X \oplus A$ and $\varepsilon_A(X) = X \ominus A$.

6.16 Theorem. *Let $\mathcal{L}$ be a complete lattice for which the Basic Assumption holds. Then the classes of adjunctional and structural $\mathbb{T}$-openings coincide. To be precise,*

$$\alpha_A^{\mathbb{T}}(X) = (X \ominus A) \oplus A, \tag{6.25}$$

for $A, X \in \mathcal{L}$.

Proof. The invariance domain of the adjunctional opening $\delta_A\varepsilon_A$ equals $\mathrm{Ran}(\delta_A) = \{X \oplus A \mid X \in \mathcal{L}\} = \{A \oplus X \mid X \in \mathcal{L}\}$. This is the subset of $\mathcal{L}$

which contains A and is $\mathbb{T}$-invariant and sup-closed. In other words, $\mathrm{Inv}(\delta_A \varepsilon_A) = \mathrm{Inv}(\alpha_A^{\mathbb{T}})$. But since, by Corollary 6.2, an opening is uniquely determined by its invariance domain it follows that $\alpha_A^{\mathbb{T}} = \delta_A \varepsilon_A$, which was to be proved. □

A similar result holds for closings if the dual Basic Assumption holds; see Remark 5.11. As we pointed out at the end of Section 5.2, however, $\mathbb{T}$-adjunctions use different Minkowski additions and subtractions in this case.

Now we present two examples (for closings) which show that the (dual) Basic Assumption is indeed essential.

6.17 Example. (The closed sets $\mathcal{F}(\mathbb{R}^d)$)
This example is a continuation of Section 5.5.2. Take $\mathcal{L} = \mathcal{F}(\mathbb{R}^d)$, ℓ the singletons, and $\mathbb{T}$ the translation group. It has been demonstrated that every $\mathbb{T}$-adjunction on $\mathcal{F}(\mathbb{R}^d)$ is of the form $(\varepsilon_A, \delta_A)$, where

$$\delta_A(X) = X \overline{\oplus} A = \overline{\bigcup_{a \in A} X_a},$$
$$\varepsilon_A(X) = X \overline{\ominus} A = \bigcap_{a \in A} X_{-a}.$$

Let ψ be the structural closing defined in (5.50), that is, $\psi = \beta_C^{\mathbb{T}}$ where $C = \{x \in \mathbb{R}^d \mid |x| \geq 1\}$. Suppose that ψ is an adjunctional closing; in other words, ψ is of the form $\varepsilon_A \delta_A$ for some $A \in \mathcal{F}(\mathbb{R}^d)$. Then

$$\psi(X) = (X \overline{\oplus} A) \overline{\ominus} A = \bigcap_{h \in A} (X \overline{\oplus} A)_{-h} = \bigcap_{h \in A} (X \overline{\oplus} A_{-h});$$

in other words, $\psi = \bigwedge_{h \in A} \delta_{A_{-h}}$. In Section 5.5.2, however, it has been argued that ψ cannot be represented as an intersection of dilations, and we conclude that ψ is not an adjunctional closing. The same argument shows that it is also not possible to decompose ψ as an infimum of adjunctional closings.

6.18 Example. (Convex sets)
Consider the complete lattice $\mathcal{C}(\mathbb{R}^2)$ of convex subsets of $\mathbb{R}^2$; let ℓ be the singletons and $\mathbb{T}$ the translation group. In Section 5.5.3, it was proved that every $\mathbb{T}$-adjunction has the form $(\varepsilon_A, \delta_A)$ given by

$$\delta_A(X) = X \oplus A\,, \qquad \varepsilon_A(X) = X \ominus A,$$

where $\oplus, \ominus$ denote the ordinary Minkowski sum and difference and where $A \subseteq \mathbb{R}^2$ is convex. Let L be the nonnegative x-axis; the closing $\beta = \beta_L^{\mathbb{T}}$ maps a subset X onto $\mathbb{R}^2$ if X is not contained inside some translate of L. Suppose there exists a convex set A such that

$$\beta(X) = (X \oplus A) \ominus A,$$

for every convex set X. It is clear that A must be nonempty and that A cannot be the entire space $\mathbb{R}^2$. Let D be the unit disk; then $\beta(D) = \mathbb{R}^2$, i.e., $(D \oplus A) \ominus A = \mathbb{R}^2$. This means $((D \oplus A) \ominus A) \oplus A = D \oplus A = \mathbb{R}^2$. This is possible only if $A = \mathbb{R}^2$, a contradiction. Therefore, β is not an adjunctional closing.

6.19 Example. We show that every opening α on a complete lattice $\mathcal{L}$ can be obtained as a composition $\delta\varepsilon$ where (ε, δ) is an adjunction between $\mathcal{L}$ and some other complete lattice $\mathcal{M}$. More specifically, we choose $\mathcal{M} = \mathcal{P}(\mathrm{Inv}(\alpha))$ and define $\varepsilon : \mathcal{L} \to \mathcal{M}$ by

$$\varepsilon(X) = \{A \in \mathrm{Inv}(\alpha) \mid A \leq X\}$$

and $\delta : \mathcal{M} \to \mathcal{L}$ by

$$\delta(\mathcal{H}) = \bigvee \mathcal{H}.$$

It is easy to check that

$$\delta(\mathcal{H}) \leq X \iff \mathcal{H} \subseteq \varepsilon(X),$$

for $X \in \mathcal{L}$ and $\mathcal{H} \subseteq \mathrm{Inv}(\alpha)$. This shows that indeed (ε, δ) defines an adjunction. Using (6.12), it is obvious that $\delta\varepsilon(X) = \bigvee\{A \in \mathrm{Inv}(\alpha) \mid A \leq X\} = \alpha(X)$. For closings an analogous result holds.

6.4. $\mathbb{T}$-openings: the nonabelian case

Section 5.8 contains a comprehensive discussion of $\mathbb{T}$-adjunctions on $\mathcal{P}(E)$ under the assumption that the pair $(\mathbb{T}, E)$ constitutes a homogeneous space, that is, $\mathbb{T}$ is an automorphism group on $\mathcal{P}(E)$ which acts transitively on E; cf. Assumption 5.31. The present section examines $\mathbb{T}$-openings (and $\mathbb{T}$-closings), in particular, the relation between structural and adjunctional $\mathbb{T}$-openings.

In Section 6.1 we have not imposed any restriction on the automorphism group, so all results obtained there apply to the present situation as well. In particular, Theorem 6.10 gives that every $\mathbb{T}$-opening can be represented as a supremum of structural $\mathbb{T}$-openings. Concerning adjunctional $\mathbb{T}$-openings, recall from Theorem 5.43 that every $\mathbb{T}$-adjunction on $\mathcal{P}(E)$ is of the form $(\varepsilon_A, \delta_A)$, where

$$\delta_A(X) = \pi\big[\lambda(X)\,\overline{\oplus}\,\lambda(A)\big], \tag{6.26}$$

$$\varepsilon_A(X) = \pi_e\big[\lambda(X)\,\overline{\ominus}\,\lambda(A)\big], \tag{6.27}$$

and where $A \subseteq E$ is Σ-invariant. In operator notation:

$$\delta_A = \pi\,\Delta_{\lambda(A)}\lambda, \qquad \varepsilon_A = \pi_e\mathcal{E}_{\lambda(A)}\lambda. \tag{6.28}$$

The corresponding adjunctional opening is given by

$$\begin{aligned}\delta_A\varepsilon_A &= \pi\,\Delta_{\lambda(A)}\lambda\pi_e\mathcal{E}_{\lambda(A)}\lambda = \pi\,\Delta_{\lambda(A)}\mathcal{E}_\Sigma\mathcal{E}_{\lambda(A)}\lambda\\ &= \pi\,\Delta_{\lambda(A)}\mathcal{E}_{\Sigma\overline{\oplus}\lambda(A)}\lambda = \pi\,\Delta_{\lambda(A)}\mathcal{E}_{\lambda(\Sigma A)}\lambda;\end{aligned}$$

here we have used (5.80), (5.69), and (5.95), respectively. Using that A is Σ-invariant, it follows

$$\delta_A\varepsilon_A = \pi\,\Delta_{\lambda(A)}\mathcal{E}_{\lambda(A)}\lambda. \tag{6.29}$$

Let $\mathcal{A}_{\mathbb{A}}^{\mathbb{T}}$ be the structural $\mathbb{T}$-opening on $\mathcal{P}(\mathbb{T})$ by the structuring element $\mathbb{A} \subseteq \mathbb{T}$, that is,

$$\mathcal{A}_{\mathbb{A}}^{\mathbb{T}}(\mathbb{X}) = \bigcup\{\tau(\mathbb{A}) \mid \tau \in \mathbb{T}\,,\ \tau(\mathbb{A}) \subseteq \mathbb{X}\}. \tag{6.30}$$

Exactly the same proof as for Theorem 6.16 shows

$$\mathcal{A}_{\mathbb{A}}^{\mathbb{T}} = \Delta_{\mathbb{A}}\mathcal{E}_{\mathbb{A}}. \tag{6.31}$$

In other words, on $\mathcal{P}(\mathbb{T})$ adjunctional and structural $\mathbb{T}$-openings coincide. Combination of (6.29) and (6.31) leads to

$$\delta_A\varepsilon_A = \pi\,\mathcal{A}_{\lambda(A)}^{\mathbb{T}}\lambda.$$

So, for every $X \subseteq E$,

$$\delta_A\varepsilon_A(X) = \pi\Big(\bigcup\{\tau\lambda(A) \mid \tau\lambda(A) \subseteq \lambda(X)\}\Big)$$

$$= \bigcup \{\tau\pi\lambda(A) \mid \tau\lambda(A) \subseteq \lambda(X)\};$$

here we have used (6.30) and the fact that π is a dilation (see Proposition 5.37(b)) which commutes with elements of $\mathbb{T}$. From (5.78) and the observation that $\tau\lambda(A) \subseteq \lambda(X)$ if and only if $\tau(A) \subseteq X$, one gets

$$\delta_A\varepsilon_A(X) = \bigcup \{\tau(A) \mid \tau(A) \subseteq X\}.$$

This expression coincides with the expression for a structural $\mathbb{T}$-opening. This suggests that, as in Theorem 6.16, the classes of adjunctional and structural openings coincide. This, however, is not true for the following reason. In our computations we have used explicitly that A is Σ-invariant, that is, $A = \Sigma A$. If one does not want to make any a priori assumption on A, one must replace A by ΣA in the formula for $\delta_A\varepsilon_A$.

6.20 Proposition. *Let A be an arbitrary subset of E. The adjunctional opening $\delta_A\varepsilon_A$ is given by*

$$\delta_A\varepsilon_A(X) = \bigcup \{\tau(\Sigma A) \mid \tau(\Sigma A) \subseteq X\}. \tag{6.32}$$

The adjunctional and structural $\mathbb{T}$-opening by A coincide; i.e., $\delta_A\varepsilon_A = \alpha_A^{\mathbb{T}}$, if and only if A is Σ-invariant.

We point out that the inequality

$$\delta_A\varepsilon_A \leq \alpha_A^{\mathbb{T}}$$

always holds. In fact, the latter proposition says that the class of structural openings is larger than the class of adjunctional openings, assuming that $\Sigma \neq \{\mathrm{id}\}$. The difference between both classes of openings is nicely illustrated in Example 6.21. Before we present this example, we give a different characterization of the structural opening $\alpha_A^{\mathbb{T}}$, namely, as the opening which derives from an adjunction $(\varepsilon^{\uparrow}, \delta^{\downarrow})$ between $\mathcal{P}(E)$ and $\mathcal{P}(\mathbb{T})$. In fact, the foregoing computations show that every structural $\mathbb{T}$-opening on $\mathcal{P}(E)$ has the form

$$\alpha_A^{\mathbb{T}} = \pi \mathcal{A}_{\lambda(A)}^{\mathbb{T}} \lambda,$$

where $A \subseteq E$. Using (6.31), this gives

$$\alpha_A^{\mathbb{T}} = \pi \Delta_{\lambda(A)} \mathcal{E}_{\lambda(A)} \lambda. \tag{6.33}$$

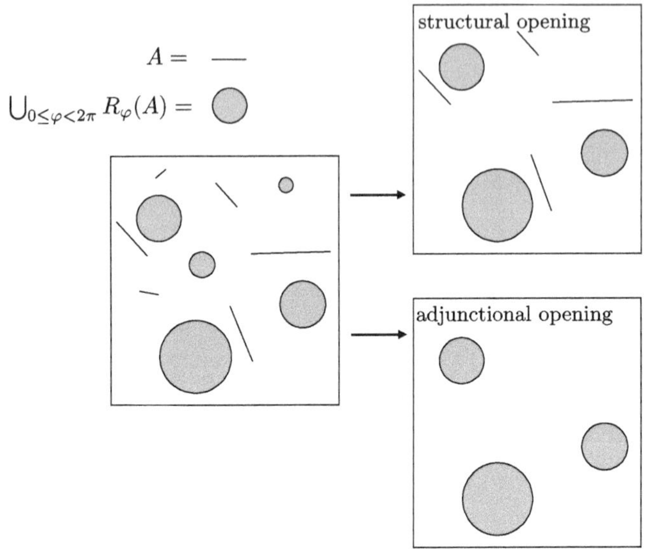

Figure 6.3 Structural and adjunctional opening by a line segment under translation–rotation invariance.

With Proposition 5.37(b), it follows immediately that the pair $(\varepsilon^{\uparrow}, \delta^{\downarrow}) = (\mathcal{E}_{\lambda(A)}\lambda, \pi\Delta_{\lambda(A)})$ defines an adjunction between $\mathcal{P}(E)$ and $\mathcal{P}(\mathbb{T})$, and it is obvious that $\alpha_A^{\mathbb{T}} = \delta^{\downarrow}\varepsilon^{\uparrow}$.

6.21 Example. (Translation–rotation morphology on $\mathbb{R}^2$)
Consider the framework described in Section 5.9; in particular, $\mathbb{T}$ is the group of translations and rotations on $\mathbb{R}^2$. Let A be a line segment as in Fig. 6.3.

The opening $\alpha_A^{\mathbb{T}}(X)$ of the set X is the union of all rotated and translated copies of A which fit inside X. Let X be a collection of line segments of varying lengths; the structural opening $\alpha_A^{\mathbb{T}}(X)$ preserves all line segments whose length is not smaller than that of A. However, the adjunctional opening $\delta_A\varepsilon_A$ coincides with the structural opening by ΣA, which is the disk with diameter equal to the length of A; see Fig. 6.3. Thus $\alpha_A^{\mathbb{T}}$ can be expressed as

$$\alpha_A^{\mathbb{T}}(X) = \bigcup_{0\le\varphi<2\pi} X \circ R_\varphi(A),$$

where R_φ is the rotation around the origin about an angle φ and $X \circ B$ denotes the translation invariant opening defined in (4.49). Observe also that

$\alpha_A^{\mathbb{T}}$ coincides with the opening $\overline{\alpha}$ obtained from (5.105) if α is the translation invariant opening $\alpha(X) = X \circ A$. On the other hand, the adjunctional opening $\delta_A \varepsilon_A$ is given by

$$\delta_A \varepsilon_A(X) = X \circ \Big(\bigcup_{0 \le \varphi < 2\pi} R_\varphi(A) \Big).$$

We leave the derivation of these formulas as an exercise to the reader.

6.5. Annular openings

In Section 4.4 we have introduced the *annular opening* on $\mathcal{P}(\mathbb{E}^d)$. For convenience we recall its definition here. Let $A \subseteq \mathbb{E}^d$ be a symmetric structuring element; then the operator $\mathrm{id} \wedge \delta_A : X \mapsto X \cap (X \oplus A)$ is an opening, the annular opening by A. Its invariance domain is generated by all pairs $\{x, x + a\}$ where $x \in \mathbb{E}^d, a \in A$. In this section it is shown how this notion can be extended to complete lattices for which the infinite supremum distributive law (2.5) holds. Suppose that $\mathcal{L}$ is a complete lattice for which the distributive law

$$X \wedge (Y \vee Z) = (X \wedge Y) \vee (X \wedge Z) \qquad (*)$$

holds. Assume that τ is an automorphism on $\mathcal{L}$ with inverse τ^{-1}, and define the dilation δ by

$$\delta = \tau \vee \tau^{-1}.$$

It is obvious that $(\mathrm{id} \wedge \delta)^2 \le \mathrm{id} \wedge \delta$. If one can show that $\delta(\mathrm{id} \wedge \delta) \ge \mathrm{id} \wedge \delta$, then $(\mathrm{id} \wedge \delta)^2 = \mathrm{id} \wedge \delta \wedge \delta(\mathrm{id} \wedge \delta) \ge \mathrm{id} \wedge \delta$, whence it follows that $\mathrm{id} \wedge \delta$ is idempotent, and therefore an opening. Now

$$\begin{aligned}
\delta(\mathrm{id} \wedge \delta) &= (\tau \vee \tau^{-1})(\mathrm{id} \wedge \delta) = \tau(\mathrm{id} \wedge \delta) \vee \tau^{-1}(\mathrm{id} \wedge \delta) \\
&= (\tau \wedge \tau\delta) \vee (\tau^{-1} \wedge \tau^{-1}\delta) \ge (\tau \wedge \mathrm{id}) \vee (\tau^{-1} \wedge \mathrm{id}) \\
&= \mathrm{id} \wedge (\tau \vee \tau^{-1}) = \mathrm{id} \wedge \delta,
\end{aligned}$$

where we have used the distributive law $(*)$ in the penultimate equality. This result is generalized in the following proposition.

6.22 Proposition. *Let $\mathcal{L}$ be a complete lattice satisfying the infinite supremum distributive law* (2.5)*, and let $\mathbb{A}$ be a collection of automorphisms which satisfy*

$$\forall \sigma \in \mathbb{A}\ \exists \sigma' \in \mathbb{A} : \quad \sigma\sigma' \ge \mathrm{id}. \qquad (6.34)$$

Then $\mathrm{id} \wedge \bigvee \mathbb{A}$ is an opening.

Proof. Let $\delta = \bigvee \mathbb{A}$. The foregoing argument implies that it suffices to show $\delta(\mathrm{id} \wedge \delta) \geq \mathrm{id} \wedge \delta$. By the infinite supremum distributivity,

$$\mathrm{id} \wedge \delta = \bigvee_{\sigma' \in \mathbb{A}} (\mathrm{id} \wedge \sigma');$$

hence,

$$\delta(\mathrm{id} \wedge \delta) = \bigvee_{\sigma \in \mathbb{A}} \sigma\Big[\bigvee_{\sigma' \in \mathbb{A}} \mathrm{id} \wedge \sigma'\Big] = \bigvee_{\sigma \in \mathbb{A}} \bigvee_{\sigma' \in \mathbb{A}} (\sigma \wedge \sigma\sigma').$$

From (6.34) it follows that $\bigvee_{\sigma' \in \mathbb{A}} (\sigma \wedge \sigma\sigma') \geq \sigma \wedge \mathrm{id}$. Using the infinite supremum distributivity once more, one gets

$$\delta(\mathrm{id} \wedge \delta) \geq \bigvee_{\sigma \in \mathbb{A}} (\sigma \wedge \mathrm{id}) = \mathrm{id} \wedge \delta.$$

This completes the proof. □

Of particular interest is the case where $\mathbb{T}$ is an abelian automorphism group and ℓ is a sup-generating family in $\mathcal{L}$ such that the Basic Assumption 5.10 holds. In particular, this implies that one can define an abelian group operation $+$ on ℓ (after fixing an origin $o \in \ell$). Let, for $A \in \mathcal{L}$, δ_A be the $\mathbb{T}$-dilation given by (5.20) and $\ell(A) = \{a \in \ell \mid a \leq A\}$. The following theorem is the main result of this section.

6.23 Theorem. *Let $\mathcal{L}$ be a complete lattice which satisfies the infinite supremum distributive law* (2.5) *and which contains a sup-generating family ℓ. Furthermore, let $\mathbb{T}$ be an abelian automorphism group on $\mathcal{L}$ such that the Basic Assumption holds. If $A \in \mathcal{L}$ satisfies the condition*

$$\forall a \in \ell(A)\ \exists a' \in \ell(A) : \quad a + a' \geq o, \tag{6.35}$$

then $\mathrm{id} \wedge \delta_A$ is a $\mathbb{T}$-opening. Its invariance domain is the sup-closed family generated by $x \vee (x - a)$, where $x \in \ell$ and $a \in \ell(A)$.

Proof. Let τ_a be the unique element of $\mathbb{T}$ which maps o onto a (see Section 5.2). Evidently, $a + a' \geq o$ in (6.35) is equivalent to $\tau_a\tau_{a'} \geq \mathrm{id}$. This implies that condition (6.34) holds for $\mathbb{A} = \{\tau_a \mid a \in \ell(A)\}$. Since $\bigvee \mathbb{A} = \delta_A$, the first assertion is proved.

Take $x \in \ell$ and $a \in \ell(A)$. We show that $x \vee (x - a) \in \mathrm{Inv}(\mathrm{id} \wedge \delta_A)$, or equivalently, that $\delta_A(x \vee (x - a)) \geq x \vee (x - a)$. By (6.35) there is an $a' \in \ell(A)$ such that $a' \geq -a$; hence,

$$\delta_A(x \vee (x - a)) = \delta_A(x) \vee \delta_A(x - a) \geq (x + a') \vee (x - a + a) \geq (x - a) \vee x.$$

Since by Proposition 6.7(a), the invariance domain of an opening is closed under suprema, the supremum of a given family of elements $x \vee (x - a)$, with $x \in \ell$ and $a \in \ell(A)$, lies in $\mathrm{Inv}(\mathrm{id} \wedge \delta_A)$.

Conversely, assume $X \in \mathrm{Inv}(\mathrm{id} \wedge \delta_A)$; we show that X can be written as a supremum of elements $x \vee (x - a)$, where $x \in \ell$ and $a \in \ell(A)$. Define $X' = \bigvee_{a \in \ell(A)} \bigvee_{x \in \ell(X \wedge X_a)} (x \vee (x - a))$. Since $x \le X$ and $x - a \le X$ for $x \in \ell(X \wedge X_a)$, it is obvious that $X' \le X$. From $X \in \mathrm{Inv}(\mathrm{id} \wedge \delta_A)$ and infinite supremum distributivity, we derive

$$\begin{aligned} X = X \wedge (X \oplus A) = X \wedge \bigvee_{a \in \ell(A)} X_a = \bigvee_{a \in \ell(A)} (X \wedge X_a) \\ = \bigvee_{a \in \ell(A)} \bigvee_{x \in \ell(X \wedge X_a)} x \le \bigvee_{a \in \ell(A)} \bigvee_{x \in \ell(X \wedge X_a)} (x \vee (x - a)) = X', \end{aligned}$$

whence equality follows. This proves the result. □

Note that in the Euclidean case condition (6.35) says that $-a \in A$ if $a \in A$, that is, A is symmetric. In the following example we apply the theorem to grey-scale functions.

6.24 Example. (Grey-scale functions)
Consider the complete lattice $\mathrm{Fun}(\mathbb{E}^d, \overline{\mathbb{R}})$ of grey-scale functions. In Example 2.18(c) we have seen that this lattice satisfies the infinite supremum distributive law. The pulse functions $\ell = \{f_{x,t} \mid x \in \mathbb{E}^d\ ,\ t \in \mathbb{R}\}$ constitute a sup-generating family on which the group of horizontal and vertical translations acts transitively. In Section 5.7 it has been demonstrated that every dilation is of the form

$$\Delta_G(F)(x) = \bigvee_{x \in \mathbb{E}^d} \big[F(x - h) + G(h)\big].$$

Recall from (4.134) that the domain $\mathrm{dom}(G)$ of a function G is the set of all x for which $G(x) > -\infty$. Theorem 6.23 states that $\mathrm{id} \wedge \Delta_G$ is an annular opening on $\mathrm{Fun}(\mathbb{E}^d)$ if for every pulse function $f_{x,t} \le G$ there exists a pulse function $f_{x',t'} \le G$ such that $f_{x,t} + f_{x',t'} \ge f_{0,0}$, that is, $f_{x+x',t+t'} \ge f_{0,0}$. The latter means that $x + x' = 0$ and $t + t' \ge 0$. Furthermore, $f_{x,t} \le G$ is equivalent to $G(x) \ge t$. Therefore, $t + t' \ge 0$ implies $G(x) + G(x') \ge 0$. On the other hand, $G(x) + G(x') \ge 0$ implies that there exist t, t' such that $G(x) \ge t$, $G(x') \ge t'$ and $t + t' \ge 0$; for one may choose $t = G(x)$ and $t' = G(x')$. Thus, under the given assumptions, condition (6.35) amounts to the following conditions on the structuring function G:

(i) $\mathrm{dom}(G)$ is symmetric;
(ii) $G(x) + G(-x) \geq 0$ for every $x \in \mathrm{dom}(G)$.
It is not difficult to show that in the multiplicative case (cf. Section 5.7.2) condition (ii) has to be replaced by:
(ii′) $G(x)G(-x) \geq 1$ for every $x \in \mathrm{dom}(G)$.
More results about annular openings for grey-scale functions can be found in Example 11.36.

In a similar way, one can also construct *annular closings*, i.e., closings of the form $\mathrm{id} \vee \varepsilon$ for some erosion ε. In this case one needs a complete lattice which satisfies the infinite infimum distributive law (2.6). Then, if $\mathbb{A}$ is a collection of automorphisms such that

$$\forall \sigma \in \mathbb{A}\ \exists \sigma' \in \mathbb{A}: \quad \sigma\sigma' \leq \mathrm{id},$$

the operator $\mathrm{id} \vee \bigwedge \mathbb{A}$ is a closing. Likewise one can formulate the analogue of Theorem 6.23 according to the Duality Principle; this involves, among other things, the dual Basic Assumption. It is also possible, however, to formulate a result for annular closings which utilizes the Basic Assumption itself. If one makes the same assumptions as in Theorem 6.23, except that the infinite supremum distributive law (2.5) is replaced by the infinite infimum distributive law (2.6); then $\mathrm{id} \vee \varepsilon_A$ is an opening. The proof is left as an exercise to the reader.

6.6. Openings from inf-overfilters

If (ε, δ) is an adjunction between $\mathcal{L}$ and $\mathcal{M}$, then $\delta\varepsilon$ defines an opening on $\mathcal{L}$. Suppose now that both (ε, δ) and (ε', δ') are adjunctions between $\mathcal{L}$ and $\mathcal{M}$ such that $\delta' \geq \delta$, and hence $\varepsilon' \leq \varepsilon$. Then $\delta'\varepsilon$ need not be anti-extensive. In particular, $\delta'\varepsilon$ is in general not an opening. The operator $\mathrm{id} \wedge \delta'\varepsilon$ is anti-extensive, however, and moreover it defines an opening on $\mathcal{L}$. To see this, consider the expression $\delta'\varepsilon(\mathrm{id} \wedge \delta'\varepsilon)$. Since $\delta' \geq \delta$, it follows that

$$\delta'\varepsilon(\mathrm{id} \wedge \delta'\varepsilon) \geq \delta'\varepsilon(\mathrm{id} \wedge \delta\varepsilon) = \delta'\varepsilon\delta\varepsilon = \delta'\varepsilon.$$

Here we have used that $\delta\varepsilon \leq \mathrm{id}$ and $\varepsilon\delta\varepsilon = \varepsilon$; cf. Proposition 3.14. This means in particular that $(\mathrm{id} \wedge \delta'\varepsilon)^2 \geq \mathrm{id} \wedge \delta'\varepsilon$. As the reverse inequality is trivial, $\mathrm{id} \wedge \delta'\varepsilon$ is idempotent and hence an opening. The operator $\delta'\varepsilon$ introduced here is a prototypical example of what is called an inf-overfilter.

6.25 Definition.

(a) An increasing operator ψ on $\mathcal{L}$ is called an *inf-overfilter* if

$$\psi(\mathrm{id} \wedge \psi) = \psi. \tag{6.36}$$

(a′) An increasing operator ψ on $\mathcal{L}$ is called a *sup-underfilter* if

$$\psi(\mathrm{id} \vee \psi) = \psi. \tag{6.37}$$

Note that inf-overfilters and sup-underfilters are dual notions in the sense of the Duality Principle; this section considers only inf-overfilters, but the Duality Principle says that for every statement concerning openings and inf-overfilters, there exists a dual counterpart concerning closings and sup-underfilters.

Since for an increasing operator we always have $\psi(\mathrm{id} \wedge \psi) \leq \psi$, we must, in order to prove that ψ is an inf-overfilter, only show that the inequality $\psi(\mathrm{id} \wedge \psi) \geq \psi$ holds. If ψ is an inf-overfilter, then

$$\psi^2 \geq \psi(\mathrm{id} \wedge \psi) = \psi.$$

Increasing operators with the property $\psi^2 \geq \psi$ are called *overfilters*. Chapter 12 contains a comprehensive discussion of overfilters, inf-overfilters, and other classes of morphological operators which are relevant in the context of morphological filtering. A motivation for introducing inf-overfilters in this chapter on openings is given by the following important result.

6.26 Theorem.

(a) *If ψ is an inf-overfilter, then* $\mathrm{id} \wedge \psi$ *is an opening.*

(a′) *If ψ is a sup-underfilter, then* $\mathrm{id} \vee \psi$ *is a closing.*

Proof. Let ψ be an inf-overfilter. Then $(\mathrm{id} \wedge \psi)^2 = (\mathrm{id} \wedge \psi) \wedge \psi(\mathrm{id} \wedge \psi) = \mathrm{id} \wedge \psi \wedge \psi = \mathrm{id} \wedge \psi$. This shows that $\mathrm{id} \wedge \psi$ is idempotent, and since the operator is also anti-extensive, it is an opening. □

It is trivial that every extensive increasing operator is an inf-overfilter. Since the opening derived from them by Theorem 6.26(a) is the identity operator, this class of inf-overfilters is not of great interest.

6.27 Proposition. *The inf-overfilters constitute a sup-closed family in $\mathcal{O}^+(\mathcal{L})$.*

Proof. Let ψ_i be an inf-overfilter for every $i \in I$; define $\psi = \bigvee_{i \in I} \psi_i$. Then $\psi(\mathrm{id} \wedge \psi) \geq \psi_i(\mathrm{id} \wedge \psi_i) = \psi_i$. Therefore, $\psi(\mathrm{id} \wedge \psi) \geq \bigvee_{i \in I} \psi_i = \psi$, which shows that ψ is an inf-overfilter. □

The next result shows how to build inf-overfilters.

6.28 Proposition.

(a) *If ψ is an inf-overfilter and ϕ an increasing operator such that $\phi \geq \mathrm{id} \wedge \psi$, then $\phi\psi$ is an inf-overfilter.*

(b) *If (ε, δ) is an adjunction between $\mathcal{L}$ and $\mathcal{M}$ and $\phi : \mathcal{M} \to \mathcal{L}$ is an increasing operator with $\phi \geq \delta$, then $\phi\varepsilon$ is an inf-overfilter on $\mathcal{L}$.*

Proof. (a): Using $\psi^2 \geq \psi$, one gets

$$\begin{aligned}\phi\psi(\mathrm{id} \wedge \phi\psi) &\geq \phi\psi(\mathrm{id} \wedge (\mathrm{id} \wedge \psi)\psi) = \phi\psi(\mathrm{id} \wedge \psi \wedge \psi^2)\\ &\geq \phi\psi(\mathrm{id} \wedge \psi) = \phi\psi.\end{aligned}$$

(b): From $\varepsilon\delta\varepsilon = \varepsilon$ it follows that

$$\phi\varepsilon(\mathrm{id} \wedge \phi\varepsilon) \geq \phi\varepsilon(\mathrm{id} \wedge \delta\varepsilon) = \phi(\varepsilon \wedge \varepsilon\delta\varepsilon) = \phi\varepsilon.$$

This concludes the proof. □

From (a) it follows immediately that ψ^2 is an inf-overfilter if ψ is one. Furthermore, since every opening α is an inf-overfilter, (a) means that $\phi\alpha$ is an inf-overfilter if $\phi \geq \alpha$.

6.29 Example. (Rank-max openings)

Consider the complete Boolean lattice $\mathcal{P}(\mathbb{Z}^2)$. Let $A \subseteq \mathbb{Z}^2$ be the octagon depicted in Fig. 6.4 and let $\alpha_A(X) = X \circ A$. Fig. 6.4 illustrates the following properties:

- *α_A is sensitive to noise.* Let X be an infinite strip with a thickness of five pixels. Then X is invariant under α_A. Removal of one pixel from X may have the effect that a "considerable" portion of the original is removed by α_A.
- *α_A is shape-dependent.* Concretely, if X has a more curved boundary than A (see Fig. 6.4), then α_A removes parts of this boundary and imposes the shape of A on X.

We look for an opening α which is more flexible than α_A. In fact this opening should satisfy the following requirement: the portion $X \cap A_h$ lies within $\alpha(X)$ if $X \cap A_h$ does not differ too much from A_h. Observe that $\alpha_A(X)$ consists of all portions $X \cap A_h$ that coincide with A_h. To quantify the condition "does not differ too much from", assume that there exist $B^1, B^2, \ldots, B^m \subseteq A$ which can be interpreted as the minimal sets which

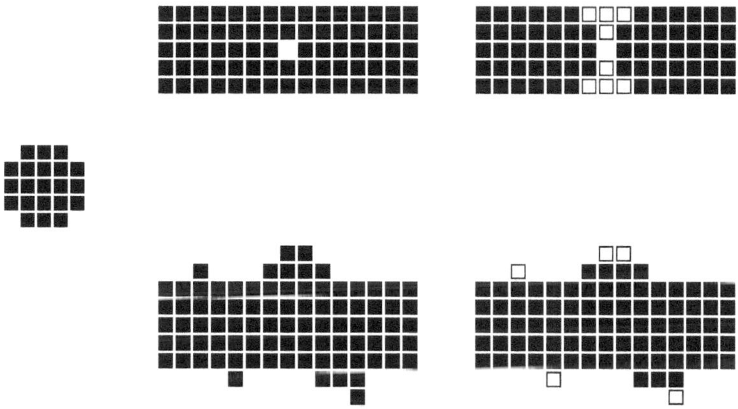

Figure 6.4 A strip (of infinite length) which is 5 pixels thick is invariant under opening by an octagon (left). Top: removal of one pixel has a drastic effect on the opening. Bottom: the opening removes thin parts of the boundary.

"differ not too much from A". Now $\alpha(X)$ consists of all portions $X \cap A_h$ such that $B^i_h \subseteq X$ is satisfied for some $i = 1, 2, \dots, m$. That is,

$$\alpha(X) = \bigcup_{i=1}^{m} \Big[\bigcup \{X \cap A_h \mid B^i_h \subseteq X\} \Big].$$

To rewrite this expression, observe that

$$\begin{aligned} \bigcup \{X \cap A_h \mid B^i_h \subseteq X\} &= X \cap \big[\bigcup \{A_h \mid B^i_h \subseteq X\} \big] \\ &= X \cap \big[\{h \in \mathbb{Z}^2 \mid B^i_h \subseteq X\} \oplus A \big] \\ &= X \cap \big[(X \ominus B^i) \oplus A \big], \end{aligned}$$

for every $i = 1, 2, \dots, m$. Using that $\oplus$ distributes over union, we get

$$\alpha(X) = X \cap \Big[\Big(\bigcup_{i=1}^{m} (X \ominus B^i) \Big) \oplus A \Big]. \tag{6.38}$$

In other words,

$$\alpha = \mathsf{id} \wedge \bigvee_{i=1}^{m} \delta_A \varepsilon_{B^i} = \mathsf{id} \wedge \delta_A \bigvee_{i=1}^{m} \varepsilon_{B^i}. \tag{6.39}$$

As outlined at the beginning of this section, $\delta_A \varepsilon_{B^i}$ is an inf-overfilter since $\delta_A \geq \delta_{B^i}$. Now, from the fact that inf-overfilters constitute a sup-closed family (Proposition 6.27) and Theorem 6.26, we conclude that α is indeed an opening.

If one takes for B^i all subsets of A with k points, where $1 \leq k \leq \operatorname{card}(A)$, then $\bigvee_{i=1}^{m} \varepsilon_{B^i}$ is the *rank operator* $\rho_{A,k}$; cf. Section 4.5. Openings of the form $\mathsf{id} \wedge \delta_A \rho_{A,k}$ are called *rank-max openings* since the underlying inf-overfilter $\delta_A \rho_{A,k}$ is the composition of a rank operator and a dilation; the latter is also called max operator in the literature. Note that for $k = \operatorname{card}(A)$, the rank operator $\rho_{A,k}$ coincides with the erosion ε_A, and so in that case the opening α coincides with the structural opening α_A. Illustrations of rank-max openings are given in Fig. 6.5.

In the remainder of this section we look for a complete characterization of inf-overfilters and the invariance domain of the associated openings.

6.30 Theorem. *Let J, K be nonempty index sets, let $(\varepsilon_j, \delta_j)$ be an adjunction on $\mathcal{L}$ for $j \in J$, and let ϕ_{kj} be an increasing operator on $\mathcal{L}$ for $j \in J$, $k \in K$. Assume that $\phi_{kj} \geq \delta_j$ for all j, k. Then ψ given by*

$$\psi = \bigwedge_{k\in K} \bigvee_{j\in J} \phi_{kj}\varepsilon_j \tag{6.40}$$

is an inf-overfilter. The invariance domain of the opening $\mathsf{id} \wedge \psi$ comprises all $X \in \mathcal{L}$ such that for every $j \in J$ there is some $Y_j \in \mathcal{L}$ with

$$\bigvee_{j\in J} \delta_j(Y_j) \leq X \leq \bigwedge_{k\in K} \bigvee_{j\in J} \phi_{kj}(Y_j). \tag{6.41}$$

If, in addition, $\mathcal{L}$ satisfies the infinite supremum distributive law (2.5), then the invariance domain of $\mathsf{id} \wedge \psi$ comprises all $X \in \mathcal{L}$ such that for every $k \in K$ and $j \in J$ there exist $X_{kj}, Y_j \in \mathcal{L}$ with

$$\delta_j(Y_j) \leq X_{kj} \leq \phi_{kj}(Y_j) \quad \text{and} \quad X = \bigwedge_{k\in K} \bigvee_{j\in J} X_{kj}. \tag{6.42}$$

Proof. For $k \in K$, we define $\psi_k = \bigvee_{j\in J} \phi_{kj}\varepsilon_j$. Since $\phi_{kj} \geq \delta_j$, we get $\psi_k \geq \bigvee_{j\in J} \delta_j\varepsilon_j$. The operator $\alpha = \bigvee_{j\in J} \delta_j\varepsilon_j$ is an opening, since $\delta_j\varepsilon_j$ is an opening and a supremum of openings is again an opening. For every $j \in J$ we have $\psi_k\alpha \geq \phi_{kj}\varepsilon_j\delta_j\varepsilon_j = \phi_{kj}\varepsilon_j$, and therefore $\psi_k\alpha \geq \bigvee_{j\in J} \phi_{kj}\varepsilon_j = \psi_k$. It is evident that $\psi_k\alpha \leq \psi_k$, and it follows that $\psi_k\alpha = \psi_k$. Now $\psi = \bigwedge_{k\in K} \psi_k$ and

$$\psi(\mathsf{id} \wedge \psi) = \bigwedge_{k\in K} \psi_k(\mathsf{id} \wedge \bigwedge_{k\in K} \psi_k) \geq \bigwedge_{k\in K} \psi_k(\mathsf{id} \wedge \alpha)$$

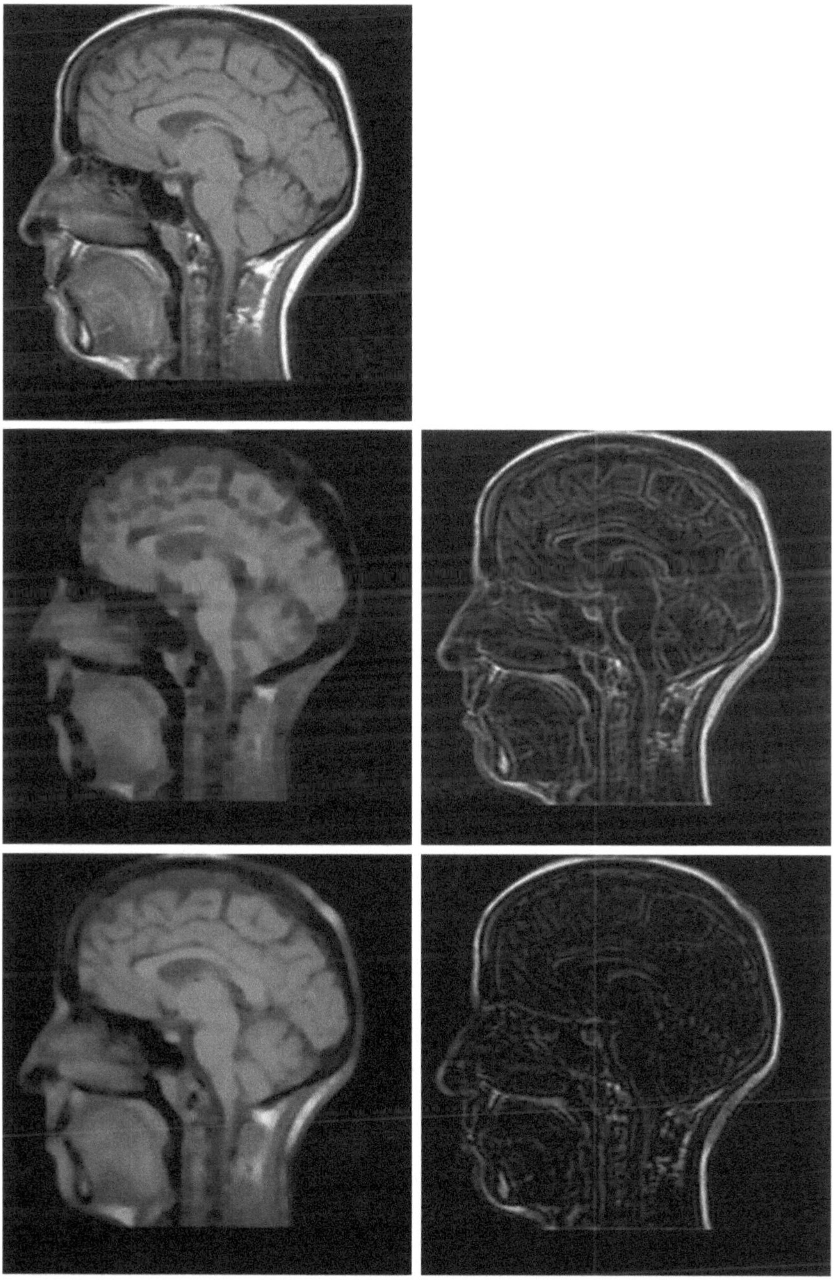

Figure 6.5 Rank-max opening and top-hat transforms. The structuring element A is the 7×7 square. Top: original image F. Middle: opening $F \circ A$ and top-hat $F - F \circ A$. Bottom: rank-max opening $F \wedge \delta_A \rho_{A,40}(F)$ and top-hat $F - (F \wedge \delta_A \rho_{A,40}(F))$.

$$= \bigwedge_{k\in K} \psi_k \alpha = \bigwedge_{k\in K} \psi_k = \psi;$$

this shows that ψ is an inf-overfilter.

As $\mathrm{id} \wedge \psi = \mathrm{id} \wedge (\bigwedge_{k\in K} \psi_k) = \bigwedge_{k\in K} (\mathrm{id} \wedge \psi_k)$ and each $\mathrm{id} \wedge \psi_k$ is anti-extensive, X is a fixpoint of $\mathrm{id} \wedge \psi$ if and only if it is a fixpoint of each $\mathrm{id} \wedge \psi_k$. Define $Y_j = \varepsilon_j(X)$; by adjunction, $\delta_j(Y_j) \le X$, and hence $\bigvee_{j\in J} \delta_j(Y_j) \le X$. Thus the first inequality in (6.41) follows. Furthermore,

$$X = (\mathrm{id} \wedge \psi_k)(X) \le \psi_k(X) = \bigvee_{j\in J} \phi_{kj}\varepsilon_j(X) = \bigvee_{j\in J} \phi_{kj}(Y_j),$$

for every $k \in K$, and the second inequality in (6.41) follows.

Conversely, suppose that (6.41) holds. Then $\delta_j(Y_j) \le X$ for every $j \in J$; so $Y_j \le \varepsilon_j(X)$ by adjunction. Combining this with the second inequality in (6.41),

$$X \le \bigwedge_{k\in K} \bigvee_{j\in J} \phi_{kj}(Y_j) \le \bigwedge_{k\in K} \bigvee_{j\in J} \phi_{kj}\varepsilon_j(X) = \bigwedge_{k\in K} \psi_k(X) = \psi(X);$$

hence $(\mathrm{id} \wedge \psi)(X) = X$.

Finally, assume that the infinite supremum distributive law (2.5) holds. It is obvious that (6.42) implies (6.41). Suppose, conversely, that (6.41) holds. For every $j \in J$, we define $X_{kj} = X \wedge \phi_{kj}(Y_j)$; obviously, $X_{kj} \le \phi_{kj}(Y_j)$. By the first inequality in (6.41) it follows that $\delta_j(Y_j) \le X$; as $\phi_{kj} \ge \delta_j$, we derive

$$X_{kj} = X \wedge \phi_{kj}(Y_j) \ge X \wedge \delta_j(Y_j) = \delta_j(Y_j).$$

As $X \le \bigwedge_{k\in K} \bigvee_{j\in J} \phi_{kj}(Y_j)$, by the second inequality in (6.41), we get the following by (2.5):

$$X = X \wedge \bigwedge_{k\in K} \Big(\bigvee_{j\in J} \phi_{kj}(Y_j)\Big) = \bigwedge_{k\in K} \bigvee_{j\in J} \big(X \wedge \phi_{kj}(Y_j)\big) = \bigwedge_{k\in K} \bigvee_{j\in J} X_{kj}.$$

Hence (6.42) holds. □

To illustrate these abstract results we return to Example 6.29. Note that $\mathcal{P}(\mathbb{Z}^2)$ obeys the infinite supremum distributive law (2.5). It is obvious that the opening given by formula (6.39) is of the form (6.40) (where K contains only one element). Consequently, the invariance domain contains all $X \subseteq \mathbb{Z}^2$ of the form $X = \bigcup_{j=1}^{m} X_j$, where $X_j \subseteq \mathbb{Z}^2$ satisfies

$$Y_j \oplus B^j \subseteq X_j \subseteq Y_j \oplus A.$$

Choosing $Y_j = \{h\}$, one gets that every set X with $B^j_h \subseteq X \subseteq A_h$ lies in the invariance domain. Moreover, in this example the invariance domain is generated by all sets X with this property. For the rank-max opening $\mathrm{id} \wedge \delta_A \rho_{A,k}$, where $1 \leq k \leq \mathrm{card}(A)$, the invariance domain is sup-generated by all translates of subsets of A which contain at least k elements.

So far we have not made any particular assumption concerning $\mathbb{T}$-invariance of the underlying operators. The reader may easily verify that all the results of this section obtained so far remain valid if we assume that the operators involved are $\mathbb{T}$-invariant.

We now present conditions under which each $\mathbb{T}$-inf-overfilter (i.e., $\mathbb{T}$-invariant inf-overfilter) takes the form (6.40).

6.31 Proposition. *Suppose that $\mathbb{T}$ is an automorphism group on the complete lattice $\mathcal{L}$ such that*

(1) *every $\mathbb{T}$-opening is a supremum of adjunctional $\mathbb{T}$-openings;*

(2) *every increasing $\mathbb{T}$-operator fixing O is an infimum of $\mathbb{T}$-dilations.*

Let $\psi \neq o$ be a $\mathbb{T}$ operator such that $\psi(O) = O$. Then ψ is a $\mathbb{T}$-inf-overfilter if and only if there exists a nonempty family of $\mathbb{T}$-adjunctions $(\varepsilon_j, \delta_j)$, $j \in J$, on $\mathcal{L}$ and a nonempty family of $\mathbb{T}$-dilations δ_{kj}, $j \in J$, $k \in K$, on $\mathcal{L}$ such that

$$\forall j \in J \ \forall k \in K: \quad \delta_{kj} \geq \delta_j \quad \text{and} \quad \psi = \bigwedge_{k \in K} \bigvee_{j \in J} \delta_{kj} \varepsilon_j. \tag{6.43}$$

Proof. As (6.43) is a particular case of (6.40), every operator having this form is a $\mathbb{T}$-inf-overfilter. Suppose that ψ is a $\mathbb{T}$-inf-overfilter; then $\mathrm{id} \wedge \psi$ is an opening; so by (1) there exists a decomposition $\mathrm{id} \wedge \psi = \bigvee_{j \in J} \delta_j \varepsilon_j$ where $J \neq \emptyset$. As ψ is an increasing $\mathbb{T}$-operator and $\psi(O) = O$, it follows from (2) that there exists a decomposition $\psi = \bigwedge_{k \in K} \delta'_k$, where $K \neq \emptyset$. Thus

$$\psi = \psi(\mathrm{id} \wedge \psi) = \bigwedge_{k \in K} \delta'_k \bigvee_{j \in J} \delta_j \varepsilon_j = \bigwedge_{k \in K} \bigvee_{j \in J} \delta'_k \delta_j \varepsilon_j,$$

because every dilation δ'_k distributes over the supremum. For every $j \in J$ and $k \in K$ we have $\delta'_k \geq \psi \geq \mathrm{id} \wedge \psi \geq \delta_j \varepsilon_j$, so $\delta'_k \delta_j \geq \delta_j \varepsilon_j \delta_j = \delta_j$. Putting $\delta_{kj} = \delta'_k \delta_j$, we arrive at the expression for ψ in (6.43). □

From the introduction of Section 6.3 and Theorem 5.2(a′) it follows that assumptions (1) and (2) are satisfied if $\mathbb{T} = \{\mathrm{id}\}$. If $\mathbb{T}$ is nontrivial, then (1) is true if the Basic Assumption is satisfied (Theorems 6.16 and 6.10), whereas (2) holds if the dual Basic Assumption is satisfied (see the paragraph following Theorem 5.22). These two facts can be combined if we assume that $\mathcal{L}$ possesses a $\mathbb{T}$-compatible negation. Let ε_A, δ_A be defined by (5.20). The following result holds.

6.32 Theorem. *Assume that $\mathcal{L}$ is a complete lattice with a sup-generating family ℓ, and let $\mathbb{T}$ be an automorphism group on $\mathcal{L}$ such that the Basic Assumption holds. Assume further that $\mathcal{L}$ has an negation ν which is $\mathbb{T}$-compatible. Then every $\mathbb{T}$-inf-overfilter on $\mathcal{L}$ has the form*

$$\psi = \bigwedge_{k \in K} \bigvee_{j \in J} \delta_{A_{kj}} \varepsilon_{B_j},$$

where A_{kj}, B_j are elements of $\mathcal{L}$ such that $B_j \subseteq A_{kj}$ for $k \in K, j \in J$.

Note that every increasing $\mathbb{T}$-operator ψ with $\psi(O) \neq O$ is identically I and hence an inf-overfilter. Therefore the condition $\psi(O) = O$ is superfluous in this case.

6.7. Granulometries

This section presents a formal treatment of granulometries on complete lattices. The definitions given here are straightforward generalizations of concepts defined in Section 4.4. More concrete results for binary continuous images can be found in Section 9.6. Furthermore, Section 11.11 discusses granulometries for grey-scale images.

6.33 Definition. A *granulometry* on the complete lattice $\mathcal{L}$ is a parametrized family of openings $\{\alpha_r \mid r \in \mathcal{R}\}$, where $\mathcal{R}$ is a poset, which satisfy

$$\alpha_s \leq \alpha_r, \qquad s \geq r. \tag{6.44}$$

Condition (6.44) is equivalent to

$$\operatorname{Inv}(\alpha_s) \subseteq \operatorname{Inv}(\alpha_r), \qquad s \geq r, \tag{6.45}$$

and also to

$$\alpha_r \alpha_s = \alpha_s \alpha_r = \alpha_s, \qquad s \geq r. \tag{6.46}$$

This follows immediately from Corollary 6.2. If no confusion about the poset $\mathcal{R}$ is possible, we will use the shorter notation $\{\alpha_r\}$ instead of $\{\alpha_r \mid r \in \mathcal{R}\}$.

6.34 Proposition. *Let $\mathcal{R}$ be a poset; if $\{\alpha_r^i \mid r \in \mathcal{R}\}$ is a granulometry on $\mathcal{L}$ for every i in the index set I, then $\{\bigvee_{i\in I} \alpha_r^i \mid r \in \mathcal{R}\}$ is a granulometry as well.*

Proof. To prove this result, we use Theorem 6.3(a), which states that an arbitrary supremum of openings is again an opening. Now, since $\alpha_s^i \leq \alpha_r^i$ if $s \geq r$ for every $i \in I$, the same holds for their supremum over i. This yields the result. □

The following result is obvious.

6.35 Proposition. *Suppose that α_r is an opening for every r in the poset $\mathcal{R}$; then $\{\tilde{\alpha}_r\}$ given by $\tilde{\alpha}_r = \bigvee_{s\geq r} \alpha_s$ is a granulometry.*

The remainder of this section deals with a generalization of Minkowski granulometries; cf. Section 4.4. Toward that goal, assume in addition that both $\mathbb{T}$ and $\mathcal{R}$ are abelian automorphism groups on $\mathcal{L}$. We refer to elements of $\mathbb{T}$ as translations and to elements of $\mathcal{R}$ as multiplications. Note that, by the previous assumption, $\mathcal{R}$ also carries a partial ordering. Henceforth we write rX instead of $r(X)$; we denote the unit of $\mathcal{R}$ by 1, that is, $1X = X$ for $X \in \mathcal{L}$. In all examples encountered in this book, the multiplication group $\mathcal{R}$ can be parametrized by the positive reals. To be precise, $\mathcal{R} = \{r(v) \mid v > 0\}$ such that $r(v)^{-1} = r(1/v)$ and $r(v) \leq r(w)$ if and only if $v \leq w$. Here r^{-1} denotes the inverse multiplication of r. In order to get useful results one has to make an assumption concerning compatibility of translations and multiplications.

6.36 Compatibility Assumption. For every $\tau \in \mathbb{T}$ and $r \in \mathcal{R}$, the composition $r\tau r^{-1}$ is an element of $\mathbb{T}$.

Define the mapping $\cdot$ from $\mathcal{R} \times \mathbb{T}$ into $\mathbb{T}$ by

$$r \cdot \tau = r\tau r^{-1}, \tag{6.47}$$

for $r \in \mathcal{R}$ and $\tau \in \mathbb{T}$. Note that in the Euclidean case where $\mathcal{L} = \mathcal{P}(\mathbb{R}^d)$, $\mathbb{T}$ is the translation group and $\mathcal{R}$ the group of scalar positive multiplications, $r\big[(r^{-1}X)_h\big] = X_{rh}$; this means that $r \cdot \tau$ is a translation over rh if τ is a translation over h. One can easily derive the following properties:

$$s \cdot (r \cdot \tau) = (sr) \cdot \tau, \tag{6.48}$$

$$r \cdot (\tau\sigma) = (r \cdot \tau)(r \cdot \sigma), \tag{6.49}$$

for $r, s \in \mathcal{R}$, and $\tau, \sigma \in \mathbb{T}$. Define, for $\mathcal{M} \subseteq \mathcal{L}$ and $r \in \mathcal{R}$,

$$r\mathcal{M} = \{rX \mid X \in \mathcal{M}\}.$$

The next result shows that $\mathbb{T}$-openings are compatible with multiplications.

6.37 Proposition. *Assume that the Compatibility Assumption 6.36 holds; for* $\mathcal{M} \subseteq \mathcal{L}$,

$$\alpha^{\mathbb{T}}(r\mathcal{M})(rX) = r\alpha^{\mathbb{T}}(\mathcal{M})(X), \tag{6.50}$$

if $r \in \mathcal{R}$ *and* $X \in \mathcal{L}$.

Proof. It has to be demonstrated that $r^{-1}\alpha^{\mathbb{T}}(r\mathcal{M})(rX) = \alpha^{\mathbb{T}}(\mathcal{M})(X)$ for $r \in \mathcal{R}$, $\mathcal{M} \subseteq \mathcal{L}$, and $X \in \mathcal{L}$. By (6.2),

$$\begin{aligned} r^{-1}\alpha^{\mathbb{T}}(r\mathcal{M})(rX) &= r^{-1}\Big(\bigvee\{\tau(rA) \mid A \in \mathcal{M}\ ,\ \tau \in \mathbb{T}\ ,\ \tau(rA) \leq rX\}\Big) \\ &= \bigvee\{r^{-1}\tau(rA) \mid A \in \mathcal{M}\ ,\ \tau \in \mathbb{T}\ ,\ r^{-1}\tau(rA) \leq X\} \\ &= \bigvee\{(r^{-1}\cdot\tau)(A) \mid A \in \mathcal{M},\ \tau \in \mathbb{T}\ ,\ (r^{-1}\cdot\tau)(A) \leq X\}. \end{aligned}$$

It follows from (6.48) that $r^{-1}\cdot\tau$ runs over $\mathbb{T}$ if τ does; from this the identity follows. □

6.38 Definition. A granulometry $\{\alpha_r\}$ on $\mathcal{L}$ is called a $(\mathbb{T},\mathcal{R})$-*Minkowski granulometry* if every α_r is a $\mathbb{T}$-opening and

$$\alpha_r(rX) = r\alpha_1(X), \tag{6.51}$$

for $X \in \mathcal{L}$ and $r \in \mathcal{R}$.

A Minkowski granulometry is completely characterized by its invariance domain.

6.39 Proposition. *Assume that the Compatibility Assumption 6.36 holds. A granulometry* $\{\alpha_r\}$ *on* $\mathcal{L}$ *is a* $(\mathbb{T},\mathcal{R})$*-Minkowski granulometry if and only if*

$$\mathrm{Inv}(\alpha_r) = r\,\mathrm{Inv}(\alpha_1), \qquad r > 0, \tag{6.52}$$

and $\mathrm{Inv}(\alpha_1)$ *is* $\mathbb{T}$*-invariant.*

Proof. The only if-statement is straightforward. We prove the if-part. The assertion about $\mathbb{T}$-invariance is easy. Now assume that (6.52) holds; we prove (6.51). From the observation that $\alpha_r(rX) \in \mathrm{Inv}(\alpha_r) = r\,\mathrm{Inv}(\alpha_1)$, it follows that $\alpha_1(r^{-1}\alpha_r(rX)) = r^{-1}\alpha_r(rX)$. Thus

$$\alpha_1(r^{-1}rX) \geq \alpha_1(r^{-1}\alpha_r(rX)) = r^{-1}\alpha_r(rX),$$

or equivalently, $r\alpha_1(X) \geq \alpha_r(rX)$. On the other hand, $\alpha_1(X) \in \mathrm{Inv}(\alpha_1)$, and so $r\alpha_1(X) \in \mathrm{Inv}(\alpha_r)$; this gives $\alpha_r(rX) \geq \alpha_r(r\alpha_1(X)) = r\alpha_1(X)$. Therefore, $\alpha_r(rX) = r\alpha_1(X)$, and the proof is finished. □

If α_r is a $(\mathbb{T}, \mathcal{R})$-Minkowski granulometry, then $\mathrm{Inv}(\alpha_r) = r\mathcal{C}$, where $\mathcal{C} = \mathrm{Inv}(\alpha_1)$. Since $\mathrm{Inv}(\alpha_s) \subseteq \mathrm{Inv}(\alpha_r)$ for $s \geq r$, it follows that $s\mathcal{C} \subseteq r\mathcal{C}$ if $s \geq r$. This is equivalent to

$$r\mathcal{C} \subseteq \mathcal{C} \qquad \text{for } r \geq 1. \tag{6.53}$$

We conclude that $\mathcal{C}$ is closed under suprema (since α_1 is an opening), $\mathbb{T}$-invariant (since α_1 is $\mathbb{T}$-invariant), and invariant under multiplications $r \geq 1$. In fact, these observation provide a basis for the construction of Minkowski granulometries. Let $\mathcal{M}$ be an arbitrary subset of $\mathcal{L}$, and define $\mathcal{C}$ as the smallest subset of $\mathcal{L}$ that contains $\mathcal{M}$ and which is closed under suprema, $\mathbb{T}$-invariant, and invariant under multiplications $r \geq 1$. Let α_r be the opening generated by $r\mathcal{C}$, that is, $\alpha_r = \alpha^{\mathbb{T}}(r\mathcal{C})$; then $\{\alpha_r\}$ defines a $(\mathbb{T}, \mathcal{R})$-Minkowski granulometry. Moreover, every $(\mathbb{T}, \mathcal{R})$-Minkowski granulometry is of this form. The results are summarized in the following theorem.

6.40 Theorem. *Assume that the Compatibility Assumption 6.36 holds. The family $\{\alpha_r\}$ defines a $(\mathbb{T}, \mathcal{R})$-Minkowski granulometry if and only if there is a family $\mathcal{M} \subseteq \mathcal{L}$ such that*

$$\alpha_r = \bigvee_{s \geq r} \alpha^{\mathbb{T}}(s\mathcal{M}). \tag{6.54}$$

If $\mathcal{M}$ contains one element, say A, then the expression reduces to $\alpha_r = \bigvee_{s \geq r} \alpha^{\mathbb{T}}_{sA}$. The granulometry $\{\alpha_r\}$ is called a *structural granulometry* if every opening α_r is a structural $\mathbb{T}$-opening. In the foregoing expression one must get rid of the supremum over $s \geq r$ in order to obtain a structural granulometry. Proposition 6.11(a) shows that such reduction is possible if sA is rA-open with respect to $\mathbb{T}$, or equivalently, if

$$rA \text{ is } A\text{-open for } r \geq 1. \tag{6.55}$$

Section 9.6 discusses (Minkowski) granulometries for continuous binary images. In that case it is possible to give a complete characterization of the structuring elements A which satisfy (6.55); see Theorem 9.57. In Section 11.11 we apply the abstract results derived here to grey-scale images.

In fact, we consider two different multiplication groups there, T-scalings and H-scalings.

Finally, we point out that it is also possible to use families of dilations for the construction of granulometries on $\mathcal{L}$. We restrict to the case where $\mathcal{R} = (0, \infty)$.

6.41 Proposition. *Let $\mathcal{L}, \mathcal{M}$ be complete lattices, and let, for $r > 0$, $(\varepsilon_r, \delta_r)$ be an adjunction between $\mathcal{L}$ and $\mathcal{M}$ such that the semigroup property*

$$\delta_r \delta_s = \delta_{r+s}, \qquad r, s > 0, \tag{6.56}$$

holds. Then the openings $\alpha_r = \delta_r \varepsilon_r$ define a granulometry on $\mathcal{L}$.

Proof. From (6.56) and Proposition 3.16(c) it is easy to derive that $\varepsilon_r \varepsilon_s = \varepsilon_{r+s}$ for $r, s > 0$. Then, if $s \geq r$,

$$\alpha_s \alpha_r = \delta_s \varepsilon_s \delta_r \varepsilon_r = \delta_s \varepsilon_{s-r} \varepsilon_r \delta_r \varepsilon_r = \delta_s \varepsilon_{s-r} \varepsilon_r = \delta_s \varepsilon_s = \alpha_s.$$

This proves the result. □

6.8. Dominance and incidence structures

This chapter as well as the previous one is concerned with the extension of mathematical morphology to complete lattices. It has been argued that the main building blocks for a general theory of morphology are adjunctions between two lattices. This section presents an alternative approach to this fundamental notion.

6.42 Definition. Given two complete lattices $\mathcal{L}$ and $\mathcal{M}$, the binary relation $\Subset$ on $\mathcal{M} \times \mathcal{L}$ is called a *dominance relation* if for two arbitrary families $\{X_i \mid i \in I\} \subseteq \mathcal{L}$ and $\{Y_j \mid j \in J\} \subseteq \mathcal{M}$ the following holds:

$$\bigvee_{j \in J} Y_j \Subset \bigwedge_{i \in I} X_i \iff \forall\, i, j:\ Y_j \Subset X_i. \tag{6.57}$$

If one takes for I the empty set, one gets

$$Y \Subset I_{\mathcal{L}}, \qquad \text{for } Y \in \mathcal{M};$$

dually, if one takes for J the empty set, one gets

$$O_{\mathcal{M}} \Subset X, \qquad \text{for } X \in \mathcal{L}.$$

Furthermore, if $Y \Subset X$, $Y' \le Y$ and $X \le X'$, then $Y' \Subset X'$. If $(\mathcal{L}, \le)$ is a complete lattice, then $\le$ defines a dominance relation on $\mathcal{L} \times \mathcal{L}$. As the following result shows, there exists a one–one relation between dominance relations on $\mathcal{M} \times \mathcal{L}$ and adjunctions between $\mathcal{L}$ and $\mathcal{M}$.

6.43 Proposition. *Let $\Subset$ be a dominance relation on $\mathcal{M} \times \mathcal{L}$; the pair of operators (ε, δ) given by*

$$\varepsilon(X) = \bigvee \{Y \in \mathcal{M} \mid Y \Subset X\}, \tag{6.58}$$

$$\delta(Y) = \bigwedge \{X \in \mathcal{L} \mid Y \Subset X\} \tag{6.59}$$

defines an adjunction between $\mathcal{L}$ and $\mathcal{M}$. For all $X \in \mathcal{L}$ and $Y \in \mathcal{M}$ the following relations hold:

$$Y \Subset X \iff Y \le \varepsilon(X) \iff \delta(Y) \le X \tag{6.60}$$

and

$$\varepsilon(X) \Subset X \qquad \text{and} \qquad Y \Subset \delta(Y). \tag{6.61}$$

Conversely, if (ε, δ) is an adjunction between $\mathcal{L}$ and $\mathcal{M}$, then $\Subset$ defined by (6.60) defines a dominance structure on $\mathcal{M} \times \mathcal{L}$ and (6.58) (6.59) and (6.61) are satisfied.

Proof. Let $\Subset$ be a dominance relation, and let δ, ε be the operators given by (6.58)–(6.59). The relations in (6.61) follow immediately from (6.57). It is obvious that $Y \Subset X$ implies that $Y \le \varepsilon(X)$ and $\delta(Y) \le X$. Conversely, if $Y \le \varepsilon(X)$, then $Y \Subset X$ since $Y \le \varepsilon(X) \Subset X$. With these observations the first part of the proposition is proved.

Let (ε, δ) be an adjunction, and define $\Subset$ by (6.60). Suppose $\bigvee_{j \in J} Y_j \Subset \bigwedge_{i \in I} X_i$; then $\bigvee_{j \in J} \delta(Y_j) = \delta(\bigvee_{j \in J} Y_j) \le \bigwedge_{i \in I} X_i$; this gives $\delta(Y_j) \le X_i$ for every i, j. Thus $Y_j \Subset X_i$ for every i, j. The reverse implication in (6.57) follows by a similar argument. The properties in (6.61) follow immediately from the definition of a dominance relation. □

Assume throughout the remainder of this section that $\mathcal{L}$ is self-dual and that ν is a negation on $\mathcal{L}$; as usual, $X^* = \nu(X)$. We define an *incidence relation* $\sim$ on $\mathcal{M} \times \mathcal{L}$ as follows:

$$Y \sim X \iff Y \not\Subset X^*. \tag{6.62}$$

From Definition 6.42 one easily derives that for arbitrary families X_i, $i \in I$, in $\mathcal{L}$ and Y_j, $j \in J$, in $\mathcal{M}$ one has

$$\bigvee_{j\in J} Y_j \sim \bigvee_{i\in I} X_i \iff \exists\, i,j:\ Y_j \sim X_i. \tag{6.63}$$

Conversely, if $\sim$ is a binary relation on $\mathcal{M} \times \mathcal{L}$ with property (6.63), then $\Subset$ given by

$$Y \Subset X \iff Y \not\sim X^*, \tag{6.64}$$

defines a dominance relation on $\mathcal{M} \times \mathcal{L}$.

From (6.58)–(6.59) it follows easily that the corresponding adjunction can be written in terms of the incidence relation as follows:

$$\varepsilon(X) = \bigvee\{Y \in \mathcal{M} \mid Y \not\sim X^*\}, \tag{6.65}$$

$$\delta(Y) = \left[\bigvee\{X \in \mathcal{L} \mid Y \not\sim X\}\right]^*. \tag{6.66}$$

To obtain, e.g., the second relation observe that

$$\begin{aligned}[\delta(Y)]^* &= \left[\bigwedge\{X \in \mathcal{L} \mid Y \Subset X\}\right]^* = \left[\bigwedge\{X \in \mathcal{L} \mid Y \not\sim X^*\}\right]^* \\ &= \bigvee\{X^* \mid X \in \mathcal{L},\ Y \not\sim X^*\} = \bigvee\{X' \mid X' \in \mathcal{L},\ Y \not\sim X'\};\end{aligned}$$

here we have substituted $X' = X^*$. This gives the result. We conclude this section with two examples.

6.44 Example. (Translation invariant morphology)
Let $\mathcal{L} = \mathcal{M} = \mathcal{P}(\mathbb{R}^d)$ and $A \subseteq \mathbb{R}^d$. The pair (ε, δ) given by $\varepsilon(X) = X \ominus A$, $\delta(Y) = Y \oplus A$ is an adjunction. The corresponding dominance relation can easily be found if one uses (6.60):

$$Y \Subset X \iff Y \subseteq X \ominus A.$$

Thus, by (6.64), the corresponding incidence relation is given by

$$Y \sim X \iff Y \not\Subset X^c \iff Y \cap (X^c \ominus A)^c \neq \varnothing.$$

Since, by (4.40), $(X^c \ominus A)^c = X \oplus \check{A}$, one gets

$$Y \sim X \iff Y \cap (X \oplus \check{A}) \neq \varnothing.$$

6.45 Example. (Kendall's random set theory)
Kendall (1974) developed a theory for random sets $X \subseteq E$, where E is a given nonvoid space. His construction is based on the following simple idea: check for a given collection $\mathcal{T}$ of test sets, called *traps*, which cover the entire space E, whether they intersect X or not. The *incidence function* of a set X is the function $I_X : \mathcal{T} \to \{0, 1\}$ with $I_X(T) = 1$ if $X \cap T \neq \varnothing$ and 0 otherwise. Such an approach leads to the problem of characterizing all functions $f : \mathcal{T} \to \{0, 1\}$ which are incidence functions, i.e., which have the form $f = I_X$. We use the previous theory of incidence structures to deal with this problem. Let $\mathcal{M} = \mathcal{P}(E)$, and let $\mathcal{L}$ be the lattice of all functions $f : \mathcal{T} \to \{0, 1\}$. Since $\mathcal{L}$ is isomorphic to $\mathcal{P}(\mathcal{T})$, it is a Boolean lattice with complement $f^* = 1 - f$. Define the relation $\sim$ on $\mathcal{M} \times \mathcal{L}$ by

$$X \sim f \iff \exists\, T \in \mathcal{T} : \; X \cap T \neq \varnothing \text{ and } f(T) = 1.$$

The reader should have no difficulty in verifying that $\sim$ indeed defines an incidence relation. The associated adjunction (ε, δ) between $\mathcal{L}$ and $\mathcal{M}$ can be derived from (6.65)–(6.66). The erosion ε is given by

$$\varepsilon(f) = \bigcup \{X \subseteq E \mid X \not\sim f^*\}.$$

We derive

$$\begin{aligned} X \not\sim f^* &\iff \forall\, T : \; X \cap T = \varnothing \text{ or } f(T) = 1 \\ &\iff \forall\, T : \; f(T) = 0 \Rightarrow X \cap T = \varnothing \\ &\iff \forall\, T : \; f(T) = 0 \Rightarrow X \subseteq T^{\mathrm{c}}. \end{aligned}$$

This means

$$\varepsilon(f) = \bigcap \{T^{\mathrm{c}} \mid T \in \mathcal{T} \text{ and } f(T) = 0\}.$$

It is easy to show that

$$\delta(X) = I_X.$$

Consider the opening $\delta\varepsilon$ on $\mathcal{L}$. A function f is open with respect to this opening, that is, $\delta\varepsilon(f) = f$, if and only if

$$f(T) = 0 \iff [\delta\varepsilon(f)](T) = 0.$$

Since $\delta\varepsilon(f) = I_{\varepsilon(f)}$, the latter condition is equivalent to $\varepsilon(f) \cap T = \varnothing$. Using

$$\varepsilon(f) = [\bigcup_{f(T')=0} T']^c,$$

one arrives at the following characterization: a function f is open with respect to $\delta\varepsilon$ if and only if

$$f(T) = 0 \iff T \subseteq \bigcup_{f(T')=0} T'.$$

A function f with this property has been called a *strong incidence function* by Kendall (1974). Therefore, the functions open with respect to $\delta\varepsilon$ are precisely the strong incidence functions. On the other hand, we know from Theorem 3.25 that $\mathrm{Inv}(\delta\varepsilon) = \mathrm{Ran}(\delta)$. The latter set contains all incidence functions I_X. This shows that the equation $I_X = f$ has a solution if and only if f is a strong incidence function.

6.9. Bibliographical notes

It is impossible to discuss all morphological literature where openings and closings occur, and we shall not try to do so. A comprehensive discussion of openings for binary Euclidean images can be found in Matheron (1975) and Serra (1982). In the first reference, one also finds the classical version of Theorem 6.10 about the representation of translation invariant openings and closings. The utility of openings and closings with respect to cleaning of noisy images has been emphasized by Serra (1982). Song and Delp (1990) propose an algorithm for noise removal by a translation invariant opening which is the supremum of openings with eight different structuring elements.

The study of openings and closings on complete lattices was initiated by Matheron and Serra in the early eighties; an account of their work can be found in Serra (1988), in particular, in Chapters 5–8. A systematic study of $\mathbb{T}$-invariant openings and closings on arbitrary complete lattices was first made by Ronse and Heijmans (1991); in fact, the theory developed in Sections 6.1–6.3 is based mainly on their work. Ronse (1994b) presents a comprehensive account of openings on complete lattices and generalizes some of the results obtained by Ronse and Heijmans (1991). In fact, Proposition 6.8 is closely related to Proposition 2.7 in Ronse (1994b).

Serra and co-workers (Serra, 1988) speak of "morphological openings and closings" instead of "adjunctional openings and closings". Since these operators derive from adjunctions, however, the nomenclature "adjunctional" seems more appropriate. Another instance where it is necessary to distinguish between structural and adjunctional openings is in graph morphology; see Heijmans et al. (1992).

The definitions of the lower and the upper envelope $\check{\psi}$ and $\hat{\psi}$ of an operator ψ are originally due to Matheron (Serra, 1988, Chapter 6). The openings in Example 6.12 are called *trivial openings* by Serra and Vincent (1992).

The concept of a polar opening results from the work of the author (Heijmans, 1987); the example in Example 6.14 is due to Roerdink and Heijmans (1988).

The exposition in Section 6.4 on $\mathbb{T}$-openings on $\mathcal{P}(E)$, where $(\mathbb{T}, E)$ is a homogeneous space, is based on work by Roerdink (1989, 1990a, 1990c, 1993).

Annular openings occur for the first time in the work of Serra (1988). The results in Section 6.5 are due to Ronse and Heijmans (1991); see also Ronse (1994b).

Overfilters, inf-overfilters, as well as their dual notions underfilters and sup-underfilters, have been invented by Matheron (Serra, 1988, Chapter 6). The characterization of inf-overfilters in (6.40) is due to Serra (1988, Theorem 9.7). Ronse and Heijmans (1991, Subsection 4.1) extended Matheron's and Serra's work on inf-overfilters and gave complete characterizations of inf-overfilters which are $\mathbb{T}$-invariant; see in particular Proposition 6.31 and Theorem 6.32. To a large extent, the exposition in Section 6.6 is based on the account given in Ronse and Heijmans (1991). Similar results can be found in Ronse (1994b). Rank-max openings were introduced by Ronse (1988), where they were used for the extraction of blood vessels in x-ray angiographic images; some of Ronse's results were reproduced in Ronse and Heijmans (1991).

The theory of granulometries on complete lattices in Section 6.7 has not appeared elsewhere. Note, however, that the definition of a granulometry given in Definition 6.33 is nothing but a straightforward generalization of Matheron's original definition (Matheron, 1975).

Incidence and dominance structures have been studied by Baddeley and Heijmans (1992). In that paper one can also find several other applications, including in stochastic geometry and robot path planning.

A useful method to construct openings and closings, as well as other morphological filters, is through iteration of an increasing operator. This method is discussed in great detail in Chapter 13. There it is shown that, under rather mild conditions on ψ, iterates of the anti-extensive operator $\mathrm{id} \wedge \psi$ converge toward the lower envelope $\check{\psi}$.

Before we conclude this section we mention one final application of openings, namely, as a tool for shape decomposition. In the literature, one can find two different approaches, one by Pitas and Venetsanopoulos (1990) and one by Ronse (1992). A brief overview of both approaches can be found in Heijmans (1994b).

In Section 9.7 we discuss another important class of openings, the *connectivity openings*.

CHAPTER SEVEN

Hit-or-miss topology and semi-continuity

Henk J.A.M. Heijmans
Formerly Centre for Mathematics and Computer Science, Amsterdam, Netherlands

Contents

In image analysis in general and in mathematical morphology in particular, topology may enter at two levels. First, topology can be used as a modeling tool. In many cases, $\mathcal{P}(E)$ is not an appropriate model for binary images. Using topological concepts, this class can be restricted; one can, for instance, consider closed sets, compact sets, or connected sets. But topology can also be useful on a higher level. In this chapter we consider the closed subsets of a topological space as a model for binary images. We will supply this class of images with a topology, the so-called hit-or-miss topology.

So far, the theory developed in this book concerns mainly algebraic properties of morphological operators. In certain situations, however, topological properties may be equally important. An interesting example will be discussed in the next chapter, where it is shown that (semi-) continuity properties of morphological operators are indispensable for their discretization. The present chapter presents an elementary discussion of topological and metrical spaces. However, the emphasis is on the discussion of the hit-or-miss topology on the one hand and the Hausdorff metric and myope

Advances in Imaging and Electron Physics, Volume 216
ISSN 1076-5670
https://doi.org/10.1016/bs.aiep.2020.07.007

topology (for compact sets) on the other. Finally, some words are devoted to semi-continuity properties of morphological operators.

7.1. Topology: basic concepts

This section collects some basic definitions and results about general topological spaces which are used elsewhere in this book. The material covered in this section can be found in any good textbook on topology; whenever possible we adopt the terminology used in the book by Dugundji (1966).

7.1 Definition. Given a set E, a family $\mathcal{G}$ of subsets of E is called a *topology* on E if $\varnothing$ and E are members of G and if every finite intersection as well as every union of members of $\mathcal{G}$ is again a member of $\mathcal{G}$.

Note that this definition does not require that an infinite intersection of members of $\mathcal{G}$ is again contained in $\mathcal{G}$. If $\mathcal{G}$ is a topology on E, then the pair $(E, \mathcal{G})$ is called a *topological space*. If there is no danger of confusion, we say that E is a topological space. Sometimes we write $\mathcal{G}(E)$ instead of $\mathcal{G}$. The members of the family $\mathcal{G}$ are called the *open sets*. Given a point $x \in E$, a set $U \subseteq E$ is called a *neighbourhood* of x if U is an open set which contains x. A set is called *closed* if its complement (relative to E) is open. One derives easily that $\varnothing$ and E are closed, that a finite union of closed sets is closed, and that an arbitrary intersection of closed sets is closed. The closed sets are denoted by $\mathcal{F}(E)$, or just $\mathcal{F}$. Let $(E, \mathcal{G})$ be a topological space and $E' \subseteq E$; then

$$\mathcal{G}(E') = \{X \cap E' \mid X \in \mathcal{G}(E)\}$$

defines a topology on E', called the *relative topology* of E'.

7.2 Examples.

(a) Let E be an arbitrary set; then $\mathcal{G} = \{\varnothing, E\}$ defines a topology, called the *indiscrete topology*.

(b) If E is a set and $\mathcal{G} = \mathcal{P}(E)$, the family of all subsets of E, then $\mathcal{G}$ defines a topology on E, called the *discrete topology*. With respect to this topology, every set is open; the smallest neighbourhood containing a point x is the singleton $\{x\}$.

(c) Let $E = \mathbb{R}^d$. We denote by B° the open ball with centre 0 and radius 1, that is, $B^\circ = \{x \in \mathbb{R}^d \mid \|x\| < 1\}$. Furthermore, we denote by $B^\circ(x, r)$ the ball with centre x and radius r; note that $B^\circ(x,r) = (rB^\circ)_x$. Call a set $X \subseteq \mathbb{R}^d$ open if for every $x \in X$ there is an $r > 0$ such that $B^\circ(x,r) \subseteq X$. With this definition, $\mathbb{R}^d$ becomes a topological space; this topology is called the *Euclidean topology*.

Let E be a topological space and $X \subseteq E$. By the *interior* of X, written X°, is meant the largest open set contained in X, i.e.,

$$X^\circ = \bigcup\{G \in \mathcal{G} \mid G \subseteq X\}.$$

The *closure* $\overline{X}$ of X is the smallest closed set that contains X, i.e.,

$$\overline{X} = \bigcap\{F \in \mathcal{F} \mid X \subseteq F\}.$$

Note the duality between these two definitions.

The *boundary* ∂X of a set X is the set of all points x such that every neighbourhood of x has nonempty intersection with both X and its complement X^c. It is easy to see that

$$\partial X = \overline{X} \setminus X^\circ.$$

7.3 Example. (Closure versus closing)
Kuratowski (1972) defines a topological space by means of a *closure* operation. His definition can be expressed conveniently in morphological terms. A set E is called a topological space if there exists an operator κ on $\mathcal{P}(E)$ such that the following axioms hold:

(i) $\kappa(\varnothing) = \varnothing$;
(ii) κ is extensive;
(iii) κ is idempotent;
(iv) $\kappa(X \cup Y) = \kappa(X) \cup \kappa(Y)$.

The set $\kappa(X)$ is called the closure of X and a set X is closed if $\kappa(X) = X$. We show that this definition coincides with the definition of a topological space just given. It suffices to show that an arbitrary intersection of closed sets is closed. From (iv) it is obvious that κ is increasing; in combination with (ii) and (iii), this means that κ is a closing. The invariance domain is, by definition, the family of closed sets. By Lemma 3.22(a′), this family is closed under intersections. This shows that the axioms (in terms of the closed sets) of a topology are satisfied.

7.4 Definition. Let $(E, \mathcal{G})$ be a topological space. A family $\mathcal{B} \subseteq \mathcal{G}$ is called a *basis* for $\mathcal{G}$ if every open set can be written as a union of members of $\mathcal{B}$.

A family $\mathcal{C} \subseteq G$ is called a *subbasis* for $\mathcal{G}$ if the family obtained by taking all finite intersections of members of $\mathcal{C}$ is a basis for $\mathcal{G}$.

7.5 Examples.

(a) Let E be a set; the family of all singletons $\{\{x\} \mid x \in E\}$ is a basis for the discrete topology on E.

(b) The family of all open balls $\{B^\circ(x, r) \mid r > 0,\ \ x \in \mathbb{R}^d\}$ is a basis for the Euclidean topology on $\mathbb{R}^d$. Actually, this topology also possesses a countable basis, namely, $\{B^\circ(x, r) \mid r > 0 \text{ is rational and } x \in \mathbb{R}^d \text{ has rational coordinates}\}$.

A topological space satisfies the *second axiom of countability* if it has a countable basis. A topological space E satisfies the *first axiom of countability* if for every $x \in E$ there is countable family U_n of neighbourhoods of x such that every neighbourhood V of x is contained within some U_n. It is obvious that the second axiom of countability implies the first; the converse is not true, however.

The concept of a subbasis provides a general method for defining a topology. In fact, let $\mathcal{C}$ be any family of subsets of E; define $\mathcal{G}$ as the smallest collection of subsets of E such that (i) $\varnothing, E \in \mathcal{G}$, (ii) $\mathcal{C} \subseteq \mathcal{G}$, (iii) $\mathcal{G}$ is closed under arbitrary unions and finite intersections. Then $(E, \mathcal{G})$ is a topological space which has $\mathcal{C}$ as a subbasis. Section 7.4 contains an application of the subbasis approach.

7.6 Definition. A topological space $(E, \mathcal{G})$ is *Hausdorff* if any two distinct points $x, y \in E$ possess neighbourhoods with empty intersection.

Obviously, the Euclidean space $\mathbb{R}^d$ is Hausdorff: in fact, let $x, y \in \mathbb{R}^d$ and $r = \|x - y\|$. Then $B^\circ(x, r/3)$ and $B^\circ(y, r/3)$ are neighbourhoods of x and y with empty intersection.

Let E be a topological space; call $\mathcal{H} \subseteq \mathcal{G}$ an *open covering* of E if the union of all sets contained in $\mathcal{H}$ is E. The space E is said to be *compact* if every open covering $\mathcal{H}$ of E has a finite subcovering, that is, if there exist $H_1, \ldots, H_n \in \mathcal{H}$ such that $E = H_1 \cup \cdots \cup H_n$. The subset $X \subseteq E$ is called compact if the space X with the relative topology is compact. Finite subsets are always compact. The Euclidean space $\mathbb{R}^d$ is not compact since the open covering $\{B^\circ(0, r) \mid r > 0\}$ does not have a finite subcovering. A classical result, known as the *Heine–Borel theorem*, says that a set $X \subseteq \mathbb{R}^d$ is compact if and only if it is closed and bounded. A subset $X \subseteq E$ is called *relatively compact* if $\overline{X}$ is compact.

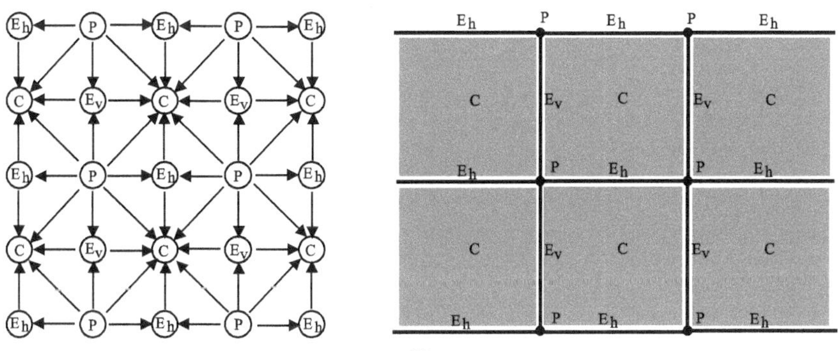

Figure 7.1 Khalimskii topology in two dimensions. (a) The adjacency graph; an arrow from x to y means that $x \rhd y$. (b) Representation of the Khalimskii topology in $\mathbb{R}^2$.

7.7 Definition. A Hausdorff space is *locally compact* if each point has a neighbourhood which is relatively compact. It is called *separable* if it contains a countable subset the closure of which is the entire space.

A locally compact Hausdorff space has a basis comprising relatively compact open sets. If, in addition, the space has a countable basis, then it has a countable basis of relatively compact open sets.

7.8 Example. (Adjacency and topology)
Consider a nonempty set E and a reflexive relation $\rhd$ on $E \times E$. Recall that reflexivity means that $x \rhd x$ for every $x \in E$. If $x \rhd y$, we say that x is *adjacent* to y. This adjacency relation induces a topology on E as follows: a set $X \subseteq E$ is called open if $x \in X$ implies that elements to which x is adjacent, i.e., elements of the set $\{y \mid x \rhd y\}$, are contained in X. It is easy to show that an infinite intersection of open sets is open. Furthermore, if there exist two distinct points x, y in E such that $x \rhd y$, then the space E is not Hausdorff, for every neighbourhood of x contains y. An interesting example is depicted in Fig. 7.1.

The topology illustrated here is known as the *Khalimskii topology*; see Khalimskii (1969). In this example one distinguishes four types of members, referred to as C, P, E_h, and E_v; see Fig. 7.1(a). The representation of this structure in Fig. 7.1(b) gives more insight. Moreover, this representation shows the relevance of this topology with respect to the fields of image processing and computer graphics. The C-members are represented by open squares, called *cells*, the E_h, E_v-members by the horizontal and vertical straight *edges*, respectively, and the P-members by *points*. The adjacency relation can be captured as follows: $x \triangleright y$ if x is contained in the boundary of y. It is obvious that every point is adjacent to four cells and four edges, every edge is adjacent to two cells, and a cell is not adjacent to any other element.

Let E be a topological space and x_n a sequence in E. We say that x_n *converges to* $x \in E$ if for every neighbourhood U of x there is an N such that $x_n \in U$ for every $n \geq N$; in this case we write $x_n \to x$. We say that x is an *accumulation point* of x_n if for every neighbourhood U of x and every integer N there exists an $n \geq N$ such that $x_n \in U$; in other words, x is an accumulation point of x_n if there is a subsequence n_k of n such that $x_{n_k} \to x$ as $k \to \infty$. More generally, x is an accumulation point of the set X if every neighbourhood of x contains infinitely many points of X.

In topological spaces satisfying the first axiom of countability, sequential convergence is adequate to express all topological concepts. For instance, the closure $\overline{X}$ of a set $X \subseteq E$ comprises all points $x \in E$ for which there is a sequence in X which converges toward x.

Let E_1, E_2 be topological spaces, and let ψ be a mapping from E_1 into E_2. We say that ψ is *continuous at the point* $x \in E_1$ if for each neighbourhood V of $\psi(X)$ in E_2 there is a neighbourhood U of x in E_1 such that $\psi(U) \subseteq V$. A mapping is called *continuous* if the inverse image of an open set in E_2 is an open set in E_1. It is rather easy to show that ψ is continuous if and only if ψ is continuous at each point in E_1. If E_1 satisfies the first axiom of countability, then a mapping $\psi : E_1 \to E_2$ is continuous at x if and only if $\psi(x_n) \to \psi(x)$ for every sequence x_n with $x_n \to x$.

7.9 Example. (Boolean lattices and regular closed sets)
Recall from Section 2.3 that a subset X of a topological space E is called a regular closed set if $X = X^{\circ -}$ or, alternatively, if $X = X^{\perp\perp}$, where $X^{\perp} = X^{c-}$. We have shown that the regular closed subsets $\mathcal{R}(E)$ of E constitute a complete Boolean lattice. Ronse (1990a) has proved that every complete Boolean lattice is isomorphic to the lattice of regular closed subsets of some topological space E. Observe that for a set E endowed with the discrete topology, $\mathcal{R}(E) = \mathcal{P}(E)$. To prove the result by Ronse one needs the following auxiliary result.

Given a set E and a complete Boolean lattice $\mathcal{R} \subseteq \mathcal{P}(E)$ ordered by inclusion, such that

(i) $\varnothing, E \in \mathcal{R}$;
(ii) $X, Y \in \mathcal{R} \Rightarrow X \cup Y \in \mathcal{R}$.

If $\mathcal{F}$ contains all subsets of E obtained by intersections of members of $\mathcal{R}$, then E is a topological space with closed subsets $\mathcal{F}(E) = \mathcal{F}$ and regular closed sets $\mathcal{R}(E) = \mathcal{R}$.

Proof. To show that E is a topological space, one must prove that $\varnothing, E \in \mathcal{F}$, that an intersection of members of $\mathcal{F}$ lies in $\mathcal{F}$, and that a finite union of members of $\mathcal{F}$ lies in $\mathcal{F}$. These facts are straightforward. The difficult part of the proof consists in showing that $\mathcal{R}$ contains the regular closed sets. Suppose one can show that for each family $X_i \in \mathcal{R}$, $i \in I$, the following holds:

$$(\bigwedge_{i\in I} X_i)^* = (\bigcap_{i\in I} X_i)^{c-}. \qquad (\dagger)$$

Then, in particular, $X^* = X^{c-}$ for $X \in \mathcal{R}$. As $X^{**} = X$, this means that $X^{\circ-} = X^{c-c-} = X$, and so every member of $\mathcal{R}$ defines a regular set. We show that the converse also holds; from (†) we derive

$$\bigwedge_{i\in I} X_i = (\bigwedge_{i\in I} X_i)^{**} = (\bigcap_{i\in I} X_i)^{c-c-} = (\bigcap_{i\in I} X_i)^{\circ-}.$$

By definition, every closed set X is of the form $X = \bigcap_{i\in I} X_i$, where $X_i \in \mathcal{R}$. The set X is regular if $X = X^{\circ-}$, that is,

$$\bigcap_{i\in I} X_i = \bigwedge_{i\in I} X_i;$$

but this implies that $X \in \mathcal{R}$. Therefore, it remains to prove (†). Take $Y \in \mathcal{R}$; by infinite infimum distributivity,

$$\begin{aligned}
(\bigwedge_{i\in I} X_i)^* \subseteq Y &\iff Y \vee (\bigwedge_{i\in I} X_i) = E \\
&\iff \bigwedge_{i\in I}(Y \cup X_i) = E \\
&\iff Y \cup X_i = E, \qquad \text{for every } i \in I, \\
&\iff Y \cup (\bigcap_{i\in I} X_i) = E \\
&\iff (\bigcap_{i\in I} X_i)^c \subseteq Y \\
&\iff (\bigcap_{i\in I} X_i)^{c-} \subseteq Y,
\end{aligned}$$

where, in the final equivalence, the fact that Y is closed has been used. This means that $(\bigwedge_{i\in I} X_i)^*$ is the smallest $Y \in \mathcal{R}$ such that

$(\bigcap_{i\in I} X_i)^{c-} \subseteq Y$. By the very definition of $\mathcal{F}$, $(\bigcap_{i\in I} X_i)^{c-}$ is the intersection of all such sets Y, and it follows that both sets are equal. This proves the result. □

Now let $\mathcal{L}$ be an arbitrary complete Boolean lattice. Define $E = \mathcal{L} \setminus \{I\}$. Let, for $A \in \mathcal{L}$, the set $[A] \subseteq E$ be given by

$$[A] = \{X \in E \mid X \not\geq A\}.$$

Define $\mathcal{R} \subseteq \mathcal{P}(E)$ as the family of all sets $[A]$, where $A \in \mathcal{L}$. It is not difficult to prove that the mapping $A \mapsto [A]$ from $\mathcal{L}$ into $\mathcal{R}$ defines a lattice isomorphism. From $[O] = \varnothing$, $[I] = E$, and the identity

$$[A \vee B] = [A] \cup [B],$$

it follows that the conditions in the preceding auxiliary result are satisfied. Therefore, $\mathcal{R}$ contains the regular closed sets of the topological space E.

7.2. Metric spaces

It is the purpose of this section to discuss briefly some concepts from the theory of metric spaces. Some further results can be found in Chapter 9.

7.10 Definition. A function $d : E \times E \to \mathbb{R}_+$ is called a *metric* or *distance function* if it satisfies the following axioms:

(D1) $d(x, y) = 0$ if and only if $x = y$;

(D2) $d(x, y) = d(y, x)$;

(D3) $d(x, z) \leq d(x, y) + d(y, z)$.

This last property is called the *triangle inequality*. The pair (E, d) is called a *metric space*.

We call $d(x, y)$ the *distance* between the points x and y. If $r > 0$ and $x \in E$, then $B(x, r) = \{y \in E \mid d(x, y) \leq r\}$ and $B^\circ(x, r) = \{y \in E \mid d(x, y) < r\}$ are the closed and open ball with radius r and centre x, respectively.

7.11 Examples.

(a) Let E be an arbitrary set. Given $x, y \in E$, define $d(x, y) = 1$ if $x \neq y$ and 0 otherwise. It is trivial that d defines a metric on E.

(b) On the space $\mathbb{R}^d$ one can define a family of metrics d_k, $k = 1, 2, \ldots, \infty$, as follows:

$$d_k(x, y) = \left\{\Sigma_{i=1}^{d} |x_i - y_i|^k\right\}^{1/k},$$

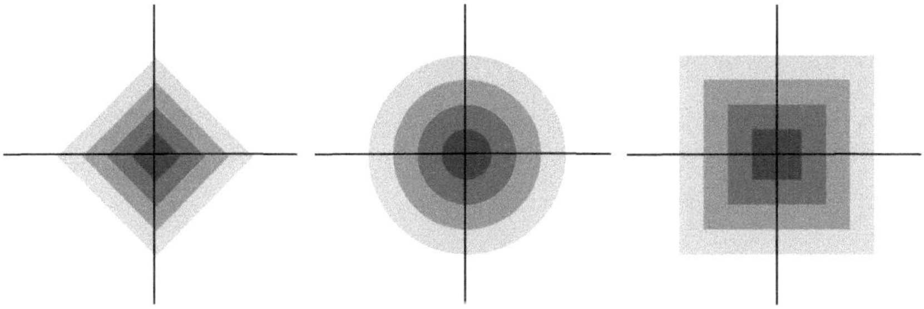

Figure 7.2 The closed balls with respect to the metric d_k for $k = 1, 2, \infty$.

if $1 \leq k < \infty$ and

$$d_\infty(x, y) = \max_{1 \leq i \leq d} |x_i - y_i|.$$

The metric d_2 corresponds to the well-known *Euclidean distance*. In Section 9.2 it is explained that these metrics derive from a norm. The discrete variants of d_1 and d_∞ are called the *city block distance* and *chessboard distance*, respectively; see Example 9.73. In Fig. 7.2 we have depicted the closed balls for $k = 1, 2, \infty$.

On every metric space (E, d) one can define a topology as follows: call a set X open if for every $x \in X$ there is an $r > 0$ such that $B^\circ(x, r) \subseteq X$. One can easily verify that this defines a topology. The family of open balls $B^\circ(x, r)$ where $x \in E$ and $r > 0$ defines a basis for this topology. A topological space E is called *metrizable* if its topology is induced by a metric on E. Every metric space satisfies the first axiom of countability since $\{B^\circ(x, r) \mid r > 0 \text{ rational}\}$ is a countable basis at x. Furthermore, a metric space is separable if and only if it has a countable basis (Dugundji, 1966, Theorem IX.5.6).

Two metrics on the set E are said to be *topologically equivalent* when they induce the same topology on E. One can show that the metrics of Example 7.11(b) induce the same topology, namely, the Euclidean topology; see also Section 9.2.

Let the function $f : \mathbb{R}_+ \to \mathbb{R}_+$ be nondecreasing and satisfy $f(u) = 0$ iff $u = 0$ and $f(u+v) \leq f(u) + f(v)$. If d is a metric on E and $d'(x, y) = f(d(x, y))$, then d' is also a metric. A well-known example is given by the function $f(u) = u/(1 + u)$. In this particular case the two metrics are topologically equivalent.

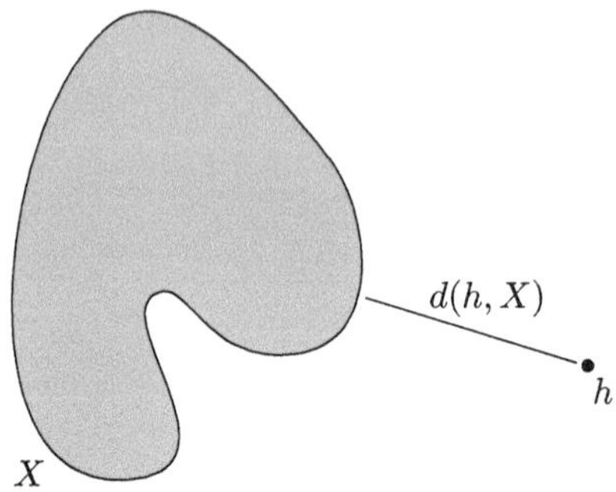

Figure 7.3 The distance between a point and a set.

Define the distance between a point $h \in E$ and a nonempty set $X \subseteq E$ by

$$d(h, X) = \inf_{x \in X} d(h, x); \tag{7.1}$$

see Fig. 7.3.

It is obvious that $d(h, X) = 0$ iff $h \in \overline{X}$. This distance concept plays an important role in the following section, where the Hausdorff metric will be discussed.

7.12 Proposition. *Let (E, d) be a metric space. Then*

$$|d(h, X) - d(k, X)| \leq d(h, k), \tag{7.2}$$

for $h, k \in E$ and $X \subseteq E$.

Proof. For $x \in X$ one has, by the triangle inequality $d(h, x) \leq d(h, k) + d(k, x)$. Taking the infimum over all x, first on the left-hand side, subsequently on the right-hand side, one gets $d(h, X) \leq d(h, k) + d(k, X)$. Analogously, $d(k, X) \leq d(h, X) + d(h, k)$. Combining both estimates gives the result. □

We need some further terminology.

7.13 Definition. Let (E, d) be a metric space; a sequence $\{x_n\}$ in E is called a *Cauchy sequence* if, for every $\epsilon > 0$, there is an integer $N = N(\epsilon)$ such that, for all $n, m \geq N$, we have $d(x_n, x_m) \leq \epsilon$.

Every convergent sequence is a Cauchy sequence. If the converse holds, that is, if every Cauchy sequence is convergent, then the space (E, d) is called *complete*.

The previous concepts, although they concern convergence of sequences, refer to metric properties and not topological ones. By this, we mean that the property may be true with respect to one metric on E but not with respect to another one, even when both metrics induce the same topology. It is, e.g., possible that of two metrics which induce the same topology, only one is complete.

A subset X of a metric space (E, d) is *bounded* if there is a real number $L > 0$ such that $d(x, y) \leq L$ for every pair $x, y \in X$. A well-known result in topology says that every compact set is closed and bounded.

7.14 Definition. A metric space (E, d) is called *finitely compact* if every infinite bounded subset has at least one accumulation point.

Finitely compact metric spaces play a central role in Chapter 9. This section concludes with some basic properties. For their proofs we refer to Busemann (1955) or the excellent monograph by Rinow (1961). First observe that $\mathbb{R}^d$ is finitely compact; in fact this is an alternative formulation of the Bolzano–Weierstrass theorem in analysis.

7.15 Proposition. *Let (E, d) be a metric space. The following two assertions are equivalent:*

(i) *(E, d) is finitely compact;*

(ii) *a set $X \subseteq E$ is compact iff it is closed and bounded.*

That finite compactness is a metric property and not a topological one is easily understood if one recalls that with every metric d one can associate the bounded metric $d' = d/(1 + d)$, which is topologically equivalent.

7.16 Proposition.

(a) *Every compact metric space is finitely compact.*

(b) *Every finitely compact metric space is complete, locally compact, and separable.*

7.3. Hausdorff metric

One of the main goals of this chapter is to provide the closed subsets of a topological space E with a topology. First consider the case where E is a metric space. Following Hausdorff (1962) we define the *upper* and *lower (closed) limit* of a sequence of sets in E. Hausdorff also introduced a metric, nowadays called the *Hausdorff metric*, on the space of bounded closed subsets of E and pointed out the connection with the upper and lower limit. Later, Busemann (1955) extended this metric to the space of all closed sets.

Section 7.3.1 presents definitions of the upper and lower limit of a set and introduces the Hausdorff–Busemann metric for closed sets. Subsequently, Section 7.3.2 considers the original Hausdorff metric for bounded closed sets. Throughout this section it is assumed that (E, d) is a finitely compact metric space.

7.3.1 Closed sets

To get some intuition we start with an example. Consider the increasing family of disks in $\mathbb{R}^2$,

$$(x - r)^2 + y^2 \leq r^2.$$

Does there exist a notion of convergence under which this family tends to the right half-plane as $r \to \infty$?

Given a sequence X_n of nonempty subsets of E, define the *upper limit* $\overline{\lim}\, X_n$ as the collection of all accumulation points of sequences $x_n \in X_n$. In other words, $x \in \overline{\lim}\, X_n$ if there exists a subsequence n_k and points $x_{n_k} \in X_{n_k}$ such that $x_{n_k} \to x$. Furthermore, define the *lower limit* $\underline{\lim}\, X_n$ as the set of points x for which there is a sequence $x_n \in X_n$ which tends to x. It is obvious that

$$\underline{\lim}\, X_n \subseteq \overline{\lim}\, X_n. \tag{7.3}$$

The sequence X_n is said to converge to X (notation $\lim X_n = X$) if $\underline{\lim}\, X_n = \overline{\lim}\, X_n = X$. The set $\lim X_n$, if it exists, is called the *closed limit* of X_n. The following result is easily established.

7.17 Proposition. *Suppose that E is a finitely compact metric space and that $X_n \subseteq E$.*

(a) $\overline{\lim}\, X_n$ *and* $\underline{\lim}\, X_n$ *are closed sets.*

(b) $\overline{\lim}\, X_n = \overline{\lim}\, \overline{X}_n$ *and* $\underline{\lim}\, X_n = \underline{\lim}\, \overline{X}_n$.

7.18 Examples.

(a) Let $X_n = X$ for every n; then $\lim X_n = \overline{X}$.

(b) Let $X_n = \{x_n\}$, where $x_n \in E$. Then $\overline{\lim}\, X_n$ is the set of accumulation points of the sequence x_n. Furthermore, $\underline{\lim}\, X_n$ is the singleton containing the limit of the sequence x_n if it exists; otherwise it is the empty set.

(c) Let $X_n \subseteq \mathbb{R}^2$ be the horizontal line $y = n$. Then $\overline{\lim}\, X_n = \underline{\lim}\, X_n = \varnothing$; hence $\lim X_n = \varnothing$.

(d) If X_n is the right closed half-plane in $\mathbb{R}^2$ if n is odd and the left closed half-plane if n is even, then $\overline{\lim}\, X_n = \mathbb{R}^2$ and $\underline{\lim}\, X_n$ is the y-axis.

There exists a distance function D on $\mathcal{P}(E)$ such that

$$\lim X_n = \overline{X} \iff D(X_n, X) \to 0.$$

In fact, let o be a fixed element of E, and define

$$D(X, Y) = \sup_{h \in E} |d(h, X) - d(h, Y)| e^{-d(o,h)}, \tag{7.4}$$

for nonempty $X, Y \subseteq E$; here $d(h, X)$ is given by (7.1). It is easy to see that $|d(h, X) - d(h, Y)| \leq d(o, X) + d(o, Y) + 2d(o, h)$. Since $d(o, h)e^{-d(o,h)} \leq 1$ for all h, the value given by (7.4) is finite. From $d(h, X) = d(h, \overline{X})$ it follows that

$$D(X, \overline{X}) = 0.$$

It is obvious that $D(X, Y) = D(Y, X)$; furthermore,

$$D(X, Y) = 0 \iff \overline{X} = \overline{Y}. \tag{7.5}$$

In fact, if $\overline{X} = \overline{Y}$, then $d(h, X) = d(h, \overline{X}) = d(h, \overline{Y}) = d(h, Y)$ for every $h \in E$, and it follows that $D(X, Y) = 0$. Conversely, if $D(X, Y) = 0$, then $d(h, X) = d(h, Y)$ for every $h \in E$. If $h \in \overline{X} \setminus \overline{Y}$, then $d(h, X) = 0$, and hence $d(h, Y) = 0$. This gives $h \in \overline{Y}$, a contradiction. The same conclusion can be reached for $h \in \overline{Y} \setminus \overline{X}$; therefore, $\overline{X} = \overline{Y}$.

Next, we prove the triangle inequality

$$D(X, Z) \leq D(X, Y) + D(Y, Z), \qquad \text{for } X, Y, Z \subseteq E.$$

For every $\epsilon > 0$ there is a point $h \in E$ such that

$$\begin{aligned} D(X, Z) &\leq |d(h, X) - d(h, Z)| e^{-d(o,h)} + \epsilon \\ &\leq \Big[|d(h, X) - d(h, Y)| + |d(h, Y) - d(h, Z)| \Big] e^{-d(o,h)} + \epsilon \\ &\leq D(X, Y) + D(Y, Z) + \epsilon. \end{aligned}$$

As this holds for every $\epsilon > 0$, the assertion follows. Thus the following result has been proved.

7.19 Proposition. *Let E be a finitely compact metric space. The nonempty closed sets in E provided with the distance D given by* (7.4) *constitute a metric space.*

We call D the *Hausdorff–Busemann metric.* The next result establishes the connection between the closed limit just introduced and the metric D. We use the following convention. If we say "there is a subsequence $x_{n_k} \in X_{n_k}$ such that ...", then we mean "there is a subsequence n_k and a sequence $x_{n_k} \in X_{n_k}$ such that ...".

7.20 Theorem. *Suppose that E is a finitely compact metric space. Let X, X_n be nonempty closed sets in E; then $\lim X_n = X$ if and only if $D(X_n, X) \to 0$.*

Proof. "if": Assume that $D(X_n, X) \to 0$; for every $x \in X$,

$$D(X_n, X) \geq |d(x, X_n) - d(x, X)|e^{-d(o,x)} = d(x, X_n)e^{-d(o,x)}.$$

This means that $d(x, X_n) \to 0$ as $n \to \infty$. So there is a point $x_n \in X_n$ such that $d(x, x_n) \to 0$; therefore, $X \subseteq \underline{\lim}\, X_n$. On the other hand, if $x \in \overline{\lim}\, X_n$, then there is a subsequence $x_{n_k} \in X_{n_k}$ such that $x_{n_k} \to x$. It follows that

$$\begin{aligned} D(X_{n_k}, X)e^{d(o,x)} &\geq |d(x, X_{n_k}) - d(x, X)| \\ &\geq d(x, X) - d(x_{n_k}, X_{n_k}) - d(x, x_{n_k}). \end{aligned}$$

Letting k tend to ∞ one gets $d(x, X) = 0$; hence $x \in X$. Therefore, $\overline{\lim}\, X_n \subseteq X$, and we may conclude that $\lim X_n = X$.

"only if": Assume that $\lim X_n = X$; it must be demonstrated that $D(X_n, X) \to 0$. If not, there is an $\epsilon > 0$ and a subsequence n_k such that $D(X_{n_k}, X) \geq \epsilon$. Without loss of generality, we may assume $D(X_n, X) \geq \epsilon$. Thus there exist points h_n such that

$$|d(h_n, X) - d(h_n, X_n)|e^{-d(o,h_n)} \geq \frac{\epsilon}{2}. \tag{7.6}$$

Given $x \in X$, there exists $x_n \in X_n$ such that $x_n \to x$. From the estimates

$$\begin{aligned} \frac{\epsilon}{2}e^{d(o,h_n)} &\leq |d(h_n, X) - d(h_n, X_n)| \leq d(h_n, x_n) + d(h_n, x) \\ &\leq 2d(o, h_n) + d(o, x_n) + d(o, x), \end{aligned}$$

one deduces easily that h_n is bounded. From (7.6) it follows that there are the following two possibilities: either there is a subsequence n_k such that

$$d(h_{n_k}, X_{n_k}) - d(h_{n_k}, X) \geq \frac{\epsilon}{2}, \tag{7.7}$$

or there is a subsequence n_l such that

$$d(h_{n_l}, X) - d(h_{n_l}, X_{n_l}) \geq \frac{\epsilon}{2}. \tag{7.8}$$

In the first case, choose $y_{n_k} \in X$ such that

$$d(h_{n_k}, X) \geq d(h_{n_k}, y_{n_k}) - \frac{\epsilon}{4}.$$

Then y_{n_k} has an accumulation point $y \in X$; here we have used that E is finitely compact. It is evident that $y \in \overline{\lim}\, X_{n_k}$. Then, by (7.7),

$$\begin{aligned} d(y_{n_k}, X_{n_k}) &\geq d(h_{n_k}, X_{n_k}) - d(y_{n_k}, h_{n_k}) \\ &\geq d(h_{n_k}, X_{n_k}) - d(h_{n_k}, X) - \frac{\epsilon}{4} \geq \frac{\epsilon}{4}. \end{aligned}$$

Thus, for large k, $d(y, X_{n_k}) \geq \epsilon/8$. But this contradicts $y \in \overline{\lim}\, X_{n_k}$.

In the second case, pick $y_{n_l} \in X_{n_l}$ such that

$$d(h_{n_l}, X_{n_l}) \geq d(h_{n_l}, y_{n_l}) - \frac{\epsilon}{4}.$$

Using that h_n is bounded, one concludes easily that y_{n_l} is bounded. Hence this sequence has an accumulation point $y \in X$. Using (7.8), it follows

$$\begin{aligned} d(y_{n_l}, X) &\geq d(h_{n_l}, X) - d(h_{n_l}, y_{n_l}) \\ &\geq d(h_{n_l}, X) - d(h_{n_l}, X_{n_l}) - \frac{\epsilon}{4} \geq \frac{\epsilon}{4}. \end{aligned}$$

But this contradicts the fact that y_{n_l} has an accumulation point in X. □

7.3.2 Compact sets

If one restricts attention to nonempty compact sets, denoted by $\mathcal{K}'(E)$ or just $\mathcal{K}'$, then the exponential factor $e^{-d(o,h)}$ in the expression for D can be omitted. Define

$$D_H(X, Y) = \sup_{h \in E} |d(h, X) - d(h, Y)|, \tag{7.9}$$

for $X, Y \in \mathcal{K}'(E)$. Here the subscript "H" indicates that D_H is the original *Hausdorff metric*; see Hausdorff (1962). Analogously to Proposition 7.19 one has the following result.

7.21 Proposition. *If E is a finitely compact metric space, then D_H defines a metric on $\mathcal{K}'(E)$.*

Define

$$\hat{D}(X, Y) = \sup_{x \in X} d(x, Y), \tag{7.10}$$

for two nonempty sets X, Y. The next result contains an alternative characterization of the Hausdorff metric.

7.22 Proposition. *Suppose that E is a finitely compact metric space. For $X, Y \in \mathcal{K}'$,*

$$D_H(X, Y) = \max\{\hat{D}(X, Y), \hat{D}(Y, X)\}. \tag{7.11}$$

Proof. Let $X, Y \in \mathcal{K}'$; for every $\epsilon > 0$, there is an $x \in X$ such that $d(x, Y) \geq \hat{D}(X, Y) - \epsilon$. Since $d(x, X) = 0$, we get

$$\hat{D}(X, Y) - \epsilon \leq |d(x, Y) - d(x, X)| \leq D_H(X, Y).$$

A similar expression holds for $\hat{D}(Y, X)$, and we conclude $D_H(X, Y) \geq \max\{\hat{D}(X, Y), \hat{D}(Y, X)\}$.

To prove the reverse inequality, take $h \in E$. We have

$$d(h, X) \leq d(y, X) + d(h, y),$$

for arbitrary y. Choose $y \in Y$ such that $d(h, y) \leq d(h, Y) + \epsilon$; this gives

$$\begin{aligned} d(h, X) &\leq d(y, X) + d(h, Y) + \epsilon \\ &\leq \hat{D}(Y, X) + d(h, Y) + \epsilon \\ &\leq \max\{\hat{D}(X, Y), \hat{D}(Y, X)\} + d(h, Y) + \epsilon. \end{aligned}$$

We find a similar inequality for $d(h, Y)$, and thus

$$|d(h, X) - d(h, Y)| \leq \max\{\hat{D}(X, Y), \hat{D}(Y, X)\} + \epsilon.$$

Taking the supremum over h, we get

$$D_H(X, Y) \leq \max\{\hat{D}(X, Y), \hat{D}(Y, X)\} + \epsilon.$$

As this holds for every $\epsilon > 0$, the inequality also holds for $\epsilon = 0$, and the result follows. □

The Hausdorff metric is illustrated in Fig. 7.4.

Yet a third characterization of the Hausdorff metric can be found in Example 9.38.

As $e^{-d(o,h)} \leq 1$ for every $h \in E$,

$$D(X, Y) \leq D_H(X, Y), \qquad X, Y \in \mathcal{K}'.$$

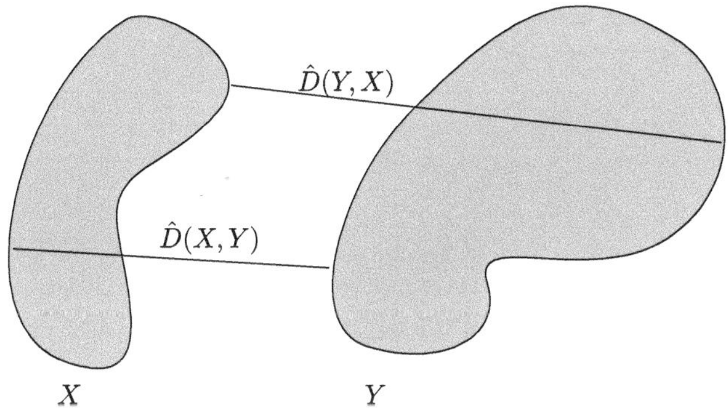

Figure 7.4 The Hausdorff distance between X and Y equals $\hat{D}(Y,X)$ because $\hat{D}(X,Y) < \hat{D}(Y,X)$.

Therefore, $D_H(X_n, X) \to 0$ implies $D(X_n, X) \to 0$. That the converse is not true can be seen from the following counterexample. Take for $X_n \subseteq \mathbb{R}$ the set $X_n = \{0, n\}$. It is easy to see that $\lim X_n = \{0\}$; hence, by Theorem 7.20, $D(X_n, X) \to 0$, where $X = \{0\}$. However, $D_H(X_n, X) = n$; in particular, $D_H(X_n, X)$ does not tend to zero. This observation means that the topology on $\mathcal{K}'$ induced by the metric D is not equivalent to the topology induced by D_H; see also Section 7.6.

7.4. Hit-or-miss topology

Throughout this section E is a locally compact Hausdorff space with countable basis unless stated otherwise. For our purposes, the main example is $E = \mathbb{R}^d$, but the theory is also valid for more general spaces. Let $\mathcal{F}(E)$, $\mathcal{G}(E)$, and $\mathcal{K}(E)$ be the closed, open, and compact subsets of E, respectively. When no confusion about the underlying space E seems possible, the argument E is suppressed in our the notation. We define a topology on $\mathcal{F}(E)$, called the *hit-or-miss topology*. When E is a finitely compact metric space, restriction of this topology to $\mathcal{F}'(E) = \mathcal{F}(E) \setminus \{\varnothing\}$ coincides with the topology induced by the Hausdorff–Busemann metric discussed in the previous section.

Define, for $A \subseteq E$,

$$\mathcal{F}^A = \{X \in \mathcal{F} \mid X \cap A = \varnothing\},$$
$$\mathcal{F}_A = \{X \in \mathcal{F} \mid X \cap A \neq \varnothing\}.$$

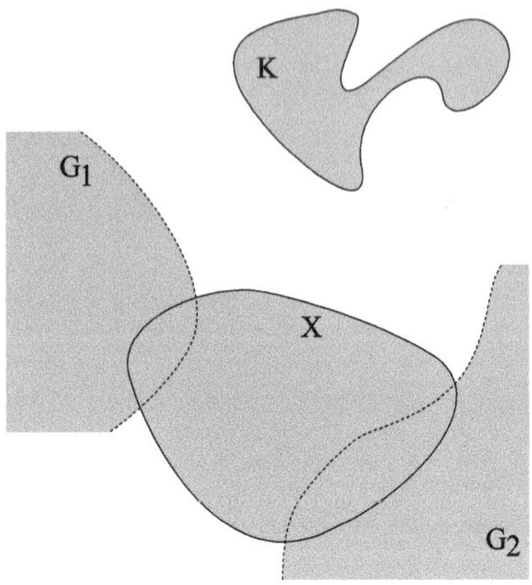

Figure 7.5 The set X is contained in $\mathcal{F}^K_{G_1,G_2}$.

For an arbitrary family $\{A_i \mid i \in I\}$ of subsets of E,

$$\bigcap_{i \in I} \mathcal{F}^{A_i} = \mathcal{F}^{\bigcup_{i\in I} A_i}, \tag{7.12}$$

$$\bigcup_{i \in I} \mathcal{F}_{A_i} = \mathcal{F}_{\bigcup_{i\in I} A_i}. \tag{7.13}$$

7.23 Definition. The *hit-or-miss topology* is the topology generated by the subbasis $\{\mathcal{F}^K \mid K \in \mathcal{K}\} \cup \{\mathcal{F}_G \mid G \in \mathcal{G}\}$.

One obtains a basis for this topology by taking all finite intersections of subbasis members. From (7.12) it follows

$$\bigcap_{i=1}^{n} \mathcal{F}^{K_i} = \mathcal{F}^K;$$

here $K = K_1 \cup \cdots \cup K_n$ is a compact set. Thus a basis for the hit-or-miss topology is constituted by the sets

$$\mathcal{F}^K_{G_1,G_2,\ldots,G_n} = \mathcal{F}^K \cap \mathcal{F}_{G_1} \cap \cdots \cap \mathcal{F}_{G_n}, \tag{7.14}$$

where K is compact and $G_1, \ldots, G_n$ are open; refer to Fig. 7.5 for an illustration.

It is allowed that $n = 0$, in which case (7.14) reduces to $\mathcal{F}^K$. Furthermore, $\mathcal{F}^K_G = \mathcal{F}_G$ if $K = \varnothing$.

7.24 Theorem. *The space $\mathcal{F}$ with the hit-or-miss topology is Hausdorff, compact, and has a countable basis.*

Proof. "Hausdorff": Given $X, Y \in \mathcal{F}$ such that $X \neq Y$, there is a point x which lies in X but not in Y. Thus there exists a neighbourhood U of x with compact closure such that $Y \cap \overline{U} = \varnothing$. This means that $X \in \mathcal{F}_U$, $Y \in \mathcal{F}^{\overline{U}}$, and $\mathcal{F}_U \cap \mathcal{F}^{\overline{U}} = \varnothing$.

"Compactness": Observe first that in the proof of compactness of a topological space one may restrict attention to open coverings the members of which are contained in the subbasis of the topology. Assume that there exist K_i $(i \in I)$ in $\mathcal{K}$ and G_j $(j \in J)$ in $\mathcal{G}$ such that

$$(\bigcup_{i\in I} \mathcal{F}^{K_i}) \cup (\bigcup_{j\in J} \mathcal{F}_{G_j}) = \mathcal{F}.$$

Taking complements with respect to $\mathcal{F}$ gives

$$(\bigcap_{i\in I} \mathcal{F}_{K_i}) \cap (\bigcap_{j\in J} \mathcal{F}^{G_j}) = \varnothing. \qquad (*)$$

Put $G = \bigcup_{j\in J} G_j$; then by (7.12), $\bigcap_{j\in J} \mathcal{F}^{G_j} = \mathcal{F}^G$. Suppose that $K_i \not\subseteq G$ for every $i \in I$; then $G^c \cap \overline{(\bigcup_{i\in I} K_i)}$ is a closed set which is disjoint from G but intersects every K_i, meaning that $(*)$ is not satisfied. Thus we may conclude that $K_{i_0} \subseteq G$ for some i_0. As K_{i_0} is compact, there exist finitely many indices $j_1, j_2, \ldots, j_p$ in J such that $K_{i_0} \subseteq G_{j_1} \cup G_{j_2} \cup \cdots \cup G_{j_p}$. Then

$$\mathcal{F}_{K_{i_0}} \cap \mathcal{F}^{G_{j_1}} \cap \mathcal{F}^{G_{j_2}} \cap \cdots \cap \mathcal{F}^{G_{j_p}} = \varnothing;$$

hence, taking complements,

$$\mathcal{F}^{K_{i_0}} \cup \mathcal{F}_{G_{j_1}} \cup \mathcal{F}_{G_{j_2}} \cup \cdots \cup \mathcal{F}_{G_{j_p}} = \mathcal{F}.$$

Thus we have obtained a finite subcovering.

"Countable basis": Since the topology on E has a countable basis and is locally compact, it has a countable basis $\mathcal{B}$ comprising relatively compact open sets. Consider the countable family of subsets $\mathcal{F}_{B_1,\ldots,B_n}^{\overline{C_1}\cup\cdots\cup\overline{C_m}}$, where $B_1, \ldots, B_n, C_1, \ldots, C_m \in \mathcal{B}$. We claim that this collection defines a basis for the hit-or-miss topology. To prove this, we show that for every $X \in \mathcal{F}$ and for every basis member $\mathcal{F}_{G_1,G_2,\ldots,G_n}^K$ that contains X there is a member $\tilde{\mathcal{F}}$ in our new family such that $X \in \tilde{\mathcal{F}} \subseteq \mathcal{F}_{G_1,G_2,\ldots,G_n}^K$. We only consider the case where $X \neq \varnothing$; for $X = \varnothing$, the proof is left to the reader. So let

$\varnothing \neq X \in \mathcal{F}^K_{G_1,G_2,\ldots,G_n}$. Take $x_i \in X \cap G_i$; then $x \notin K$, and hence $x \in G_i \cap K^c$. Choose $B_i \in \mathcal{B}$ such that

$$x \in B_i \subseteq \overline{B}_i \subseteq G_i \cap K^c;$$

apparently, $\mathcal{F}_{B_i} \subseteq \mathcal{F}_{G_i}$. Let $k \in K$; then $k \in X^c$. There exists a $C(k) \in \mathcal{B}$ such that $k \in C(k) \subseteq \overline{C(k)} \subseteq X^c$. Thus $\{C(k) \mid k \in K\}$ is an open covering of K, and, since K is compact, it has a finite subcovering $C_1, \ldots, C_m$. From $\overline{C_j} \subseteq X^c$ it follows that

$$K \subseteq \overline{C_1} \cup \cdots \cup \overline{C_m} \subseteq X^c.$$

Thus we have shown

$$X \in \mathcal{F}^{\overline{C_1}\cup\cdots\cup\overline{C_m}}_{B_1,\ldots,B_n} \subseteq \mathcal{F}^K_{G_1,\ldots,G_n}.$$

This proves the result. □

Since the space $\mathcal{F}$ has a countable basis, sequential convergence is adequate for topological concepts, in particular for the characterization of (semi-) continuity of operators. Henceforth $X_n \xrightarrow{\mathcal{F}} X$ means that $X_n, X \in \mathcal{F}$ and X_n converges to X with respect to the hit-or-miss topology. It will be shown that convergence in $\mathcal{F}$ with respect to the hit-or-miss topology can be characterized in terms of the upper and lower limit introduced in the previous section. Before making this explicit, observe that the definitions of upper and lower limit carry over to an arbitrary Hausdorff space; furthermore, the results in Proposition 7.17 remain valid in this case. In what follows, "eventually" means "for n sufficiently large".

7.25 Proposition. *Let X_n be a sequence in $\mathcal{F}$ and $X \in \mathcal{F}$. Then $X_n \xrightarrow{\mathcal{F}} X$ iff the following two rules are obeyed:*

(i) *if $X \cap G \neq \varnothing$, then $X_n \cap G \neq \varnothing$ eventually, for every open set G;*

(ii) *if $X \cap K = \varnothing$, then $X_n \cap K = \varnothing$ eventually, for every compact set K.*

Proof. The proof of the only if-statement is straightforward.

"if": Assume that (i) and (ii) hold; we show that $X_n \xrightarrow{\mathcal{F}} X$. Let $\mathcal{F}_0 \subseteq \mathcal{F}$ be open with respect to the hit-or-miss topology and $X \in \mathcal{F}_0$. There exist open sets $G_1, G_2, \ldots, G_m$ and a compact set K such that $X \in \mathcal{F}^K_{G_1,\ldots,G_n} \subseteq \mathcal{F}_0$. In other words, $X \cap K = \varnothing$ and $X \cap G_i \neq \varnothing$, $i = 1, \ldots, m$. From rules (i)–(ii) one derives that $X_n \cap K = \varnothing$ and $X_n \cap G_i \neq \varnothing$, $i = 1, \ldots, m$, eventually. This means that $X_n \in \mathcal{F}^K_{G_1,\ldots,G_n} \subseteq \mathcal{F}_0$ eventually; therefore $X_n \xrightarrow{\mathcal{F}} X$. □

7.26 Proposition. *Let X_n be a sequence in $\mathcal{F}$ and $X \in \mathcal{F}$. Rules* (i) *and* (ii) *in Proposition 7.25 are, respectively, equivalent to*

(i) $X \subseteq \underline{\lim}\, X_n$;

(ii) $\overline{\lim}\, X_n \subseteq X$.

In particular, $X_n \xrightarrow{\mathcal{F}} X$ if (i)–(ii) *are satisfied.*

Proof. We prove only that rule (i) in Proposition 7.25 is equivalent to (i) here. For rule (ii) the proof is very similar.

$\Rightarrow$: Assume that $x \in X$. To show that $x \in \underline{\lim}\, X_n$ one can use the following property: in a locally compact Hausdorff space with countable basis, every point x has a system of relatively compact neighbourhoods U_n such that $U_{n+1} \subseteq U_n$ and $\bigcap_{n\geq 1} U_n = \{x\}$. Since $X \cap U_p \neq \varnothing$, it follows that $X_n \cap U_p \neq \varnothing$ eventually, say for $n \geq N_p$. Let $x_n \in X_n \cap U_p$ for $n = N_p, N_p + 1, \ldots, N_{p+1} - 1$; it is obvious that this sequence converges toward x. Therefore, $x \in \underline{\lim}\, X_n$.

$\Leftarrow$: If $X = \varnothing$, there is nothing to be proved. Assume that $X \neq \varnothing$ and that G is an open set such that $X \cap G \neq \varnothing$. Then $\underline{\lim}\, X_n \cap G \neq \varnothing$; so there must be a sequence $x_n \in X_n$ converging to an element $x \in G$. This means, however, that $x_n \in G$ eventually; that is, $X_n \cap G \neq \varnothing$ eventually. □

This result suggests alternative characterizations of the upper and lower limit. In fact, $\underline{\lim}\, X_n$ is the largest closed set that satisfies Proposition 7.25(i) and $\overline{\lim}\, X_n$ is the smallest closed set that satisfies Proposition 7.25(ii); see Fig. 7.6 for an illustration.

Combination of Proposition 7.26 and Theorem 7.20 leads to the following result.

7.27 Corollary. *Let (E, d) be a finitely compact metric space. The topology on $\mathcal{F}'(E)$ induced by the Hausdorff–Busemann metric coincides with the relative hit-or-miss topology.*

Observe that a finitely compact metric space is Hausdorff, locally compact and has a countable basis. The last property is a consequence of the fact that a finitely compact space is separable (cf. Proposition 7.16).

For future reference we mention the following alternative expression for the upper limit:

$$\overline{\lim}\, X_n = \bigcap_{N\geq 1} \overline{\bigcup_{n\geq N} X_n}. \tag{7.15}$$

The inclusion $\subseteq$ is straightforward. To prove $\supseteq$, let $x \in \bigcap_{N\geq 1} \overline{\bigcup_{n\geq N} X_n}$. Let U_n be a decreasing sequence of neighbourhoods of x with intersection $\{x\}$. We construct a sequence x_{m_k} such that $x_{m_k} \in X_{m_k} \cap U_k$. As $x \in \overline{\bigcup_{n\geq N} X_n}$

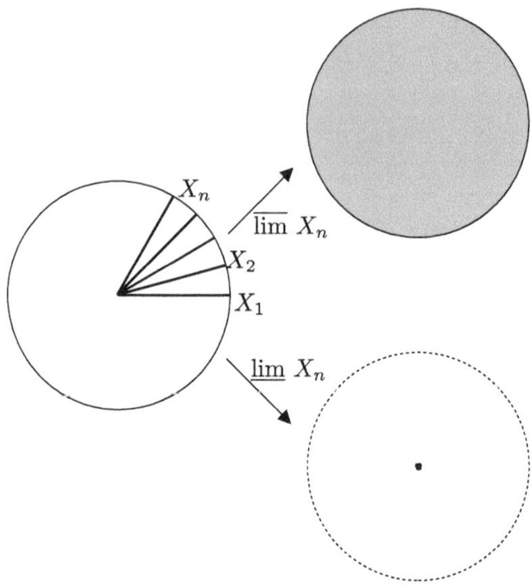

Figure 7.6 Let X_n be the radii of the closed unit disk which form an angle $n\vartheta$ with the positive real axis. If $\vartheta/2\pi$ is irrational, then $\overline{\lim}\ X_n$ is the closed unit disk, whereas $\underline{\lim}\ X_n = \{(0,0)\}$.

for every $N \geq 1$, there is a sequence x_k^N in $\bigcup_{n\geq N} X_n$ which converges to x as $k \to \infty$. Choose $x_{k_1}^1 \in U_1$. Clearly, $x_{k_1}^1 \in X_{m_1}$ for some $m_1 \geq 1$; now $x_{m_1} = x_{k_1}^1$. Next, choose $x_{k_2}^{m_1+1}$ inside U_2. Obviously, $x_{k_2}^{m_1+1} \in X_{m_2}$ for some $m_2 > m_1$; put $x_{m_2} = x_{k_2}^{m_1+1}$. One can repeat this procedure for every k and obtain a sequence $x_{m_k} \in X_{m_k} \cap U_k$. From the fact that $x_{m_k} \in U_k$, it follows that $x_{m_k} \to x$; hence $x \in \overline{\lim}\ X_n$, which had to be proved.

In Section 3.1 two different notions of monotone convergence were introduced. These concepts are not a priori related to some topology but make use of a partial ordering relation. Here we point out the connection between monotone convergence on $\mathcal{P}(E)$ and convergence with respect to the hit-or-miss topology.

Let X_n be subsets of E; we write

$$X_n \downarrow X \quad \text{if} \quad X_1 \supseteq X_2 \supseteq \cdots \qquad \text{and} \qquad X = \bigcap_{n\geq 1} X_n,$$

and we say that X_n decreases toward X. We write

$$X_n \uparrow X \quad \text{if} \quad X_1 \subseteq X_2 \subseteq \cdots \qquad \text{and} \qquad X = \bigcup_{n\geq 1} X_n,$$

and we say that X_n increases toward X.

7.28 Proposition. *Assume E is a locally compact Hausdorff space with countable basis; let $X_n, X \in \mathcal{F}(E)$ and $Y \in \mathcal{P}(E)$. Then*

(a) $X_n \downarrow X \Rightarrow X_n \xrightarrow{\mathcal{F}} X$;

(b) $X_n \uparrow Y \Rightarrow X_n \xrightarrow{\mathcal{F}} \overline{Y}$.

Proof. (a): Assume that $X_n \downarrow X$; from (7.15) it follows immediately that $\overline{\lim}\, X_n = X$. Assume $x \in X$; then $x \in X_n$ for every n, and hence $x \in \underline{\lim}\, X_n$. This means that $\overline{\lim}\, X_n - X \subseteq \underline{\lim}\, X_n$. But thc inclusion $\overline{\lim}\, X_n \supseteq \underline{\lim}\, X_n$ holds trivially, and it follows that both sets are equal; so $X_n \xrightarrow{\mathcal{F}} X$.

(b): If $X_n \uparrow Y$, then $\overline{\bigcup_{n \geq N} X_n} = \overline{\bigcup_{n \geq 1} X_n} = \overline{Y}$; now (7.15) gives $\overline{\lim}\, X_n = \overline{Y}$. Conversely, if $y \in Y$, then $y \in X_N$ for some N. This means, however, that $y \in X_n$ for $n \geq N$; hence $y \in \underline{\lim}\, X_n$. Thus $Y \subseteq \underline{\lim}\, X_n$, and as $\underline{\lim}\, X_n$ is closed, $\overline{Y} \subseteq \underline{\lim}\, X_n$. Therefore, $\overline{\lim}\, X_n = \overline{Y} \subseteq \underline{\lim}\, X_n$, and a similar argument as in (a) gives that both sets equal $\overline{Y}$. This means that $X_n \xrightarrow{\mathcal{F}} \overline{Y}$. □

Section 13.3 discusses convergence of arbitrary sequences in $\mathcal{P}(E)$ and the relation with the hit-or-miss topology.

7.5. Myope topology

Throughout this section it is assumed that E is a locally compact Hausdorff space with countable basis. Denote by $\mathcal{K} = \mathcal{K}(E)$ the compact subsets of E and by $\mathcal{K}'(E)$ the nonempty compact subsets. The families $\mathcal{K}^A$ and $\mathcal{K}_A$ are respectively defined by

$$\mathcal{K}^A = \{X \in \mathcal{K} \mid X \cap A = \varnothing\}, \qquad \mathcal{K}_A = \{X \in \mathcal{K} \mid X \cap A \neq \varnothing\}.$$

The families $\mathcal{K}^F$ with F closed and $\mathcal{K}_G$ with G open constitute a subbasis of a topology on $\mathcal{K}$, called the *myope topology*. A basis of this topology is constituted by the sets

$$\mathcal{K}^F_{G_1,\ldots,G_n} = \mathcal{K}^F \cap \mathcal{K}_{G_1} \cap \cdots \cap \mathcal{K}_{G_n}.$$

If E is compact, then $\mathcal{F}(E) = \mathcal{K}(E)$ and the myope topology coincides with the hit-or-miss topology. The relative hit-or-miss topology on $\mathcal{K}$ is generated by the families $\mathcal{F}^K \cap \mathcal{K} = \mathcal{K}^K$ (K compact) and $\mathcal{F}_G \cap \mathcal{K} = \mathcal{K}_G$ (G open). By definition, these families are open for the myope topology. If F is a noncompact closed set, however, then $\mathcal{K}^F$ is open for the myope topology but not for the relative hit-or-miss topology. This means that the myope topology is finer (i.e., has more open sets) than the relative hit-or-miss topology if E is not compact. This is nicely illustrated by the following example in $\mathbb{R}^2$. The closed balls K_n with radius 1 and centre $(n, 0)$ converge to the empty set with respect to the hit-or-miss topology. These balls do not converge myopically, however. For suppose that K is their limit; there is a vertical closed right half-plane H which has empty intersection with K, i.e., $K \in \mathcal{K}^H$. The balls K_n lie completely inside H if n is large enough, however, in particular, $K_n \notin \mathcal{K}^H$.

7.29 Lemma. *If $\mathcal{C}$ is a subset of $\mathcal{K}$ which is compact for the myope topology, then $\mathcal{C}$ is also compact for the hit-or-miss topology, and the two topologies are identical on $\mathcal{C}$.*

Proof. The first assertion is an immediate consequence of the fact that the myope topology is finer than the (relative) hit-or-miss topology. Obviously, every subset of $\mathcal{C}$ closed for the hit-or-miss topology is also closed for the myope topology. On the other hand, a subset $\mathcal{C}' \subseteq \mathcal{C}$ which is closed for the myope topology is compact for the myope topology. This implies that it is compact (and therefore closed) for the hit-or-miss topology. This proves the result. □

Consider the subset of $\mathcal{K}$ given by $\mathcal{C} = \{X \in \mathcal{K} \mid X \subseteq K\}$, where K is a compact set. Then $\mathcal{C} = \mathcal{K}^{K^c} = \mathcal{F}^{K^c}$ is a closed (hence compact) subset of $\mathcal{F}$. The relative hit-or-miss topology and the myope topology on $\mathcal{C}$ coincide and we conclude that $\mathcal{C}$ is compact for the myope topology.

7.30 Lemma. *A set $\mathcal{C} \subseteq \mathcal{K}$ is compact for the myope topology if and only if $\mathcal{C}$ is closed in $\mathcal{F}$, and there exists a $K \in \mathcal{K}$ such that $X \subseteq K$ for every $X \in \mathcal{C}$.*

Proof. "if": Follows easily from the preceding arguments.

"only if": Let $\mathcal{C} \subseteq \mathcal{K}$ be compact for the myope topology. The previous lemma says that $\mathcal{C}$ is also compact (and hence closed) in $\mathcal{F}$ for the hit-or-miss topology. Define $\mathcal{B}$ as the collection of open sets in E with compact closure. For every compact set $K \subseteq E$ there exists a set $B \in \mathcal{B}$ such that $K \subseteq B$. In fact, for every $k \in K$ there exists a relatively compact open neighbourhood $B(k)$ of k. These open neighbourhoods are an open covering of K. Since K is compact there exists a finite subcovering $B(k_1), \dots, B(k_p)$; the union of these sets lies in $\mathcal{B}$ and contains K. Now $\{\mathcal{K}^{B^c} \mid B \in \mathcal{B}\}$ defines an open covering of $\mathcal{C}$; since $\mathcal{C}$ is compact, it has a finite subcovering, that is,

$$\mathcal{C} \subseteq \bigcup_{i=1}^{n} \mathcal{K}^{B_i{}^c} \subseteq \mathcal{K}^{\cap(B_i^c)} = \mathcal{K}^{B^c},$$

where $B = \bigcup_{i=1}^{n} B_i$. This implies that every $X \in \mathcal{C}$ satisfies $X \cap B^c = \varnothing$, i.e., $X \subseteq B$. In particular, $X \subseteq \overline{B}$ for $X \in \mathcal{C}$. This proves the result. □

We can now prove the following result.

7.31 Theorem. *$\mathcal{K}$ with the myope topology is Hausdorff, locally compact, and has countable basis.*

Proof. The proof that the myope topology has the Hausdorff property and that it possesses a countable basis is very much the same as for the hit-or-miss topology: see Theorem 7.24. It remains to prove local compactness. Given $K \in \mathcal{K}$, one must show that there exists a neighbourhood of K with compact closure. Since the underlying space E is locally compact, there exists an open relatively compact set $B \subseteq E$ such that $K \subseteq B$ (cf. the proof of Lemma 7.30). Then $\mathcal{K}^{B^c}$ is a neighbourhood of K whose closure is contained within $\mathcal{K}^{\overline{B}^c}$. From Lemma 7.30 we know that $\mathcal{K}^{\overline{B}^c}$ is compact; therefore, the closure of $\mathcal{K}^{B^c}$ is compact as well. □

7.32 Remark. The finite sets (including $\varnothing$) lie dense in $\mathcal{K}$ with respect to the myope topology. In fact, let $K \in \mathcal{K}$ be nonempty; for every $n = 1, 2, \ldots$ the open balls $B^\circ(k, 1/n)$, where k runs over K, are an open covering of K. Since K is compact there is a finite subcovering $B^\circ(k_1, 1/n), \ldots, B^\circ(k_{p_n}, 1/n)$. Define $K_n = \{k_1, k_2, \ldots, k_{p_n}\}$; it is easy to show that $K_n \to K$ myopically.

Another immediate consequence of Lemma 7.30 is the following proposition.

7.33 Proposition. *A sequence X_n converges to X in $\mathcal{K}$ if and only if the following two conditions are satisfied:*

(i) *there exists $K \in \mathcal{K}$ such that $X_n \subseteq K$ for each n;*

(ii) $X_n \xrightarrow{\mathcal{F}} X$.

By $X_n \xrightarrow{\mathcal{K}} X$ we mean that $X, X_n \in \mathcal{K}$ and X_n converges to X with respect to the myope topology. The following result expresses the relation between monotone convergence and myopic convergence. It follows without serious effort from Propositions 7.28 and 7.33.

7.34 Proposition. *Let X_n be a sequence in $\mathcal{K}$; then*

(a) $X_n \downarrow X \Rightarrow X_n \xrightarrow{\mathcal{K}} X$;

(b) $X_n \uparrow Y \Rightarrow X_n \xrightarrow{\mathcal{K}} \overline{Y}$ *if $\overline{Y}$ is compact.*

Finally, one can establish the following connection between the myope topology and the Hausdorff metric.

7.35 Proposition. *Let (E, d) be a finitely compact metric space. The topology on $\mathcal{K}'(E)$ induced by the Hausdorff metric is equivalent to the relative myope topology.*

Proof. Let $o \in E$ be fixed, and let $Y \subseteq E$; it is easy to show that the following assertions are equivalent:

(i) Y is bounded;

(ii) $\overline{Y}$ is bounded;

(iii) $\{d(o, y) \mid y \in Y\}$ is bounded.

Assume that $X_n \to X$ in $\mathcal{K}'$ myopically. Proposition 7.33 gives that $X_n \subseteq K$ for every n and $X_n \xrightarrow{\mathcal{F}} X$. Furthermore, Corollary 7.27 gives that $D(X_n, X) \to 0$. But it is rather easy to prove that for subsets $Y, Z \subseteq K$ one has $D_H(Y, Z) \leq cD(Y, Z)$ for some constant $c > 0$. This means that $D_H(X, X_n) \to 0$ as well.

Conversely, assume $D_H(X_n, X) \to 0$. Suppose one can show that $Y = \bigcup_{n \geq 1} X_n$ is bounded. By the preceding observation, $K = \overline{Y}$ is bounded as well, and Proposition 7.15 says that K is compact. Thus $X_n \subseteq K$ and $D(X_n, X) \leq D_H(X_n, X) \to 0$. By Corollary 7.27, $X_n \xrightarrow{\mathcal{F}} X$, and Proposition 7.33 gives that $X_n \xrightarrow{\mathcal{K}} X$. Therefore, it remains to prove that Y is bounded, or equivalently, that $\{d(o, y) \mid y \in Y\}$ is bounded. Suppose that $d(o, x_{n_k}) \geq k$ for some sequence $x_{n_k} \in X_{n_k}$. If $x \in X$; then $d(x_{n_k}, x) \geq d(x_{n_k}, o) - d(o, x)$; this leads to

$$d(x_{n_k}, X) \geq \inf_{x \in X} [d(x_{n_k}, o) - d(o, x)] \geq k - \sup_{x \in X} d(o, x).$$

Since X is compact, it is bounded; putting $L = \sup_{x \in X} d(o, x)$, we find $d(x_{n_k}, X) \geq k - L$. Thus we get

$$D_H(X_{n_k}, X) \geq \hat{D}(X_{n_k}, X) \geq d(x_{n_k}, X) \geq k - L.$$

But this contradicts $D_H(X_n, X) \to 0$. The proof is finished. □

7.6. Semi-continuity

Throughout this section E is a locally compact Hausdorff space with countable basis. Let S be an arbitrary topological space and ψ a mapping from S into $\mathcal{F}$. Given $\mathcal{H} \subseteq \mathcal{F}$, denote by $\psi^{-1}(\mathcal{H})$ the subset of S containing all elements mapped into $\mathcal{H}$ by ψ.

7.36 Definition. The mapping $\psi : S \to \mathcal{F}$ is *upper semi-continuous* (*u.s.c.*) if $\psi^{-1}(\mathcal{F}^K)$ is open in S for each $K \in \mathcal{K}$. If, on the other hand, $\psi^{-1}(\mathcal{F}_G)$ is open in S for each $G \in \mathcal{G}$, then ψ is *lower semi-continuous* (*l.s.c.*). The mapping ψ is *continuous* if it is both u.s.c. and l.s.c.

7.37 Proposition. *Let S be a topological space with countable basis, and let ψ be a mapping from S into $\mathcal{F}$.*

(a) *ψ is u.s.c. iff $s_n \to s$ implies that $\overline{\lim}\ \psi(s_n) \subseteq \psi(s)$.*

(b) *ψ is l.s.c. iff $s_n \to s$ implies that $\psi(s) \subseteq \underline{\lim}\ \psi(s_n)$.*

Proof. We prove only (a) and leave (b) as an exercise to the reader.

"only if": Upper semicontinuity of ψ means that $\psi^{-1}(\mathcal{F}^K)$ is open in S for every compact set K. Let $s_n \to s$ in S; we show that $\overline{\lim}\ \psi(s_n) \subseteq \psi(s)$. By Proposition 7.26 it suffices to show that $\psi(s) \cap K = \varnothing$ implies that $\psi(s_n) \cap K = \varnothing$ eventually, for every compact set K. But the latter is equivalent to the assertion that $\psi^{-1}(\mathcal{F}^K)$ is open.

"if": Analogous proof. □

Since $\mathcal{F}$ is a complete lattice, the set of mappings $\psi : S \to \mathcal{F}$ has a complete lattice structure. Let, for each i in some index set I, ψ_i be a mapping from S into $\mathcal{F}$. Define $\bigwedge_{i\in I} \psi_i$ and $\bigvee_{i\in I} \psi_i$ as the mappings from S into $\mathcal{F}$ given, respectively, by

$$(\bigwedge_{i\in I} \psi_i)(s) = \bigcap_{i\in I} \psi_i(s),$$

$$(\bigvee_{i\in I} \psi_i)(s) = \overline{\bigcup_{i\in I} \psi_i(s)}.$$

7.38 Proposition. *Let S be a topological space, and let $\psi_i : S \to \mathcal{F}$ for every $i \in I$.*

(a) *If every mapping ψ_i is u.s.c., then $\bigwedge_{i\in I} \psi_i$ is u.s.c. as well.*

(b) *If every mapping ψ_i is l.s.c., then $\bigvee_{i\in I} \psi_i$ is l.s.c. as well.*

Proof. (a): Let ψ_i be u.s.c., and let $s_n \to s$ in S; then $\overline{\lim}\ \psi_i(s_n) \subseteq \psi_i(s)$ for every $i \in I$. So $\overline{\lim} \bigwedge_{i\in I} \psi_i(s_n) \subseteq \psi_i(s)$ for $i \in I$. But this implies $\overline{\lim} \bigwedge_{i\in I} \psi_i(s_n) \subseteq \bigwedge_{i\in I} \psi_i(s)$, and the result follows.

(b): Analogous proof. □

Of particular interest is the case $S = \mathcal{F}$. As far as increasing operators are concerned, one can characterize upper semi-continuity in terms of monotone convergence. Recall from Definition 3.7 that an operator is $\downarrow$-continuous if $X_n \downarrow X$ implies $\psi(X_n) \downarrow \psi(X)$.

7.39 Proposition. *Let $\psi : \mathcal{F} \to \mathcal{F}$ be an increasing operator. The following conditions are equivalent:*

(i) *ψ is u.s.c.;*

(ii) *ψ is $\downarrow$-continuous;*

(iii) *$\overline{\lim}\, \psi(X_n) \subseteq \psi(\overline{\lim}\, X_n)$ for every sequence X_n in $\mathcal{F}$.*

Proof. (i) $\Rightarrow$ (ii): Assume that ψ is u.s.c. and that $X_n \downarrow X$. Then $X_n \xrightarrow{\mathcal{F}} X$ by Proposition 7.28; hence $\overline{\lim}\, \psi(X_n) \subseteq \psi(X)$. Since $X \subseteq X_n$, $\psi(X) \subseteq \underline{\lim}\, \psi(X_n)$, and it follows that $\psi(X_n) \xrightarrow{\mathcal{F}} \psi(X)$. Since ψ is increasing, one gets $\psi(X_n) \downarrow \bigcap_{n\geq 1} \psi(X_n)$; application of Proposition 7.28 gives $\psi(X_n) \xrightarrow{\mathcal{F}} \bigcap_{n\geq 1} \psi(X_n)$. But this is possible only if $\psi(X) = \bigcap_{n\geq 1} \psi(X_n)$, i.e., $\psi(X_n) \downarrow \psi(X)$.

(ii) $\Rightarrow$ (iii): Let X_n be a sequence in $\mathcal{F}$ and define $Y_n = \overline{\bigcup_{k\geq n} X_k}$. By (7.15), $Y_n \downarrow \overline{\lim}\, X_n$, and we conclude from (ii) that $\psi(Y_n) \downarrow \psi(\overline{\lim}\, X_n)$, that is,

$$\bigcap_{n\geq 1} \psi(\overline{\bigcup_{k\geq n} X_k}) = \psi(\overline{\lim}\, X_n).$$

Obviously, $\psi(\overline{\bigcup_{k\geq n} X_k}) \supseteq \overline{\bigcup_{k\geq n} \psi(X_k)}$, whence it follows

$$\bigcap_{n\geq 1} \overline{\bigcup_{k\geq n} \psi(X_k)} \subseteq \psi(\overline{\lim}\, X_n).$$

Applying (7.15) once again leads to

$$\overline{\lim}\, \psi(X_n) \subseteq \psi(\overline{\lim}\, X_n),$$

which was to be proved.

(iii) $\Rightarrow$ (i): Suppose that $X_n \xrightarrow{\mathcal{F}} X$; then $\overline{\lim}\, X_n = X$. Now (iii) implies that $\overline{\lim}\, \psi(X_n) \subseteq \psi(X)$, and so ψ is u.s.c. □

Proposition 7.39 expresses that for increasing operators on $\mathcal{F}$ the lattice-theoretical notion of upper semi-continuity and the topological notion of upper semi-continuity coincide. An extension of this result for nonincreasing operators is stated in Proposition 13.14.

7.40 Corollary. *Assume that S is a topological space with countable basis; let $\psi_1 : S \to \mathcal{F}$ be a mapping which is u.s.c., and let $\psi_2 : \mathcal{F} \to \mathcal{F}$ be a mapping which is u.s.c. and increasing. Then $\psi_2\psi_1$ is u.s.c. as well.*

Proof. If $s_n \to s$ in S, then $\overline{\lim}\ \psi_1(s_n) \subseteq \psi_1(s)$. Since ψ_2 is increasing,

$$\psi_2(\overline{\lim}\ \psi_1(s_n)) \subseteq \psi_2\psi_1(s).$$

Furthermore, Proposition 7.39 gives

$$\overline{\lim}\ \psi_2\psi_1(s_n) \subseteq \psi_2(\overline{\lim}\ \psi_1(s_n)) \subseteq \psi_2\psi_1(s);$$

so $\psi_2\psi_1$ is u.s.c. □

A function $f : E \to E$ is called a *homeomorphism* if it is continuous, bijective, and has a continuous inverse. Such functions can be extended to $\mathcal{F}(E)$ by putting

$$f(X) = \{f(x) \mid x \in X\}.$$

Note that we use the same notation for this extension.

7.41 Proposition. *Let f be a homeomorphism on E. Then the operator $f : \mathcal{F} \to \mathcal{F}$ is continuous.*

The proof follows easily if one uses that for all x, x_n in $\mathbb{R}^d$ one has $x_n \to x$ if and only if $f(x_n) \to f(x)$. This implies

$$\underline{\lim}\, f(X_n) = f(\underline{\lim}\ X_n) \qquad \text{and} \qquad \overline{\lim}\, f(X_n) = f(\overline{\lim}\ X_n),$$

for every sequence X_n in $\mathcal{F}$. This result has some useful implications in the case where $E = \mathbb{R}^d$. Throughout the remainder of this section we restrict ourselves to this case. Proposition 7.41 implies that translations $X \mapsto X_h$, multiplications $X \mapsto rX$, and rotations are continuous operators. Of particular interest are the basic morphological operators, viz. dilations, erosions, closings, and openings.

7.42 Lemma. *Let $X \subseteq \mathbb{R}^d$ be closed, and let $A \subseteq \mathbb{R}^d$ be nonempty and compact. Then the sets $X \oplus A$, $X \ominus A$, $X \bullet A$, $X \circ A$ are closed as well.*

Proof. First we show that $X \oplus A$ is closed. Given $y_n \in X \oplus A$ and $y_n \to y$, we show that $y \in X \oplus A$. Every y_n is of the form $y_n = x_n + a_n$, where $x_n \in X$ and $a_n \in A$. Since A is compact, the sequence a_n must have a convergent subsequence in A. Without loss of generality, one can assume that a_n converges, say toward a. Then $x_n = y_n - a_n \to y - a$, and since X is closed, $y - a \in X$. Therefore, $y = (x - a) + a \in X \oplus A$.

Furthermore, $X \ominus A = \bigcap_{a \in A} X_{-a}$ is an intersection of closed sets and as such it is closed as well.

Now $X \bullet A = Y \ominus A$, where $Y = X \oplus A$. If X is closed, then Y is closed; hence $X \bullet A$ is closed as well. A similar argument shows that $X \circ A$ is closed. □

Consider the product space $\mathcal{F} \times \mathcal{K}$ endowed with the product topology (Dugundji, 1966, Chapter IV). In this topology, which also has a countable basis, a sequence (X_n, A_n) converges toward (X, A) if X_n converges toward X in $\mathcal{F}$ and A_n converges toward A in $\mathcal{K}$.

7.43 Proposition.

(a) *The operator $(X, A) \mapsto X \oplus A$ from $\mathcal{F} \times \mathcal{K}$ into $\mathcal{F}$ is continuous.*

(b) *The operator $(X, A) \mapsto X \ominus A$ from $\mathcal{F} \times \mathcal{K}$ into $\mathcal{F}$ is u.s.c.*

Proof. First we show that $A_n \xrightarrow{\mathcal{K}} \varnothing$ implies that $A_n = \varnothing$ eventually. Let $A_n \xrightarrow{\mathcal{K}} \varnothing$; by Proposition 7.33, $A_n \subseteq A_0$ for some compact set A_0 and $A_n \xrightarrow{\mathcal{F}} \varnothing$. Since $A_0 \cap \varnothing = \varnothing$, we get that $A_n \cap A_0 = \varnothing$ eventually; this means, however, that $A_n = \varnothing$ eventually.

(a): Let $X_n \xrightarrow{\mathcal{F}} X$ and $A_n \xrightarrow{\mathcal{K}} A$. Since $X \oplus \varnothing = \varnothing$ for every closed set X, the proof is straightforward if $A = \varnothing$. So assume from now on that $A \neq \varnothing$. We use Proposition 7.26 to show that $X_n \oplus A_n \xrightarrow{\mathcal{F}} X \oplus A$. First we verify rule (i). Let $y = x + a \in X \oplus A$, where $x \in X$ and $a \in A$. Since $X_n \xrightarrow{\mathcal{F}} X$ and $A_n \xrightarrow{\mathcal{F}} A$ (see Proposition 7.33), there exist $x_n \in X_n$ and $a_n \in A_n$ such that $x_n \to x$ and $a_n \to a$. Then $x_n + a_n \in X_n \oplus A_n$ and $x_n + a_n \to x + a$; therefore $x + a \in \underline{\lim}\,(X_n \oplus A_n)$. This proves that $X \oplus A \subseteq \underline{\lim}\,(X_n \oplus A_n)$. To prove rule (ii) of Proposition 7.26, let $y \in \overline{\lim}\,(X_n \oplus A_n)$, and let $y_{n_i} = x_{n_i} + a_{n_i}$, with $x_{n_i} \in X_{n_i}, a_{n_i} \in A_{n_i}$, be a sequence tending to y as $i \to \infty$. Since, by Proposition 7.33, $A_n \subseteq A_0$ for some compact set A_0, it follows that a_{n_i} has a convergent subsequence. Without loss of generality, one may assume that $a_{n_i} \to a$ where a lies in A. Then $x_{n_i} = y_{n_i} - a_{n_i} \to y - a$ and $y - a \in X$; so $y = y - a + a \in X \oplus A$. This proves (a).

(b): Let $X_n \xrightarrow{\mathcal{F}} X$ and $A_n \xrightarrow{\mathcal{K}} A$; we show $\overline{\lim}\, X_n \ominus A_n \subseteq X \ominus A$. If $A = \varnothing$, then $X \ominus A = \mathbb{R}^d$ and the result holds trivially. Take $y \in \overline{\lim}\, X_n \ominus A_n$; there is a subsequence $y_{n_i} \in X_{n_i} \ominus A_{n_i}$ such that $y_{n_i} \to y$. We must show that $y \in X \ominus A$, or in other words, that $y + a \in X$ for $a \in A$. Since $y_{n_i} \in X_{n_i} \ominus A_{n_i}$, we have $(A_{n_i})_{y_{n_i}} \subseteq X_{n_i}$. There is a subsequence m_i of n_i and elements $a_{m_i} \in A_{m_i}$ such that $a_{m_i} \to a$ as $i \to \infty$. Then $y_{m_i} + a_{m_i} \in (A_{m_i})_{y_{m_i}} \subseteq X_{m_i}$ and $y_{m_i} + a_{m_i} \to y + a$. From $X_n \xrightarrow{\mathcal{F}} X$ it follows that $y + a \in X$, which concludes the proof. □

This result, in combination with Corollary 7.40, implies that all basic morphological operators are u.s.c.

7.44 Corollary. *Let A be a compact structuring element.*

(a) *The dilation $X \mapsto X \oplus A$ on $\mathcal{F}(\mathbb{R}^d)$ is continuous*

(b) *The erosion $X \mapsto X \ominus A$ on $\mathcal{F}(\mathbb{R}^d)$ is u.s.c.*

(c) *The closing $X \mapsto X \bullet A$ on $\mathcal{F}(\mathbb{R}^d)$ is u.s.c.*

(d) *The opening $X \mapsto X \circ A$ on $\mathcal{F}(\mathbb{R}^d)$ is u.s.c.*

Boundedness of A is essential as shown by the following example. Assume that A contains a sequence a_n which converges to infinity. Take $X_n = \{-a_n\}$; then $X_n \xrightarrow{\mathcal{F}} \varnothing$. However, $0 \in \overline{\lim}\,(X_n \oplus A)$ since $0 = -a_n + a_n \in X_n \oplus A$. Therefore, the dilation $X \mapsto X \oplus A$ is not u.s.c.

It is easy to find examples which show that erosion is not l.s.c. in general. Take for X_n the closed ball with radius $1 - 1/n$; it is evident that X_n converges toward B, the closed unit ball. Furthermore, $X_n \ominus B = \varnothing$, whereas $B \ominus B = \{0\}$. This implies that $X \mapsto X \ominus B$ is not l.s.c.

The *kernel* $\mathcal{V}(\psi)$ of an operator $\psi : \mathcal{F}(\mathbb{R}^d) \to \mathcal{F}(\mathbb{R}^d)$ is defined by

$$\mathcal{V}(\psi) = \{X \in \mathcal{F}(\mathbb{R}^d) \mid 0 \in \psi(X)\};$$

cf. Definition 4.8.

7.45 Proposition. *The translation invariant operator ψ on $\mathcal{F}(\mathbb{R}^d)$ is u.s.c. if and only if its kernel $\mathcal{V}(\psi)$ is closed with respect to the hit-or-miss topology.*

Proof. "if": Suppose that ψ is a translation invariant operator on $\mathcal{F}(\mathbb{R}^d)$ with closed kernel. Assume $X_n \xrightarrow{\mathcal{F}} X$; we must show $\overline{\lim}\,\psi(X_n) \subseteq \psi(X)$. Let $h \in \overline{\lim}\,\psi(X_n)$; there is a sequence $h_{n_k} \in \psi(X_{n_k})$ which tends to h. From the translation invariance of ψ we deduce that $0 \in \psi\big((X_{n_k})_{-h_{n_k}}\big)$; hence $(X_{n_k})_{-h_{n_k}} \in \mathcal{V}(\psi)$. Since $(X_{n_k})_{-h_{n_k}} \xrightarrow{\mathcal{F}} X_{-h}$ and $\mathcal{V}(\psi)$ is closed, it follows that $X_{-h} \in \mathcal{V}(\psi)$. But this means that $h \in \psi(X)$, which had to be proved.

"only if": Assume that ψ is u.s.c. Given a sequence $X_n \in \mathcal{V}(\psi)$ converging to $X \in \mathcal{F}$, we must show that $X \in \mathcal{V}(\psi)$. But this is obvious as $0 \in \psi(X_n)$ for every n; hence $0 \in \overline{\lim}\,\psi(X_n) \subseteq \psi(X)$. □

7.46 Proposition. *Given an operator $\psi : \mathcal{F}(\mathbb{R}^d) \to \mathcal{F}(\mathbb{R}^d)$ which is increasing and translation invariant; then ψ is u.s.c. if and only if there exists a family $\mathcal{A} \subseteq \mathcal{K}(\mathbb{R}^d)$ of structuring elements which is closed with respect to the myope topology such that $\psi(X) = \bigcap_{A \in \mathcal{A}} X \oplus A$.*

Proof. "if": Follows immediately from Proposition 7.38(a).

"only if": Extend ψ to an increasing operator $\psi_{\mathcal{P}}$ on $\mathcal{P}(\mathbb{R}^d)$ by putting $\psi_{\mathcal{P}}(X) = \psi(\overline{X})$, $X \subseteq \mathbb{R}^d$. By Theorem 4.15, $\psi_{\mathcal{P}}(X) = \bigcap_{A \in \mathcal{V}(\psi^*_{\mathcal{P}})} X \oplus \check{A}$. Later in the proof it is shown that for every $A \in \mathcal{V}(\psi^*_{\mathcal{P}})$ there exists a compact set $K_A \subseteq A$ such that $K_A \in \mathcal{V}(\psi^*_{\mathcal{P}})$ as well. Putting $\mathcal{A} = \{K_A \mid A \in \mathcal{V}(\psi^*_{\mathcal{P}})\}$, one gets $\psi_{\mathcal{P}}(X) = \bigcap_{K \in \mathcal{A}} X \oplus \check{K}$; to understand this, observe that $X \oplus \check{A} \supseteq X \oplus \check{K}_A$ for every $A \in \mathcal{V}(\psi^*_{\mathcal{P}})$. Since $\psi(X) = \psi_{\mathcal{P}}(X)$ if X is closed, this means $\psi(X) = \bigcap_{K \in \mathcal{A}} X \oplus \check{K}$ for $X \in \mathcal{F}(\mathbb{R}^d)$. Furthermore, if $K_n \in \mathcal{A}$ and $K_n \xrightarrow{\mathcal{K}} K$, then $X \oplus \check{K}_n \xrightarrow{\mathcal{F}} X \oplus \check{K}$. From $\psi(X) \subseteq X \oplus \check{K}_n$ it follows that $\psi(X) \subseteq X \oplus \check{K}$ as well. Therefore, one may take the closure of $\mathcal{A}$ in the myopic sense. Thus it remains to be shown that for every $A \in \mathcal{V}(\psi^*_{\mathcal{P}})$ one can find a compact set $K \in \mathcal{V}(\psi^*_{\mathcal{P}})$ with $K \subseteq A$. Note that $A \in \mathcal{V}(\psi^*_{\mathcal{P}})$ iff $0 \notin \psi_{\mathcal{P}}(\overline{A^c})$. Take $K_n = \{a \in A \mid d(0,a) \le n\}$, where d denotes Euclidean distance. Then $\overline{K_n^c} = \{a \in A \mid d(0,a) \ge n\} \cup \overline{A^c}$, and it is obvious that $\overline{K_n^c} \xrightarrow{\mathcal{F}} \overline{A^c}$. From the upper semi-continuity of ψ it follows that $0 \notin \psi_{\mathcal{P}}(\overline{K_n^c})$ eventually, saying that $K_n \in \mathcal{V}(\psi^*_{\mathcal{P}})$. This concludes the proof. □

7.7. Basis representations

In Section 5.5.2 it was argued that every increasing translation invariant operator ψ on $\mathcal{F}(\mathbb{R}^d)$ can be written as a supremum of erosions:

$$\psi(X) = \overline{\bigcup_{A \in \mathcal{V}(\psi)} X \ominus A}. \tag{7.16}$$

Recall that a set $A \in \mathcal{V}(\psi)$ is called a *minimal kernel element* if $\mathcal{V}(\psi)$ does not contain a kernel element which is strictly smaller than A. The family of minimal kernel elements is called the *basis* of ψ and is denoted by $\mathcal{V}_b(\psi)$. Theorem 5.27 applied to $\mathcal{F}(\mathbb{R}^d)$ says that $\mathcal{V}(\psi)$ may be replaced by $\mathcal{V}_b(\psi)$ in (7.16), assuming that the operator ψ is l.u.s.c.; cf. Definition 3.6. We prove the following extension of Proposition 7.39.

7.47 Proposition. *An increasing operator ψ on $\mathcal{F}(\mathbb{R}^d)$ is l.u.s.c. if and only if it is u.s.c.*

Proof. "only if": If ψ is l.u.s.c., then it is also $\downarrow$-continuous; now Proposition 7.39 implies that ψ is u.s.c.

"if": Assume ψ is u.s.c.; suppose that every chain $\mathcal{C}$ in $\mathcal{F}$ contains a decreasing sequence X_n such that $\bigcap\{C \mid C \in \mathcal{C}\} = \bigcap_{n \ge 1} X_n$. By Lemma 3.8(a), the operator ψ is l.u.s.c. iff it is $\downarrow$-continuous. Since, by Proposition 7.39,

$\downarrow$-continuity is equivalent to upper semi-continuity, this would give the result. Let $\mathcal{C}$ be a chain in $\mathcal{F}$; we show that there exists a sequence $X_n \in \mathcal{C}$ with $\bigcap_{n\geq 1} X_n = X$, where $X = \bigcap\{C \mid C \in \mathcal{C}\}$. If $X \in \mathcal{C}$, then we can take the sequence constantly X. Assume therefore that $X \notin \mathcal{C}$. First we prove that X is a limit point of $\mathcal{C}$ relative to the hit-or-miss topology. It suffices to show that every basis member $\mathcal{F}^K_{G_1,G_2,\dots,G_n}$ of the hit-or-miss topology which contains X has nonempty intersection with $\mathcal{C}$. By definition, $X \in \mathcal{F}^K_{G_1,G_2,\dots,G_n}$ means that $X \cap G_i \neq \varnothing$ $(i = 1, 2, \dots, n)$ and $X \cap K = \varnothing$. Since $X \subseteq C$ for every $C \in \mathcal{C}$ we conclude that $C \cap G_i \neq \varnothing$ for $i = 1, 2, \dots, n$. It remains to show that there is a $C \in \mathcal{C}$ such that $C \cap K = \varnothing$. Note that $X \cap K = \varnothing$ is equivalent to $K \subseteq X^c = \bigcup\{C^c \mid C \in \mathcal{C}\}$. Thus the family $\{C^c \mid C \in \mathcal{C}\}$ is an open covering of K, and since K is compact it must possess a finite subcovering $\{C_1^c, C_2^c, \dots, C_q^c\}$. Since $\mathcal{C}$ is a chain the collection $C_1, C_2, \dots, C_q$ has a least element say C_j. Then $K \subseteq C_j^c$, that is, $K \cap C_j = \varnothing$, which had to be shown. Thus we have demonstrated that X is a limit point of $\mathcal{C}$. Since $\mathcal{F}(\mathbb{R}^d)$ has a countable basis, there must exist a sequence X_n in $\mathcal{C}$ such that $X_n \xrightarrow{\mathcal{F}} X$. Now, for every $n \geq 1$ there is an $m > n$ such that $X_m \subseteq X_n$; for otherwise there would exist an n such that $X_n \subseteq X_m$ for $m \geq n$. This, however, would imply that $X_n \subseteq X$, and hence $X = X_n \in \mathcal{C}$, a contradiction. It follows that we can choose a sequence $n_1 < n_2 < n_3 < \cdots$ such that $X_{n_1} \supseteq X_{n_2} \supseteq \cdots$. Then $X_{n_i} \xrightarrow{\mathcal{F}} \bigcap_{i\geq 1} X_{n_i}$. But also $X_{n_i} \xrightarrow{\mathcal{F}} X$, and it follows that $\bigcap_{i\geq 1} X_{n_i} = X$. This completes the proof. □

7.48 Corollary. *Every increasing translation invariant operator ψ on $\mathcal{F}(\mathbb{R}^d)$ which is u.s.c. has nonempty basis $\mathcal{V}_b(\psi)$ and can be decomposed as*

$$\psi(X) = \overline{\bigcup_{A \in \mathcal{V}_b(\psi)} X \ominus A}, \tag{7.17}$$

for $X \in \mathcal{F}(\mathbb{R}^d)$.

7.8. Bibliographical notes

There exist several elementary books on topology; we mention in particular the treatises by Hausdorff (1962), Kuratowski (1972), and Dugundji (1966). Applications of image processing, computational geometry, and computer graphics are partially responsible for a revival of interest in topology, especially topology for discrete structures. In particular, the Khalimskii topology, briefly discussed in Example 7.8, was "invented" in 1970 by Khalimskii (1969), but has recently been "reinvented" by Kovalevsky (1989). Since then it has been studied by several authors; refer to Kong et al. (1991), Kong and Rosenfeld (1989), and the references mentioned therein for a nice overview. The true bibliophiles may find it interesting that Matheron (1974) made a detailed study of finite topologies already in 1974.

Two standard references on metric spaces are Busemann (1955) and Rinow (1961). The definition of a finitely compact metric space is due to Busemann (1955). As the name indicates, the Hausdorff metric originates from Hausdorff (1962). The modification to closed sets discussed in Section 7.3.1 appears for the first time in Busemann (1955); this explains our terminology "Hausdorff–Busemann metric". For a comprehensive discussion of both metrics refer to Rinow (1961). The proof of Theorem 7.20 is due to Busemann (1955).

The hit-or-miss topology seems to have appeared for the first time in a paper by Fell (1962). Note, however, that Fell called this topology the H-topology. The nomenclature "hit-or-miss topology" was invented by Matheron (1975), who examined this topology in great detail; apparently, Matheron was not aware of Fell's pioneering work. The contents of Sections 7.4–7.6 are based on Matheron's work. Note, however, that Matheron does not discuss the relation between the hit-or-miss topology and the topology induced by the Hausdorff–Busemann metric. Michael (1951) examines the topology on $\mathcal{F}(E)$ with basis elements $\mathcal{F}^F_{G_1,\dots,G_n}$, where F is closed, and $G_1, \dots, G_n$ are open. This topology is sometimes called Vietoris topology; see, e.g., Kuratowski and Mostowski (1976). Some further discussions on these topologies can be found in Baddeley (1991); Beer (1985, 1987); Goutsias (1993); Vervaat (1988). Finally, we point out that the hit-or-miss topology is a special case of the Lawson topology in continuous lattices (with reverse partial ordering); see Gierz et al. (1980).

Beer (1985, 1987) discusses the class of metric spaces in which the relation $\lim X_n = X$ for closed sets is equivalent to pointwise convergence of $d(\cdot, X_n)$ to $d(\cdot, X)$. He shows that this class includes the finitely compact metric spaces. Beer (1987) also points out the relation with the hit-or-miss topology. It seems, however, that Beer was unaware of the work by Busemann (1955) and Matheron (1975). For a comprehensive discussion of the myope topology and its relation with the Hausdorff metric refer again to Matheron (1975).

Matheron is primarily interested in a theory of random closed sets; in this theory, semi-continuous operators are important because they are measurable; application of a measurable operator to a random closed set yields a random closed set. An elementary introduction to random set theory and semi-continuous operators is presented by Goutsias (1993).

Proposition 7.46 is due to Matheron (1975, Section 8.2); however, the proof given here is different.

The basis representation theorem in Corollary 7.48 is due to Maragos (1989b); note that Maragos proves an analogous result for u.s.c. functions.

In their book entitled *Set-Valued Analysis*, Aubin and Frankowska (1990) provide many examples where set-valued maps occur naturally. In the first chapter they present a stimulating discussion of semi-continuity properties of such maps.

CHAPTER EIGHT

Discretization

Henk J.A.M. Heijmans
Formerly Centre for Mathematics and Computer Science, Amsterdam, Netherlands

Contents

This chapter examines a discretization procedure for binary images and binary image operators. The procedure is based on a morphological sampling strategy and uses the hit-or-miss topology to formulate the concept of approximation.

8.1. Statement of the problem

Images processed by a computer are discrete (or digital) images: they comprise a finite number of elements. Usually these picture elements (pixels) are arranged regularly in a grid, e.g., the square or hexagonal grid. The ultimate goal of a computerized image analysis system is the extraction of useful information from an image. Cell biologists, for example, are often interested in the statistics of certain cell characteristics such as the mean curvature of the cell boundary or the area of the cell nucleus. To extract such parameters one requires analytical methods deriving, e.g., from integral geometry. Morphological algorithms are developed for discrete images, however, because in practice one has a finite amount of data at one's disposal and because the computers on which these data are to be processed are digital. Thus there exists a gap between the mathematical tools, which are intrinsically continuous, and the practical solution, which requires discrete computations.

The following example illustrates this point. Suppose one is interested in the perimeter of a set $X \subseteq \mathbb{R}^2$. One can attempt to approximate this

Advances in Imaging and Electron Physics, Volume 216
ISSN 1076-5670
https://doi.org/10.1016/bs.aiep.2020.07.008

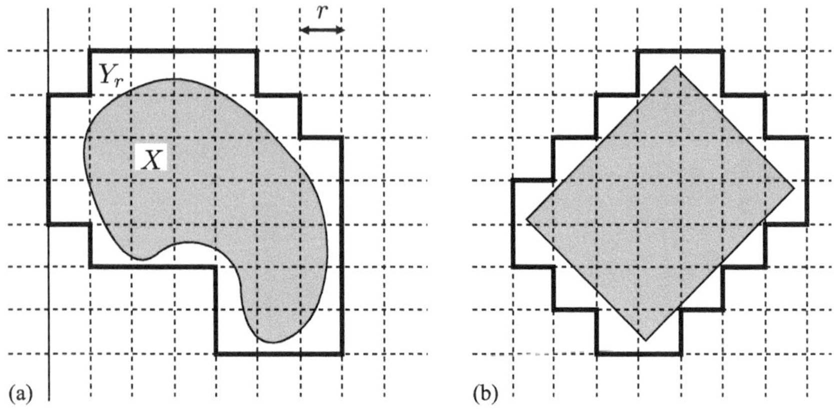

Figure 8.1 Discretization of the perimeter.

quantity by choosing a square grid in $\mathbb{R}^2$ with mesh width r and replacing X by Y_r, the set of all pixels that intersect X; see Fig. 8.1(a). Does the perimeter of Y_r, which is a multiple of r, converge to the perimeter of X as r tends to zero?

The example in Fig. 8.1(b) shows that in general, the answer is negative: here the perimeter of Y_r converges to $\sqrt{2}$ times the perimeter of X.

The gap between continuous and discrete morphology needs to be bridged by a discretization procedure for images and image operators. This procedure must contain the following three ingredients.

- *Sampling of continuous images:* Section 8.2 discusses a sampling strategy transforming a continuous image into a discrete one comprising at most countably many picture elements; in practice, the original image is bounded and the discretized image is finite. To compare the discretized image with the original one, one has to embed the discrete images into the space of continuous ones; this can be achieved by means of a representation operator.
- *Topology:* To formalize the concept of approximation one requires a topology on the space of images. We use the hit-or-miss topology discussed in the previous chapter for that purpose. If one samples an image at finer and finer scales, one obtains a sequence of discretizations which converges toward the original image.
- *Approximation of image operators:* Finally, the theory must deal with the question of under which conditions a continuous morphological operator can be approximated by a sequence of discrete ones. At this point one has to face the following dilemma: should one require that the

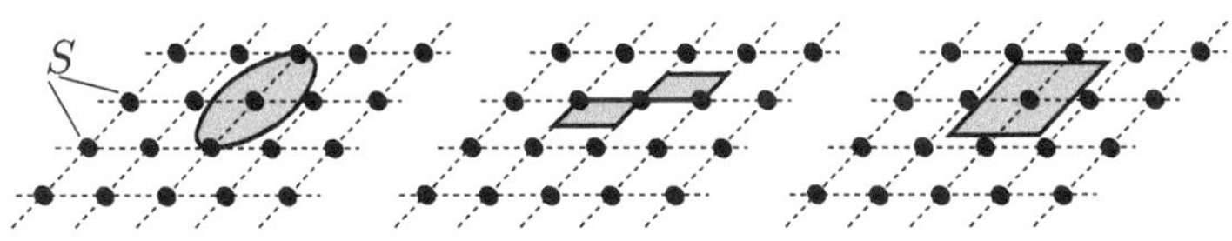

Figure 8.2 Some sampling strategies. The grey-shaded region denotes the sampling element *C*.

discretized operator has essentially the same structure as the original one? For example, must the discretization of a dilation be again a dilation? It is apparent that such additional requirements make the problem considerably more difficult.

8.2. Morphological sampling

A first step in any discretization procedure is the reduction of a continuous image to some mosaic of pixels. A typical way to achieve this is through sampling. This section discusses one specific sampling strategy. Since the hit-or-miss topology plays an important role in this chapter, the space $\mathcal{F}(\mathbb{R}^d)$ is used as a model for continuous binary images. Let S be a regular grid in $\mathbb{R}^d$ given by

$$S = \{k_1 u_1 + \cdots + k_d u_d \mid k_i \in \mathbb{Z}\}, \tag{8.1}$$

where u_i are linearly independent vectors in $\mathbb{R}^d$. We call S the *sampling grid*. Let C be an open subset of $\mathbb{R}^d$, called the *sampling element*. Without loss of generality, we may assume

$$0 \in C.$$

Some sampling strategies are depicted in Fig. 8.2.

Let the function $C^* : \mathbb{R}^d \to \mathcal{P}(S)$ be defined by

$$C^*(x) = \{s \in S \mid x \in C_s\}, \tag{8.2}$$

where C_s is the translation of C over s. It is easy to check that $s \in C^*(x)$ if and only if $x \in C_s$. Note that $C^*(x) = \check{C}_x \cap S$, where $\check{C}$ is the reflected set $\{-c : c \in C\}$. Refer to Fig. 8.3 for an illustration.

Assume that the collection of all translates C_s, $s \in S$, spans the entire space $\mathbb{R}^d$, that is,

$$S \oplus C = \mathbb{R}^d. \tag{8.3}$$

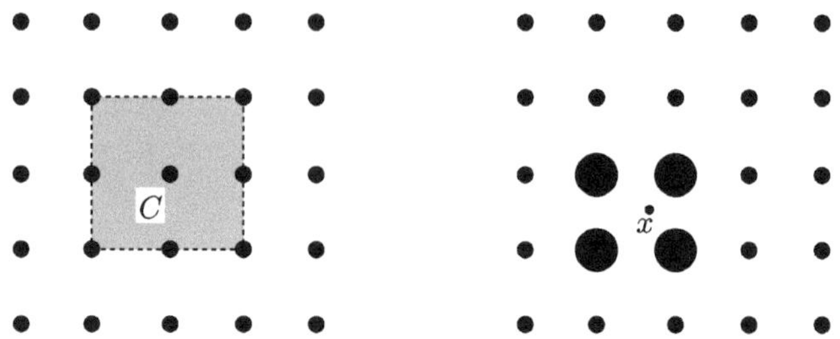

Figure 8.3 A sampling element C (left) and the corresponding set $C^*(x)$ (right).

This assumption henceforth referred to as *Covering Assumption*, is equivalent to

$$C^*(x) \neq \emptyset. \tag{8.4}$$

The sampling strategies illustrated in Fig. 8.2 obey this Covering Assumption. In practice one takes $C = \{x_1 u_1 + \cdots + x_d u_d \mid -a < x_i < a\}$, where $a > \frac{1}{2}$ in order that (8.3) is satisfied. The third example in Fig. 8.2 is of this kind. Define the *sampling operator* $\sigma : \mathcal{F}(\mathbb{R}^d) \to \mathcal{P}(S)$ by

$$\sigma(X) = \{s \in S : C_s \cap X \neq \emptyset\} = (X \oplus \check{C}) \cap S. \tag{8.5}$$

Refer to Fig. 8.4 for an illustration.

8.1 Lemma. *For every $V \subseteq S$, the set $\{x \in \mathbb{R}^d \mid C^*(x) \subseteq V\}$ is closed.*

Proof. Let $x_n \to x$ and $C^*(x_n) \subseteq V$; we must show that $C^*(x) \subseteq V$. Assume $s \in C^*(x)$, that is, $x \in C_s$. Since C_s is open, $x_n \in C_s$ for large enough n. But this implies that $s \in C^*(x_n) \subseteq V$ for such n. This gives the result. □

Define, for $V \subseteq S$,

$$\rho(V) = \{x \in \mathbb{R}^d \mid C^*(x) \subseteq V\}. \tag{8.6}$$

The previous lemma means that ρ maps $\mathcal{P}(S)$ into $\mathcal{F}(\mathbb{R}^d)$.

8.2 Proposition. *The pair (ρ, σ) defines an adjunction between $\mathcal{P}(S)$ and $\mathcal{F}(\mathbb{R}^d)$.*

Proof. We show that $\sigma(X) \subseteq V$ iff $X \subseteq \rho(V)$ for $X \in \mathcal{F}(\mathbb{R}^d)$ and $V \in \mathcal{P}(S)$.

$\Rightarrow$: Let $\sigma(X) \subseteq V$ and $x \in X$. We must show that $x \in \rho(V)$, i.e., $C^*(x) \subseteq V$. Let $s \in C^*(x)$; then $x \in C_s$, and so $C_s \cap X \neq \emptyset$; this gives $s \in \sigma(X)$, and hence $s \in V$.

'⇐': Let $X \subseteq \rho(V)$ and $s \in \sigma(X)$; we must show that $s \in V$. From $s \in \sigma(X)$ we conclude $C_s \cap X \neq \varnothing$. Let $x \in C_s \cap X$; this implies that $x \in \rho(V)$, and hence that $C^*(x) \subseteq V$. From $x \in C_s$ it follows that $s \in C^*(x)$, and so $s \in V$. □

One has the following alternative characterization of ρ:

$$\rho(V) = \Big[\bigcup_{s \in S \setminus V} C_s\Big]^c, \tag{8.7}$$

for $V \subseteq S$. To prove $\subseteq$ assume $x \in \rho(V)$ and $x \in C_s$ for some $s \in S \setminus V$. Then $C^*(x) \subseteq V$ and, since $s \in C^*(x)$, also $s \in V$, which is a contradiction. To prove $\supseteq$ assume that $x \notin \rho(V)$; then $C^*(x) \nsubseteq V$. So there exists $s \in C^*(x)$ such that $s \notin V$. In other words, $x \in C_s$ for some $s \in S \setminus V$; this means that $x \notin \big[\bigcup_{s \in S \setminus V} C_s\big]^c$.

We call ρ the *representation operator*. The composition

$$\pi = \rho\sigma \tag{8.8}$$

is called *approximation operator*. Since (ρ, σ) is an adjunction, π defines a closing on $\mathcal{F}(\mathbb{R}^d)$, that is, $\pi^2 = \pi$ and $X \subseteq \pi(X)$; cf. Theorem 3.25. Moreover, the adjunction property means that resampling of the approximated image gives the same outcome as the original sampling of X. In other words, $\sigma\pi = \sigma$; cf. Proposition 3.14. The sampling and approximation procedure are illustrated in Fig. 8.4.

From (8.5)–(8.6) one obtains

$$\begin{aligned}
\pi(X) = \rho\sigma(X) &= \{x \in \mathbb{R}^d : C^*(x) \subseteq (X \oplus \check{C}) \cap S\} \\
&\subseteq \{x \in \mathbb{R}^d : (\check{C}_x \cap S) \cap ((X \oplus \check{C}) \cap S) \neq \emptyset\} \\
&\subseteq \{x \in \mathbb{R}^d : \check{C}_x \cap (X \oplus \check{C}) \neq \emptyset\} \\
&= (X \oplus \check{C}) \oplus C = X \oplus L,
\end{aligned}$$

where

$$L = C \oplus \check{C}. \tag{8.9}$$

This leads to the following estimates for π:

$$X \subseteq \pi(X) \subseteq X \oplus L, \tag{8.10}$$

for every set $X \subseteq \mathbb{R}^d$.

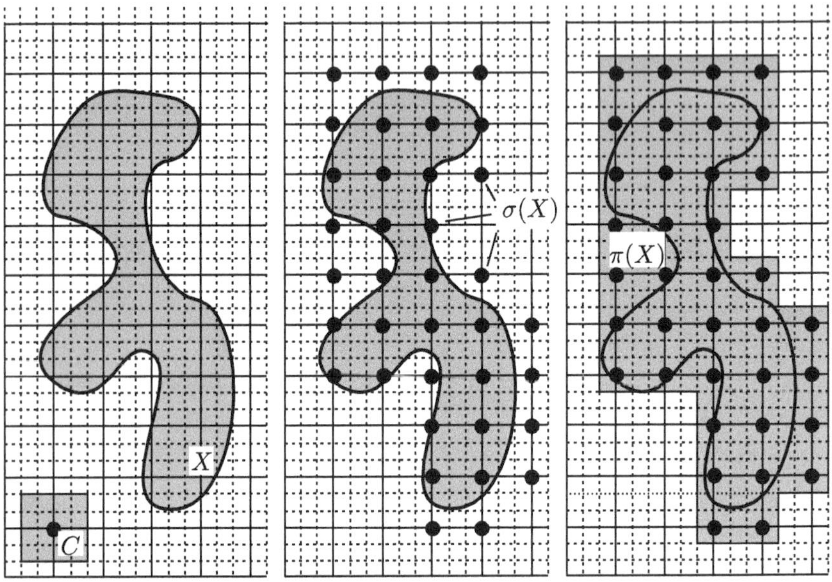

Figure 8.4 Sampling and approximation.

8.3. Discretization of images

By a discretization of a binary image $X \subseteq \mathbb{R}^d$ is meant an approximation of X by a sequence $X_n \subseteq \mathbb{R}^d$, $n \geq 1$, where X_n has a discrete representation: see Definition 8.4 for a precise characterization. The topology exploited here is the hit-or-miss topology discussed in the previous chapter. Hereafter the following property of this topology will be used.

8.3 Lemma. *Let U_n, V_n, X_n be sequences in $\mathcal{F}(\mathbb{R}^d)$ such that $U_n \subseteq X_n \subseteq V_n$ and $U_n \xrightarrow{\mathcal{F}} X$, $V_n \xrightarrow{\mathcal{F}} X$. Then also $X_n \xrightarrow{\mathcal{F}} X$.*

This follows easily from the observation that $\underline{\lim}\, U_n \subseteq \underline{\lim}\, X_n \subseteq \underline{\lim}\, V_n$ and $\overline{\lim}\, U_n \subseteq \overline{\lim}\, X_n \subseteq \overline{\lim}\, V_n$. We now present a formal definition of a discretization.

8.4 Definition. A *discretization* $\mathcal{D}$ on $\mathcal{F}(\mathbb{R}^d)$ is a sequence of triples (S_n, σ_n, ρ_n), where S_n is a countable set, σ_n is an operator from $\mathcal{F}(\mathbb{R}^d)$ to $\mathcal{P}(S_n)$, and ρ_n is an operator from $\mathcal{P}(S_n)$ to $\mathcal{F}(\mathbb{R}^d)$ such that $\rho_n\sigma_n(X) \xrightarrow{\mathcal{F}} X$, for every $X \in \mathcal{F}(\mathbb{R}^d)$. The notation $\mathcal{D} = \{S_n, \sigma_n, \rho_n\}_{n\geq 1}$ is used to denote such discretizations.

Note that although $\rho_n\sigma_n(X)$ lies in the original space $\mathcal{F}(\mathbb{R}^d)$, the operator $\rho_n\sigma_n$ assumes only values in a "discrete" subspace of $\mathcal{F}(\mathbb{R}^d)$.

This section discusses a discretization procedure based on the sampling strategy described in the previous section. Throughout the remainder of this section the following assumptions are made.

8.5 Assumption. C_n and S_n are sequences of sampling elements and sampling grids, respectively, such that the following properties hold:

(i) for every n relation (8.3) holds, that is, $S_n \oplus C_n = \mathbb{R}^d$;

(ii) every C_n is an open set;

(iii) $\overline{C_n}$ converges myopically to $\{0\}$.

8.6 Example. Let S be a regular grid as in (8.1), and let r_n be a sequence of real positive numbers tending to 0. Define $S_n = r_n S$ and $C_n = r_n C$ where $C = \{x_1 u_1 + \cdots + x_d u_d \mid -a < x_i < a\}$; here $a > \frac{1}{2}$ is such that $S_n \oplus C_n = \mathbb{R}^d$. It is obvious that $\overline{C_n} \xrightarrow{\mathcal{K}} \{0\}$; hence Assumption 8.5 is satisfied.

In what follows the notation $(C_n)_s$ and $C_n(s)$ is used interchangeably. Define

$$L_n = C_n \oplus \check{C}_n.$$

It is evident that L_n is open and $0 \in L_n$.

8.7 Lemma. *Suppose that Assumption 8.5 is satisfied; then* $\overline{L_n} \xrightarrow{\mathcal{K}} \{0\}$.

Proof. First note that $\overline{L_n} = \overline{\check{C}_n} \oplus \overline{C_n}$. It is obvious that $\overline{\check{C}_n} \to \{0\}$ both with respect to the hit-or-miss and the myope topology. Thus Proposition 7.43(a) leads to $\overline{L_n} = \overline{C_n} \oplus \check{C}_n \to \{0\} \oplus \{0\} = \{0\}$ with respect to the hit-or-miss topology. The fact that $\overline{C_n} \xrightarrow{\mathcal{K}} \{0\}$ in combination with Proposition 7.33 implies that $\overline{C_n} \subseteq K$ for all n, hence $\overline{L_n} \subseteq K \oplus \check{K}$ for all n. Applying Proposition 7.33 once more, one concludes that $\overline{L_n} \to \{0\}$ with respect to the myope topology. □

Let σ_n be the sampling operator given by (8.5) with $S = S_n$, that is,

$$\sigma_n(X) = \{s \in S_n \mid C_n(s) \cap X \neq \varnothing\};$$

let ρ_n be the adjoint erosion given by (8.6),

$$\rho_n(V) = \{x \in \mathbb{R}^d \mid C_n^*(x) \subseteq V\}.$$

Finally, let π_n be the approximation operator

$$\pi_n = \rho_n \sigma_n.$$

8.8 Proposition. $\pi_n(X) \xrightarrow{\mathcal{F}} X$ *for every* $X \in \mathcal{F}(\mathbb{R}^d)$.

Proof. Let $X \in \mathcal{F}(\mathbb{R}^d)$; from (8.10) it follows that

$$X \subseteq \pi_n(X) \subseteq X \oplus L_n \subseteq X \oplus \overline{L_n}.$$

Now Lemma 8.7 and Proposition 7.43(a) imply $X \oplus \overline{L_n} \xrightarrow{\mathcal{F}} X$; using Lemma 8.3, one gets $\pi_n(X) \xrightarrow{\mathcal{F}} X$. □

8.9 Corollary. *The sequence* $\mathcal{D} = \{S_n, \sigma_n, \rho_n\}_{n\geq 1}$ *defines a discretization on* $\mathcal{F}(\mathbb{R}^d)$.

In general, the convergence of the sequence $\pi_n(X)$ of approximations toward X is not monotone. An interesting problem is to find conditions which guarantee

$$\cdots \subseteq \pi_{n+1}(X) \subseteq \pi_n(X) \subseteq \pi_{n-1}(X) \subseteq \cdots,$$

and hence that $\pi_n(X) \downarrow X$. In that case, the approximations $\pi_n(X)$ approach X from above in a monotone fashion, and $\mathcal{D}$ is called a *constricting discretization*. Note that $X \subseteq \pi_n(X)$ is satisfied automatically since every π_n is a closing. It turns out that one can give a complete solution to this problem in terms of the following condition:

$$\forall x \in \mathbb{R}^d \ \forall s \in C_n^*(x) \ \exists s' \in C_{n+1}^*(x): \quad C_{n+1}(s') \subseteq C_n(s). \tag{8.11}$$

8.10 Proposition. *The discretization* $\mathcal{D} = \{S_n, \sigma_n, \rho_n\}_{n\geq 1}$ *is constricting if and only if condition* (8.11) *is satisfied.*

Proof. Recall that $\sigma_n(X) = \{s \in S_n \mid C_n(s) \cap X \neq \varnothing\}$ and that $\pi_n(X) = \{x \in \mathbb{R}^d \mid C_n^*(x) \subseteq \sigma_n(X)\}$.

"if": Let (8.11) hold, and suppose $x \in \pi_{n+1}(X)$; it has to be demonstrated that $x \in \pi_n(X)$, that is, $C_n^*(x) \subseteq \sigma_n(X)$. If $s \in C_n^*(x)$, then, by (8.11), there is an $s' \in C_{n+1}^*(x)$ such that $C_{n+1}(s') \subseteq C_n(s)$. Now $C_{n+1}(s') \cap X \neq \varnothing$, and therefore $C_n(s) \cap X \neq \varnothing$.

"only if": Assume that $\pi_{n+1}(X) \subseteq \pi_n(X)$; let $x \in \mathbb{R}^d$ and $s \in C_n^*(x)$. Define $X = [C_n(s)]^c$; then $s \notin \sigma_n(X)$. Therefore, $C_n^*(x) \nsubseteq \sigma_n(X)$, and thus

$$\begin{array}{ccc} \mathcal{F}(\mathbb{R}^d) & \xrightarrow{\psi} & \mathcal{F}(\mathbb{R}^d) \\ \rho_n \uparrow \downarrow \sigma_n & & \rho_n \uparrow \downarrow \sigma_n \\ \mathcal{P}(S_n) & \xrightarrow{\psi_n} & \mathcal{P}(S_n) \end{array}$$

Figure 8.5 Intertwining diagram for the discretization of an operator.

$x \notin \pi_n(X)$. This implies that $x \notin \pi_{n+1}(X)$, and so $C^*_{n+1}(x) \not\subseteq \sigma_{n+1}(X)$. Thus there must exist an $s' \in C^*_{n+1}(x)$ such that $s' \notin \sigma_{n+1}(X)$. The latter means that $C_{n+1}(s') \cap X = \varnothing$, that is, $C_{n+1}(s') \subseteq C_n(s)$, which was to be proved. □

Section 8.5 is entirely devoted to a regular sampling strategy for which condition (8.11) is satisfied. Furthermore, this section contains some illustrations which may help the reader to get some intuition for this problem.

8.4. Discretization of operators

Now that we know how to approximate closed sets by discrete ones, we may consider the problem of approximating morphological operators on $\mathcal{F}(\mathbb{R}^d)$ by discrete operators. The following definition formalizes this problem.

8.11 Definition. Let $\mathcal{D} = \{S_n, \sigma_n, \rho_n\}_{n\geq 1}$ be a discretization of $\mathcal{F}(\mathbb{R}^d)$, and let ψ be an operator mapping $\mathcal{F}(\mathbb{R}^d)$ into itself. If ψ_n is a sequence of operators on $\mathcal{P}(S_n)$ such that

$$\rho_n \psi_n \sigma_n(X) \xrightarrow{\mathcal{F}} \psi(X),$$

for every $X \in \mathcal{F}(\mathbb{R}^d)$, then $\{\psi_n\}_{n\geq 1}$ is called a *discretization of ψ with respect to $\mathcal{D}$* or, alternatively, a *$\mathcal{D}$-discretization of ψ*.

The question whether a given operator is discretizable depends on the discretization $\mathcal{D}$ of the space $\mathcal{F}(\mathbb{R}^d)$. The foregoing definition allows that ψ is discretizable with respect to one discretization $\mathcal{D}$, but not with respect to another $\mathcal{D}'$. In Theorem 8.12, for example, it will be shown that constricting discretizations allow a larger class of discretizable operators than discretizations which lack this property. Now suppose that $\mathcal{D} = \{S_n, \sigma_n, \rho_n\}_{n\geq 1}$ is a discretization of $\mathcal{F}(\mathbb{R}^d)$, and let ψ be an operator on $\mathcal{F}(\mathbb{R}^d)$. The diagram in Fig. 8.5 indicates how to obtain a sequence of discrete operators ψ_n.

Defining

$$\psi_n = \sigma_n \psi \rho_n,$$

one finds

$$\rho_n \psi_n \sigma_n = \pi_n \psi \pi_n,$$

where π_n is the approximation operator $\rho_n \sigma_n$.

8.12 Theorem. *Let $\mathcal{D} = \{S_n, \sigma_n, \rho_n\}_{n \geq 1}$ be a discretization of $\mathcal{F}(\mathbb{R}^d)$, let ψ be an operator on $\mathcal{F}(\mathbb{R}^d)$, and let $\psi_n : \mathcal{P}(S_n) \to \mathcal{P}(S_n)$ be given by*

$$\psi_n = \sigma_n \psi \rho_n.$$

Then $\{\psi_n\}_{n \geq 1}$ is a $\mathcal{D}$-discretization of ψ in either of the following two cases:

(a) *ψ is continuous;*

(b) *$\mathcal{D}$ is a constricting discretization and ψ is increasing and u.s.c.*

Proof. The proof of (a) is trivial.

(b): For $X \in \mathcal{F}(\mathbb{R}^d)$ one has $\pi_n(X) \downarrow X$, and so $\psi \pi_n(X) \downarrow \psi(X)$ since ψ is u.s.c. We show

$$\bigcap_{n \geq 1} \pi_n \psi \pi_n(X) = \psi(X).$$

The inclusion $\supseteq$ follows from the fact that every π_n is a closing. On the other hand, for $m \geq 1$,

$$\bigcap_{n \geq 1} \pi_n \psi \pi_n(X) \subseteq \bigcap_{n \geq 1} \pi_n \psi \pi_m(X) = \psi \pi_m(X).$$

Take the intersection over $m \geq 1$ on the right-hand side and use the upper semi-continuity of ψ; this gives

$$\bigcap_{n \geq 1} \pi_n \psi \pi_n(X) \subseteq \psi(X),$$

which proves the result. □

Although this theorem provides only a partial answer to the question of which morphological operators are discretizable, it applies to some of the most important morphological transformations such as dilations (which are continuous), erosions, closings, and openings (which are u.s.c. and increasing): see Corollary 7.44.

8.5. Covering discretization

This section considers a regular discretization procedure which has some particularly nice properties; in particular, it is constricting. Let $u_1, \ldots, u_d$ be independent vectors in $\mathbb{R}^d$, and let S be the grid spanned by these vectors. Take

$$C = (-u_1, u_1) \oplus \cdots \oplus (-u_d, u_d), \tag{8.12}$$

or alternatively, $C = \{x_1 u_1 + \cdots + x_d u_d \mid x_i \in (-1, 1)\}$. It is obvious that $C \oplus S = \mathbb{R}^d$; furthermore,

$$C \cap S = \{0\}. \tag{8.13}$$

Let σ, ρ be as before, that is,

$$\sigma(X) = \{s \in S \mid C_s \cap X \neq \varnothing\}, \qquad X \in \mathcal{F}(\mathbb{R}^d),$$
$$\rho(V) = \{x \in \mathbb{R}^d \mid C^*(x) \subseteq V\}, \qquad V \in \mathcal{P}(S).$$

Let $\pi = \rho\sigma$, and denote the range of π by $\mathcal{F}_\pi$,

$$\mathcal{F}_\pi = \{\pi(X) \mid X \in \mathcal{F}(\mathbb{R}^d)\};$$

note that $\mathcal{F}_\pi$ is a subset of $\mathcal{F}(\mathbb{R}^d)$.

8.13 Lemma. *For every $X \in \mathcal{F}_\pi$ one has the following expressions:*
(a) $\sigma(X) = X \cap S$;
(b) $X = \rho(X \cap S)$.

Proof. First we prove that $C^*(s) = \{s\}$ for $s \in S$. It is obvious that $s \in C^*(s)$. Assume $s' \in C^*(s)$; then $s \in C_{s'}$, and hence $s - s' \in C$. But $s - s' \in S$ and $C \cap S = \{0\}$, which implies $s = s'$.

(a): Let $X \in \mathcal{F}_\pi$; then $X = \pi(X) = \rho\sigma(X)$. Thus

$$X \cap S = \rho\sigma(X) \cap S = \{s \in S \mid C^*(s) \subseteq \sigma(X)\} = \sigma(X).$$

(b): Let $X \in \mathcal{F}_\pi$; then, by (a), $\sigma(X) = X \cap S$. So $\rho\sigma(X) = \pi(X) = X = \rho(X \cap S)$. □

Note that this lemma uses only that $C \cap S = \{0\}$. For the next lemma it is explicitly needed that C is of the form described in (8.12).

8.14 Lemma. *If $X, Y \in \mathcal{F}_\pi$, then $X \oplus Y, X \ominus Y \in \mathcal{F}_\pi$ as well, and*

$$(X \cap S) \oplus (Y \cap S) = (X \oplus Y) \cap S.$$

Proof. To prove this lemma, note first that

$$C^*(x+y) \subseteq C^*(x) \oplus C^*(y),$$

for $x, y \in \mathbb{R}^d$. We give a sketch of the proof of this fact for the case $d = 2$, but the argument carries over to the general case easily. We consider only the situation where neither x nor y is positioned on one of the grid lines. Then both $C^*(x)$ and $C^*(y)$ contain four grid points, namely, those grid points which are closest to x and y, respectively. Now $C^*(x) \oplus C^*(y)$ contains nine grid points, which constitute a three by three square; $x + y$ lies in the convex hull formed by these nine points. Thus it is obvious that $C^*(x+y) \subseteq C^*(x) \oplus C^*(y)$.

We are now ready to prove the lemma. First observe that the inclusion $\subseteq$ is trivial; we prove $\supseteq$. Let $s \in (X \oplus Y) \cap S$; hence $s = x + y$ for some $x \in X$ and $y \in Y$. From the fact that $X \in \mathcal{F}_\pi$ and Lemma 8.13(b), it follows that $x \in \rho(X \cap S)$, that is, $C^*(x) \subseteq X \cap S$. A similar argument gives $C^*(x) \subseteq Y \cap S$. Then

$$\{s\} = C^*(s) = C^*(x+y) \subseteq C^*(x) \oplus C^*(y) \subseteq (X \cap S) \oplus (Y \cap S).$$

This concludes the proof. □

One can define a discretization in the following way. Let $S_1 = S$ and $C_1 = C$, and define

$$S_{n+1} = \frac{1}{2} S_n, \qquad C_{n+1} = \frac{1}{2} C_n; \tag{8.14}$$

see Fig. 8.6(a) for an example.

It is obvious that Assumption 8.5 is satisfied; furthermore, one can show that condition (8.11) is satisfied as well. In fact, let $x \in \mathbb{R}^d$ and take $s \in C_n^*(x)$. We must show that there exists an $s' \in C_{n+1}^*(x)$ such that $C_{n+1}(s') \subseteq C_n(s)$. We restrict ourselves to $d = 2$ for the sake of exposition. From Fig. 8.6(b) it is immediately clear how to choose s' in this case. If x does not lie on one of the grid lines of S_n through s, then x is contained in one of the cells $C_{n+1}(s_1)$, $C_{n+1}(s_2)$, $C_{n+1}(s_3)$, $C_{n+1}(s_4)$. If x does lie on of these grid lines, however, then one has to choose one of the other four $(n+1)$-cells contained in $C_n(s)$.

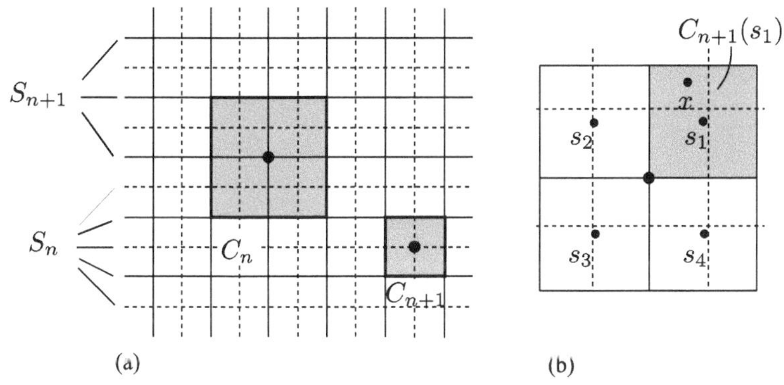

Figure 8.6 (a) A discretization scheme for which the constriction property is satisfied. (b) If $x \in C_n(s)$, then there exists an $s' \in C^*_{n+1}(x)$ such that $C_{n+1}(s') \subseteq C_n(s)$. In the case here $s' = s_1$. See also text.

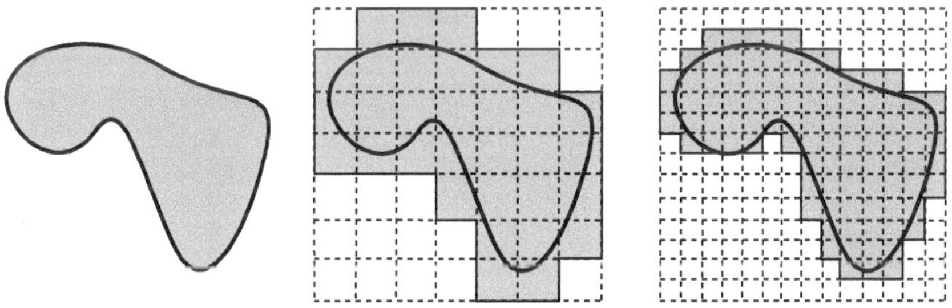

Figure 8.7 Two successive steps of the covering discretization.

We point out that one may also choose a multiplication factor $1/p$ (where p is a positive integer) instead of $1/2$ in (8.14) without destroying the validity of condition (8.11). For other multiplication factors, however, condition (8.11) is not satisfied. The resulting discretization procedure is called *covering discretization*; refer to Fig. 8.7 for an illustration.

The decreasingness property $\pi_1(X) \subseteq \pi_2(X) \subseteq \cdots \subseteq X$ in combination with the properties of the range space $\mathcal{F}_\pi$ mentioned in Lemmas 8.13 and 8.14 suggest yet another discretization procedure for increasing translation invariant operators. Consider, as a first step, the dilation operator. In contrast to the discretization discussed in the previous section, the covering discretization of a dilation is again a dilation.

8.15 Proposition. *Consider the dilation $\delta(X) = X \oplus A$ on $\mathcal{F}(\mathbb{R}^d)$, where $A \subseteq \mathbb{R}^d$ is a compact structuring element. Define the dilation δ_n on $\mathcal{P}(S_n)$ by*

$\delta_n(V) = V \oplus \sigma_n(A)$. Then δ_n is a discretization of δ with respect to the covering discretization of $\mathcal{F}(\mathbb{R}^d)$.

Proof. One gets

$$\rho_n \delta_n \sigma_n(X) = \rho_n\big[\sigma_n(X) \oplus \sigma_n(A)\big].$$

Using that $\sigma_n \pi_n = \sigma_n$, one obtains

$$\begin{aligned}
\rho_n \delta_n \sigma_n(X) &= \rho_n\big[\sigma_n \pi_n(X) \oplus \sigma_n \pi_n(A)\big] \\
&= \rho_n\big[(\pi_n(X) \cap S_n) \oplus (\pi_n(A) \cap S_n)\big] \\
&= \rho_n\big[(\pi_n(X) \oplus \pi_n(A)) \cap S_n\big] \\
&= \pi_n(X) \oplus \pi_n(A).
\end{aligned}$$

Here we have successively used Lemmas 8.13(a), 8.14, and 8.13(b). From the fact that the covering discretization is constricting we get that $\pi_n(X) \downarrow X$ and $\pi_n(A) \downarrow A$; hence $\pi_n(X) \xrightarrow{\mathcal{F}} X$ and $\pi_n(A) \xrightarrow{\mathcal{K}} A$ (see Propositions 7.28(a) and 7.34(a)). Now Proposition 7.43(a) implies

$$\pi_n(X) \oplus \pi_n(A) \xrightarrow{\mathcal{F}} X \oplus A.$$

Therefore, δ_n defines a discretization of δ. □

Consider the operator ψ given by

$$\psi(X) = \bigcap_{A \in \mathcal{A}} X \oplus A, \tag{8.15}$$

where $\mathcal{A}$ is a collection of compact structuring elements. It is shown that ψ is discretizable with respect to the covering discretization already discussed. Define

$$\psi_n(V) = \bigcap_{A \in \mathcal{A}} V \oplus \sigma_n(A),$$

for $V \subseteq S_n$. Using that ρ_n is an erosion, one derives

$$\begin{aligned}
\rho_n \psi_n \sigma_n(X) &= \rho_n\Big(\bigcap_{A \in \mathcal{A}} \sigma_n(X) \oplus \sigma_n(A)\Big) \\
&= \bigcap_{A \in \mathcal{A}} \rho_n\big(\sigma_n(X) \oplus \sigma_n(A)\big)
\end{aligned}$$

$$= \bigcap_{A \in \mathcal{A}} \pi_n(X) \oplus \pi_n(A).$$

Here the same argument as in the proof of the previous proposition has been used. It is obvious that this expression defines a decreasing sequence and, using Proposition 8.15,

$$\begin{aligned}\bigcap_{n \geq 1} \rho_n \psi_n \sigma_n(X) &= \bigcap_{A \in \mathcal{A}} \bigcap_{n \geq 1} \pi_n(X) \oplus \pi_n(A) \\ &= \bigcap_{A \in \mathcal{A}} X \oplus A \\ &= \psi(X).\end{aligned}$$

Thus ψ_n defines a discretization of ψ. The operator ψ in (8.15) is increasing, translation invariant, and u.s.c. (by the fact that an arbitrary intersection of u.s.c. operators is u.s.c.). On the other hand, Proposition 7.46 states that an increasing translation invariant operator $\psi : \mathcal{F} \to \mathcal{F}$ is u.s.c. if and only if it admits the representation (8.15) for a family $\mathcal{A} \subseteq \mathcal{K}$ which is closed with respect to the myope topology. These results are listed in the following proposition.

8.16 Proposition. *Let $\psi : \mathcal{F} \to \mathcal{F}$ be an increasing translation invariant u.s.c. operator, and let ψ be represented as*

$$\psi(X) = \bigcap_{A \in \mathcal{A}} X \oplus A,$$

where $\mathcal{A} \subseteq \mathcal{K}$ is myopically closed. Furthermore, let $\mathcal{D} = \{S_n, \sigma_n, \rho_n\}$ be the covering discretization of $\mathcal{F}$. Then the operators

$$\psi_n(V) = \bigcap_{A \in \mathcal{A}} V \oplus \sigma_n(A)$$

define a $\mathcal{D}$-discretization of ψ.

8.6. Bibliographical notes

The theory developed in this chapter, published before in Heijmans (1992a), is to a large extent motivated by the work of Serra (1982, Chapter VII). Serra's work can be regarded as a first attempt to formalize the concept of discretization in the context of mathematical morphology. In fact, the basic idea underlying the approach discussed in this chapter can be found in Serra (1982). Our contribution consists hereof that we put our ideas in a rigorous algebraic and topological framework, which enables us to develop a concise and rather general theory of discretization and to avoid some of the ambiguities occurring in Serra (1982). Heijmans (1991a) discuss a procedure for the discretization of sets which is slightly more general in the sense that we do not restrict ourselves a priori to the case where S_n in Definition 8.4 is a grid. The covering discretization has also been discussed by Serra (1982). We point out that our definition is different from the one by Serra, although this distinction is rather subtle. One of the differences is that in our case the approximation operator π may yield isolated grid points and grid segments, whereas in Serra's approach the discretization comprises only closed cells.

In the literature one finds many discrete approximations of the perimeter. The quality of these approximations depends on the regularity of the sets under consideration, among other things. This subject, however, falls outside the scope of this chapter, and we refer the interested reader to Serra (1982, Section VII.F.2) or Dorst and Smeulders (1987).

Dougherty and Giardina (1987a) make estimates of the following kind:

$$\|f(\cdot, X) - f_n(\cdot, X_n)\| \le e(X, n),$$

where $f(r, \cdot)$ is a parametrized family of Euclidean functionals (e.g., size distributions) with discretizations $f_n(r, \cdot)$ and $\|\cdot\|$ is a norm on the set of all distribution functions $f(r)$, viz. a sup-norm or L^1-norm. In their analysis Dougherty and Giardina restrict to size distributions obtained from dilations, erosions, or openings by a parametrized family of (convex) structuring elements. Furthermore, they use a discretization very similar to the covering discretization discussed in Section 8.5.

Computer graphics is another field where discretization (or digitization) is of great importance. We refer, in particular, to the work of van Lierop (1987). The goal there, however, is to develop a consistent theory of discrete geometrical transformations. In contrast, we are primarily interested in Euclidean image transformations and functionals and regard their discrete versions only as a method to obtain approximations.

CHAPTER NINE

Convexity, distance, and connectivity

Henk J.A.M. Heijmans
Formerly Centre for Mathematics and Computer Science, Amsterdam, Netherlands

Contents

This chapter explains why mathematical morphology is considered to be a geometrical approach in image processing. It shows how to employ basic geometrical notions, like convexity and distance, for the construction of operators which can be utilized to extract geometric information from an image. The main operators treated in this chapter are metric dilations, distance transforms, granulometries, skeletons, and geodesic operators. This final class requires the notion of *geodesic distance*, a notion which will also be treated in this chapter.

9.1. Convexity

This section deals with convex sets, in particular, with those properties which are relevant in the context of mathematical morphology. Unless stated otherwise, it is assumed that E is a vector space (or linear space) over the real numbers.

9.1 Definition. A set $X \subseteq E$ is *convex* if $rx + (1 - r)y \in X$ for $x, y \in X$ and $0 \leq r \leq 1$.

Advances in Imaging and Electron Physics, Volume 216
ISSN 1076-5670
https://doi.org/10.1016/bs.aiep.2020.07.009

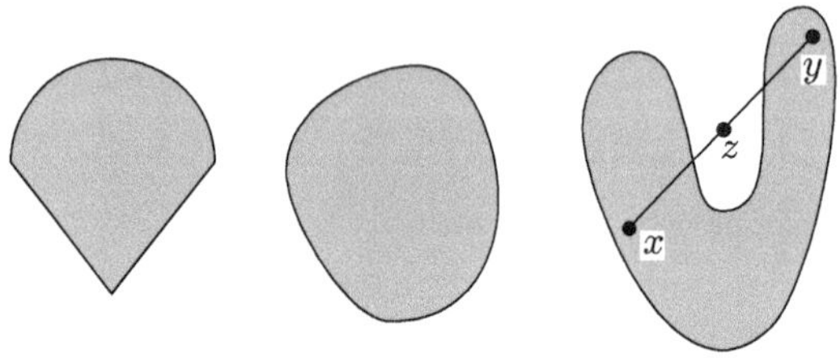

Figure 9.1 Three sets, the first two of which are convex. The set at the right is not convex: x and y lie in X, but the convex combination z lies outside X.

The vector $\sum_{i=1}^{n} r_i x_i$ is called a convex combination of the vectors x_i if all coefficients r_i are nonnegative and $\sum_{i=1}^{n} r_i = 1$. If X is convex, then every convex combination of elements in X lies in X again. See Fig. 9.1 for an illustration.

9.2 Proposition. *Let E be a vector space. A set $X \subseteq E$ is convex if and only if*

$$rX \oplus sX = (r+s)X,$$

for every $r, s \geq 0$.

Proof. "if": Assume that the identity holds and that $x, y \in X$ and $0 \leq r \leq 1$; then $rx + (1-r)y \in rX \oplus (1-r)X = X$. Therefore, X is convex.

"only if": Assume that X is convex and that $r, s \geq 0$. Then, for every $x, y \in X$,

$$rx + sy = (r+s)(\frac{r}{r+s}x + \frac{s}{r+s}y) \in (r+s)X,$$

which proves $rX \oplus sX \subseteq (r+s)X$. The reverse inclusion is trivial. □

The intersection of an arbitrary collection of convex sets is convex. The *convex hull* of a set $X \subseteq E$ is the intersection of all convex sets containing X and is denoted by $\mathrm{co}(X)$; see Fig. 9.2. Clearly, X is convex if and only if $X = \mathrm{co}(X)$.

It is obvious that

$$\mathrm{co}(rX) = r\,\mathrm{co}(X), \tag{9.1}$$

for $X \subseteq E$ and $r \in \mathbb{R}$.

The next proposition is a classical result due to Caratheodory. Its proof can be found in any good textbook on convex sets, e.g., Valentine (1964, Theorem 1.20).

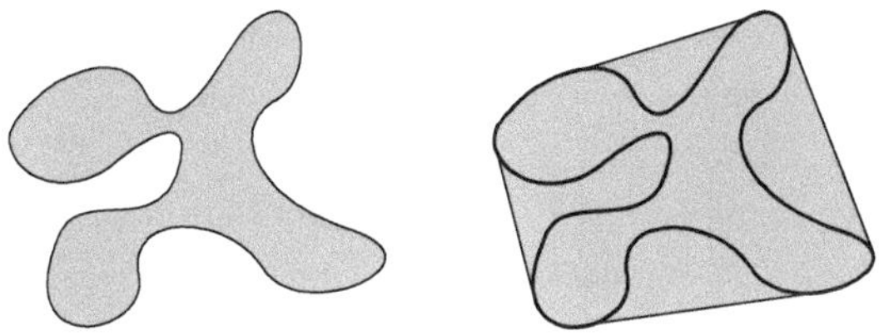

Figure 9.2 A set and its convex hull.

9.3 Proposition. (Caratheodory's Theorem) *Let E be a vector space. The convex hull of a set $X \subseteq \mathbb{R}^d$ comprises all points $\sum_{i=1}^{d+1} r_i x_i$ where $x_i \in X$, $r_i \in [0, 1]$, and $\sum_{i=1}^{d+1} r_i = 1$.*

9.4 Proposition. *Let E be a vector space. For $X, Y \subseteq E$,*

$$\mathrm{co}(X \oplus Y) = \mathrm{co}(X) \oplus \mathrm{co}(Y). \tag{9.2}$$

In particular, if X, Y are both convex, then $X \oplus Y$ is convex too.

Proof. $\subseteq$: Caratheodory's theorem says that elements of $\mathrm{co}(X \oplus Y)$ are of the form $\sum_{i=1}^{d+1} r_i(x_i + y_i)$ where $x_i \in X$, $y_i \in Y$ and $\sum_{i=1}^{d+1} r_i = 1$. But this sum can be decomposed as $\sum_{i=1}^{d+1} r_i x_i + \sum_{i=1}^{d+1} r_i y_i$, which, again by Caratheodory's theorem, is an element of $\mathrm{co}(X) \oplus \mathrm{co}(Y)$.

$\supseteq$: Let $x \in \mathrm{co}(X)$ and $y \in \mathrm{co}(Y)$. Then x, y can be written as $x = \sum_{i=1}^{d+1} r_i x_i$ and $y = \sum_{j=1}^{d+1} s_j y_j$, where $x_i \in X$ and $y_j \in Y$ and where the coefficients r_i and s_j sum up to 1. Then $x + y = \sum_{i=1}^{d+1} \sum_{j=1}^{d+1} r_i s_j (x_i + y_j)$; this latter sum is a convex combination of elements in $X \oplus Y$ and therefore is contained in $\mathrm{co}(X \oplus Y)$. □

A *topological vector space* is a vector space with a Hausdorff topology such that vector addition and scalar multiplication are continuous operations with respect to all variables. The Euclidean space $\mathbb{R}^d$ is a topological vector space. In fact, every d-dimensional topological vector space is linearly isomorphic to $\mathbb{R}^d$.

9.5 Proposition. *Let E be a topological vector space. If $X \subseteq E$ is convex, then both X° and $\overline{X}$ are convex.*

Proof. Assume that $x, y \in X^\circ$ and $r \in [0, 1]$. There is a neighbourhood U of 0 such that $U_x \subseteq X$. Then $rU_x + (1 - r)y \subseteq X$, that is, $(rU)_{rx+(1-r)y} \subseteq X$. This means that $rx + (1 - r)y \in X^\circ$.

Assume that $x, y \in \overline{X}$ and $r \in [0, 1]$. Let U be a neighbourhood of 0. There is another neighbourhood V of 0 such that $V \oplus V \subseteq U$. One can find points $x_0, y_0 \in X$ such that $r(x_0 - x), (1 - r)(y_0 - y) \in V$. Then $r(x_0 - x) + (1 - r)(y_0 - y) \in V \oplus V \subseteq U$. This means that $rx_0 + (1 - r)y_0 - (rx + (1 - r)y) \in U$. Since X is convex, $rx_0 + (1 - r)y_0 \in X$. This implies that every neighbourhood of $rx + (1 - r)y$ contains a point of X, and therefore $rx + (1 - r)y \in \overline{X}$. □

Let E be a topological vector space. Define the *closed convex hull* $\overline{\text{co}}(X)$ of a set X as the intersection of all closed convex sets containing X. It is easy to see that $\overline{\text{co}}(X) = \overline{\text{co}(X)}$.

A *linear functional* on E is a mapping $f : E \to \mathbb{R}$ which is additive and homogeneous, that is, $f(rx + sy) = rf(x) + sf(y)$ for $x, y \in E$ and $r, s \in \mathbb{R}$.

9.6 Definition. Let E be a topological vector space, f a continuous linear functional on E, and $c \in \mathbb{R}$. The set $\{x \in E \mid f(x) = c\}$ is called a *hyperplane* in E and the set $\{x \in E \mid f(x) \leq c\}$ is called a *closed half-space*; the set $\{x \in E \mid f(x) < c\}$ is called an *open half-space*.

Most of the following results are valid for the Euclidean space $\mathbb{R}^d$. A number of them, however, can be extended rather easily to so-called locally convex topological vector spaces. A closed convex set $X \subseteq \mathbb{R}^d$ and a point h which lies outside X can be separated by a half-space H, that is, $X \subseteq H$ and $h \notin H$. This is a consequence of the separation theorem in convex analysis; refer to Valentine (1964, Section II.A) or Marti (1977, Section II.4).

9.7 Proposition. *The closed convex hull of a set $X \subseteq \mathbb{R}^d$ is the intersection of all closed half-spaces that contain X.*

Proof. It is evident that $\overline{\text{co}}(X)$ is a subset of the intersection of all half-spaces that contain X. Assume, on the other hand, that $h \notin \overline{\text{co}}(X)$. Because of the separation result already mentioned, there is a closed half-space H such that $\overline{\text{co}}(X) \subseteq H$ and $h \notin H$. Since $X \subseteq H$, h does not lie in the intersection of closed half-spaces including X. □

9.8 Proposition. *If X is a closed subset and A a bounded subset of $\mathbb{R}^d$, then $X \bullet A \subseteq \text{co}(X)$. In particular, if X is convex, then $X \bullet A = X$.*

Proof. Assume that $h \notin \text{co}(X)$. There is a half-space H such that $X \subseteq H$ and $h \notin H$. Since A is bounded, there is a translate $\check{A}_z$ of $\check{A}$ with $\check{A}_z \cap H = \varnothing$ and $h \in \check{A}_z$. Using the expression for $H \bullet A$ given by (4.57), it follows that $h \notin H \bullet A$, and hence $h \notin X \bullet A$. This concludes the proof. □

9.9 Proposition. *The convex hull of a compact set is again compact, and the mapping* $\mathrm{co}(\cdot) : \mathcal{K}(\mathbb{R}^d) \to \mathcal{K}(\mathbb{R}^d)$ *is continuous.*

Proof. Let $X \subseteq \mathbb{R}^d$ be compact. We show that $\mathrm{co}(X)$ is compact as well. It is obvious that $\mathrm{co}(X)$ is bounded, and it remains to be shown that $\mathrm{co}(X)$ is closed. Assume that $x_n \in \mathrm{co}(X)$ and $x_n \to x$; we show that $x \in \mathrm{co}(X)$. By Proposition 9.3, x_n is of the form $x_n = \sum_{i=1}^{d+1} r_n^i x_n^i$ where $r_n^i \geq 0$, $\sum_{i=1}^{d+1} r_n^i = 1$, and $x_n^i \in X$ for every n. Let i be fixed for the moment. Since X is compact, the sequence $\{x_n^i\}$ has a convergent subsequence. A similar remark applies to the sequences $\{r_n^i\}$. Without loss of generality, we may assume that the entire sequences are convergent. That is, $x_n^i \to x^i$, where $x^i \in X$ and $r_n^i \to r^i$. It is clear that $r^i \geq 0$ and that $\sum_{i=1}^{d+1} r^i = 1$. This means that $x = \sum_{i=1}^{d+1} r^i x^i$, and hence that $x \in \mathrm{co}(X)$.

To prove the continuity of $\mathrm{co}(\cdot)$, let $\{X_n\}$ be a sequence in $\mathcal{K}$ which converges myopically toward $X \in \mathcal{K}$. By Proposition 7.33 there is a $K \in \mathcal{K}$ such that $X_n \subseteq K$ for every n. This gives that $\mathrm{co}(X) \subseteq \mathrm{co}(K)$. Using Proposition 7.33 once more, we conclude that we are finished if we can show that $\mathrm{co}(X_n) \to \mathrm{co}(X)$ with respect to the hit-or-miss topology. Let G be open and $\mathrm{co}(X) \cap G \neq \varnothing$. By Proposition 9.3 there exist $0 \leq r^i \leq 1$ which sum to 1 and $x^i \in X$ such that $\sum_{i=1}^{d+1} r^i x^i \in G$. Since G is open, there is an $\epsilon > 0$ such that for all y^i with distance to x^i less than ϵ we have $\sum_{i=1}^{d+1} r^i y^i \in G$. Because $X_n \to X$ and $X \cap B^\circ(x^i, \epsilon) \neq \varnothing$, it follows that $X_n \cap B^\circ(x^i, \epsilon) \neq \varnothing$ eventually. If x_n^i lies in the intersection, then $\sum_{i=1}^{d+1} r^i x_n^i \in G$. This shows that $\mathrm{co}(X_n) \cap G \neq \varnothing$ eventually. By Proposition 7.26 it follows that $\mathrm{co}(X) \subseteq \underline{\lim}\, \mathrm{co}(X_n)$.

It remains to be demonstrated that $\overline{\lim}\, \mathrm{co}(X_n) \subseteq \mathrm{co}(X)$. We use that $\mathrm{co}(X)$ is the intersection of all $\overline{H}$ where H is an open half-space containing X. In fact, for every such open half-space $X \cap H^c = \varnothing$, and so $X_n \cap H^c = \varnothing$ eventually. This means that $X_n \subseteq H$, and so $\mathrm{co}(X_n) \subseteq H$ eventually. Suppose now that $x_{n_k} \in \mathrm{co}(X_{n_k})$ converges toward x; then $x \in \overline{H}$. As this holds for every H, also $x \in \mathrm{co}(X)$. This finishes the proof. □

9.10 Definition.

(a) Assume that E is a topological vector space. A compact subset $A \subseteq E$ which is convex and has nonempty interior is called *convex body*.

(b) A convex body $A \subseteq \mathbb{R}^d$ is said to be *smooth* if $A \circ \epsilon B = A$ for some $\epsilon > 0$. Here B is the closed unit ball.

9.11 Lemma. *Let* A *be a smooth compact convex body in* $\mathbb{R}^d$*. Assume that* h_n *is a sequence in* $\mathbb{R}^d$, $x \in (nA)_{h_n}$, *and* $(nA)_{h_n} \xrightarrow{\mathcal{F}} X$, *where* X *is a nonempty closed set. Then* X *is a closed half-space or* $X = \mathbb{R}^d$.

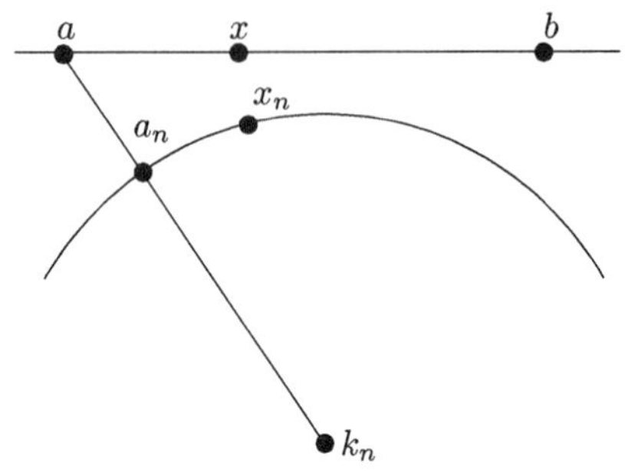

Figure 9.3 Ball notation.

Proof. It is easy to see that X is closed and convex. Since $A \circ \epsilon B = A$, we may conclude that there is a sequence k_n in $\mathbb{R}^d$ such that $x \in (n\epsilon B)_{k_n} \subseteq (nA)_{h_n}$. Suppose we can show that $(n\epsilon B)_{k_n}$, or a subsequence, converges to a closed half-space; this half-space is included in X, and the assertion follows easily from the fact that X is convex.

We may assume without loss of generality that $\epsilon = 1$. Since $\mathcal{F}$ is compact, the sequence $(nB)_{k_n}$ (or a subsequence) converges to a closed set Y. It is easy to see that Y is convex. Suppose we can prove that Y^c is convex, too. By the separation theorem for disjoint convex sets (Valentine, 1964, Theorem 2.9) Y and Y^c can be separated by a hyperplane. Thus it follows immediately that Y is a closed half-space. Therefore, it remains to show that Y^c is convex.

Assume that the points a, b are contained in Y^c and that the convex combination $x = ra + (1 - r)b$ lies in Y. By the separation theorem there is a hyperplane which separates $\{a\}$ and Y and a hyperplane which separates $\{b\}$ and Y. If these hyperplanes were different, then the closed balls $(nB)_{k_n}$ would converge to the empty set. Therefore, there is a hyperplane through a, b, and x such that Y is contained in one of the corresponding half-spaces, say H. There is a sequence $x_n \in (nB)_{k_n}$ which converges to x; the preceding observation shows that we may assume that the balls $(nB)_{k_n}$ lie in H. We may also assume that x_n lies on the surface of $(nB)_{k_n}$. The intersection point of the straight line segment between k_n and a with the surface of the ball $(nB)_{k_n}$ is denoted by a_n; see Fig. 9.3.

It is easy to check that $a_n \to a$, and so $a \in Y$, a contradiction. This concludes the proof. □

9.12 Proposition. *Let A be a smooth convex body in $\mathbb{R}^d$. For every compact set $X \subseteq \mathbb{R}^d$,*

$$\lim X \bullet rA = \overline{(\bigcup_{r>0} X \bullet rA)} = \mathrm{co}(X), \qquad as\ r \to \infty;$$

here the limit is taken with respect to the myope topology of $\mathcal{K}$.

Proof. Without loss of generality, it may be assumed that $0 \in A$. From Proposition 9.8 it follows that $X \bullet rA \subseteq \mathrm{co}(X)$ for every $r > 0$, and thus $(\bigcup_{r>0} X \bullet rA)^- \subseteq \mathrm{co}(X)$. To prove that $\mathrm{co}(X) \subseteq (\bigcup_{r>0} X \bullet rA)^-$, define $K = (\bigcup_{r>0} X \bullet rA)^-$; then K is compact. Let $h \in K^c$; we show that $h \notin \mathrm{co}(X)$. For sufficiently small $\epsilon > 0$, we have $h \in (\epsilon \check{A})_h \subseteq K^c$. Using (1.55), one gets

$$h \in (\epsilon \check{A})_h \subseteq (X \bullet rA)^c = X^c \circ r\check{A},$$

for every $r > 0$. This means that

$$\begin{aligned} h &\in \left[(X^c \ominus r\check{A}) \oplus r\check{A}\right] \ominus \epsilon\check{A} \\ &= \left[(r-\epsilon)\check{A} \oplus \left((X^c \ominus (r-\epsilon)\check{A}) \ominus \epsilon\check{A}\right) \oplus \epsilon\check{A}\right] \ominus \epsilon\check{A} \\ &= (X^c \ominus r\check{A}) \oplus (r-\epsilon)\check{A}, \end{aligned}$$

if $r \geq \epsilon$. In other words, $h \in (X^c \ominus (r+\epsilon)\check{A}) \oplus r\check{A} = (X^c \ominus \epsilon\check{A}) \circ r\check{A}$, for $r \geq 0$. Thus, for every integer n there is an x_n such that

$$h \in (n\check{A})_{x_n} \subseteq X^c \ominus \epsilon\check{A}.$$

This gives

$$h \in (\epsilon\check{A})_h \subseteq ((n+\epsilon)\check{A})_{x_n} \subseteq X^c.$$

The sequence $((n+\epsilon)\check{A})_{x_n}$ has an accumulation point F in $\mathcal{F}(\mathbb{R}^d)$ with $h \in F \subseteq X^c$, that is, $F \cap X = \varnothing$. From the previous lemma it follows that F is a closed half-space. Now $X \subseteq F^c$ implies $\mathrm{co}(X) \subseteq F^c$ since F^c is an open half-space. Therefore, $F \cap \mathrm{co}(X) = \varnothing$, and in particular, $h \notin \mathrm{co}(X)$, which was to be proved. □

Proposition 9.2 implies that every convex set A satisfies

$$A = (A/n) \oplus (A/n) \oplus \cdots \oplus (A/n) \quad [n \text{ terms}]. \tag{9.3}$$

We denote this as $A = (A/n)^{\oplus n}$.

9.13 Definition. The set $A \subseteq E$ is *infinitely divisible* for Minkowski sum if for every integer $n \geq 1$ there exists a set B_n such that

$$A = B_n^{\oplus n}. \tag{9.4}$$

The identity in (9.3) means that every convex set is infinitely divisible. Convexity is not a prerequisite for infinite divisibility, however. For example, the set $A \subseteq \mathbb{R}^2$ containing the first and third quadrants satisfies $A = A^{\oplus n}$ for every $n \geq 1$, and, therefore, this set is infinitely divisible. Yet one can establish the following result.

9.14 Proposition. *A compact set $A \subseteq \mathbb{R}^d$ is infinitely divisible for the Minkowski sum if and only if it is convex.*

Proof. It suffices to prove the only if-part. Let A be compact and assume that $A = B_n^{\oplus n}$. Obviously, B_n is bounded. Since $\overline{B_n}^{\oplus n} = \overline{B_n^{\oplus n}} = \overline{A} = A$, we may assume without loss of generality that B_n is closed. There is a subsequence of $\{nB_n\}$ which converges in $\mathcal{F}$. Without loss of generality, we may assume that the whole sequence converges, say toward A'. Since $nB_n \subseteq B_n^{\oplus n} = A$, Proposition 7.33 gives that $\{nB_n\}$ also converges in $\mathcal{K}$ toward A'. Furthermore, it follows from (9.2) and Proposition 9.2 that

$$\mathrm{co}(A) = \mathrm{co}(B_n^{\oplus n}) = [\mathrm{co}(B_n)]^{\oplus n} = n\,\mathrm{co}(B_n) = \mathrm{co}(nB_n).$$

Using that $\mathrm{co}(\cdot)$ is continuous on $\mathcal{K}$ (Proposition 9.9), one finds

$$\mathrm{co}(A) = \mathrm{co}(A').$$

If one can show that $\mathrm{co}(A') \subseteq A$, then $\mathrm{co}(A) = A$, meaning that A is convex. Take $x \in \mathrm{co}(A')$; then $x = \sum_{i=1}^{d+1} r_i x_i$, where $x_i \in A'$, $0 \leq r_i \leq 1$, and $\sum_{i=1}^{d+1} r_i = 1$. For every i there is a sequence $\{b_n^i\}$ with $b_n^i \in B_n$ such that $nb_n^i \to x^i$. It is easy to see that one can find integers p_n^i such that

$$\sum_{i=1}^{d+1} p_n^i = n \qquad \text{and} \qquad \frac{1}{n} p_n^i \to r_i \text{ as } n \to \infty.$$

Define $x_n = \sum_{i=1}^{d+1} p_n^i b_n^i$; it is obvious that $x_n \in B_n^{\oplus n} = A$. Moreover,

$$x_n = \sum_{i=1}^{d+1} \frac{p_n^i}{n} nb_n^i \to \sum_{i=1}^{d+1} r_i x_i = x;$$

this means that $x \in A$. □

9.15 Proposition. *Let A be a nonempty compact subset of $\mathbb{R}^d$. Then $\{(A/n)^{\oplus n}\}$ converges to $\mathrm{co}(A)$ in $\mathcal{K}(\mathbb{R}^d)$.*

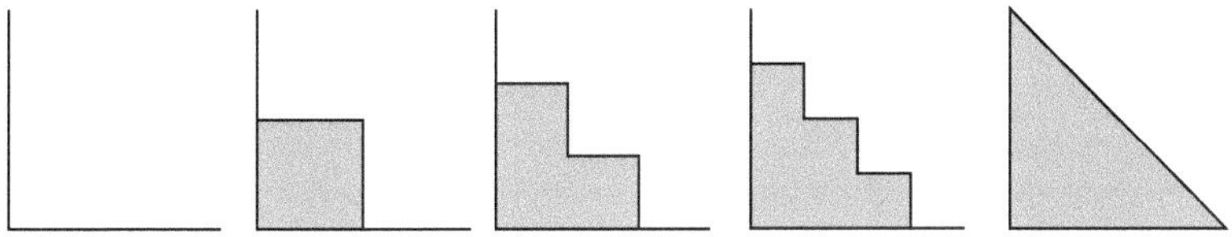

Figure 9.4 The sets A, $(A/n)^{\oplus n}$ for $n = 2, 3, 4$, and the limit $\mathrm{co}(A)$.

Proof. Since $(A/n)^{\oplus n} \subseteq \mathrm{co}(A)$ it suffices, by Proposition 7.33, to show that $(A/n)^{\oplus n} \to \mathrm{co}(A)$ in $\mathcal{F}$. It is evident that $\overline{\lim}\,(A/n)^{\oplus n} \subset \mathrm{co}(A)$; thus it remains to prove that $\mathrm{co}(A) \subseteq \underline{\lim}\,(A/n)^{\oplus n}$. Every $x \in \mathrm{co}(A)$ can be written as $x = \sum_{i=1}^{d+1} r_i x_i$, where $x_i \in A$, $0 \le r_i \le 1$, and $\sum r_i = 1$. As in the proof of the previous result there exist integers p_n^i such that

$$\sum_{i=1}^{d+1} p_n^i = n \qquad \text{and} \qquad \frac{1}{n} p_n^i \to r_i \text{ as } n \to \infty.$$

Define $x_n = \sum_{i=1}^{d+1} p_n^i x_i / n$; then $x_n \in (A/n)^{\oplus n}$ and $x_n \to x$. This means that $x \in \underline{\lim}\,(A/n)^{\oplus n}$, and the proof is finished. □

This result is illustrated in Fig. 9.4.

In Proposition 9.2 we have seen that A is convex iff $rA \oplus sA = (r+s)A$. We have the following interesting generalization of this result.

9.16 Theorem. *A continuous mapping $A : \mathbb{R}_+ \to \mathcal{K}'(\mathbb{R}^d)$ satisfies the semigroup property*

$$A(r) \oplus A(s) = A(r+s), \qquad r, s \ge 0, \tag{9.5}$$

if and only if $A(r) = rA$ for some compact convex set $A \subseteq \mathbb{R}^d$.

Proof. The if-part follows from Proposition 9.2. Assume on the other hand that $A(\cdot)$ satisfies the assumptions of the theorem. Then $A(r) = A(r/n)^{\oplus n}$, meaning that $A(r)$ is infinitely divisible and therefore convex (Proposition 9.14). From the semigroup property and the convexity of $A(r)$, we derive

$$A\left(\frac{k}{n}\right) = \frac{k}{n} A(1), \qquad \text{for } k \ge 0,\ n \ge 1.$$

The continuity of $A(\cdot)$ gives that $A(r) = rA(1)$ for every $r > 0$. Putting $A = A(1)$, we have proved the assertion. □

Recall from Proposition 4.22 that $X \circ A \subseteq X \circ B$ for every $X \subseteq \mathbb{R}^d$ if and only if A is B-open. If A is convex and $r \geq 1$, then, by Proposition 9.2,

$$rA \circ A = \big[((r-1)A \oplus A) \ominus A\big] \oplus A = (r-1)A \oplus A = rA,$$

that is, rA is A-open. The next theorem shows that for compact A the converse is also true.

9.17 Theorem. *Let $A \subseteq \mathbb{R}^d$ be compact; then rA is A-open for every $r \geq 1$ if and only if A is convex.*

The proof of this result requires some further preparations. If $A_k \subseteq \mathbb{R}^d$ for $k = 1, 2, \ldots, n$, then $\bigoplus_{k=1}^{n} A_k$ denotes $A_1 \oplus \cdots \oplus A_n$, and $\bigoplus_{k=1}^{\infty} A_k$ denotes the set comprising all points $\sum_{k=1}^{\infty} a_k$ with $a_k \in A_k$; here it is tacitly assumed that all sums converge.

9.18 Lemma. *Let $A \subseteq \mathbb{R}^d$ be compact, and assume that $0 \leq r < 1$. Then $\bigoplus_{k=0}^{n} r^k A$ converges myopically toward $\bigoplus_{k=0}^{\infty} r^k A$ in $\mathcal{K}$.*

Proof. For $A = \varnothing$ the result is trivial, so we assume that A is nonempty. First observe that

$$\bigoplus_{k=0}^{n} r^k A \subseteq \mathrm{co}\Big(\bigoplus_{k=0}^{n} r^k A\Big) = \sum_{k=0}^{n} r^k \,\mathrm{co}(A) \subseteq \frac{1}{1-r}\,\mathrm{co}(A).$$

Here we have used Proposition 9.4. Then, by Proposition 7.33, it suffices to consider convergence in $\mathcal{F}$. Define $S_n = \bigoplus_{k=0}^{n} r^k A$ and $S = \overline{\bigoplus_{k=0}^{\infty} r^k A}$. Assume first that $0 \in A$. Then $\{S_n\}$ is increasing and $\overline{\bigcup_{n \geq 1} S_n} = S$; thus Proposition 7.28 implies that $S_n \xrightarrow{\mathcal{F}} S$. Now, if $0 \notin A$, then $\{S_n\}$ is in general not increasing. Choose $a \in A$ arbitrary and define $A' = A_{-a}$. Then $0 \in A'$ and $S_n = \big[\bigoplus_{k=0}^{n} r^k A'\big]_{a_n}$, where $a_n = (1 + r + r^2 + \cdots + r^n)a$. Since, by the preceding argument, $\bigoplus_{k=0}^{n} r^k A' \to \bigoplus_{k=0}^{\infty} r^k A'$ and $a_n \to (1-r)^{-1} a$, it follows from Proposition 7.43(a) that $S_n \to S$. □

Proof of Theorem 9.17. We need only demonstrate the only if-part. Take $\epsilon > 0$; then $(1+\epsilon)A \circ A = (1+\epsilon)A$. In other words, $(1+\epsilon)A = [(1+\epsilon A) \ominus A] \oplus A$. Putting $B_\epsilon = \epsilon^{-1}((1+\epsilon)A \ominus A)$, we get

$$(1+\epsilon)A = A \oplus \epsilon B_\epsilon. \tag{9.6}$$

Taking the convex hull on both sides and using (9.2) gives

$$(1+\epsilon)\,\mathrm{co}(A) = \mathrm{co}(A) \oplus \epsilon\,\mathrm{co}(B_\epsilon).$$

Since co(A) is convex, we have $(1+\epsilon)\,\mathrm{co}(A)=\mathrm{co}(A)\oplus\epsilon\,\mathrm{co}(A)$. Taking the erosion by co(A) on both sides leads to $\epsilon\,\mathrm{co}(A)\bullet\mathrm{co}(A)=\epsilon\,\mathrm{co}(B_\epsilon)\bullet\mathrm{co}(A)$. From Proposition 9.8 we derive

$$\mathrm{co}(A)=\mathrm{co}(B_\epsilon).$$

In particular, this implies

$$B_\epsilon\subseteq\mathrm{co}(A). \tag{9.7}$$

Repeated application of (9.6) gives that

$$A=\frac{A}{(1+\epsilon)^n}\oplus\Big(\bigoplus_{k=1}^{n}\frac{\epsilon B_\epsilon}{(1+\epsilon)^k}\Big). \tag{9.8}$$

From (9.7) and Proposition 9.2 we derive that

$$\begin{aligned}\bigoplus_{k=1}^{n}\frac{\epsilon B_\epsilon}{(1+\epsilon)^k}&\subseteq\bigoplus_{k=1}^{n}\frac{\epsilon\,\mathrm{co}(A)}{(1+\epsilon)^k}=\Big(\sum_{k=1}^{n}\frac{\epsilon}{(1+\epsilon)^k}\Big)\,\mathrm{co}(A)\\&\subseteq\Big(\sum_{k=1}^{\infty}\frac{\epsilon}{(1+\epsilon)^k}\Big)\,\mathrm{co}(A)=\mathrm{co}(A).\end{aligned}$$

Lemma 9.18 implies that the sequence $\bigoplus_{k=1}^{n}\epsilon(1+\epsilon)^{-k}B_\epsilon$ converges toward $\bigoplus_{k=1}^{\infty}\epsilon(1+\epsilon)^{-k}B_\epsilon$ in $\mathcal{K}$. Furthermore, $A/(1+\epsilon)^n\xrightarrow{\mathcal{K}}\{0\}$, and we get

$$A=\bigoplus_{k=1}^{\infty}\frac{\epsilon B_\epsilon}{(1+\epsilon)^k}$$

from (9.8). We write this as

$$A=\bigoplus_{i=1}^{N}\bigoplus_{k=0}^{\infty}\frac{\epsilon B_\epsilon}{(1+\epsilon)^{i+kN}}=B_1(\epsilon)\oplus\cdots\oplus B_N(\epsilon), \tag{9.9}$$

where

$$B_i(\epsilon)=\bigoplus_{k=0}^{\infty}\frac{\epsilon B_\epsilon}{(1+\epsilon)^{i+kN}},\qquad i=1,\ldots,N;$$

here $N\geq 1$ is arbitrary. Observe that $B_i(\epsilon)=(1+\epsilon)^{-i+1}B_1(\epsilon)$. Using a similar argument as before, we infer that every $B_i(\epsilon)$ is compact and that

$$B_i(\epsilon)\subseteq\frac{\epsilon(1+\epsilon)^{N-i}}{(1+\epsilon)^N-1}\,\mathrm{co}(A).$$

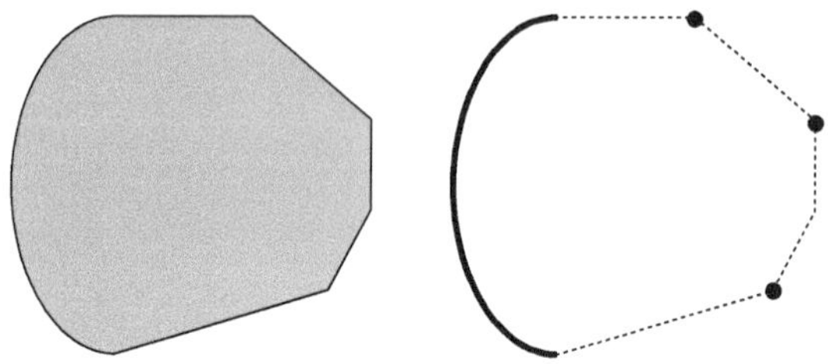

Figure 9.5 The extreme points (right) of a set (left).

So there is a sequence $\epsilon_n \downarrow 0$ such that $B_1(\epsilon_n) \xrightarrow{\mathcal{K}} B$ as $n \to \infty$. Then $B_i(\epsilon_n) \xrightarrow{\mathcal{K}} B$ for every $i = 1, 2, \ldots, N$. In combination with (9.9) and the continuity of Minkowski addition this means that $A = B^{\oplus N}$. We can carry out this procedure for every integer N, and so we may conclude that A is infinitely divisible. But then A is convex by Proposition 9.14. □

We end this section with some material which we will need later.

9.19 Definition. Let E be a topological vector space. A convex set X is called *strictly convex* if it contains no straight line segment in its boundary, that is, if $x, y \in X, x \neq y$ implies that $rx + (1 - r)y \in X^\circ$ for $0 < r < 1$.

The second set in Fig. 9.1 is strictly convex, but the first one is not.

9.20 Definition. Let X be a convex set in a vector space E. A point $e \in X$ is called an *extreme point* of X if $e = rx + (1 - r)y$ with $x, y \in X$ implies $e = x$ or $e = y$.

For example, the extreme points of a square in $\mathbb{R}^2$ are the four corner points. The extreme points of a disk are the points on the circle. Another example is illustrated in Fig. 9.5.

We conclude this section with a famous result, the Krein–Milman Theorem. A proof is found in Valentine (1964, Theorem 11.4).

9.21 Theorem. (Krein–Milman Theorem)
If $X \subseteq \mathbb{R}^d$ is a compact set, then $\overline{\text{co}}(X)$ is equal to the closed convex hull of the extreme points of X.

9.2. Geodesic distance and M-convexity

This section discusses several well-known results from the theory of metric spaces. In particular, it introduces the notions of geodesic distance and metric convexity (M-convexity). Most of the results will be given without proof. Readers interested in technical details should refer to Blumenthal (1953), Busemann (1955), and Rinow (1961). Recall from Definition 7.10 that (E, d) is a *metric space* if $d : E \times E \to \mathbb{R}_+$ is a function satisfying the axioms

(D1) $d(x, y) = 0$ if and only if $x = y$;
(D2) $d(x, y) = d(y, x)$;
(D3) $d(x, z) \leq d(x, y) + d(y, z)$.

The last property is usually called triangle inequality. One calls $d(x, y)$ the distance between x and y. Section 7.2 concerns mainly topological properties of metric spaces. The present chapter, however, concentrates on geometrical properties such as convexity and path-connectedness.

The closed and open balls centred at x with radius r are denoted by $B(x, r)$ and $B^\circ(x, r)$, respectively:

$$B(x, r) = \{y \in E \mid d(x, y) \leq r\}, \qquad B^\circ(x, r) = \{y \in E \mid d(x, y) < r\}.$$

9.22 Definition. Given a metric space (E, d) and three distinct points $u, x, y \in E$, we say that u is *between* x and y if $d(x, y) = d(x, u) + d(u, y)$.

The best known example of a metric space is the Euclidean space $\mathbb{R}^d$; here the distance between two points $x, y \in \mathbb{R}^d$ is given by

$$d_2(x, y) = \|x - y\| = \Big(\sum_{k=1}^{d} (x_k - y_k)^2\Big)^{1/2}. \tag{9.10}$$

One calls $\|x\|$ the Euclidean norm of the vector x; see below. The validity of the triangle inequality (in most cases, of the three axioms characterizing a metric this is the most difficult one to prove) follows from standard arguments. One can show that $d_2(x, z) = d_2(x, y) + d_2(y, z)$ if and only if y lies on the straight line segment between x and y. The Euclidean metric, as well as the other metrics d_k introduced in Example 7.11(b), are metrics of the so-called Minkowski type; see Example 9.34.

This section continues with a brief discussion of curves in metric spaces. A comprehensive treatment can be found in Rinow (1961). A *curve* is a continuous mapping f from a closed interval $[a, b]$ into the metric space (E, d). Note that a curve may cross itself or even repeat itself. Consider the partition π of $[a, b]$ given by $a = t_0 \leq t_1 \leq \cdots \leq t_n = b$, and define

$$L_\pi(f) = \sum_{k=1}^{n} d(f(t_{k-1}), f(t_k)).$$

Define the *length* $L(f)$ of the curve f as the least upper bound of $L_\pi(f)$ for all possible partitions π of $[a, b]$:

$$L(f) = \sup_{\pi} L_\pi(f).$$

If this quantity is finite, then the curve represented by f is called *rectifiable*. Note that $L(f) = 0$ if and only if f is constant on $[a, b]$. Let $[t_1, t_2]$ be a subinterval of $[a, b]$, and denote by $L(f; t_1, t_2)$ the length of the restriction of f to $[t_1, t_2]$. It is apparent that

$$L(f; t_1, t_2) \geq d(f(t_1), f(t_2)). \tag{9.11}$$

There exists a representation of f which uses curve length as parameter. Let $a \leq t \leq b$, and put

$$\ell(t) = L(f; a, t).$$

It is not difficult to show that ℓ is an increasing continuous function. Given $0 \leq s \leq L = L(f)$, there exists a t such that $\ell(t) = s$. In general, t is not unique; however, for all solutions t the point $f(t)$ is the same. Putting

$$c(s) = f(t),$$

we get that $c(o) = f(a)$, $c(L) = f(b)$, and the point sets $\{c(s) \mid 0 \leq s \leq L\}$ and $\{f(t) \mid a \leq t \leq b\}$ are the same. The function c is called the *normal representation* of f. Furthermore, $c(0)$ and $c(L)$ are called *endpoints* of the curve. From now on we think of a curve as the equivalence class of all functions f with the same normal representation. A curve with endpoints x and y is called a *path* between x and y. A curve is called an *arc* if it has a parametrization f such that $f(t_1) \neq f(t_2)$ when $t_1 \neq t_2$.

9.23 Definition. A rectifiable path whose length is not greater than the length of any other path with the same endpoints is called a *geodesic path*.

Every geodesic path is an arc, for otherwise it would contain loops, which could then be removed. Fig. 9.6 shows two paths between x and y; only one of them is a geodesic path.

9.24 Proposition. *Let E be a finitely compact metric space. If the points $x, y \in E$ can be connected by a rectifiable path, then there exists a geodesic path between x and y.*

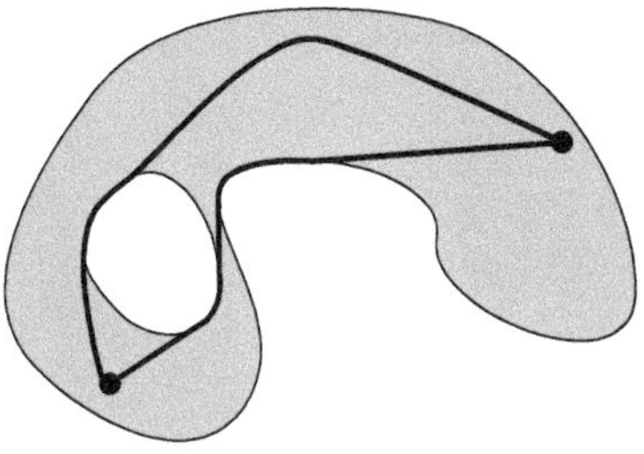

Figure 9.6 Two paths between x and y. The lower one is a geodesic path.

9.25 Definition. The space E is called *path-connected* if every two points can be joined by a path. If, moreover, every two points can be joined by a rectifiable path, then E is called *finitely path-connected.*

Henceforth we write "connected" whenever we mean "path-connected".

9.26 Remark. We make the following cautionary remark. In topology there exists the following definition of connectivity: a space E is connected if it is not the union of two nonempty disjoint open sets. Every path-connected space is connected, but not conversely. Refer to Dugundji (1966, Chapter V) for further details.

Assume that two points x, y in a finitely compact metric space E can be joined by a rectifiable path. Then, by Proposition 9.24 there exists a geodesic path between x and y. The length of this path is denoted by $d_E(x, y)$ and is called the *geodesic distance* between x and y. If there exists no rectifiable path between x and y, then we put $d_E(x, y) = \infty$. For an arbitrary metric space, $d_E(x, y)$ is defined as the infimum of $L(f)$, where f ranges over all paths between x and y. In this case, finiteness of $d_E(x, y)$ does not always mean that there exists a path between x and y with length $d_E(x, y)$. One calls d_E the *geodesic metric* or *intrinsic metric*, and one says that (E, d) has an intrinsic metric if $d = d_E$. Strictly speaking, d_E is not a metric since it may attain the value ∞. It is easy to show that d_E satisfies the three properties of a metric, however, that is,

$$d_E(x, y) = 0 \iff x = y; \tag{9.12}$$

$$d_E(x, y) = d_E(y, x); \tag{9.13}$$

$$d_E(x, z) \le d_E(x, y) + d_E(y, z). \tag{9.14}$$

Furthermore, it follows immediately from (9.11) that

$$d(x, y) \leq d_E(x, y), \qquad \text{for all } x, y \in E. \tag{9.15}$$

A path between x and y whose length equals $d(x, y)$ is called a *metric segment* between x and y. Because of (9.15), every metric segment is a geodesic path between its endpoints; the converse is not true in general. Note that a space (E, d) with an intrinsic metric is finitely path-connected; between every two points x, y there exists a rectifiable path with length $d(x, y)$. Note also that every geodesic path in a space with an intrinsic metric is a metric segment.

9.27 Proposition. *Assume that the metric space (E, d) is finitely path-connected. Every rectifiable curve in (E, d) is a rectifiable curve in (E, d_E), and their respective lengths are the same.*

This result implies in particular that (E, d_E) has an intrinsic metric if E is finitely path-connected.

9.28 Definition. A subset X of a metric space E is called *metrically convex*, or *M-convex*, if for every two distinct points $x, y \in X$ there exists a third point $z \in X$ which lies between x and y. It is called *continuously M-convex* if for every two distinct points there exists a geodesic path between them. If, moreover, such geodesic paths are unique, then the set X is called *simply M-convex*.

It is not difficult to show that every continuously M-convex space with intrinsic metric is M-convex. The converse is not true in general, however. For instance, the set of rational numbers provided with the metric $d(x, y) = |x - y|$ is M-convex but not continuously M-convex.

9.29 Proposition. *Assume that the metric space (E, d) is finitely compact. Then E is M-convex if and only if it has an intrinsic metric. In that case, the space is also continuously M-convex.*

To distinguish between metric convexity and the convexity notion defined in the previous section, we refer to the latter as *linear convexity*. The next example discusses a metric on $\mathbb{R}^d$ which allows only trivial M-convex sets. But Example 9.34 will show that the converse is also possible: there do exist simply M-convex sets which are not linearly convex.

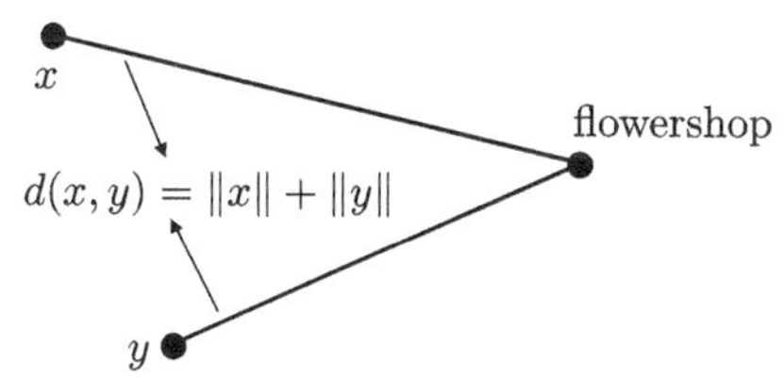

Figure 9.7 Flowershop distance between the points x and y.

9.30 Example. (Flowershop distance)
Let $\|\cdot\|$ denote the Euclidean norm on $\mathbb{R}^d$; see (9.10). Define a distance d by

$$d(x, y) = \begin{cases} 0, & \text{if } x = y, \\ \|x\| + \|y\|, & \text{if } x \neq y. \end{cases}$$

Note that $d(x, 0) = \|x\|$. It is easy to verify that d is a distance. The distance d is called *flowershop distance*. To understand this nomenclature, interpret 0 as the position of a flowershop. A person living at x who is visiting his girlfriend at y has to make a detour to the flowershop and therefore walk a distance $\|x\| + \|y\|$; this distance is illustrated in Fig. 9.7.

It is evident that 0 is the only point between two distinct points x and y different from 0. Furthermore, there exist no curves besides the constant curves residing at one point. As a result, a set is M-convex if it is a singleton, or if it contains the origin. Furthermore, the only continuously M-convex sets are the singletons.

Assume that E is a real vector space. A function $p : E \to \mathbb{R}_+$ is called a *norm* if the following axioms are satisfied:
(N1) $p(x) = 0$ if and only if $x = 0$;
(N2) $p(rx) = |r|p(x)$ for $r \in \mathbb{R}$ and $x \in E$;
(N3) $p(x + y) \leq p(x) + p(y)$ for $x, y \in E$.
With every norm one can associate a metric d by putting

$$d(x, y) = p(x - y), \qquad x, y \in E.$$

Obviously, every metric deriving from a norm is translation invariant, i.e.,

$$d(x + h, y + h) = d(x, y), \qquad x, y, h \in E,$$

and homogeneous, i.e.,

$$d(rx, ry) = |r|d(x, y), \qquad r \in \mathbb{R},\ x, y \in E.$$

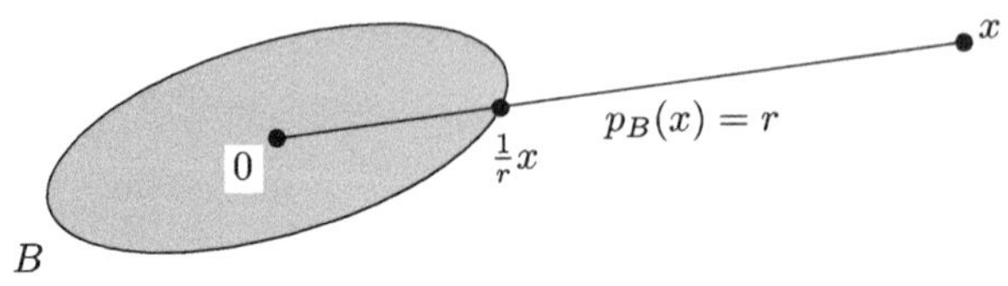

Figure 9.8 The Minkowski functional $p_B(x)$.

Conversely, with every homogeneous translation invariant metric d on a vector space E one can associate a norm p by putting

$$p(x) = d(x, 0), \qquad x \in E.$$

It is obvious that the unit ball $B = \{x \mid p(x) \leq 1\}$ is convex and that the interior of B contains all points rx where $|r| < 1$ and $p(x) \leq 1$. Furthermore, B is reflection symmetric with respect to the origin. The ball with centre x and radius r is given by $B(x, r) = (rB)_x$.

9.31 Definition. A finite-dimensional normed vector space is called a *Minkowski space.*

The norm in a Minkowski space is uniquely determined by the closed unit ball B. Suppose, namely, that E is a finite-dimensional topological vector space and that B is a convex body with 0 in its interior which is symmetric with respect to 0. Define

$$p_B(x) = \inf\{r > 0 \mid \frac{1}{r}x \in B\}.$$

See Fig. 9.8 for an illustration.

It is evident that p_B satisfies axioms (N1)–(N2). To prove (N3), assume that $x, y \in E$, and put $r_0 = p_B(x)$, $s_0 = p_B(y)$. So if $r > r_0$, then $x/r \in B$ and if $s > s_0$, then $y/s \in B$. Since B is convex,

$$\frac{1}{r+s}(x+y) = \frac{r}{r+s} \cdot \frac{1}{r}x + \frac{s}{r+s} \cdot \frac{1}{s}y \in B.$$

This means that $p_B(x+y) \leq r+s$, and as this holds for all $r > r_0$ and $s > s_0$ it follows that $p_B(x+y) \leq r_0 + s_0 = p_B(x) + p_B(y)$. Thus (N3) holds, and p_B defines a norm on E. It is evident that the unit ball with respect to this metric is B. The function p_B is called the *Minkowski functional*, or *gauge functional.*

Assume that the Minkowski space E is spanned by the linearly independent vectors $e_1, e_2, \dots, e_d$. Every $x \in E$ can be uniquely written as $x = x_1 e_1 + \cdots + x_d e_d$. There exist constants $C_p, c_p > 0$ depending on the norm p such that

$$c_p \|x\| \leq p(x) \leq C_p \|x\|,$$

where $\|x\|$ is the Euclidean norm $\|x\| = (|x_1|^2 + \cdots + |x_d|^2)^{1/2}$. If p and p' are norms on the finite dimensional vector space E, then p, p' are equivalent in the sense that

$$cp(x) \leq p'(x) \leq Cp(x),$$

where $c = c_{p'}/C_p$ and $C = C_{p'}/c_p$. In particular, this implies that the topology induced on a Minkowski space is the Euclidean topology.

9.32 Proposition. *A Minkowski space is finitely compact and has an intrinsic metric.*

Proof. Finite compactness is obvious. For every two points $x, y \in E$ the straight segment $c(t) = tx + (1 - t)y$, $0 \leq t \leq 1$, defines a geodesic path with endpoints x and y, and its length is equal to the distance between x and y. Therefore, $d = d_E$. □

By Proposition 9.29 every Minkowski space is continuously M-convex. The next result describes the relation between M-convexity and linear convexity when the underlying space is of Minkowski type.

9.33 Proposition. *Let E be a Minkowski space with norm p and unit ball B. The following are equivalent:*

- **(i)** *E is simply M-convex;*
- **(ii)** *$p(x + y) = p(x) + p(y)$ implies that x, y are linearly dependent;*
- **(iii)** *B is strictly convex.*

Proof. (i) ⇒ (ii): Assume that $p(x + y) = p(x) + p(y)$. One can construct a geodesic path from 0 to $x + y$ by taking the union of the straight line segment from 0 to x and the segment from x to $x + y$. Since E is simply M-convex, this path must coincide with the straight line segment from 0 to $x + y$. This means that x and y are linearly dependent.

(ii) ⇒ (i): Let $x, y \in E$; then the straight line segment between x and y is a geodesic path. On the other hand, every point z that lies on a geodesic path between x and y lies between these two points: $d(x, y) = d(x, z) + d(z, y)$.

This implies $p(x-z)+p(z-y)=p(x-y)$. By (ii) one finds that $x-z$ and $z-y$ are linearly dependent, whence it follows that z lies on the line segment between x and y. This means that this line segment is the only geodesic path from x to y.

(ii) $\Rightarrow$ (iii): Let $p(x)=1$ and $p(y)=1$, and assume that $p(rx+(1-r)y)=1$ for some $r\in(0,1)$. Then rx and $(1-r)y$ must be linearly dependent by (ii); this gives $x=y$. Thus B is strictly convex.

(iii) $\Rightarrow$ (ii): Similar. □

9.34 Example. In Example 7.11(b) we have introduced the metrics d_k, $1\le k\le\infty$, on $\mathbb{R}^d$ given by

$$d_k(x,y)=\{\sum_{i=1}^{d}|x_i-y_i|^k\}^{1/k},$$

if $1\le k<\infty$ and

$$d_\infty(x,y)=\max_{1\le i\le d}|x_i-y_i|.$$

These metrics are generated by the norms

$$p_k(x)=\{\sum_{i=1}^{d}|x_i|^k\}^{1/k},$$

if $1\le k<\infty$ and

$$p_\infty(x)=\max_{1\le i\le d}|x_i|.$$

It is obvious that each of these metrics generates the Euclidean topology. If $1<k<\infty$, then the corresponding Minkowski space is simply M-convex because every geodesic path is a straight line segment. This is no longer true, however, when $k=1$ or ∞. In fact, in these two cases the unit balls are not strictly convex; see Fig. 7.2. Alternatively, this follows from the fact that rule (ii) of Proposition 9.33 is not satisfied in these two cases. Hereafter we restrict ourselves to the 2-dimensional case. For $k=1$ as well as for $k=\infty$ one finds that $p_k((2,1))=p_k((1,1)+(1,0))$, though $(1,1)$ and $(1,0)$ are not linearly dependent. Furthermore, it is easy to see that geodesic paths between two fixed points are in general not unique; see Fig. 9.9.

Every straight line segment is a geodesic path, but there may be other geodesic paths with the same endpoints. If $k=1$, then the path consisting

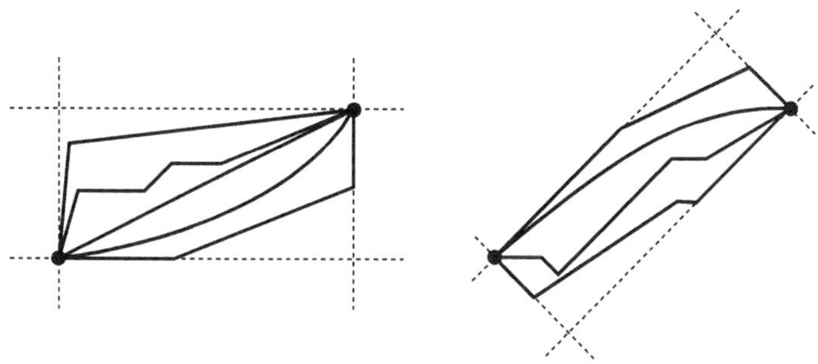

Figure 9.9 Left: geodesic paths in $(\mathbb{R}^2, d_1)$ Right: geodesic paths in $(\mathbb{R}^2, d_\infty)$.

of the horizontal line segment from $(0,0)$ to $(1,0)$ and the vertical line segment from $(1,0)$ to $(1,1)$ is a geodesic path between $(0,0)$ and $(1,1)$. Like every geodesic path, this path is simply M-convex; however, it is not linearly convex.

9.3. Metric dilations

Assume that (E, d) is a metric space. Define for $r \geq 0$ the *metric dilation* δ^r on $\mathcal{P}(E)$ by

$$\delta^r(X) = \bigcup_{x \in X} B(x, r). \tag{9.16}$$

By the fact that $B(x, 0) = \{x\}$, one has $\delta^0 = \mathsf{id}$. In general, $\delta^r(X)$ is a subset of the set of points whose distance to X is not greater than r. However, the following result holds.

9.35 Proposition. *Let the metric space* (E, d) *be finitely compact. If* $X \subseteq E$ *is closed, then*

$$\delta^r(X) = \{h \in E \mid d(h, X) \leq r\}, \tag{9.17}$$

where $d(h, X) = \inf_{x \in X} d(h, x)$.

Proof. The inclusion $\subseteq$ is trivial, and we only have to prove the inclusion $\supseteq$. Suppose that $h \in E$ and $d(h, X) \leq r$. There is a sequence $x_n \in X$ such that $d(h, x_n) \leq r + 1/n$. Obviously, this sequence is bounded. Because E is finitely compact, there is an $x \in E$ and a subsequence n_k such that $x_{n_k} \to x$.

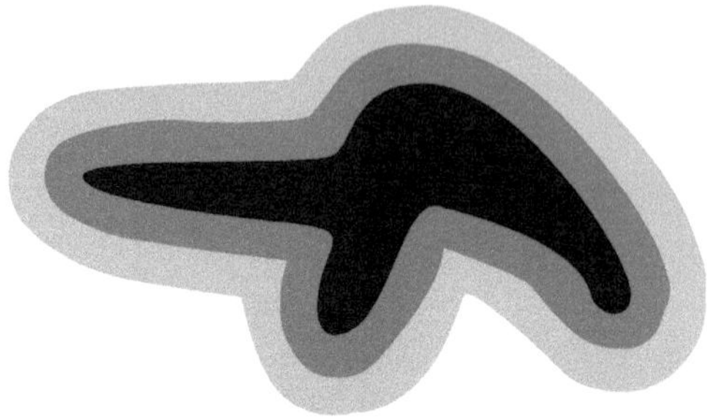

Figure 9.10 A set X (black) and its metric dilates $\delta^1(X)$ (dark grey) and $\delta^2(X)$ (light grey) with respect to Euclidean distance.

Since X is closed, $x \in X$. Then

$$d(h, x) \le d(h, x_{n_k}) + d(x_{n_k}, x) \le r + \frac{1}{n_k} + d(x_{n_k}, x).$$

As this holds for every k, it follows that $d(h, x) \le r$. In other words, $h \in B(x, r) \subseteq \delta^r(X)$. □

Two metric dilations with respect to the Euclidean distance are illustrated in Fig. 9.10.

The family δ^r, $r \ge 0$, satisfies the following properties:

(Δ0) $\delta^0 = \mathrm{id}$;
(Δ1) $\delta^s \ge \delta^r$ if $s \ge r$ and $\bigcap_{s>r} \delta^s(\{x\}) = \delta^r(\{x\})$, for every $x \in E$;
(Δ2) $x \in \delta^r(\{y\}) \iff y \in \delta^r(\{x\})$, for every $x, y \in E$;
(Δ3) $\delta^r \delta^s \le \delta^{r+s}$, $r, s \ge 0$;
(Δ4) $\bigcup_{r \ge 0} \delta^r(\{x\}) = E$.

It is important to remark that (Δ1) does not imply that $\bigwedge_{s>r} \delta^s = \delta^r$. One can recover the distance d from the expression

$$d(x, y) = \inf\{r > 0 \mid y \in \delta^r(\{x\})\}. \tag{9.18}$$

Assume, on the other hand, that δ^r, $r \ge 0$, is a family of dilations which satisfies the conditions (Δ0)–(Δ4). Define d by (9.18); we show that d is a metric on E. Note that $d(x, y)$ is finite by (Δ4). We must show that the axioms (D1)–(D3) of a metric space mentioned at the beginning of the previous section are satisfied. To prove (D1), assume that $d(x, y) = 0$. Then $y \in \delta^r(\{x\})$ for every $r > 0$. Using (Δ0) and (Δ1), one infers that $y \in \bigcap_{r>0} \delta^r(\{x\}) = \delta^0(\{x\}) = \{x\}$; hence $y = x$. Axiom (D2) concerning the symmetry follows immediately from (Δ2). To prove the triangle inequality

(D3) we use (Δ3). Assume that $d(x, y) = r$ and $d(y, z) = s$. Then $y \in \delta^r(\{x\})$ and $z \in \delta^s(\{y\})$. Therefore, $z \in \delta^s\delta^r(\{x\}) \subseteq \delta^{r+s}(\{x\})$; this gives $d(x, z) \leq r + s$. Finally, it is not difficult to show that the closed balls associated with this metric d coincide with $\delta^r(\{x\})$. We summarize our conclusions in the following proposition.

9.36 Proposition. *To every metric d on E there corresponds a unique one-parameter family of dilations δ^r, $r \geq 0$, which satisfies the axioms (Δ0)–(Δ4). The relation between the distance function d and the metric dilations δ^r is described by* (9.16) *and* (9.18).

Assume that E is a Minkowski space with unit ball B; then $B(x, r) = (rB)_x$ and

$$\delta^r(X) = X \oplus rB.$$

Since B is convex one may invoke Proposition 9.2, and thus one finds

$$\delta^r\delta^s(X) = X \oplus (sB \oplus rB) = X \oplus [(s + r)B] = \delta^{r+s}(X).$$

The next result shows that this fact is nothing but a special case of a more general statement.

9.37 Theorem. *Let (E, d) be a metric space which is continuously M-convex. The metric dilations δ^r on $\mathcal{P}(E)$ given by* (9.16) *satisfy the semigroup relation*

$$\delta^r\delta^s = \delta^{r+s}, \qquad r, s \geq 0. \tag{9.19}$$

Conversely, if (E, d) is finitely compact and if the semigroup relation (9.19) *holds, then (E, d) is continuously M-convex.*

Proof. Assume that (E, d) is continuously M-convex. It is sufficient to prove that $\delta^{r+s} \leq \delta^r\delta^s$, since the opposite relation is always satisfied. Suppose that $y \in \delta^{r+s}(\{x\})$; this means that $d(x, y) \leq r + s$. Evidently, there exist $r' \leq r$ and $s' \leq s$ such that $d(x, y) = r' + s'$. Since E is continuously M-convex, there is a geodesic path between x and y. On this path lies a point z with $d(x, z) = s'$ and $d(z, y) = r'$. Then $z \in \delta^{s'}(\{x\}) \subseteq \delta^s(\{x\})$ and $y \in \delta^{r'}(\{z\}) \subseteq \delta^r(\{z\})$; therefore, $y \in \delta^r\delta^s(\{x\})$.

To prove the second assertion, it suffices to show that (E, d) is M-convex; cf. Proposition 9.29. Assume that the semigroup relation (9.19) holds. Take $x, y \in E$, and let $r = d(x, y)$. We show that there exists a point between x and y. If $0 < s < r$, then $\delta^r = \delta^{r-s}\delta^s$; in particular, $y \in \delta^{r-s}\delta^s(\{x\})$.

So there is a $z \in \delta^s(\{x\})$ with $y \in \delta^{r-s}(\{z\})$. This means that $d(x, z) \le s$ and $d(y, z) \le r - s$. Since $r = d(x, y) \le d(x, z) + d(z, y) \le r$ it follows that $d(x, z) = s$ and $d(y, z) = r - s$. Thus z lies between x and y. □

Invoking Proposition 9.29 one concludes that the metric dilations satisfy the semigroup property (9.19) if the underlying space E is finitely compact and has an intrinsic metric. In particular, this holds when E is a Minkowski space.

9.38 Example. (The Hausdorff metric revisited)
Proposition 7.22 gives the following characterization for the Hausdorff metric D_H on the nonempty compact subsets $\mathcal{K}'(E)$ of a finitely compact metric space E:

$$D_H(X, Y) = \max\{\hat{D}(X, Y), \hat{D}(Y, X)\},$$

where $\hat{D}(X, Y) = \sup_{x \in X} d(x, Y)$. At the end of this example it is demonstrated that

$$\hat{D}(X, Y) = \inf\{r \ge 0 \mid X \subseteq \delta^r(Y)\}. \tag{9.20}$$

From this and an analogous identity for $\hat{D}(Y, X)$ it follows that

$$D_H(X, Y) = \inf\{r \ge 0 \mid X \subseteq \delta^r(Y) \text{ and } Y \subseteq \delta^r(X)\}. \tag{9.21}$$

For the space $\mathbb{R}^d$ with the Euclidean metric this leads to the following formula:

$$D_H(X, Y) = \inf\{r \ge 0 \mid X \subseteq Y \oplus rB \text{ and } Y \subseteq X \oplus rB\}, \tag{9.22}$$

where B is the closed unit ball.

It remains to prove identity (9.20). Put $r_0 = \inf\{r \ge 0 \mid X \subseteq \delta^r(Y)\}$. If $r > r_0$, then $X \subseteq \delta^r(Y)$, and so $d(x, Y) \le r$ if $x \in X$ by (9.17); this means that $\hat{D}(X, Y) \le r$. As this holds for every $r > r_0$, it follows that $\hat{D}(X, Y) \le r_0$.

On the other hand, putting $r = \hat{D}(X, Y)$, it holds that $d(x, Y) \le r$ if $x \in X$. Therefore, $X \subseteq \delta^r(Y)$ by Proposition 9.35. Thus $r_0 \le r$, which concludes the proof.

9.4. Distance transform

Throughout this section it is assumed that (E, d) is a finitely compact space.

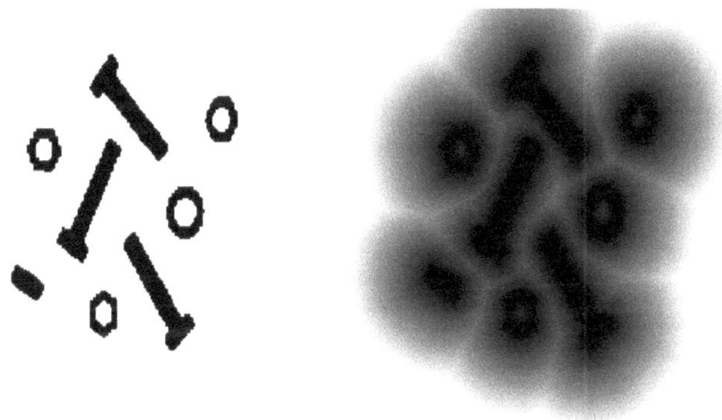

Figure 9.11 Distance transform (right) of a set X (left) based on the 5-7-11 chamfer metric; cf. Section 9.9. The distance transform $\mathbf{\Delta}(X)$ is shown as a grey-scale image: the brighter the image at the point h, the higher the value $\mathbf{\Delta}(X, h)$.

9.39 Definition. The *distance transform* $\mathbf{\Delta}(X)$ of a set $X \subseteq E$ is the function given by

$$[\mathbf{\Delta}(X)](h) = d(h, X), \qquad h \in E,$$

with the convention that $d(h, \varnothing) = \infty$ for $h \in E$.

Usually we write $\mathbf{\Delta}(X, h)$ for $[\mathbf{\Delta}(X)](h)$. Obviously, $\mathbf{\Delta}(X, h) = 0$ if $h \in X$. An example is depicted in Fig. 9.11.

It is worthwhile to notice that often in the literature the function $h \mapsto d(h, X^c)$ is called the distance transform of X. It depends on the circumstances which definition is more appropriate.

The distance transform is an operator from $\mathcal{P}(E)$ to $\mathcal{H} = \mathrm{Fun}(E, \overline{\mathbb{R}}_+)$. Utilizing Proposition 9.35 one gets immediately the following relations between the distance transform and the metric dilations δ^r.

9.40 Proposition. *Let (E, d) be a finitely compact metric space. If $X \subseteq E$ is closed, then*

$$\mathbf{\Delta}(X, h) = \inf\{r \geq 0 \mid h \in \delta^r(X)\}, \tag{9.23}$$

and also

$$\delta^r(X) = \{h \in E \mid \mathbf{\Delta}(X, h) \leq r\}, \tag{9.24}$$

for every $r \geq 0$ and $h \in E$.

Let, as usual, $\mathcal{H}'$ be the opposite lattice of $\mathcal{H}$. Denote the supremum and infimum in $\mathcal{H}'$ by $\vee'$ and $\wedge'$, respectively; $\vee$ and $\wedge$ denote the usual supremum and infimum in $\mathcal{H}$. Given a function $F : E \to \overline{\mathbb{R}}_+$, define the closed set $\mathbf{\Delta}^{\leftarrow}(F) \subseteq E$ by

$$\mathbf{\Delta}^{\leftarrow}(F) = \Big[\bigcup_{h \in E} B^{\circ}(h, F(h))\Big]^{c}. \tag{9.25}$$

Here we use the convention that $B^{\circ}(h, \infty) = E$.

9.41 Proposition. *Let (E, d) be a finitely compact metric space. The pair $(\mathbf{\Delta}^{\leftarrow}, \mathbf{\Delta})$ constitutes an adjunction between $\mathcal{H}'$ and $\mathcal{P}(E)$. The closing $\mathbf{\Delta}^{\leftarrow}\mathbf{\Delta}$ on $\mathcal{P}(E)$ is the set closure*

$$\mathbf{\Delta}^{\leftarrow}\mathbf{\Delta}(X) = \overline{X}. \tag{9.26}$$

Proof. To prove that $(\mathbf{\Delta}^{\leftarrow}, \mathbf{\Delta})$ is an adjunction between $\mathcal{H}'$ and $\mathcal{P}(E)$ we must show that for every set $X \subseteq E$ and every function $F : E \to \mathbb{R}_+$,

$$\mathbf{\Delta}(X) \geq F \iff X \subseteq \mathbf{\Delta}^{\leftarrow}(F).$$

Note that $\geq$ has the same meaning as $\leq'$.

$\Rightarrow$: $\mathbf{\Delta}(X) \geq F$ means that $F(h) \leq d(h, X)$ for every $h \in E$. We use the observation that

$$d(h, X) \geq r \iff B^{\circ}(h, r) \subseteq X^{c}, \tag{9.27}$$

for $h \in E$, $X \subseteq E$, and $r \geq 0$. Thus we get that $B^{\circ}(h, F(h)) \subseteq X^{c}$ for $h \in E$. This means that $X \subseteq \Big[\bigcup_{h \in E} B^{\circ}(h, F(h))\Big]^{c}$, which was to be proved.

$\Leftarrow$: Analogous.

Since $\mathbf{\Delta}^{\leftarrow}\mathbf{\Delta}$ is a closing, $X \subseteq \mathbf{\Delta}^{\leftarrow}\mathbf{\Delta}(X)$ for $X \subseteq E$. Since $\mathbf{\Delta}^{\leftarrow}$ maps onto closed sets, we also have $\overline{X} \subseteq \mathbf{\Delta}^{\leftarrow}\mathbf{\Delta}(X)$. To prove the opposite inclusion, take $h \in (\overline{X})^{c}$. Then $d(h, X) > 0$, and thus $h \in B^{\circ}(h, d(h, X))$. Since $\mathbf{\Delta}^{\leftarrow}\mathbf{\Delta}(X) = \Big[\bigcup_{h \in E} B^{\circ}(h, d(h, X))\Big]^{c}$, we get $\mathbf{\Delta}^{\leftarrow}\mathbf{\Delta}(X) \subseteq \Big[\bigcup_{h \in E \setminus \overline{X}} \{h\}\Big]^{c} = \overline{X}$. This finishes the proof. □

In particular, this result gives

$$\mathbf{\Delta}\Big(\bigcup_{i \in I} X_i, h\Big) = \bigwedge_{i \in I} \mathbf{\Delta}(X_i, h),$$

for every family of sets $\{X_i \mid i \in I\} \subseteq \mathcal{P}(E)$ and $h \in E$. The operator $\mathbf{\Delta}^{\leftarrow}$ can be interpreted as a kind of inverse of $\mathbf{\Delta}$, because application of $\mathbf{\Delta}^{\leftarrow}$ to the distance transform $\mathbf{\Delta}(X)$ of a set X gives a reconstruction of this set, which is exact if X is closed. The operator $\mathbf{\Delta}^{\leftarrow}$ is called the *inverse distance transform*.

We present an alternative approach to the distance transform. This approach does not start with a distance function but with a more general notion, the cost function. We outline a recursive procedure which, starting from an arbitrary cost function, converges toward a distance function and yields the associated distance transform at the same time.

9.42 Definition. Let E be an arbitrary set. A *cost function* is a function $c : E \times E \to \overline{\mathbb{R}}_+$ which satisfies the following two properties of a metric:
(D1) $c(x, y) = 0$ if and only if $x = y$;
(D2) $c(x, y) = c(y, x)$ for $x, y \in E$.

A cost function generalizes the notion of a metric: in general it does not satisfy the triangle inequality
(D3) $c(x, z) \leq c(x, y) + c(y, z)$, $x, y \in E$.
As before, let $\mathcal{H} = \text{Fun}(E, \overline{\mathbb{R}}_+)$ with the usual partial ordering. It is evident that the operator

$$\mathcal{E}_c(F)(x) = \bigwedge_{h \in E} [F(h) + c(h, x)] \tag{9.28}$$

defines an erosion on $\mathcal{H}$; in fact, the axioms (D1)–(D2) are not a prerequisite here. For arbitrary functions c_i, $i \in I$, we have

$$\mathcal{E}_{\bigwedge_{i \in I} c_i} = \bigwedge_{i \in I} \mathcal{E}_{c_i}. \tag{9.29}$$

If (D1) holds, then $\mathcal{E}_c$ is anti-extensive. Actually, one can easily check that $\mathcal{E}_c$ is anti-extensive if and only if $c(x, x) = 0$ for every $x \subset E$.

Given $X \subseteq E$, define $\mathbf{0}_X$ as the function from E to $\overline{\mathbb{R}}_+$ which is 0 on X and ∞ on X^c. By $\mathbf{0}_x$, we mean $\mathbf{0}_{\{x\}}$ if $x \in E$. Observe that

$$c(x, y) = \mathcal{E}_c(\mathbf{0}_x)(y), \tag{9.30}$$

for $x, y \in E$. If d is a metric, then the corresponding distance transform $\mathbf{\Delta}$ can be expressed in terms of the erosion $\mathcal{E}_d$ in the following way:

$$\mathbf{\Delta}(X) = \mathcal{E}_d(\mathbf{0}_X). \tag{9.31}$$

Define the *convolution* of two cost functions c, c' by

$$(c * c')(x, y) = \bigwedge_{z \in E} [c(x, z) + c'(z, y)]. \tag{9.32}$$

Without any further assumptions $c * c'$ need not be a cost function. It is easy to verify that

$$(c * c')(x, y) = (c' * c)(y, x), \tag{9.33}$$

and that convolution is associative, that is,

$$(c_1 * c_2) * c_3 = c_1 * (c_2 * c_3) \tag{9.34}$$

for arbitrary cost functions c_1, c_2, c_3. Furthermore,

$$(c * c')(x, y) \leq \inf\{c(x, y), c'(x, y)\}. \tag{9.35}$$

To show this last property one must take $z = x$ and $z = y$, respectively, in (9.32).

9.43 Proposition. *Let c, c' be arbitrary cost functions; then*

$$\mathcal{E}_{c*c'} = \mathcal{E}_{c'}\mathcal{E}_c.$$

Proof. For $F \in \mathcal{H}$ and $x \in E$,

$$\begin{aligned}
\mathcal{E}_{c*c'}(F)(x) &= \bigwedge_{h \in E} \big[F(h) + (c * c')(h, x)\big] \\
&= \bigwedge_{h \in E} \big[F(h) + \bigwedge_{z \in E} (c(h, z) + c'(z, x))\big] \\
&= \bigwedge_{h \in E} \bigwedge_{z \in E} \big[F(h) + c(h, z) + c'(z, x)\big] \\
&= \bigwedge_{z \in E} \big[\bigwedge_{h \in E} (F(h) + c(h, z)) + c'(z, x)\big] \\
&= \bigwedge_{z \in E} \big[\mathcal{E}_c(F)(z) + c'(z, x)\big] \\
&= \mathcal{E}_{c'}\mathcal{E}_c(F)(x).
\end{aligned}$$

This concludes the proof. □

9.44 Proposition. *Let c be a cost function. The following assertions are equivalent:*

(i) *c is a metric;*
(ii) $c * c = c$;
(iii) $\mathcal{E}_c$ *is idempotent.*

Proof. (i) $\Rightarrow$ (ii): If c satisfies the triangle inequality, then $(c * c)(x, y) = \bigwedge_{z \in E}[c(x, z) + c(z, y)] \geq \bigwedge_{z \in E} c(x, y) = c(x, y)$. On the other hand, $c * c \leq c$ by (9.35). Therefore, $c * c = c$.

(ii) $\Rightarrow$ (iii): Follows immediately from Proposition 9.43.

(iii) $\Rightarrow$ (i): From the fact that $\mathcal{E}_c^2(\mathbf{0}_x) = \mathcal{E}_c(\mathbf{0}_x)$ it follows that $c(x, y) = \bigwedge_{h \in E}[c(x, h) + c(h, y)]$. This implies that c satisfies the triangle inequality (D3) and is therefore a metric. □

Define the functions c^{*n}, $n \geq 1$, recursively by

$$\begin{cases} c^{*1} = c \\ c^{*(n+1)} = c * c^{*n}, & \text{if } n \geq 1. \end{cases}$$

Inequality (9.35) implies that $c^{*n}(x, x) = 0$. Using the associative law (9.34), one infers that $c^{*n} * c = c * c^{*n}$. In combination with (9.33) this implies that $c^{*n}(x, y) = c^{*n}(y, x)$ (a formal proof goes by induction). Thus c^{*n} is a cost function for every $n \geq 1$. Furthermore, (9.35) means that

$$c^{*(n+1)} \leq c^{*n}, \qquad n \geq 1,$$

and Proposition 9.43 gives

$$\mathcal{E}_c^n = \mathcal{E}_{c^{*n}}, \qquad n \geq 1.$$

Define

$$c^{*\infty} = \bigwedge_{n \geq 1} c^{*n}.$$

From (9.29) we derive that

$$\mathcal{E}_{c^{*\infty}} = \mathcal{E}_{\bigwedge_{n\geq 1} c^{*n}} = \bigwedge_{n \geq 1} \mathcal{E}_{c^{*n}} = \bigwedge_{n \geq 1} \mathcal{E}_c^n.$$

$\mathcal{E}_c$ is an erosion, and therefore it distributes over infima. Thus

$$\mathcal{E}_{c * c^{*\infty}} = \mathcal{E}_c \mathcal{E}_{c^{*}\infty} = \mathcal{E}_c\Big(\bigwedge_{n \geq 1} \mathcal{E}_c^n\Big) = \bigwedge_{n \geq 1} \mathcal{E}_c^{n+1} = \mathcal{E}_{c^{*\infty}}.$$

Therefore, $c * c^{*\infty} = c^{*\infty}$, and more generally, $c^{*n} * c^{*\infty} = c^{*\infty}$ for $n \geq 1$. Taking the infimum over all $n \geq 1$ leads to

$$c^{*\infty} * c^{*\infty} = c^{*\infty}.$$

If the property $c^{*\infty}(x, y) = 0$ iff $x = y$ holds, then the function $c^{*\infty}$ defines a metric. The following example shows that this property need not be satisfied. Let $E = \mathbb{Z}$, and define $c(x, y) = e^{-|x-y|}$ if $x \neq y$ and $c(x, x) = 0$. Then c defines a cost function. However, it follows immediately that $(c * c)(x, y) = 0$ for every x, y. This gives also that $c^{*\infty}$ is identically zero. To exclude such pathological cases, we assume that $c(x, y) \geq d(x, y)$, where d is some metric. One can easily show by induction that $c^{*n}(x, y) \geq d(x, y)$. But this implies that also $c^{*\infty}(x, y) \geq d(x, y)$, and so axiom (D1) is satisfied. We summarize our results in the following theorem.

9.45 Theorem. *Let c be a cost function on some arbitrary set E, and assume that there is a metric d such that $c(x, y) \geq d(x, y)$. Then $c^{*\infty} = \bigwedge_{n\geq 1} c^{*n}$ is a metric with $c^{*\infty} \geq d$. The corresponding erosion is given by*

$$\mathcal{E}_{c^{*\infty}} = \bigwedge_{n\geq 1} \mathcal{E}_c^n.$$

A useful choice for the lower bound d is the metric $d(x, y)$ which is constantly d_0 if $x \neq y$ and 0 if $x = y$. In combination with (9.31) this result says that one obtains a metric on a set E by repeated convolution of a cost function c. The distance transform of a set associated with this metric is obtained as the limit of the decreasing sequence $\mathcal{E}_c^n(\mathbf{0}_X)$. An application to the discrete case can be found in Section 9.9.

9.5. Geodesic and conditional operators

Assume, for the moment, that M is a finitely path-connected subset of a metric space E. Supply M with its intrinsic metric d_M, and denote the corresponding metric dilations (cf. Section 9.3) by $\delta^r(\cdot \mid M)$, that is,

$$\delta^r(X \mid M) = \bigcup_{x\in X} B_M(x, r), \tag{9.36}$$

where $B_M(x, r) = \{y \in M \mid d_M(x, y) \leq r\}$ is the *geodesic closed ball* with centre x and radius r. We call $\delta^r(\cdot \mid M)$ the *geodesic dilation* with radius r. The adjoint *geodesic erosion* is

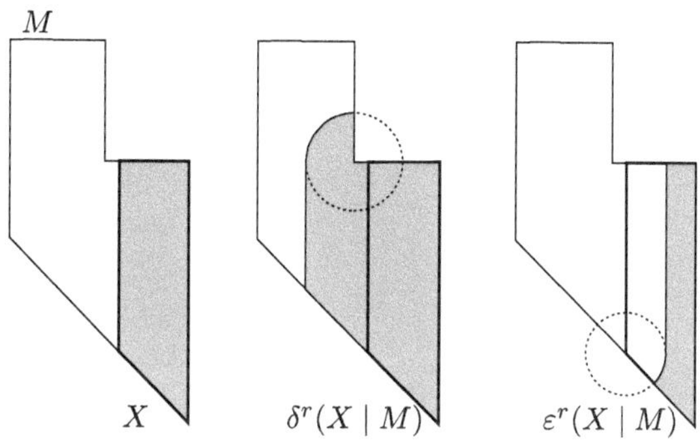

Figure 9.12 Geodesic dilation and erosion.

$$\varepsilon^r(X \mid M) = \{y \in M \mid B_M(y, r) \subseteq X\}. \tag{9.37}$$

Both operators are illustrated in Fig. 9.12 for the case $E = \mathbb{R}^2$.

To define geodesic dilations $\delta^r(\cdot \mid M)$ it is not necessary that M be finitely path-connected. One can use the same definitions when M contains more than one connected component. Note that in this case the intrinsic distance between two points which lie in different components of M is ∞. If M_i, M_j are two different components of M and if $X \subseteq M_i$, then $\delta^r(X \mid M) \cap M_j = \varnothing$.

It is easy to establish the following alternative characterization of the geodesic dilation $\delta^r(\cdot \mid M)$:

$$\delta^r(X \mid M) = \{y \in M \mid B_M(y, r) \cap X \neq \varnothing\}. \tag{9.38}$$

Geodesic dilations and erosions are complementary operators in the sense that

$$\delta^r(M \backslash X \mid M) = M \backslash \varepsilon^r(X \mid M); \tag{9.39}$$

$$\varepsilon^r(M \backslash X \mid M) = M \backslash \delta^r(X \mid M). \tag{9.40}$$

The first relation follows from (9.38):

$$\begin{aligned}
\delta^r(M \backslash X \mid M) &= \{y \in M \mid B_M(y, r) \cap M \backslash X \neq \varnothing\} \\
&= \{y \in M \mid B_M(y, r) \not\subseteq X\} \\
&= M \backslash \{y \in M \mid B_M(y, r) \subseteq X\} \\
&= M \backslash \varepsilon^r(X \mid M).
\end{aligned}$$

The second relation follows by duality.

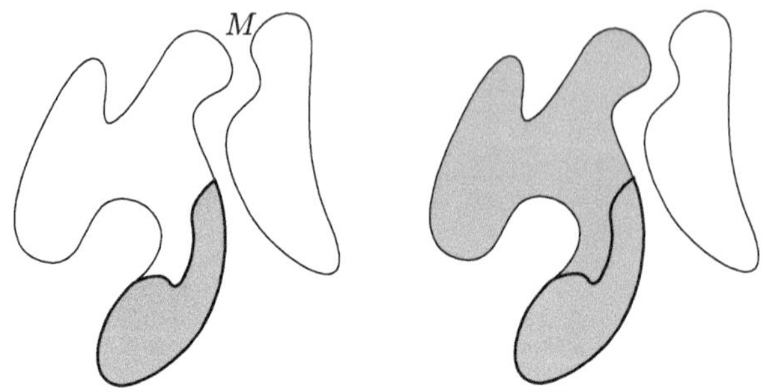

Figure 9.13 Geodesic reconstruction.

9.46 Proposition. *Let E be an arbitrary metric space, and let M be a nonempty subset of E. The geodesic dilations $\delta^r(\cdot \mid M)$ satisfy the semigroup property*

$$\delta^r(\cdot \mid M)\delta^s(\cdot \mid M) = \delta^{r+s}(\cdot \mid M), \qquad \text{for } r, s \geq 0. \tag{9.41}$$

A similar relation holds for the geodesic erosions.

This proposition is a consequence of Proposition 9.29 and Theorem 9.37.

The *geodesic opening* and *closing* are, respectively, defined by

$$\alpha_r(\cdot \mid M) = \delta^r(\cdot \mid M)\varepsilon^r(\cdot \mid M), \tag{9.42}$$

$$\beta_r(\cdot \mid M) = \varepsilon^r(\cdot \mid M)\delta^r(\cdot \mid M). \tag{9.43}$$

We prefer to use a subindex r here, since a superindex suggests that the operator with index r (r integer) can be obtained by r-fold iteration of the operator with index 1; this is true for dilation and erosion but not for opening and closing.

The *geodesic reconstruction* $\rho(X \mid M)$ is defined as the limit of the increasing family $\delta^r(X \mid M)$ as $r \to \infty$, i.e.,

$$\rho(X \mid M) = \bigcup_{r \geq 0} \delta^r(X \mid M). \tag{9.44}$$

Thus $\rho(X \mid M)$ contains all points in M from where there exists a rectifiable path to a point of X. In other words, it is the union of all components of M which have a nonempty intersection with X. See Fig. 9.13 for an example.

The reconstruction $\rho(\cdot \mid M)$ is a union of dilations and therefore a dilation. Geodesic reconstructions can be used to modify a given opening.

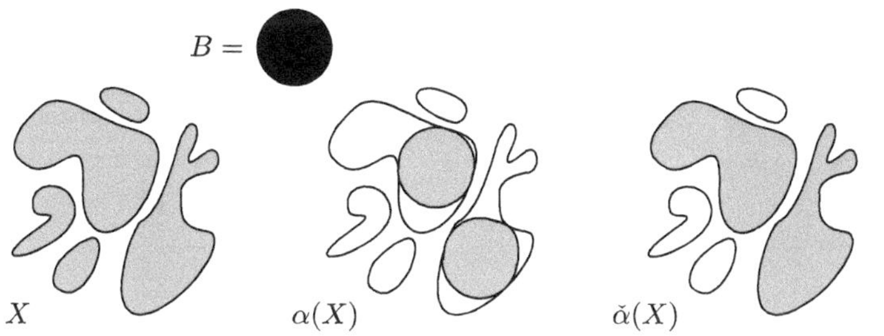

Figure 9.14 Modification $\breve{\alpha}$ of the opening $\alpha(X) = X \circ B$, where B is the closed unit disk.

9.47 Proposition. *Let E be an arbitrary metric space. If α is an opening on $\mathcal{P}(E)$, then $\breve{\alpha}$ given by*

$$\breve{\alpha}(X) = \rho(\alpha(X) \mid X), \tag{9.45}$$

is an opening as well. Furthermore, $\alpha \leq \breve{\alpha}$.

Proof. It is obvious that $\breve{\alpha}$ is increasing and anti-extensive. This means in particular that $\breve{\alpha}^2 \leq \breve{\alpha}$. On the other hand,

$$\breve{\alpha}(\breve{\alpha}(X)) = \rho(\alpha\breve{\alpha}(X) \mid \breve{\alpha}(X)) \supseteq \rho(\alpha\alpha(X) \mid \breve{\alpha}(X)) = \rho(\alpha(X) \mid \breve{\alpha}(X)).$$

By definition, this is the union of all connected components in $\breve{\alpha}(X)$ that intersect $\alpha(X)$. Since $\breve{\alpha}(X)$ is the union of all connected components of X that intersect $\alpha(X)$, it follows immediately that the last expression equals $\breve{\alpha}(X)$, so $\breve{\alpha}^2(X) \supseteq \breve{\alpha}(X)$, and the proof is completed. □

An example of such a modified opening is illustrated in Fig. 9.14; here α is the opening on $\mathcal{P}(\mathbb{R}^2)$ by the unit disk B. The modification $\breve{\alpha}$ preserves all connected components that contain a translate of the unit disk.

The next operator introduced here is the ultimate erosion. Let ε^r be the metric erosion introduced in Section 9.3. The rth ultimate erosion $\upsilon_r(X)$ comprises all points in $\varepsilon^r(X)$ that cannot be recovered from subsequent erosions $\varepsilon^s(X)$, $s > r$, by reconstruction. This is expressed by the following formula:

$$\upsilon_r(X) = \varepsilon^r(X) \setminus \bigcup_{s>0} \rho(\varepsilon^{r+s}(X) \mid \varepsilon^r(X)). \tag{9.46}$$

The *ultimate erosion* $\upsilon(X)$ is the union of all these sets:

$$\upsilon(X) = \bigcup_{r \geq 0} \upsilon_r(X). \tag{9.47}$$

In spite of the name, the ultimate erosion is not an erosion; even worse, it is not increasing.

Readers who recall the definition of the (upper) conditional dilation and erosion given in Section 3.4 will have noticed the resemblance with the geodesic operators discussed here. We will explore this relation a little deeper. For the sake of exposition we restrict ourselves to the Euclidean space $\mathbb{R}^d$. It is easy to understand, however, that the expressions for the conditional operators presented in the following are also valid in the discrete space $\mathbb{Z}^d$. Furthermore, only upper conditional operators will be considered; analogous results for lower conditional operators follow by duality.

Let $A \subseteq \mathbb{R}^d$ be a structuring element, and let $M \subseteq \mathbb{R}^d$ be a mask set. The conditional dilation $\delta_A(\cdot \mid\subseteq M)$ on $\mathcal{P}(M)$ is given by

$$\delta_A(X \mid\subseteq M) = (X \oplus A) \cap M. \tag{9.48}$$

Proposition 3.32 states that the adjoint conditional erosion is given by

$$\varepsilon_A(X \mid\subseteq M) = ((X \cup M^c) \ominus A) \cap M. \tag{9.49}$$

An illustration of the conditional dilation is given in Fig. 9.15. Geodesic dilation and conditional dilation are closely related, though they are not the same, neither from a conceptional nor from an operational point of view. To understand this, we consider the geodesic dilations associated with the Euclidean metric. The geodesic dilation $\delta^1(X \mid M)$ of a set $X \subseteq M$ contains all points $y \in M$ that can be connected to a point $x \in X$ by means of a path of length ≤ 1. On the other hand, the conditional dilation $\delta_B(X \mid\subseteq M)$, where B is the closed unit ball, contains all points whose distance to X is ≤ 1. It is evident that $\delta^1(X \mid M) \subseteq \delta_B(X \mid\subseteq M)$ and, more generally, that

$$\delta^r(X \mid M) \subseteq \delta_{rB}(X \mid\subseteq M), \tag{9.50}$$

for every $r > 0$ and $X \subseteq M$. Fig. 9.15 shows an example where the inclusion is strict.

Conditional dilations can also be used to define the conditional reconstruction. Assume that $0 \in A$; then

$$X \subseteq \delta_A(X \mid\subseteq M) = (X \oplus A) \cap M \subseteq M,$$

if $X \subseteq M$. This means that the sequence $\delta_A^n(X \mid\subseteq M)$ is increasing and has upper bound M. Define the *conditional reconstruction* $\rho_A(\cdot \mid\subseteq M)$ by

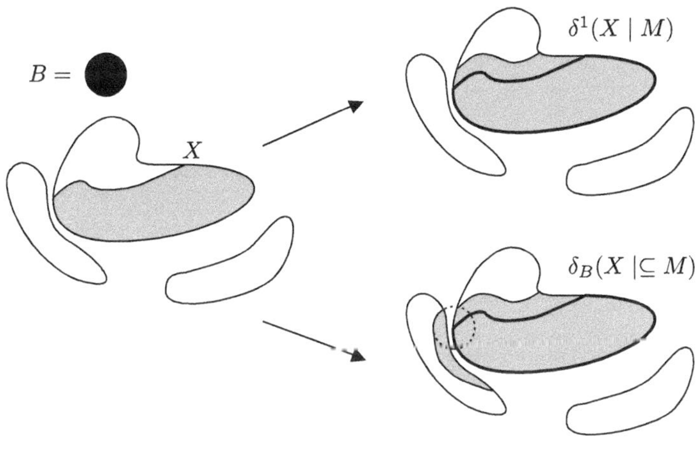

Figure 9.15 An example where $\delta^1(X \mid M)$ is strictly smaller than $\delta_B(X \mid\subseteq M)$.

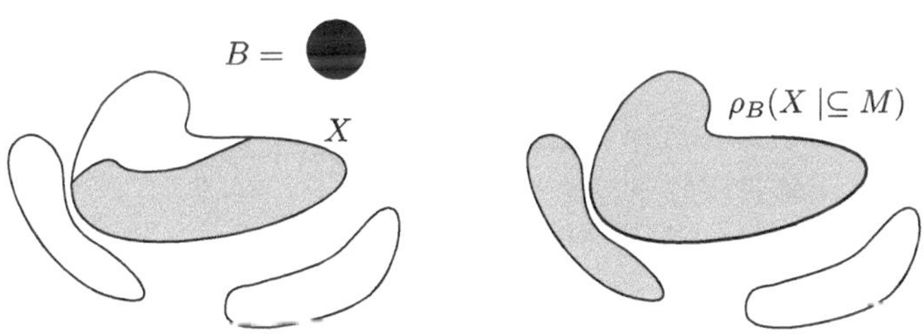

Figure 9.16 Conditional reconstruction on $\mathcal{P}(\mathbb{R}^2)$; B is the closed unit disk.

$$\rho_A(X \mid\subseteq M) = \bigcup_{n\geq 1} \delta_A^n(X \mid\subseteq M), \tag{9.51}$$

for $X \subseteq M$. See Fig. 9.16 for an illustration.

Being a union of dilations, the operator $\rho_A(\cdot \mid\subseteq M)$ is a dilation on $\mathcal{P}(M)$. In some cases it coincides with the geodesic reconstruction $\rho(\cdot \mid M)$. This happens, for instance, if 0 lies in the interior of A and $(M_i \oplus A) \cap M_j = \varnothing$ for every pair of disjoint components M_i, M_j of M. It is not difficult to think of situations where both reconstructions are substantially different, however.

9.6. Granulometries

A granulometry, which is one of the most practical tools in mathematical morphology, is an algebraic formalization of the intuitive notion of a sieving process. It enables one to compute the size distribution of the grains (or pores) in a medium. Suppose we are given a binary image representing a finite number of isolated particles. To obtain a size distribution, we put these particles through a stack of sieves with decreasing mesh widths and measure the number or the total volume of the particles remaining on a particular sieve. This results in a histogram, which may be interpreted as a size distribution. Such an intuitive approach raises immediately a number of questions. A first objection one could make is that particles are not classified according to their volume but rather according to their capacity to pass a certain mesh opening. Furthermore, one has to prescribe which motions (translations, rotations, reflections) are permitted to force a particle through a sieve. Another problem is that particles may overlap and will be classified as one large particle by the system. Thus one is led to the conclusion that the intuitive characterization of a size distribution as the outcome of a sieving process is too vague and too restricted.

Matheron (1975) was the first to realize that the concept of an opening in the morphological sense can be used to develop a theory of size distributions which is very general but also very attractive from a mathematical point of view. The key notion in his theory is a *granulometry*, meaning a tool to "measure the grains"; this notion forms the basic theme of the present section. Throughout this section we restrict ourselves to the binary image space $\mathcal{P}(\mathbb{R}^d)$.

In Section 6.7 we have discussed granulometries on complete lattices in some detail. Although a considerable part of the present section can be regarded as an application of this abstract theory, the material presented here will, to a large extent, be self-contained. When a particular result follows from the abstract theory in Section 6.7, however, its proof will be omitted.

9.48 Definition. A granulometry on $\mathcal{P}(\mathbb{R}^d)$ is a one-parameter family of openings $\{\alpha_r \mid r > 0\}$ such that

$$\alpha_s \leq \alpha_r \qquad \text{if } s \geq r. \tag{9.52}$$

Henceforth this granulometry will be denoted by $\{\alpha_r\}$.

Property (9.52) is equivalent to

$$\mathrm{Inv}(\alpha_s) \subseteq \mathrm{Inv}(\alpha_r), \qquad s \geq r. \tag{9.53}$$

From (9.52) it follows that the operators α_r obey the semigroup property

$$\alpha_r \alpha_s = \alpha_s \alpha_r = \alpha_s, \qquad s \geq r; \tag{9.54}$$

cf. Theorem 3.24. In terms of a sieving process one may think of $\alpha_r(X)$ as the particles in X which do not pass the sieve with mesh width r; see also the examples that follow.

9.49 Examples.

(a) Let α be an opening, and define $\alpha_r = \alpha$ for every $r > 0$; then α_r defines a granulometry. More generally, let α_1, α_2 be openings and $\alpha_2 \leq \alpha_1$. Take $r_1 > 0$, and define $\alpha_r = \alpha_1$ if $r \in (0, r_1]$ and $\alpha_r = \alpha_2$ if $r > r_1$; then $\{\alpha_r\}$ defines a granulometry. It is easy to extend this example to a finite collection of openings $\alpha_n \leq \alpha_{n-1} \leq \cdots \leq \alpha_1$.

(b) Given $X \subseteq \mathbb{R}^d$, denote by conn(X) the family of connected components of X. Let $A \subseteq \mathbb{R}^d$, and assume without loss of generality that $0 \in A$. Define $\alpha_r(X)$ as the union of all connected components of X that cannot be mapped inside rA by translation. In symbols,

$$\alpha_r(X) = \bigcup \{Y \in \text{conn}(X) \mid Y_h \not\subseteq rA \text{ for all } h \in \mathbb{R}^d\}.$$

Refer to Fig. 9.17(c) for an illustration.

The reader can readily verify that α_r defines a granulometry if and only if A is star-shaped with respect to 0 (i.e., $x \in A$ and $s \in [0, 1]$ implies that $sx \in A$). As a variant of this example, assume that the components Y of X may be rotated as well. Let $\mathbb{T}$ be the (nonabelian) group generated by the rotations and translations on $\mathbb{R}^d$ (see Section 5.9), and define

$$\tilde{\alpha}_r(X) = \bigcup \{Y \in \text{conn}(X) \mid \tau(Y) \not\subseteq rA \text{ for all } \tau \in \mathbb{T}\}.$$

As before, $\tilde{\alpha}_r$ defines a granulometry if A is star-shaped with respect to 0. Furthermore, $\tilde{\alpha}_r \geq \alpha_r$ for every $r > 0$. See Fig. 9.17(d) for an illustration.

(c) Let m denote Lebesgue measure on the Borel measurable subsets of $\mathbb{R}^d$. Define the image functional area : $\mathcal{P}(\mathbb{R}^d) \to [0, \infty]$ by area$(X) = m(\overline{X})$; here $\overline{X}$ denotes the closure of X. It is obvious that area$(\cdot)$ is increasing and translation invariant. If

$$\alpha_r(X) = \bigcup \{Y \in \text{conn}(X) \mid \text{area}(Y) \geq r^2\},$$

then $\{\alpha_r\}$ defines a granulometry which is invariant under translations and rotations, that is,

$$\alpha_r \tau = \tau \alpha_r, \qquad \text{for every } \tau \in \mathbb{T},$$

where $\mathbb{T}$ is the group of translations and rotations. This example is illustrated in Fig. 9.18(b).

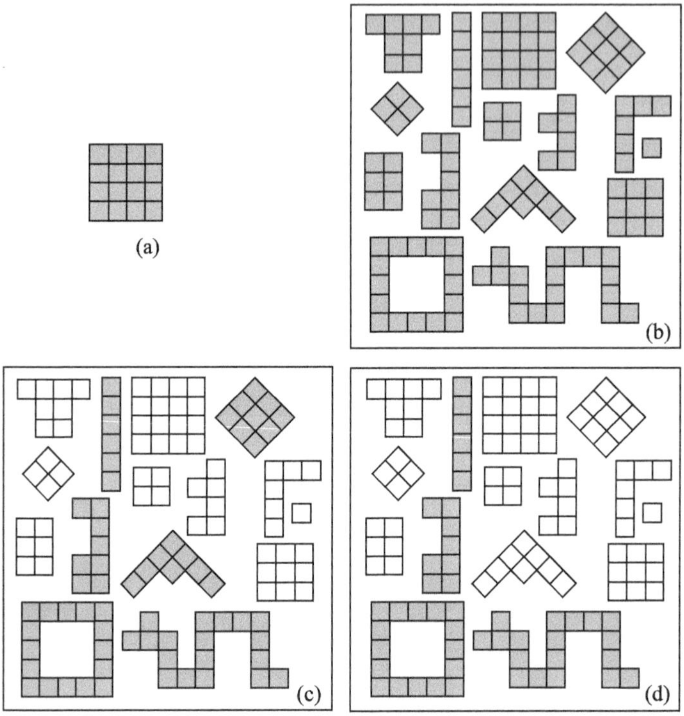

Figure 9.17 The granulometries defined in Example 9.49(b). (a) The structuring element A; (b) the original set X; (c) the set $\alpha_r(X) = \bigcup\{Y \in \mathrm{conn}(X) \mid Y_h \not\subseteq rA \text{ for all } h \in \mathbb{R}^d\}$; (d) the set $\tilde{\alpha}_r(X) = \bigcup\{Y \in \mathrm{conn}(X) \mid \tau(Y) \not\subseteq rA \text{ for all } \tau \in \mathbb{T}\}$.

(d) The following example is the most important one from a practical point of view, and it plays a prominent role hereafter. Let A be the unit square (later arbitrary convex shapes will be considered), and define

$$\alpha_r(X) = X \circ rA.$$

Since sA is rA-open if $s \geq r$, it follows immediately that $\alpha_s \leq \alpha_r$ for $s \geq r$. Refer to Fig. 9.18(c) for an illustration.

The granulometry introduced in the previous example will be examined in more detail. Suppose that Y is a component of X which contains at least one translate of rA, i.e., $Y \circ rA \neq \varnothing$. The operator $\alpha_r(X) = X \circ rA$ does not preserve the whole component Y, only the subset $Y \circ rA$. It is often desirable to retain the whole particle Y if its opening by rA is nonempty. This is achieved by introducing the following modification (see also Propo-

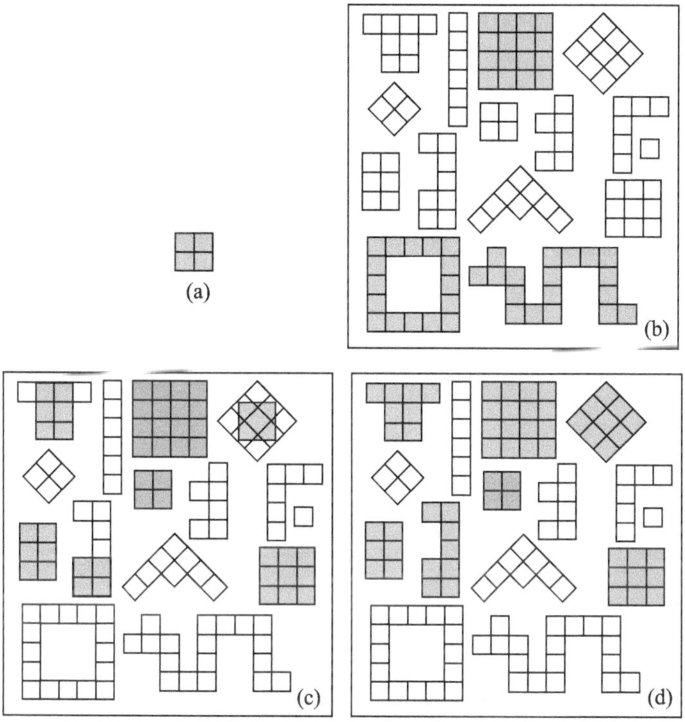

Figure 9.18 (a) Structuring element A; (b) the granulometry $\alpha_r(X) = \bigcup\{Y \in \text{conn}(X) \mid \text{area}(Y) \geq r^2\}$ with $r = 4$ (a small square corresponds to one area unit); (c) the set $X \circ rA$; (d) the set $\check{\alpha}_r(X) = \rho(X \circ rA \mid X)$.

sition 9.47):

$$\check{\alpha}_r(X) = \rho(X \circ rA \mid X),$$

where $\rho(X \mid M)$ is the geodesic reconstruction of X within the set M.

9.50 Proposition. *If $\{\alpha_r\}$ is a granulometry and if $\check{\alpha}_r$ is given by*

$$\check{\alpha}_r(X) = \rho(\alpha_r(X) \mid X), \tag{9.55}$$

then $\{\check{\alpha}_r\}$ defines a granulometry as well.

Proof. That every $\check{\alpha}_r$ is an opening is a consequence of Proposition 9.47. Furthermore, by the fact that ρ is increasing with respect to the first variable it follows that the family $\{\check{\alpha}_r\}$ obeys (9.52). □

It is easy to see that the given modification preserves translation or rotation invariance; in other words, if α_r is translation or rotation invariant, then $\breve{\alpha}_r$ is such as well. An example is depicted in Fig. 9.18(d). The next result follows from Proposition 6.34.

9.51 Proposition. *Let I be an arbitrary index set, and let $\{\alpha_r^i\}$ define a granulometry on $\mathcal{P}(\mathbb{R}^d)$ for every $i \in I$. Then $\{\bigvee_{i\in I} \alpha_r^i\}$ is a granulometry as well.*

A general method to construct granulometries is the following: if α_r is an opening for every $r > 0$, then the family $\{\tilde{\alpha}_r\}$ given by $\tilde{\alpha}_r = \bigvee_{s\geq r} \alpha_s$ defines a granulometry. Throughout the remainder of this section we restrict attention to translation invariant granulometries; these are defined as granulometries in which every opening is translation invariant.

9.52 Definition. A translation invariant granulometry $\{\alpha_r\}$ on $\mathcal{P}(\mathbb{R}^d)$ is called a *structural granulometry* if α_r is a structural opening for every $r > 0$.

If $\{\alpha_r\}$ is a structural granulometry, then $\alpha_r(X) = X \circ A(r)$ for some structuring element $A(r) \subseteq \mathbb{R}^d$. Proposition 4.22 implies that the condition that $\alpha_s \leq \alpha_r$ if $s \geq r$ is equivalent to $A(s)$ is $A(r)$-open for $s \geq r$, i.e., $A(s) \circ A(r) = A(s)$, $s \geq r$. Refer to Proposition 4.23 for a similar statement. The granulometry in Example 9.49(d) given by $\alpha_r(X) = X \circ rA$, where A is the unit square, is a structural granulometry.

9.53 Example. One can easily construct families of structuring elements $A(r)$ such that $A(s)$ is $A(r)$-open for $s \geq r$. Choose $A_1, A_2, \ldots, A_n \subseteq \mathbb{R}^d$ and $0 < r_1 < r_2 < \cdots < r_{n-1}$. Define

$$A(r) = \begin{cases} A_1, & \text{for } 0 < r \leq r_1, \\ A_1 \oplus A_2, & \text{for } r_1 < r \leq r_2, \\ \vdots & \\ A_1 \oplus A_2 \oplus \cdots \oplus A_{n-1}, & \text{for } r_{n-2} < r \leq r_{n-1}, \\ A_1 \oplus A_2 \oplus \cdots \oplus A_n, & \text{for } r_{n-1} < r. \end{cases}$$

It is easy to check that $A(s)$ is $A(r)$-open for $s \geq r$.

Recall from (4.65) that

$$rX \circ rA = r(X \circ A), \qquad r > 0.$$

This property underlies the definition of a Minkowski granulometry.

9.54 Definition. A granulometry $\{\alpha_r\}$ on $\mathcal{P}(\mathbb{R}^d)$ is called a *Minkowski granulometry* if every opening α_r is translation invariant and if the family $\{\alpha_r\}$ is compatible with scalings, i.e.,

$$\alpha_r(rX) = r\alpha_1(X), \qquad \text{for } r > 0 \text{ and } X \subseteq \mathbb{R}^d. \tag{9.56}$$

The granulometries in Examples 9.49(b)–(d) are of Minkowski type; this is not true for the granulometries in Examples 9.49(a) and 9.53. As Proposition 6.39 shows, this definition can easily be recast in terms of the invariance domain.

9.55 Proposition. *The granulometry $\{\alpha_r\}$ is of Minkowski type if and only if*

$$\mathrm{Inv}(\alpha_r) = r\,\mathrm{Inv}(\alpha_1), \qquad r > 0, \tag{9.57}$$

and $\mathrm{Inv}(\alpha_1)$ *is closed under translations.*

If α_r is a Minkowski granulometry, then $\mathrm{Inv}(\alpha_r) = r\mathcal{C}$, where $\mathcal{C} = \mathrm{Inv}(\alpha_1)$. Since $\mathrm{Inv}(\alpha_s) \subseteq \mathrm{Inv}(\alpha_r)$ for $s \geq r$, we infer that $s\mathcal{C} \subseteq r\mathcal{C}$ if $s \geq r$. This is equivalent to

$$r\mathcal{C} \subseteq \mathcal{C} \qquad \text{for } r \geq 1. \tag{9.58}$$

This shows that $\mathcal{C}$ is closed under unions (since α_1 is an opening), translations (since α_1 is translation invariant), and scalings with values $r \geq 1$. In fact, this observation forms the basis for the construction of Minkowski granulometries. Let $\mathcal{A}$ be an arbitrary collection of subsets of $\mathbb{R}^d$, and define $\mathcal{C}$ to be the smallest subset of $\mathcal{P}(\mathbb{R}^d)$ that contains $\mathcal{A}$ and is closed under unions, translations, and scalings by a factor ≥ 1. Let α_r be the opening generated by $r\mathcal{C}$. Then $\{\alpha_r\}$ defines a Minkowski granulometry. Moreover, every Minkowski granulometry is of this form. We summarize these results in the following theorem.

9.56 Proposition. *Let $\{\alpha_r\}$ be a Minkowski granulometry; then there is a family $\mathcal{A} \subseteq \mathcal{P}(\mathbb{R}^d)$ such that*

$$\alpha_r(X) = \bigcup_{s \geq r} \bigcup_{A \in \mathcal{A}} X \circ sA. \tag{9.59}$$

Conversely, if $\mathcal{A} \subseteq \mathcal{P}(\mathbb{R}^d)$, then α_r given by (9.59) *defines a Minkowski granulometry.*

For practical purposes, structural Minkowski granulometries are of particular interest. If $\mathcal{A}$ contains only one structuring element A, then the expression in (9.59) reduces to $\alpha_r(X) = \bigcup_{s \geq r} X \circ sA$. This defines a structural granulometry if and only if $X \circ sA \subseteq X \circ rA$ for $s \geq r$. This is equivalent to

$$rA \text{ is } A\text{-open} \qquad \text{for } r \geq 1. \tag{9.60}$$

Theorem 9.17 states that a compact set A satisfies this condition if and only if A is convex.

9.57 Theorem. *Assume that A is a compact structuring element. The openings $\alpha_r(X) = X \circ rA$ define a Minkowski granulometry if and only if A is convex.*

That compactness cannot be omitted here follows readily from the following example. Let $A \subseteq \mathbb{R}^2$ be the complement of the open first quadrant, that is, $A = \{(x, y) \in \mathbb{R}^2 \mid x \leq 0 \ \vee \ y \leq 0\}$. Then $rA = A$ for every $r > 0$, and thus (9.60) is satisfied. However, A is not convex.

We give an alternative formulation of Theorem 9.57. Let δ^r be the dilation given by $\delta^r(X) = X \oplus rA$, with A convex. Then the family $\{\delta^r \mid r > 0\}$ satisfies the semigroup property

$$\delta^s \delta^r = \delta^{r+s}, \qquad r, s > 0. \tag{9.61}$$

In fact, $\delta^s(\delta^r(X)) = (X \oplus rA) \oplus sA = X \oplus (rA \oplus sA) = X \oplus [(r + s)A]$. Here we have used Proposition 9.2.

Let $\{\delta^r \mid r > 0\}$ be a family of dilations, not necessarily translation invariant, satisfying semigroup property (9.61); furthermore, let ε^r be the erosion adjoint to δ^r. The openings $\alpha_r = \delta^r \varepsilon^r$ define a granulometry. In fact, for $s \geq r$,

$$\begin{aligned} \alpha_r \alpha_s &= \delta^r \varepsilon^r \delta^s \varepsilon^s = \delta^r \varepsilon^r \delta^r \delta^{s-r} \varepsilon^s \\ &= \delta^r \delta^{s-r} \varepsilon^s = \delta^s \varepsilon^s = \alpha_s, \end{aligned}$$

where we have used that $\delta^r \varepsilon^r \delta^r = \delta^r$; see (3.8). This observation applies to the geodesic openings introduced in (9.42).

9.58 Proposition. *The openings $\alpha_r(\cdot \mid M)$, $r > 0$, given by (9.42) define a granulometry on $\mathcal{P}(M)$.*

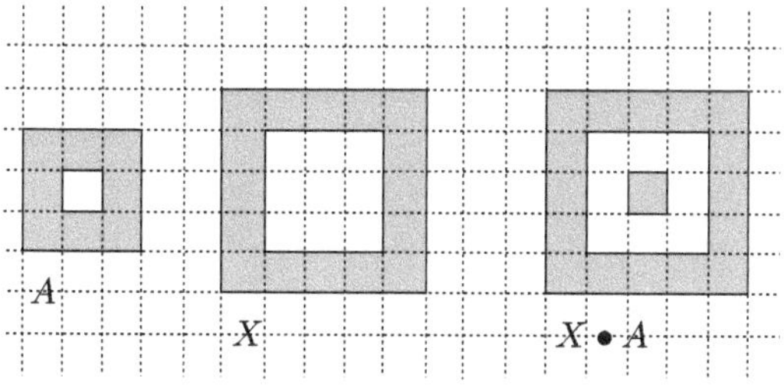

Figure 9.19 X and A are connected, but the closing $X \bullet A$ is disconnected.

9.7. Connectivity

Throughout this section we use the term *connected* although, strictly speaking, we mean *path-connected*; see also Remark 9.26. Even if connectedness is a topological rather than a geometrical notion, this chapter is the best place for a treatment of connectivity in relation to mathematical morphology.

9.59 Proposition. *Let $X, A \subseteq \mathbb{R}^d$ be connected; then $X \oplus A$ is connected as well.*

Proof. Let $x, y \in X$ and $a, b \in A$. We show that there exists a path between $x + a$ and $y + b$. Since X_a is connected, there is a path between $x + a$ and $y + a$. Furthermore, since A_y is connected, there is a path between $y + a$ and $y + b$. Joining both paths gives a path between $x + a$ and $y + b$. □

It is easy to find examples which show that a similar statement for erosions and openings is false. Fig. 9.19 contains an example which shows that it is also false for closings.

One can define connectivity on the discrete space $\mathbb{Z}^d$ by introducing a graph structure. Usually one chooses a regular grid for this purpose. It is not difficult to show that Proposition 9.59 remains valid in this case. For further results in the discrete case, refer to Section 9.9.

Serra (1988) has formalized the notion of connectivity by introducing connectivity classes.

9.60 Definition. Let E be an arbitrary set. A subset $\mathcal{C} \subseteq \mathcal{P}(E)$ is called a *connectivity class* when the following properties hold:

(i) $\varnothing \in \mathcal{C}$ and $\{x\} \in \mathcal{C}$ for every $x \in E$;

(ii) if $X_i \in \mathcal{C}$ for $i \in I$ and $\bigcap_{i \in I} X_i \neq \varnothing$, then $\bigcup_{i \in I} X_i \in \mathcal{C}$.

For example, the connected subsets of $\mathbb{R}^d$ constitute a connectivity class. Given $x \in E$, define the operator $\gamma_x : \mathcal{P}(E) \to \mathcal{P}(E)$ by

$$\gamma_x(X) = \bigcup\{C \in \mathcal{C} \mid x \in C \text{ and } C \subseteq X\}. \tag{9.62}$$

Note that, by the very definition of a connectivity class, $\gamma_x(X) \in \mathcal{C}$.

9.61 Proposition. *Let $x, y \in E$ and $X \subseteq E$.*
(a) *γ_x is an opening;*
(b) *$\gamma_x(\{x\}) = \{x\}$;*
(c) *if $x \notin X$, then $\gamma_x(X) = \varnothing$;*
(d) *either $\gamma_x(X) = \gamma_y(X)$ or $\gamma_x(X) \cap \gamma_y(X) = \varnothing$;*
(e) *$\bigcup_{y \in E} \gamma_y = \mathrm{id}$.*

Proof. (a): It is obvious that γ_x is increasing and anti-extensive; this implies $\gamma_x^2 \le \gamma_x$. To prove the reverse inequality, note that $\gamma_x(X)$ is a union of sets $C \in \mathcal{C}$ with $x \in C \subseteq X$. Every such C also satisfies $x \in C \subseteq \gamma_x(X)$; hence $\gamma_x(X) \subseteq \gamma_x(\gamma_x(X))$.

(b)–(c): Straightforward.

(d): Assume that $\gamma_x(X) \cap \gamma_y(X) \neq \varnothing$. Obviously, $\gamma_x(X), \gamma_y(X) \in \mathcal{C}$, and by property (ii) of Definition 9.60, $\gamma_x(X) \cup \gamma_y(X) \in \mathcal{C}$. Since $x \in \gamma_x(X) \cup \gamma_y(X) \subseteq X$, we conclude that $\gamma_x(X) \cup \gamma_y(X) \subseteq \gamma_x(X)$. This implies $\gamma_y(X) \subseteq \gamma_x(X)$. The other inclusion follows in a similar way, and we conclude that both sets are identical.

(e): We show that $\bigcup_{y \in E} \gamma_y(X) = X$. The inclusion $\subseteq$ holds trivially. On the other hand, assume that $x \in X$. Then

$$x \in \gamma_x(\{x\}) \subseteq \gamma_x(X) \subseteq \bigcup_{y \in E} \gamma_y(X).$$

This concludes the proof. □

The operator γ_x is called the *connectivity opening* associated with $\mathcal{C}$. Its invariance domain is $\{C \in \mathcal{C} \mid x \in C\}$.

This formal approach based on connectivity classes applies to continuous as well as discrete spaces. But it can also be used to define other types of connectivity.

9.62 Example. Let γ_x be the connectivity openings on $\mathcal{P}(\mathbb{R}^d)$ associated with the usual path-connectivity. Let $A \subseteq \mathbb{R}^d$ be a connected structuring

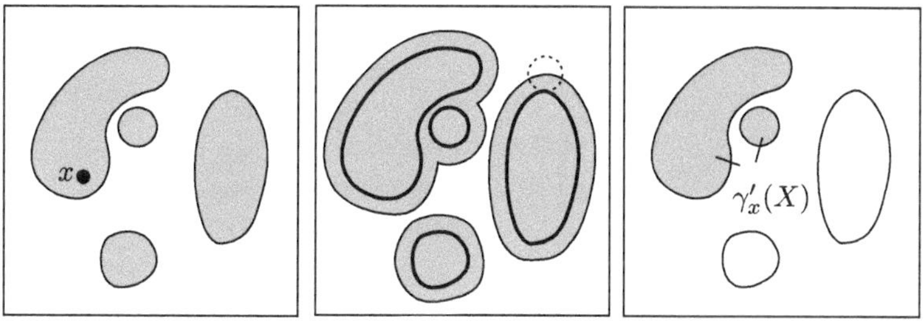

Figure 9.20 Computation of $\gamma_x'(X)$ when A is a disk. From left to right: X, $X \oplus A$, and $\gamma_x'(X)$.

element which contains the origin. For $x \in E$ we define the operator γ_x' on $\mathcal{P}(\mathbb{R}^d)$ by

$$\gamma_x'(X) = X \cap \gamma_x(X \oplus A), \qquad \text{if } x \in X,$$

and $\varnothing$ otherwise; see Fig. 9.20 for an example. We show that the γ_x' are connectivity openings associated with the connectivity class $\mathcal{C}'$ defined by

$$\mathcal{C}' = \{X \subseteq \mathbb{R}^d \mid X \subseteq C \subseteq X \oplus A \text{ for some connected set } C \subseteq \mathbb{R}^d\}.$$

If $A = \{0\}$, then $\mathcal{C}'$ is the usual connectivity class.

First we show that $\mathcal{C}'$ defines a connectivity class. Property (i) of Definition 9.60 is obvious. To prove (ii), assume that $X \subseteq C_X \subseteq X \oplus A$ and $Y \subseteq C_Y \subseteq Y \oplus A$, with C_X, C_Y connected and $X \cap Y \neq \varnothing$. Then $C_X \cap C_Y \neq \varnothing$ and $X \cup Y \subseteq C_X \cup C_Y \subseteq (X \oplus A) \cup (Y \oplus A) = (X \cup Y) \oplus A$. Therefore, $X \cup Y \in \mathcal{C}'$. It is obvious that γ_x' is anti-extensive and increasing. For idempotence it suffices to show that $(\gamma_x')^2 \geq \gamma_x'$. Since $\gamma_x' \geq \gamma_x$, this holds if $\gamma_x'\gamma_x \geq \gamma_x'$. The set $\gamma_x(X)$ is connected, and Proposition 9.59 says that $\gamma_x(X) \oplus A$ is connected as well. Thus $\gamma_x(\gamma_x(X) \oplus A) = \gamma_x(X) \oplus A$, which proves the assertion. Finally, we show that the invariance domain of γ_x' equals $\mathcal{C}_x' = \{C \in \mathcal{C}' \mid x \in C\}$. Thereto we must show that $\gamma_x'(X) = X$ if and only if

$$x \in X \subseteq C \subseteq X \oplus A, \tag{9.63}$$

for some connected set C. If (9.63) holds, then $C = \gamma_x(C) \subseteq \gamma_x(X \oplus A)$, which means that $\gamma_x'(X) = X$. Conversely, $\gamma_x'(X) = X$ implies $X \subseteq \gamma_x(X \oplus A)$, and (9.63) is satisfied if $C = \gamma_x(X \oplus A)$.

9.8. Skeleton

Skeleton transformations are a profound tool in image analysis and pattern recognition. For that reason they have received a great deal of attention in the literature. This section presents a definition of the morphological skeleton operator and some of its basic properties. In the first part of this section the continuous image space $\mathcal{P}(\mathbb{R}^d)$ will be considered. The section concludes with some remarks about the discrete case.

Throughout this section the following assumption will be made.

9.63 Assumption. A is a compact convex body in $\mathbb{R}^d$ and $0 \in A^\circ$.

The *regular* and *singular* part of a set $X \subseteq \mathbb{R}^d$ with respect to A are, respectively, defined by

$$\mathrm{reg}_A(X) = \bigcup_{r>0} X \circ rA, \tag{9.64}$$

$$\mathrm{sing}_A(X) = X \setminus \mathrm{reg}_A(X). \tag{9.65}$$

Evidently,

$$X = \mathrm{reg}_A(X) \cup \mathrm{sing}_A(X).$$

The operator $\mathrm{reg}_A(\cdot)$ is an opening on $\mathcal{P}(\mathbb{R}^d)$ which is translation invariant. It is obvious that $X^\circ \subseteq \mathrm{reg}_A(X)$, and hence that $\mathrm{sing}_A(X) \subseteq \partial X$, the boundary of X.

9.64 Definition. Let $r \geq 0$ and $h \in \mathbb{R}^d$; assume that $(rA)_h$ is contained in X. The set $(rA)_h$ is a *maximal A-shape* in X if $(rA)_h \subseteq (r'A)_{h'} \subseteq X$ for $r' \geq r$ implies that $r' = r$ and $h' = h$.

Define the rth A-skeleton subset

$$\Sigma_{A,r}(X) = \mathrm{sing}_A(X \ominus rA). \tag{9.66}$$

9.65 Lemma. *Let Assumption 9.63 be satisfied. The set $(rA)_h$ is a maximal A-shape in X if and only if $h \in \Sigma_{A,r}(X)$.*

Proof. We use that $h \in \Sigma_{A,r}(X)$ if and only if

(i) $h \in X \ominus rA$ and

(ii) $h \notin (X \ominus rA) \circ \epsilon A$, for every $\epsilon > 0$.

"if": Assume that (i) and (ii) are satisfied. We show that $(rA)_h$ is a maximal A-shape in X. Suppose that $(rA)_h \subseteq [(r+\epsilon)A]_k \subseteq X$. Then $h \in (\epsilon A)_k \subseteq X \ominus rA$, which gives that $h \in (X \ominus rA) \circ \epsilon A$, a contradiction.

"only if": Assume that $(rA)_h$ is a maximal A-shape in X. Then $h \in X \ominus rA$, i.e., (i) holds. Assume that $h \in (X \ominus rA) \circ \epsilon A$ for some $\epsilon > 0$. Then $h \in (\epsilon A)_k \subseteq (X \ominus rA)$, which implies $(rA)_h \subseteq [(r+\epsilon)A]_k \subseteq X \circ rA \subseteq X$. But then $(rA)_h$ is not a maximal A-shape, a contradiction. □

It is obvious that $\Sigma_{A,r}(X) \cap \Sigma_{A,s}(X) = \varnothing$ if $r \neq s$. Define the *A-skeleton* $\Sigma_A(X)$ of X as the (disjoint) union of all $\Sigma_{A,r}(X)$:

$$\Sigma_A(X) = \bigcup_{r \geq 0} \Sigma_{A,r}(X). \tag{9.67}$$

The *quench function* $q_A(X, \cdot)$ on $\Sigma_A(X)$ is defined by

$$q_A(X, h) = r \qquad \text{if } h \in \Sigma_{A,r}(X).$$

If A is the closed unit ball, then $q_A(X, h) = d(h, X^c)$.

9.66 Examples.

(a) Assume that A is the closed unit ball. If X is a closed ball, then the A-skeleton contains only one point, namely, the centre. If one attaches a line segment to this ball, then $\Sigma_{A,0}(X)$ is the half-open segment and $\Sigma_{A,r}(X)$, where r is the radius of the ball, is the centre of the ball; see Fig. 9.21(a). Note that in this second example $\Sigma_A(X)$ is disconnected. Furthermore, Fig. 9.21(a) also shows the A-skeleton of a rectangle.

(b) Assume that A is a square. Fig. 9.21(b) shows the A-skeleton of a ball and a rectangle, respectively.

9.67 Proposition. *If Assumption 9.63 holds, then the A-skeleton of a set has empty interior.*

Proof. Assume $h \in \Sigma_{A,r}(X)$; we show that h does not lie in the interior of $\Sigma_A(X)$. If $r = 0$, then $h \in \partial X$, and therefore it cannot lie in the interior of $\Sigma_A(X)$. Assume that $r > 0$. The set $(rA)_h$ must intersect the boundary ∂X in a point $y \neq h$. We restrict ourselves to the 2-dimensional case; for higher dimensions one can use similar arguments. We show that the half-open line segment $(h, y]$ is disjoint from $\Sigma_A(X)$. Suppose, namely, that $k \in \Sigma_{A,s}(X)$ for some $k \in (h, y]$. Then, by Lemma 9.65, $(sA)_k$ is a maximal A-shape inside X. There is a $t \in [0, 1)$ such that $k = th + (1-t)y$. It is easy to check that

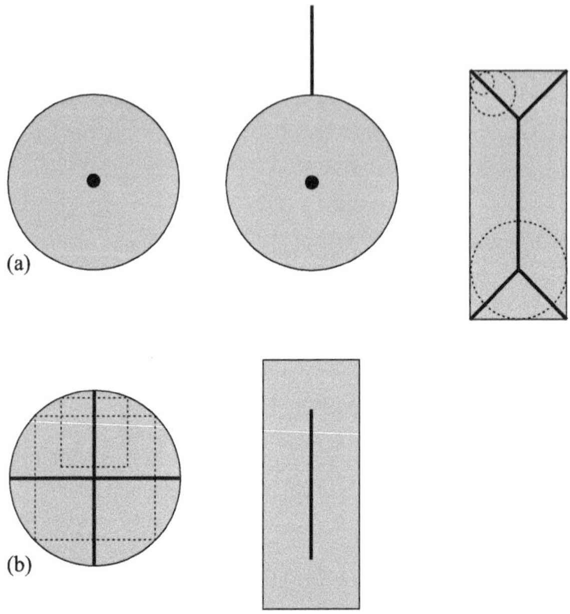

Figure 9.21 (a) The A-skeleton if A is a ball. (b) The A-skeleton if A is a square.

$s \le s_0$, where s_0 is the solution of $k + s_0 \cdot (y - h)/r = y$. A straightforward calculation shows that $s_0 = tr$, and we conclude that $s \le tr$. But it is not difficult to show that $(trA)_k \subseteq (rA)_h$; this contradicts our assumption that $(sA)_k$ is a maximal A-shape. Therefore, $(h, y] \cap \Sigma_A(X) = \varnothing$, which was to be proved. □

9.68 Proposition. *Let Assumption 9.63 be satisfied. If* $X^\circ = \varnothing$*, then* $\Sigma_A(X) = \Sigma_{A,0}(X) = X$.

This result follows immediately from the observation that rA does not fit inside X if $r > 0$. Thus if $X^\circ = \varnothing$, then $\mathrm{reg}_A(X) = \varnothing$. A combination of the former two propositions leads to the following conclusion.

9.69 Corollary. *If Assumption 9.63 holds, then the skeleton operator* $\Sigma_A : \mathcal{P}(\mathbb{R}^d) \to \mathcal{P}(\mathbb{R}^d)$ *is idempotent.*

Under certain conditions on X, the pair $(\Sigma_A(X), q_A(X, \cdot))$ contains a complete representation of X, i.e., it is possible to reconstruct X from this pair. Before we present sufficient conditions for reconstructability, we mention a class of sets for which the A-skeleton is empty.

Assume that X is an open set with convex complement, and let $r > 0$ be arbitrary. By Proposition 9.8, $X^c \bullet (r\check{A}) = X^c$; in combination with (4.55) this means that $X \circ rA = X$. Thus every $x \in X$ satisfies $x \in (rA)_h \subseteq X$ for some $h \in \mathbb{R}^d$. But this means that there is no maximal A-shape in X which contains x; hence $\Sigma_A(X) = \varnothing$.

9.70 Theorem. *Assume that A is a smooth compact convex body (see Definition 9.10(b)). If $X \subseteq \mathbb{R}^d$ is closed and contains no half-spaces, then*

$$X = \bigcup_{r \geq 0} \Sigma_{A,r}(X) \oplus rA. \tag{9.68}$$

Proof. $\supseteq$: If $h \in \Sigma_{A,r}(X)$, then $(rA)_h$ is a maximal A-shape in X by Lemma 9.65.

$\subseteq$: Assume $x \in X$; we must show that there is a maximal A-shape $(rA)_h$ in X which contains x. Assume this is not true; there is a sequence h_n such that $x \in (nA)_{h_n} \subseteq X$. (Because X is closed, it is not possible that the sequence of radii has a finite limit.) Since $\mathcal{F}(\mathbb{R}^d)$ is compact with respect to the hit-or-miss topology, the sequence $(nA)_{h_n}$ has a convergent subsequence in $\mathcal{F}$. Without loss of generality, we can assume that the entire sequence converges, say, to the closed set Y. Now Lemma 9.11 gives that $Y = \mathbb{R}^d$ or that Y is a closed half-space. From $(nA)_{h_n} \subset X$ it follows that $Y \subseteq X$. But this contradicts the assumption that X does not contain half-spaces. This concludes the proof. □

Smoothness of A is essential in Theorem 9.70. For example, if $A \subseteq \mathbb{R}^2$ is the unit square, then no maximal A-shapes lie inside a quadrant.

Note that instead of (9.68) one can also write

$$X = \bigcup_{h \in \Sigma_A(X)} (q_A(X, h)A)_h. \tag{9.69}$$

If $\Sigma_A(X)$ and $q_A(X, \cdot)$ are known, then one can also compute the eroded set $X \ominus sA$.

9.71 Proposition. *Let A be a smooth compact convex body. Assume that $X \subseteq \mathbb{R}^d$ is closed and contains no half-spaces. Then the following expressions for the erosion $X \ominus sA$ hold:*

$$\Sigma_{A,r}(X \ominus sA) = \Sigma_{A,r+s}(X) \quad \textit{and} \quad X \ominus sA = \bigcup_{r \geq 0} \Sigma_{A,r+s}(X) \oplus rA.$$

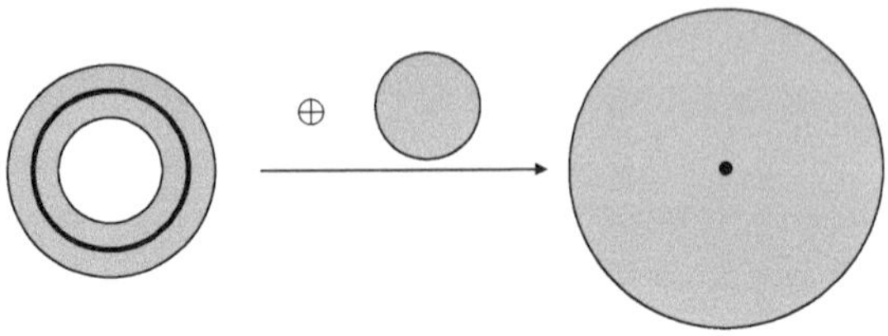

Figure 9.22 Left: the set X and its skeleton (annulus). Right: the set $X \oplus A$ and its skeleton (point).

Proof. If $s \geq 0$, then

$$\begin{aligned}\Sigma_{A,r}(X \ominus sA) &= \text{sing}_A((X \ominus sA) \ominus rA) = \text{sing}_A(X \ominus (r+s)A) \\ &= \Sigma_{A,r+s}(X).\end{aligned}$$

This proves the first identity. To prove the second, observe that under the given assumptions $X \ominus sA$ does not contain half-spaces. Now Theorem 9.70 gives the result. □

This result says that the A-skeleton of $X \ominus sA$ is a subset of the A-skeleton of X,

$$\Sigma_A(X \ominus sA) = \bigcup_{r \geq s} \Sigma_{A,r}(X).$$

It is also possible to draw conclusions about the A-skeleton of the dilated set $X \oplus sA$ and the opened set $X \circ sA$. These conclusions, however, are weaker than those for the erosion. It follows from (9.68) that

$$X \oplus sA = \bigcup_{r \geq 0} \Sigma_{A,r}(X) \oplus (r+s)A.$$

This suggests that $X \oplus sA$ has the same A-skeleton as X with quench function $q_A(X, \cdot) + s$. It is not difficult, however, to show by means of an example that this suggestion is false. Take for X the closed ring in $\mathbb{R}^2$ with inner radius 1 and outer radius 2, and let A be the closed unit ball; cf. Fig. 9.22.

The A-skeleton is the closed circle with radius $3/2$. Dilation of X by a ball with radius 1 gives a ball with radius 3. The A-skeleton of this set is just one point, namely, the centre of the ball.

Next, consider the opening $X \circ sA$. If $r \geq s$, then

$$(X \circ sA) \ominus rA = \big([(X \ominus sA) \oplus sA] \ominus sA\big) \ominus (r-s)A$$

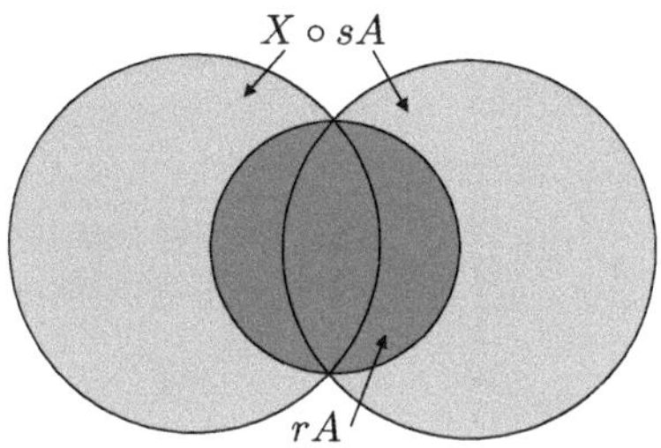

Figure 9.23 $X \circ sA$ can contain maximal A-shapes with radius $r < s$.

$$= X \ominus sA \ominus (r - s)A$$
$$= X \ominus rA.$$

Thus it follows that

$$\Sigma_{A,r}(X \circ sA) = \Sigma_{A,r}(X), \qquad \text{if } r \geq s.$$

Unfortunately, it is not possible to find an expression for $\Sigma_{A,r}(X \circ sA)$ for $r < s$. In particular, one may not conclude that $\Sigma_{A,r}(X \circ sA) = \varnothing$ if $r < s$. In Fig. 9.23 the set X is open with respect to the closed ball with radius s. However, X contains maximal balls with radius less than s.

It is possible, however, to reconstruct $X \circ sA$ from the rth A-skeleton subsets $\Sigma_{A,r}(X)$, $r \geq s$. From the expression for $X \ominus sA$ derived in Proposition 9.71, one infers that

$$X \circ sA = (X \ominus sA) \oplus sA = \Big[\bigcup_{r \geq 0} \Sigma_{A,r+s}(X) \oplus rA\Big] \oplus sA$$
$$= \bigcup_{r \geq 0} \Sigma_{A,r+s}(X) \oplus (r+s)A.$$

In other words,

$$X \circ sA = \bigcup_{r \geq s} \Sigma_{A,r}(X) \oplus rA. \tag{9.70}$$

One can use the skeletal representation of a set to define a family of morphological operators, called *quench function operators*, see Serra (1982, Exercise XI-I.8). Given a function $f : \mathbb{R}_+ \to \mathbb{R}$, define the operator $\psi_f : \mathcal{P}(\mathbb{R}^d) \to \mathcal{P}(\mathbb{R}^d)$ by

$$\psi_f(X) = \bigcup_{r \geq 0} \Sigma_{A,r}(X) \oplus f(r)A.$$

Here $f(r)A = \varnothing$ if $f(r) < 0$. For dilation $f(r) = r + s$, for erosion $f(r) = r - s$, and for the opening $f(r) = r$ if $r \geq s$ and 0 elsewhere.

This section concludes with some remarks about the discrete case. The expression for $\Sigma_{A,r}$ given in (9.66) carries over to the discrete case easily. Let $A \subseteq \mathbb{Z}^d$ be a finite structuring element, and define

$$nA = A \oplus A \oplus \cdots \oplus A \qquad [n \text{ terms}].$$

Define, for $X \subseteq \mathbb{Z}^d$ and $n \geq 0$,

$$\Sigma_{A,n}(X) = (X \ominus nA) \setminus (X \ominus nA) \circ A. \tag{9.71}$$

The discrete A-skeleton is defined by

$$\Sigma_A(X) = \bigcup_{n \geq 0} \Sigma_{A,n}(X).$$

The quench function $q_A(X, \cdot)$ becomes an integer-valued function in this case. We leave it as an exercise to the reader to show that X can be reconstructed from $\Sigma_A(X)$ and $q_A(X, \cdot)$:

$$X = \bigcup_{n \geq 0} \Sigma_{A,n}(X) \oplus nA,$$

if X is a finite set in $\mathbb{Z}^d$.

9.9. Discrete metric spaces

The notion of distance is extremely important in mathematical morphology. This final section makes some comments about the definition of metrics on discrete spaces. Furthermore, it explains how such metrics can be used to define connectivity. This opens the way to introduce discrete analogues of curves, (shortest) paths, and geodesic distance.

Basic morphological operators on $\mathcal{P}(\mathbb{Z}^d)$, like dilation and erosion, are defined in the usual way, namely, as Minkowski addition and subtraction. Chapter 5 has shown that such operators use only the group structure of $\mathbb{Z}^d$. To visualize the resulting expressions, one can choose between various representations of $\mathbb{Z}^d$ in Euclidean space. For instance, in the 2-dimensional case we can represent elements of $\mathbb{Z}^2$ as points on a square grid or points on a hexagonal grid; see Fig. 9.24. Usually, the choice between the various alternatives is made at the moment of discretization; cf. Chapter 8. It is apparent that this choice determines the operations which can be applied to the discrete image. For instance, the hexagonal grid allows six rotations, but the square grid only four. For that reason, the hexagonal grid is better suited to approximate, e.g., the Euclidean disk (evidently, this remark does not apply for the square). Furthermore, the hexagonal grid has better properties with respect to connectivity. On the other hand, current technology restricts the applicability of the hexagonal approach. The most conspicuous restriction is the fact that computers use square pixels. For the theory developed in this section the choice of the representation is, to a large extent, immaterial.

9.72 Definition. A metric space (E, d) is called *discrete* if for every $x \in E$ and for $r > 0$ sufficiently small, the closed ball $B(x, r)$ contains only finitely many points.

From now on, we assume that (E, d) is a discrete metric space. It is obvious that for every $x \in E$ there is an $r > 0$ such that $B(x, r)$ contains only the point x. Consequently, the induced topology is the discrete topology; i.e., every subset of E is open; cf. Example 7.2(b).

Two (different) points on a grid are said to be neighbours if an edge lies between them. Using this neighbour relation, we can define a path as a sequence of points such that every two subsequent points are neighbours. For two points x, y we define $d(x, y)$ as the minimal length of a path connecting these two points. It is easy to show that d defines a metric on the grid if the underlying graph structure is rich enough to construct a path between two arbitrary points.

9.73 Examples.

(a) Consider the grid constituted by the integer points in $\mathbb{R}^2$ and the horizontal and vertical edges; see Fig. 9.24(a). Two points are called *4-neighbours* if they are connected by a horizontal or vertical edge. The resulting metric is the (discrete version of the) *city block distance*:

$$d_1(x, y) = |x_1 - y_1| + |x_2 - y_2|,$$

if $x = (x_1, x_2)$ and $y = (y_1, y_2)$.

(b) If the grid contains diagonal edges also (Fig. 9.24(b)), then one obtains the *chessboard distance*

$$d_\infty(x, y) = \max\{\,|x_1 - y_1|, |x_2 - y_2|\,\},$$

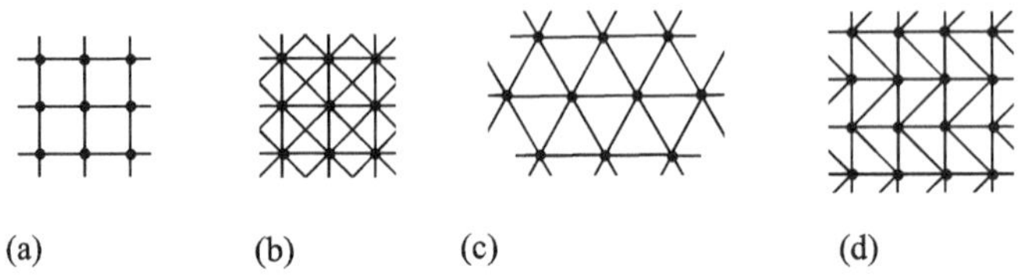

Figure 9.24 (a) 4-connected grid; (b) 8-connected grid; (c) hexagonal grid; (d) "simulated" hexagonal grid.

if $x = (x_1, x_2)$ and $y = (y_1, y_2)$. Two points are called *8-neighbours* if they are connected by a horizontal, vertical, or diagonal edge.

(c) The hexagonal grid, depicted in Fig. 9.24(c), comprises the points $nu + mv$ where $u = (1, 0)$, $v = (1/2, \sqrt{3}/2)$, and $n, m \in \mathbb{Z}$. Two points are connected by an edge if their Euclidean distance is 1. Thus every point has six neighbours. The corresponding distance is called *hexagonal distance*. Alternatively, one can impose a hexagonal grid on the integer points as in Fig. 9.24(d). Note that this "simulated" grid transforms into the "true" hexagonal grid by an affine transformation.

Discrete metrics deriving from a grid structure form a rather restricted class, and in general they give poor approximations of the Euclidean distance. This can be improved by distinguishing between the different edges constituting a path. For instance, on the 8-connected grid the horizontal and vertical edges are shorter than the diagonal ones. In fact these lengths differ by a factor $\sqrt{2}$. Merely a graph structure to model neighbour relations does not allow us to make such distinctions.

This section discusses an approach which is more general than the one just discussed. The central idea is to use a metric d on E to define neighbour relations. Recall that a point z is said to lie between x and y if $z \neq x, y$ and

$$d(x, y) = d(x, z) + d(z, y).$$

The betweenness relation can be used as a basis for the definition of neighbours.

9.74 Definition. Let (E, d) be a discrete metric space. Two points are called *d-neighbours* if there exists no point between x and y.

Note that every point is its own d-neighbour. If E contains at least two points, then every element has at least one d-neighbour other than itself. A sequence $x_1, x_2, \ldots, x_n$ in E is called a *d-path* if x_i and x_{i+1} are

d-neighbours for $i = 1, 2, \ldots, n-1$. The length of this path is given by $\sum_{i=1}^{n-1} d(x_i, x_{i+1})$. A set $X \subseteq E$ is said to be *d-connected* if for every two distinct points in X there exists a d-path between them. The reader may readily verify that this defines a connectivity class on $\mathcal{P}(E)$; see Definition 9.60. It is also obvious that for every two points x, y there must exist a shortest d-path between them; the length of this shortest path equals $d(x, y)$.

9.75 Remark. Example 7.8 introduced *adjacency relations* as a means of defining a topological space. Recall that $\triangleright$ is an adjacency relation on $E \times E$ if it is reflexive, that is, $x \triangleright x$ for $x \in E$. It is obvious that the neighbour relation just introduced is an adjacency relation, for every point is its own d-neighbour. Not every adjacency relation is induced by a discrete metric, however. In fact, the adjacency relation associated with the Khalimskii topology (see Example 7.8) is not symmetric, and thus it cannot be induced by a metric.

9.76 Example. The function $d(x, y) = \left((x_1 - y_1)^2 + (x_2 - y_2)^2\right)^{1/2}$ defines a discrete metric on $\mathbb{Z}^2$ (this follows by the fact that d is the restriction of the Euclidean metric). Two points x, y are d-neighbours if the open line segment connecting x and y does not contain points with integer coordinates. In particular, the integer points on a (possibly infinite) line segment constitute a d-connected set.

9.77 Example. (Morphology on graphs)
Let E be the set of nodes of some undirected graph G. Call $x_1, x_2, \ldots, x_n$ a path if there lies an edge between x_i and x_{i+1} for every $i = 1, 2, \ldots, n-1$. The length of this path is $n-1$. The distance $d(x, y)$ between two points x, y is the length of the shortest path between x and y. It is obvious that d defines a metric on E and that the notion of d-connectivity resulting from this metric coincides with the connectivity imposed by the graph structure. The metric dilation δ^1 maps a set X onto the set $X \cup X^1$, where X^1 contains the points in E which are connected to a point of X by some edge of the graph. Both δ^1 and its adjoint erosion ε^1 are illustrated in Fig. 9.25.

The metric dilations δ^r, with $r \geq 1$, can be obtained by r-fold iteration of δ^1. Vincent (1989) uses these definitions to extend morphological operators (viz. granulometries, skeletons, geodesic operators) to graph-based images. This theory fits perfectly well in the metric approach outlined here.

A popular class of distances in image processing are the *chamfer distances*. To introduce them we use the cost function approach of Section 9.4. Let

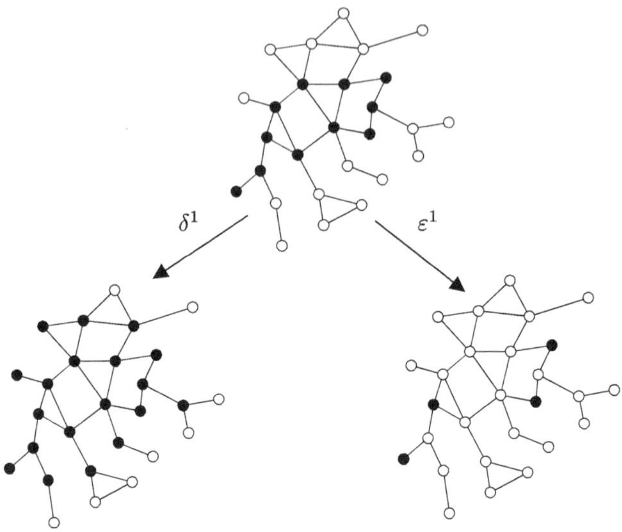

Figure 9.25 Metric dilation and erosion on a graph.

$c(x, y)$ be a cost function on $\mathbb{Z}^2$ which is translation invariant:

$$c(x+h, y+h) = c(x, y).$$

The function c is completely determined by the entries $c(0, h)$, the costs of a transition from the origin 0 to the point h. In practice one considers only cost functions with a bounded support (the value outside the support being ∞) and symmetric under rotations. Some examples are depicted in Fig. 9.26.

For computational reasons one usually restricts oneself to integer entries. The corresponding distance can be found by infinite convolution of c; that is, $d = c^{*\infty}$. An example for the 5-7-chamfer distance is given in Fig. 9.27.

Note that after division by 5 the resulting values approximate Euclidean distance. This is due to the fact that $5^2 + 5^2 = 50$, which is almost $49 = 7^2$. The reader may readily verify that in this particular example d-connectivity is the same as 8-connectivity. Furthermore, we point out that the metric dilations of Section 9.3 do not satisfy the semigroup property in this particular example. In fact, the distance between the points $(0, 0)$ and $(5, 5)$ equals 35, which is $20 + 15$. But $(5, 5) \notin \delta^{15}\big(\delta^{20}(\{(0, 0)\})\big)$, as the reader can readily check. This means also that the openings $\alpha_r = \delta^r \varepsilon^r$ cannot be used to define granulometries. Otherwise stated, the balls $B(r)$ with centre $(0, 0)$

	1	
1	0	1
	1	

1	1	1
1	0	1
1	1	1

√2	1	√2
1	0	1
√2	1	√2

4	3	4
3	0	3
4	3	4

7	5	7
5	0	5
7	5	7

	11		11	
11	7	5	7	11
	5	0	5	
11	7	5	7	11
	11		11	

Figure 9.26 Some chamfer cost functions. The first and the second one correspond to city block distance and chessboard distance, respectively.

35	33	31	29	27	25	27	29	31	33	35
33	28	26	24	22	20	22	24	26	28	33
31	26	21	19	17	15	17	19	21	26	31
29	24	19	14	12	10	12	14	19	24	29
27	22	17	12	7	5	7	12	17	22	27
25	20	15	10	5	0	5	10	15	20	25
27	22	17	12	7	5	7	12	17	22	27
29	24	19	14	12	10	12	14	19	24	29
31	26	21	19	17	15	17	19	21	26	31
33	28	26	24	22	20	22	24	26	28	33
35	33	31	29	27	25	27	29	31	33	35

Figure 9.27 5-7-chamfer distance. An approximation of the Euclidean distance is obtained after division by 5. The grey-shaded set contains an approximation of the Euclidean disk with radius 5.

and radius r do not have the property

$$B(s) \text{ is } B(r)\text{-open for } s \geq r. \tag{9.72}$$

Note, however, that $\bigvee_{s \geq r} \delta^s \varepsilon^s$ defines a granulometry. In some simple examples, e.g., the city block distance or chessboard distance, one has

$$B(r) = B(1) \oplus \cdots \oplus B(1) \qquad [r \text{ terms}],$$

in which case relation (9.72) is valid. Furthermore, a general method to obtain discrete granulometries is by means of the formula

$$\alpha_r = \delta^r \varepsilon^r,$$

where (ε, δ) is an adjunction and δ^r is the rth iterate of δ (same for ε).

If $M \subseteq X$, one can define a geodesic distance d_M on M analogously to the continuous case. Define $d_M(x, y)$ to be the length of the shortest d-path in M; if x, y lie in different d-connected components, then $d_M(x, y) = \infty$. Now, we can also introduce geodesic operators such as geodesic dilation, erosion, reconstruction, ultimate erosion, etc.

9.10. Bibliographical notes

The operators discussed in this chapter are essential for image analysis purposes. Besides publications by various members of the Fontainebleau school cited later in this section, an importance reference, especially for those readers interested in quantitative aspects of morphology, is the book by Coster and Chermant (1985).

An introduction to the theory of convex sets is given by Valentine (1964) and Marti (1977). A very enlightening, recent treatise is given by Schneider (1993).

The results stated in Proposition 9.8–Theorem 9.17 are due to Matheron (1975). Some related results in the context of integral geometry can be found in Hadwiger (1957).

Major references for the theory discussed in Section 9.2 are Blumenthal (1953), Busemann (1955) and Rinow (1961). The notion of M-convexity goes back to the work of Menger (1928). For a proof of Proposition 9.24 refer to Busemann (1955, (5.18)) and Rinow (1961, §17). Rinow calls d_E "die innere Metrik"; properties (9.12)–(9.14) can be found in Rinow (1961, §15). The proof of Proposition 9.27 can be found in Rinow (1961, §15-6). Furthermore, in Rinow (1961, §18-8,9) one finds a proof of Proposition 9.29. The flowershop distance appears for the first time in Klein (1989, pp. 24). In this reference, which comprises a study of general Voronoi diagrams, one also finds several interesting results on metric spaces.

The idea to describe a metric in terms of a one-parameter family of (metric) dilations is essentially due to Serra (1988, Section 2.4). Proposition 9.35 and Theorem 9.37 have not yet appeared elsewhere in the literature. The characterization of the Hausdorff metric given in Example 9.38 is well-known; a nice sketch appears in Hadwiger (1957, Kapitel 4), who calls $X \oplus rB$ "die äußere Parallelmenge" of X with distance r. See also Valentine (1964, Section III.B).

The distance transform is a well-known tool in image processing; it was introduced by Rosenfeld and Pfalz (1966, 1968). A detailed exposition is given by Verwer (1991). The discussion presented in Section 9.4 consists partially of new material. The construction of a metric through iteration of cost functions is based on work by Ronse and Heijmans (1991).

The utilization of geodesic and conditional operators for image analysis purposes is originally due to Lantuéjoul and Beucher (1980) and Lantuéjoul and Maisonneuve (1984). Since the appearance of these papers, these operators show up in various papers of members of the Fontainebleau school, e.g., as an auxiliary tool in segmentation algorithms; refer to Schmitt and Vincent (2010) for a recent treatment. Some theoretical results can be found in Serra (1988, Sections 4.5–4.6). A discussion of the ultimate erosion is presented in Serra (1982, Chapter XI).

An exhaustive treatment of granulometries is given by Matheron (1975). In fact, most of the theory discussed in Section 9.6 can be found there. An overview of Matheron's main results as well as some real-world applications are presented in Serra (1982, Chapter X). We point out that both Matheron and Serra use the terminology "Euclidean granulometry" instead of "Minkowski granulometry". We consider the latter terminology more appropriate, however, and propose to keep the adjective Euclidean for granulometries which use Euclidean balls as structuring elements. Maragos (1989a) uses structural granulometries $X \mapsto X \circ rA$, where A is convex, to define the *pattern spectrum*:

$$PS_X(r, A) = -\frac{d}{dr}\,\text{area}(X \circ rA), \qquad r \geq 0.$$

Using closings, this can be extended to negative r:

$$PS_X(-r, A) = \frac{d}{dr}\,\text{area}(X \bullet rA), \qquad r > 0.$$

Maragos argues that the pattern spectrum conveys various types of shape information. Mattioli (1993) and Mattioli and Schmitt (1992) provide some beautiful results about shape information contained in granulometric descriptions.

Introducing the notion of *Stieltjes–Minkowski integral* in the space of compact sets supplied with the Hausdorff metric, Matheron (1975) was able to construct a large class of families of structuring elements $\{A(r) \mid r > 0\}$ such that $A(s)$ is $A(r)$-open for $s \geq r$. The family in Example 9.53 is nothing but a special member of this class. We have seen that $\tilde{\alpha}_r = \bigvee_{s>r} \alpha_r$ defines a granulometry for an arbitrary one-parameter family of openings $\{\alpha_r\}$ on $\mathcal{P}(E)$. One can compute this opening $\tilde{\alpha}_r$ by introducing the *opening transform* $A(X)$ of a set X:

$$A(X, h) = \inf\{r > 0 \mid h \notin \alpha_r(X)\}.$$

Then the thresholded set at level r equals $\tilde{\alpha}_r(X)$, that is,

$$\tilde{\alpha}_r(X) = \{h \in E \mid A(X, h) \geq r\}.$$

A fast algorithm for computation of the opening transform in the case of the chamfer metric has been given by Nacken (1993).

Connectivity openings were introduced by Serra (1988, Section 2.6); refer to Ronse (1994b) for a more general discussion. Example 9.62 is due to Serra (1988, pp. 55ff); it has been generalized by Ronse (1994b).

There are several alternative ways to define the skeleton of a continuous set. In the literature referred to hereafter, the closed unit ball with respect to Euclidean distance is used as structuring element, and in this case we use the terminology *skeleton* instead of *A-skeleton*. In his pioneering work, Blum (1967) visualized the skeleton (or *medial axis* as he called it there; later Blum (1973) introduced the term *symmetric axis*) by means of the following metaphor: think of a set X as a dry prairie and set the grass at the edge of X afire. The resulting fires propagate toward the centre of the object X according to Huygen's principle. The medial axis is the set of points where the wavefronts intersect; the arrival times define the *medial axis function* (quench function in our terminology). The mathematical theory of skeletons was carried further in a paper by Calabi and Hartnett (1968). There one can also find the following results.

9.78 Theorem. *Let $X \subseteq \mathbb{R}^d$ be nonempty and closed. The following statements are equivalent:*

(i) *for each $x \in \mathbb{R}^d$ there is a unique nearest point on X;*

(ii) $\Sigma(X^c) = \varnothing$;

(iii) *X is convex.*

9.79 Theorem. *A closed set is uniquely determined by its convex hull and the skeletal pair $(\Sigma(X^c), q(X^c, \cdot))$ of the complement.*

The equivalence of (i) and (ii) in Theorem 9.78 is also known as *Motzkin's theorem* (Motzkin, 1935a, 1935b). Valentine (1964, Theorem 7.8) extended Motzkin's theorem to Minkowski spaces with a unit ball which is smooth and strictly convex. Observe that in both results the skeleton of the complement of X has to be computed. In the literature, this set is also known as the *exoskeleton* (Serra, 1982). The definition of the skeleton in terms of morphological operators is due to Lantuéjoul (Lantuéjoul & La, 1978); see also Serra (1982, Section XI.B). Major contributions to its theoretical foundations were made by Matheron (Serra, 1988, Chapters 11–12). Both Lantuéjoul and Matheron restrict attention to open subsets in their treatment of the skeleton. An advantage of this restriction is that the skeleton operator Σ is lower semi-continuous. From an operator-theoretic point of view, however, the restriction to the open sets has an undesirable drawback: it does not allow one to apply Σ to sets with empty interior, which are their own skeleton. In particular, Corollary 9.69 has no analogue in the framework of open sets. Recently, Serra (1991) presented a formal approach based on one-parameter families of dilations.

The definition of the discrete skeleton given at the end of Section 9.8 has some serious drawbacks. The most severe one is that it does not preserve connectivity. Note, however, that the same objection can be raised against the continuous skeleton; cf. Example 9.66. This unpleasant fact has caused an enormous proliferation of skeletonization algorithms in the literature which attempt to overcome this problem. Refer in particular to Meyer (1991) and Niblack et al. (1990). A simple (recursive) algorithm using thinnings has been given in Example 4.7.

Till recently, discrete geometry was a poorly developed area in mathematics, but with the advent of the computer era substantial progress has been made. A nice overview of the state of the art is contained in the collection (Melter et al., 1991). Our definition of d-connectivity for discrete metric spaces does not agree with the topological definition of path-connectivity. For in the discrete topology, the only subsets of E that are path-connected are the singletons. The discrete topology is not very useful for image processing. Therefore, one finds many papers in the literature containing alternative definitions. Consult Section 7.8 for some relevant literature.

For further reading on graph morphology the reader should consult the work of Vincent (1989, 1990), Heijmans et al. (1992), and the overview paper by Heijmans and Vincent (1993).

Chamfer distances were introduced by Borgefors (1984, 1986). An important contribution to this subject was made recently by Nacken (1993).

In this chapter, convexity takes a profound place. The attentive reader may have noticed that we have forsaken to define *discrete convexity*. In the literature one can find a large variety of definitions of discrete convexity. We refer to Ronse (1989a) for an extensive bibliography. Recently, Schmitt (1989) and Mattioli (1993) have obtained various interesting results on discrete convexity which are based on non-Euclidean metrics.

CHAPTER TEN

Lattice representations of functions

Henk J.A.M. Heijmans
Formerly Centre for Mathematics and Computer Science, Amsterdam, Netherlands

Contents

This chapter shows how one can extend set operators to function operators and hence apply the whole apparatus of binary morphology in the grey-scale case as well. The underlying idea is (i) to represent a function as a family of sets, (ii) to apply the set operator to this family, and (iii) to reconstruct a transformed function from the family of transformed sets; cf. Section 10.6. This chapter, which is mainly concerned with the first step, describes two power-type lattices and discusses isomorphisms between these lattices and function lattices. We point out that this procedure to extend binary morphological operators to grey-scale functions is a well-known construction, which has proved very useful in many situations.

The theory established in this chapter adopts the complete lattice framework. The advantage of such an abstract approach is that it applies to many practical examples. The domain space may be discrete as well as continuous, the grey-value set may be discrete, continuous, bounded, or unbounded. Furthermore, the given approach applies also to upper and lower semi-continuous functions. In the next chapter the abstract results will be illustrated by several concrete examples.

Advances in Imaging and Electron Physics, Volume 216
ISSN 1076-5670
https://doi.org/10.1016/bs.aiep.2020.07.010

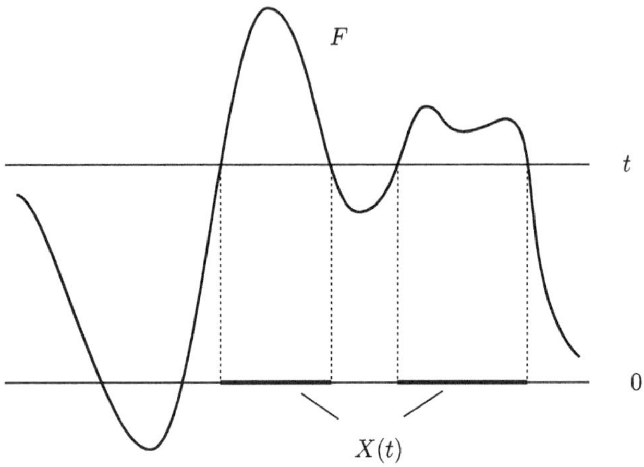

Figure 10.1 Threshold set $X(t) = \{x \in E \mid F(x) \geq t\}$.

10.1. Introduction

To get some intuition for the problems that we have addressed, consider the complete lattice $\mathrm{Fun}(E, \overline{\mathbb{R}})$. Associate with a function $F : E \to \overline{\mathbb{R}}$ its threshold sets

$$X(t) = \{x \in E \mid F(x) \geq t\}; \tag{10.1}$$

see Fig. 10.1.

These sets obey the continuity condition

$$X(t) = \bigcap_{s<t} X(s). \tag{10.2}$$

Note in particular that $X(t)$ is decreasing in t. There is a one–one correspondence between functions $F \in \mathrm{Fun}(E, \overline{\mathbb{R}})$ and families of sets $X(t)$ satisfying (10.2). In fact, the function F can be recovered from the family $X(t)$ by means of

$$F(x) = \sup\{t \in \overline{\mathbb{R}} \mid x \in X(t)\}. \tag{10.3}$$

The identities (10.1) and (10.3) define an isomorphism between the complete lattice of set-valued functions $X(\cdot)$ satisfying (10.2) and the function lattice $\mathrm{Fun}(E, \overline{\mathbb{R}})$; for a general statement we refer to Theorem 10.9. If one replaces the grey-value set $\overline{\mathbb{R}}$ by $\overline{\mathbb{Z}}$, then (10.2) is not true unless $t = \pm\infty$;

nevertheless the statement about the isomorphism between the lattice of set-valued functions and the function lattice $\mathrm{Fun}(E, \overline{\mathbb{Z}})$ remains valid. If one allows more general grey-value sets (e.g., 3-dimensional sets for the representation of colour images), the continuity requirement (10.2) gives rise to some problems. These problems will be addressed presently.

We define yet another family of threshold sets, namely,

$$Y(t) = \{x \in E \mid F(x) > t\}. \tag{10.4}$$

These sets satisfy

$$Y(t) = \bigcup_{s>t} Y(s). \tag{10.5}$$

The reader should notice carefully the duality between conditions (10.2) and (10.5). It is easy to check that F can be reconstructed from Y as follows:

$$F(x) = \inf\{t \in \overline{\mathbb{R}} \mid x \notin Y(t)\}. \tag{10.6}$$

Furthermore, X and Y are related to each other in the following sense:

$$Y(t) = \bigcup_{s>t} X(s), \tag{10.7}$$

$$X(t) = \bigcap_{s<t} Y(s). \tag{10.8}$$

Again, we point out the duality between these two relations. This chapter contains a comprehensive discussion of the mappings $X(\cdot)$ and $Y(\cdot)$ and their interrelations. We do not make any a priori assumption about the domain and range of these mappings. To begin with, X and Y are mappings from one complete lattice $\mathcal{T}$ into another, $\mathcal{L}$. It will be shown, however, that the continuity conditions (10.2) and (10.5) impose a restriction on the latticc $\mathcal{T}$.

10.2. Admissible complete lattices

What precisely is the role of the continuity condition (10.2)? If one chooses $\overline{\mathbb{Z}}$ as the set of grey-values, then it is only satisfied for $t = \pm\infty$; note that $X(-\infty) = E$, which, by definition, equals the empty intersection. If, on the other hand, the grey-value set is $\overline{\mathbb{R}}$, then (10.2) also holds for finite values of t since every such t is the supremum of all s below t. The situation becomes more complex when higher-dimensional grey-value sets, such as $\overline{\mathbb{Z}}^d$ and $\overline{\mathbb{R}}^d$ with the partial ordering of the product (see Example 2.11), are considered. Such generalizations are important if one wants to extend morphology to colour images. How to interpret (10.2) in this case? Also, for which grey-value sets does there exist a one–one correspondence between functions on the one hand and families of (threshold) sets on the other? Before we can handle such problems we need some additional concepts.

Let, throughout this chapter, $\mathcal{T}$ be a complete lattice; we refer to this lattice as *grey-value lattice* or *grey-value set*.

10.1 Definition. Given $s, t \in \mathcal{T}$, we say that s is *below* t, written $s \ll t$, if for every family t_i, $i \in I$, with $\bigvee_{i\in I} t_i \geq t$ there exists an $i \in I$ such that $t_i \geq s$. Dually, we say that s is *above* t, denoted by $s \gg t$, if for every family t_i, $i \in I$, with $\bigwedge_{i\in I} t_i \leq t$ there exists an $i \in I$ such that $t_i \leq s$.

Observe that the relation $\gg$ is the opposite of $\ll$ in the sense of the Duality Principle. Note also that

$$O \not\ll O \qquad \text{and} \qquad I \not\gg I.$$

The first relation, for example, is based on the observation that the supremum of the empty family is O and that this family does not contain an element $\geq O$. Furthermore,

$$O \ll t \text{ if } t \neq O \qquad \text{and} \qquad I \gg t \text{ if } t \neq I.$$

In the foregoing definition one can choose the family containing only t; this leads to

$$s \ll t \Rightarrow s \leq t,$$
$$s \gg t \Rightarrow s \geq t.$$

If $\mathcal{T}$ is an atomic lattice, then $s \ll t$ if and only if s is an atom and $s \leq t$.

We present some examples.

10.2 Examples.

(a) $\mathcal{T} = \overline{\mathbb{R}}$. For all $s, t \in \overline{\mathbb{R}}$:

$$s \ll t \iff t \gg s \iff s < t.$$

(b) $\mathcal{T} = \overline{\mathbb{Z}}$. For finite s, t we have $s \ll t \iff t \gg s \iff s \leq t$. Furthermore, $s \ll \infty$ for finite s, but $\infty \not\ll \infty$. Dually, $s \gg -\infty$ for finite s, but $-\infty \not\gg -\infty$

Figure 10.2 Definition of the set $\mathcal{T}$ in example (d).

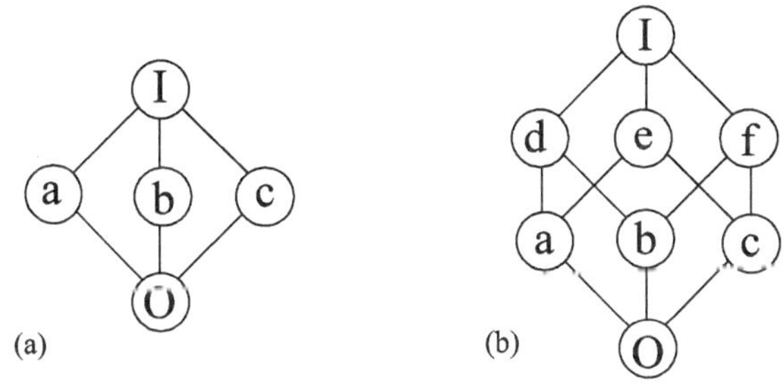

Figure 10.3 Two grey-value lattices.

(c) $\mathcal{T} = \{0, 1, \ldots, N\}$. For all s, t with $s \neq N,\ t \neq 0$ we have $s \ll t$ if and only if $t \gg s$ if and only if $s \leq t$. Furthermore, $0 \gg 0$ but $0 \not\ll 0$ and $N \ll N$ but $N \not\gg N$.

(d) $\mathcal{T} = \{-\infty\} \cup \{\ldots, -3, -2, -1\} \cup [0, \infty]$ (see Fig. 10.2). For $t < 0$ and $t > 0$ the validity of the relations $s \ll t$ and $s \gg t$ follows from the examples in (a) and (b), respectively. Consider the value $t = 0$. Since 0 is an upper but not a lower limit point, it follows that $0 \ll 0$ but $0 \not\gg 0$.

(e) $\mathcal{T} = \overline{\mathbb{R}} \times \overline{\mathbb{R}}$ with the product ordering. We have $(s_1, s_2) \ll (t_1, t_2)$ if either $s_2 = -\infty$ and $s_1 < t_1$ or $s_1 = -\infty$ and $s_2 < t_2$.

(f) $\mathcal{T} = \overline{\mathbb{Z}} \times \overline{\mathbb{Z}}$ with the product ordering. We have $(s_1, s_2) \ll (t_1, t_2)$ if either $s_2 = -\infty$ and $s_1 \leq t_1$ or $s_1 = -\infty$ and $s_2 \leq t_2$. However, $(-\infty, -\infty) \not\ll (-\infty, -\infty)$.

(g) $\mathcal{T} = \{0, 1, \ldots, N\} \times \{0, 1, \ldots, N\}$ with the product ordering. We have $(s_1, s_2) \ll (t_1, t_2)$ if either $s_2 = 0$ and $s_1 \leq t_1$ or $s_1 = 0$ and $s_2 \leq t_2$. However, $(0, 0) \not\ll (0, 0)$.

(h) $\mathcal{T} = \overline{\mathbb{R}} \times \overline{\mathbb{Z}}$ with the product ordering. We have $(s_1, s_2) \ll (t_1, t_2)$ if either $s_2 = -\infty$ and $s_1 < t_1$ or $s_1 = -\infty$ and $s_2 \leq t_2$. However, $(-\infty, -\infty) \not\ll (-\infty, -\infty)$.

(i) Consider the nondistributive lattice $\mathcal{T} = \{O, a, b, c, I\}$ represented by the Hasse diagram in Fig. 10.3(a); see also Example 2.8.

Note that in this case $a \not\ll I$. In fact, $b \vee c = I$, but neither of the elements b, c is greater than or equal to a. Analogously, $b \not\ll I$ and $c \not\ll I$. Therefore, the only element t with $t \ll I$ is $t = O$. Dually, the only element t with $t \gg O$ is $t = I$.

(j) Consider the lattice $\mathcal{T} = \{O, a, b, c, d, e, f, I\}$ represented by the Hasse diagram in Fig. 10.3(b). It is easy to show that $a, b, c \ll I$, that $d, e, f \gg O$, that $a, b \ll d$, etc. In fact, this diagram is a graphical representation of the product lattice $\{0, 1\} \times \{0, 1\} \times \{0, 1\}$.

The following properties are trivial.

10.3 Lemma. *Let* $s, s', t, t' \in \mathcal{T}$.

(a) $s' \leq s \ll t \leq t'$ *implies that* $s' \ll t'$.

(a′) $t' \geq t \gg s \geq s'$ *implies that* $t' \gg s'$.

10.4 Definition. A complete lattice $\mathcal{T}$ is called *admissible* if, for $s, t \in \mathcal{T}$,

(i) $t = \bigvee\{s \in \mathcal{T} \mid s \ll t\} = \bigwedge\{s \in \mathcal{T} \mid s \gg t\}$;

(ii) $s \ll t$ implies that $s \ll r \ll t$ for some $r \in \mathcal{T}$; $t \gg s$ implies that $t \gg r \gg s$ for some $r \in \mathcal{T}$.

If, in addition, the symmetry property

(iii) $s \ll t \iff t \gg s$, if $s \neq I$, $t \neq O$,

holds, then $\mathcal{T}$ is called *strongly admissible*.

Here $s \ll r \ll t$ means that $s \ll r$ and $r \ll t$. Property (ii) is called *interpolation property*.

All lattices discussed in Example 10.2, apart from (i), are admissible. The lattice in (i) is not admissible since $\bigvee\{s \mid s \ll I\} = O \neq I$. Furthermore, in this example, the interpolation property does not hold because $O \ll I$ but there does not exist any t with $O \ll t \ll I$. The lattices in (a)–(c) and in (j) are strongly admissible.

The following result is stated for later reference.

10.5 Lemma. *Let* $\mathcal{L}, \mathcal{T}$ *be complete lattices, and assume that* $\mathcal{T}$ *is admissible. For every mapping* $X : \mathcal{T} \to \mathcal{L}$ *the following holds:*

$$\bigwedge_{s \ll t} X(s) = \bigwedge_{s \ll t} \bigwedge_{r \ll s} X(r), \tag{10.9}$$

$$\bigvee_{s \gg t} X(s) = \bigvee_{s \gg t} \bigvee_{r \gg s} X(r). \tag{10.10}$$

Proof. Only the first identity will be proved. Since every element $X(r)$ on the right-hand side occurs also on the left-hand side, one gets immediately that $\bigwedge_{s \ll t} X(s) \leq \bigwedge_{s \ll t} \bigwedge_{r \ll s} X(r)$. If $s \ll t$, then, by the interpolation property, there is an $r \in \mathcal{T}$ such that $s \ll r \ll t$. Since $s \leq r$ we have $X(r) \leq X(s)$.

Thus to every element $X(s)$ on the left-hand side, there corresponds an element $X(r)$ on the right-hand side such that $X(r) \leq X(s)$. This proves the reverse inequality. □

10.3. Power-type lattices

Throughout this section we assume that $\mathcal{T}$ is an admissible lattice. Let $\mathcal{L}$ be an arbitrary complete lattice, and let $\mathcal{L}^{\mathcal{T}}$ be the power lattice containing all mappings from $\mathcal{T}$ into $\mathcal{L}$; we denote such mappings by bold symbols such as $\mathbf{X}, \mathbf{Y}$. The space $\mathcal{L}^{\mathcal{T}}$ with the partial ordering

$$\mathbf{X} \leq \mathbf{Y} \iff \forall t \in \mathcal{T} : \mathbf{X}(t) \leq \mathbf{Y}(t),$$

is a complete lattice under the pointwise supremum and infimum. Define the subset $\mathcal{L}\diamond\mathcal{T}$ by

$$\mathcal{L}\diamond\mathcal{T} = \{X \in \mathcal{L}^{\mathcal{T}} \mid X \text{ is decreasing}\}.$$

It is evident that $\mathcal{L}\diamond\mathcal{T}$ is a complete sublattice of $\mathcal{L}^{\mathcal{T}}$. Furthermore, one has the duality relation

$$(\mathcal{L}\diamond\mathcal{T})' \simeq \mathcal{L}'\diamond\mathcal{T}', \tag{10.11}$$

where $\mathcal{L}'$ is the opposite of $\mathcal{L}$ in the sense of the Duality Principle. Define $\mathcal{L}_{\vartriangle}\mathcal{T}$ as the set of all mappings $\mathbf{X} \in \mathcal{L}\diamond\mathcal{T}$ that satisfy

$$\mathbf{X}(\bigvee_{i\in I} t_i) = \bigwedge_{i\in I} \mathbf{X}(t_i), \tag{10.12}$$

for every family $\{t_i \mid i \in I\}$. In particular, $\mathbf{X}(O) = I$. Dually, define $\mathcal{L}_{\triangledown}\mathcal{T}$ as the set of all mappings $\mathbf{X} \in \mathcal{L}\diamond\mathcal{T}$ that satisfy

$$\mathbf{X}(\bigwedge_{i\in I} t_i) = \bigvee_{i\in I} \mathbf{X}(t_i). \tag{10.13}$$

In particular, $\mathbf{X}(I) = O$.

A mapping between two complete lattices which satisfies (10.12) (resp. (10.13)) is sometimes called an *anti-dilation* (resp. *anti-erosion*). With the partial ordering induced by $\mathcal{L}\diamond\mathcal{T}$, the set $\mathcal{L}_{\vartriangle}\mathcal{T}$ becomes a poset which has the same infimum as $\mathcal{L}\diamond\mathcal{T}$. Furthermore, the function identically I defines a greatest element; now it follows from Proposition 2.12 that $\mathcal{L}_{\vartriangle}\mathcal{T}$ is a complete lattice. Dually, $\mathcal{L}_{\triangledown}\mathcal{T}$ is a complete lattice with the same supremum as $\mathcal{L}\diamond\mathcal{T}$. Throughout this chapter, $\bigvee$ denotes the supremum on $\mathcal{L}$ or on $\mathcal{L}\diamond\mathcal{T}$ depending on whether the argument lies in $\mathcal{L}$ or in $\mathcal{L}\diamond\mathcal{T}$. Similarly, $\bigwedge$ denotes the infimum in $\mathcal{L}$ or $\mathcal{L}\diamond\mathcal{T}$. One can easily establish the duality relation

$$(\mathcal{L}_{\vartriangle}\mathcal{T})' \simeq \mathcal{L}'_{\triangledown}\mathcal{T}'.$$

So far, we have only used that $\mathcal{T}$ is a complete lattice. If also the admissibility of $\mathcal{T}$ is taken into account, a lot more can be said about the structure of the complete lattices $\mathcal{L}_{\triangledown}\mathcal{T}$ and $\mathcal{L}_{\vartriangle}\mathcal{T}$. Define the mappings $\check{\mathbf{I}}, \hat{\mathbf{I}} : \mathcal{L}\diamond\mathcal{T} \to \mathcal{L}\diamond\mathcal{T}$ by

$$(\check{\mathbf{I}}\mathbf{X})(t) = \bigvee_{s \gg t} \mathbf{X}(s), \tag{10.14}$$

$$(\hat{\mathbf{I}}\mathbf{X})(t) = \bigwedge_{s \ll t} \mathbf{X}(s). \tag{10.15}$$

In fact, $\check{\mathbf{I}}, \hat{\mathbf{I}}$ are mappings from the larger complete lattice $\mathcal{L}^{\mathcal{T}}$ to $\mathcal{L}\diamond\mathcal{T}$. For our goal, however, the lattice $\mathcal{L}^{\mathcal{T}}$ is no longer important, and henceforth we restrict ourselves to $\mathcal{L}\diamond\mathcal{T}$.

10.6 Proposition. *Assume that $\mathcal{T}$ is an admissible complete lattice.*

(a) *$\hat{\mathbf{I}}$ is a closing with invariance domain $\mathcal{L}_{\vartriangle}\mathcal{T}$.*

(a′) *$\check{\mathbf{I}}$ is an opening with invariance domain $\mathcal{L}_{\triangledown}\mathcal{T}$.*

Assume in addition that $\mathcal{T}$ is strongly admissible.

(b) *The pair of operators $(\hat{\mathbf{I}}, \check{\mathbf{I}})$ defines an adjunction on $\mathcal{L}\diamond\mathcal{T}$. Furthermore,*

$$\hat{\mathbf{I}} = \hat{\mathbf{I}}^2 = \hat{\mathbf{I}}\check{\mathbf{I}}, \tag{10.16}$$

$$\check{\mathbf{I}} = \check{\mathbf{I}}^2 = \check{\mathbf{I}}\hat{\mathbf{I}}. \tag{10.17}$$

Proof. (a): It is obvious that $\hat{\mathbf{I}}$ is increasing. Since $s \ll t$ implies that $s \le t$; hence $\mathbf{X}(s) \ge \mathbf{X}(t)$. Thus it follows immediately that $\hat{\mathbf{I}}$ is extensive. That $\hat{\mathbf{I}}^2 = \hat{\mathbf{I}}$ follows easily if one uses Lemma 10.5. This implies that $\hat{\mathbf{I}}$ is a closing.

We must show that the invariance domain of $\hat{\mathbf{I}}$ is $\mathcal{L}_{\vartriangle}\mathcal{T}$. First, take $\mathbf{X} \in \mathcal{L}_{\vartriangle}\mathcal{T}$. By the very definition of $\mathcal{L}_{\vartriangle}\mathcal{T}$ we get $\mathbf{X}(t) = \mathbf{X}(\bigvee_{s\ll t} s) = \bigwedge_{s\ll t}\mathbf{X}(s) = \hat{\mathbf{I}}\mathbf{X}(t)$. On the other hand, assume that $\hat{\mathbf{I}}\mathbf{X} = \mathbf{X}$, that is, $\bigwedge_{s\ll t}\mathbf{X}(s) = \mathbf{X}(t)$. Let $t_i \in \mathcal{T}$ for $i \in I$ and $t = \bigvee_{i\in I} t_i$. For every $s \ll t$, there is an $i(s) \in I$ such that $t_{i(s)} \ge s$. Then $\mathbf{X}(t) = \bigwedge_{s\ll t}\mathbf{X}(s) \ge \bigwedge_{s\ll t}\mathbf{X}(t_{i(s)}) \ge \bigwedge_{i\in I}\mathbf{X}(t_i)$. The reverse inequality is trivial. This concludes the proof of (a).

(b): We must show that for $\mathbf{X}, \mathbf{Y} \in \mathcal{L}\diamond\mathcal{T}$,

$$\check{\mathbf{I}}\mathbf{X} \leq \mathbf{Y} \iff \mathbf{X} \leq \hat{\mathbf{I}}\mathbf{Y}.$$

Clearly, $\check{\mathbf{I}}\mathbf{X} \leq \mathbf{Y}$ means that $\bigvee_{s \gg t} \mathbf{X}(s) \leq \mathbf{Y}(t)$, for every $t \in \mathcal{T}$. This means that $\mathbf{X}(s) \leq \mathbf{Y}(t)$ if $s \gg t$, or equivalently, if $t \ll s$. But this gives $\mathbf{X}(s) \leq \bigwedge_{t \ll s} \mathbf{Y}(t) = (\hat{\mathbf{I}}\mathbf{Y})(s)$, i.e., $\mathbf{X} \leq \hat{\mathbf{I}}\mathbf{Y}$. The other implication is proved analogously.

It remains to prove that $\hat{\mathbf{I}} = \hat{\mathbf{I}}\check{\mathbf{I}}$ and that $\check{\mathbf{I}} = \check{\mathbf{I}}\hat{\mathbf{I}}$. We only prove the first identity. Since $\check{\mathbf{I}}$ is an opening, it follows immediately that $\check{\mathbf{I}}\mathbf{X} \leq \mathbf{X}$. Since $\hat{\mathbf{I}}$ is increasing, this means $\hat{\mathbf{I}}\check{\mathbf{I}}\mathbf{X} \leq \hat{\mathbf{I}}\mathbf{X}$. On the other hand, $\hat{\mathbf{I}}\check{\mathbf{I}}$ is a closing (see Theorem 3.25), whence it follows that $\hat{\mathbf{I}}\check{\mathbf{I}}\mathbf{X} \geq \mathbf{X}$. Applying $\hat{\mathbf{I}}$ on both sides and using $\hat{\mathbf{I}}^2 = \hat{\mathbf{I}}$ gives $\hat{\mathbf{I}}\check{\mathbf{I}}\mathbf{X} \geq \hat{\mathbf{I}}\mathbf{X}$; thus equality follows. □

10.7 Corollary. *Assume that $\mathcal{T}$ is a strongly admissible complete lattice. The mapping $\check{\mathbf{I}}$ defines an isomorphism between the complete lattices $\mathcal{L}_{\vartriangle}\mathcal{T}$ and $\mathcal{L}_{\triangledown}\mathcal{T}$ with inverse $\hat{\mathbf{I}}$.*

Proof. It is evident that $\check{\mathbf{I}}$ defines an increasing mapping between $\mathcal{L}_{\vartriangle}\mathcal{T}$ and $\mathcal{L}_{\triangledown}\mathcal{T}$. Likewise, $\hat{\mathbf{I}}$ defines an increasing mapping between $\mathcal{L}_{\triangledown}\mathcal{T}$ and $\mathcal{L}_{\vartriangle}\mathcal{T}$. Furthermore, $\hat{\mathbf{I}}\check{\mathbf{I}} = \hat{\mathbf{I}}$ and $\hat{\mathbf{I}}$ coincides with the identity mapping on $\mathcal{L}_{\vartriangle}\mathcal{T}$. Dually, $\check{\mathbf{I}}\hat{\mathbf{I}} = \check{\mathbf{I}}$ defines the identity mapping on $\mathcal{L}_{\triangledown}\mathcal{T}$. This proves the result. □

Note in particular that the supremum of a collection $\mathbf{X}_i \in \mathcal{L}_{\vartriangle}\mathcal{T},\ i \in I$, is given by $\hat{\mathbf{I}}(\bigvee_{i\in I} \mathbf{X}_i)$; here $\bigvee$ denotes the supremum in $\mathcal{L}\diamond\mathcal{T}$. Dually, the infimum of this collection in $\mathcal{L}_{\triangledown}\mathcal{T}$ is given by $\check{\mathbf{I}}(\bigwedge_{i\in I} \mathbf{X}_i)$; here $\bigwedge$ denotes the infimum in $\mathcal{L}\diamond\mathcal{T}$.

10.4. Function representations

We now come to the main goal of this chapter, namely, to give a set-based representation of function lattices. In this section we consider the space Fun(E, $\mathcal{T}$) comprising all functions mapping the domain space E into the grey-value lattice $\mathcal{T}$. In the next section we shall deal with semi-continuous functions.

Define, for a function $F : E \to \mathcal{T}$, the threshold sets $\mathbf{X}^{\vartriangle}(F, t)$ and $\mathbf{X}^{\triangledown}(F, t)$ as follows:

$$\mathbf{X}^{\vartriangle}(F, t) = \{x \in E \mid F(x) \geq t\}, \tag{10.18}$$

$$\mathbf{X}^{\triangledown}(F,t)=\{x\in E \mid F(x)\not\leq t\}. \tag{10.19}$$

Let $\mathbf{X}^{\vartriangle}(F)$ and $\mathbf{X}^{\triangledown}(F)$ denote the mappings from $\mathcal{T}$ to $\mathcal{P}(E)$ given by

$$\mathbf{X}^{\vartriangle}(F)(t)=\mathbf{X}^{\vartriangle}(F,t) \qquad \text{and} \qquad \mathbf{X}^{\triangledown}(F)(t)=\mathbf{X}^{\triangledown}(F,t).$$

10.8 Proposition. *If $\mathcal{T}$ is a complete lattice, then $\mathbf{X}^{\vartriangle}(F)\in\mathcal{P}(E)_{\vartriangle}\mathcal{T}$ and $\mathbf{X}^{\triangledown}(F)\in\mathcal{P}(E)_{\triangledown}\mathcal{T}$ for every $F\in\mathrm{Fun}(E,\mathcal{T})$.*

Proof. Let $F\in\mathrm{Fun}(E,\mathcal{T})$; we show that $\mathbf{X}^{\vartriangle}(F)\in\mathcal{P}(E)_{\vartriangle}\mathcal{T}$. The second statement is proved analogously. Put $\mathbf{X}(t)=\mathbf{X}^{\vartriangle}(F,t)$, and let $t_i,\ i\in I$, be an arbitrary family in $\mathcal{T}$. We must show that

$$\mathbf{X}\Big(\bigvee_{i\in I} t_i\Big)=\bigcap_{i\in I}\mathbf{X}(t_i).$$

The inclusion $\subseteq$ follows from the decreasingness of $\mathbf{X}(\cdot)$. To prove the reverse inclusion suppose that $x\in\mathbf{X}(t_i)$ for every $i\in I$, that is, $F(x)\geq t_i$. Then $F(x)\geq\bigvee_{i\in I}t_i$, which means that $x\in\mathbf{X}(\bigvee_{i\in I}t_i)$. □

For a function $\mathbf{X}:\mathcal{T}\to\mathcal{P}(E)$, we define $\mathbf{F}^{\vartriangle}(\mathbf{X})$, $\mathbf{F}^{\triangledown}(\mathbf{X})\in\mathrm{Fun}(E,\mathcal{T})$ by

$$\mathbf{F}^{\vartriangle}(\mathbf{X})(x)=\bigvee\{t\in\mathcal{T}\mid x\in\mathbf{X}(t)\}, \tag{10.20}$$

$$\mathbf{F}^{\triangledown}(\mathbf{X})(x)=\bigwedge\{t\in\mathcal{T}\mid x\notin\mathbf{X}(t)\}. \tag{10.21}$$

The following result establishes an isomorphism between the function lattice $\mathrm{Fun}(E,\mathcal{T})$ on the one hand and the lattices $\mathcal{P}(E)_{\vartriangle}\mathcal{T}$ and $\mathcal{P}(E)_{\triangledown}\mathcal{T}$ on the other.

10.9 Theorem. *Let $\mathcal{T}$ be an arbitrary complete lattice. The complete lattices $\mathrm{Fun}(E,\mathcal{T})$, $\mathcal{P}(E)_{\vartriangle}\mathcal{T}$, and $\mathcal{P}(E)_{\triangledown}\mathcal{T}$ are all isomorphic. The mapping $\mathbf{X}^{\vartriangle}$ defines an isomorphism between $\mathrm{Fun}(E,\mathcal{T})$ and $\mathcal{P}(E)_{\vartriangle}\mathcal{T}$ with inverse $\mathbf{F}^{\vartriangle}$. Dually, the mapping $\mathbf{X}^{\triangledown}$ defines an isomorphism between $\mathrm{Fun}(E,\mathcal{T})$ and $\mathcal{P}(E)_{\triangledown}\mathcal{T}$ with inverse $\mathbf{F}^{\triangledown}$.*

Proof. We prove only the result concerning the isomorphism between $\mathrm{Fun}(E,\mathcal{T})$ and $\mathcal{P}(E)_{\vartriangle}\mathcal{T}$. It is obvious that the mappings $\mathbf{X}^{\vartriangle}$ and $\mathbf{F}^{\vartriangle}$ are increasing, so we have only to prove that they are each other's inverses.

1. Take $\mathbf{X}\in\mathcal{P}(E)_{\vartriangle}\mathcal{T}$, and define $F=\mathbf{F}^{\vartriangle}(\mathbf{X})$, i.e., $F(x)=\bigvee\{t\in\mathcal{T}\mid x\in\mathbf{X}(t)\}$. We show that $\mathbf{X}^{\vartriangle}(F)=\mathbf{X}$. If $x\in\mathbf{X}(t)$, then $F(x)\geq t$, and therefore $x\in\mathbf{X}^{\vartriangle}(F,t)$. If $x\in\mathbf{X}^{\vartriangle}(F,t)$, then $F(x)\geq t$, that is, $\bigvee\{s\in\mathcal{T}\mid x\in\mathbf{X}(s)\}\geq t$. Apply $\mathbf{X}$ to both sides and use property (10.12) together with the fact that

$\mathbf{X}(\cdot)$ is decreasing; one finds that $\bigcap\{\mathbf{X}(s) \mid s \in \mathcal{T}$ and $x \in \mathbf{X}(s)\} \subseteq \mathbf{X}(t)$. But this implies $x \in \mathbf{X}(t)$.

2. Take $F \in \text{Fun}(E, \mathcal{T})$, and let $\mathbf{X}(t) = \mathbf{X}^{\vartriangle}(F, t)$; we show that $F = \mathbf{F}^{\vartriangle}(\mathbf{X})$. Suppose first that $F(x) \geq t$; then $x \in \mathbf{X}(t)$, and so $\mathbf{F}^{\vartriangle}(\mathbf{X})(x) = \bigvee\{s \in \mathcal{T} \mid x \in \mathbf{X}(s)\} \geq t$. On the other hand, if $\mathbf{F}^{\vartriangle}(\mathbf{X})(x) \geq t$, then $\bigvee\{s \in \mathcal{T} \mid x \in \mathbf{X}(s)\} \geq t$. Applying $\mathbf{X}$ to both sides and using (10.12), one gets that $\bigcap\{\mathbf{X}(s) \mid s \in \mathcal{T}$ and $x \in \mathbf{X}(s)\} \subseteq \mathbf{X}(t)$; hence $x \in \mathbf{X}(t)$. This shows that $F(x) \geq t$. □

Define, for given $t \in \mathcal{T}$ and $X \subseteq E$, the *flat functions* $C^{\vartriangle}(X, t)$ and $C^{\triangledown}(X, t)$ by

$$C^{\vartriangle}(X, t)(x) = \begin{cases} t, & x \in X, \\ O_{\mathcal{T}}, & x \notin X, \end{cases} \tag{10.22}$$

$$C^{\triangledown}(X, t)(x) = \begin{cases} I_{\mathcal{T}}, & x \in X, \\ t, & x \notin X. \end{cases} \tag{10.23}$$

10.10 Proposition. *Let $\mathcal{T}$ be an arbitrary complete lattice.*

(a) *For every $t \in \mathcal{T}$, the pair $(\mathbf{X}^{\vartriangle}(\cdot, t), C^{\vartriangle}(\cdot, t))$ defines an adjunction between* $\text{Fun}(E, \mathcal{T})$ *and* $\mathcal{P}(E)$.

(a′) *For every $t \in \mathcal{T}$, the pair $(C^{\triangledown}(\cdot, t), \mathbf{X}^{\triangledown}(\cdot, t))$ defines an adjunction between* $\mathcal{P}(E)$ *and* $\text{Fun}(E, \mathcal{T})$.

Proof. It needs only be demonstrated that

$$C^{\vartriangle}(X, t) \leq F \iff X \subseteq \mathbf{X}^{\vartriangle}(F, t),$$

for every $F \in \text{Fun}(E, T)$ and $X \in \mathcal{P}(E)$. But one sees immediately that $C^{\vartriangle}(X, t) \leq F$ is equivalent to $F(x) \geq t$ for $x \in X$, that is, $X \subseteq \mathbf{X}^{\vartriangle}(F, t)$. This proves the result. □

10.5. Semi-continuous functions

Semi-continuous functions play a similar role for grey-scale images as the closed and open sets do for binary images. Throughout this section E is a topological space and $\mathcal{T}$ a complete lattice. As usual, $\mathcal{F}(E)$ and $\mathcal{G}(E)$ denote the closed and open subsets of E, respectively.

10.11 Definition. A function $F : E \to \mathcal{T}$ is called *upper semi-continuous* (or *u.s.c.*) if for every $t \in \mathcal{T}$ and $x \in E$ such that $t \nleq F(x)$ there exists a

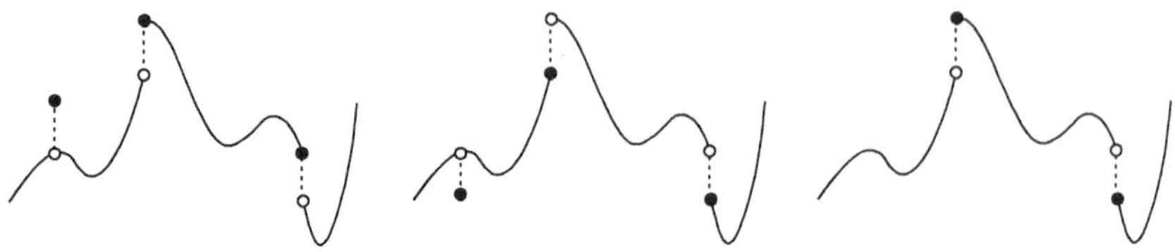

Figure 10.4 From left to right: a u.s.c. function, an l.s.c. function and a function that is neither u.s.c. nor l.s.c.

neighbourhood V of x such that $t \not\leq F(y)$ for every $y \in V$. A function $F : E \to \mathcal{T}$ is called *lower semi-continuous* (or *l.s.c.*) if for every $t \in \mathcal{T}$ and $x \in E$ such that $t \not\geq F(x)$ there exists a neighbourhood V of x such that $t \not\geq F(y)$ for every $y \in V$.

The spaces of u.s.c. and l.s.c. functions from E to $\mathcal{T}$ are denoted by $\mathrm{Fun}_u(E, \mathcal{T})$ and $\mathrm{Fun}_l(E, \mathcal{T})$, respectively. Upper semi-continuity and lower semi-continuity are dual notions in the sense of the Duality Principle: if $F : E \to \mathcal{T}$ is u.s.c., then $F : E \to \mathcal{T}'$, where $\mathcal{T}'$ is the opposite lattice of $\mathcal{T}$, is l.s.c. Fig. 10.4 shows some examples for $\mathcal{T} = \overline{\mathbb{R}}$.

If $\mathcal{T} = \{0, 1\}$, then $\mathrm{Fun}(E, \mathcal{T})$ is isomorphic to $\mathcal{P}(E)$; it is easy to show that $\mathrm{Fun}_u(E, \mathcal{T})$ is isomorphic to $\mathcal{F}(E)$ and that $\mathrm{Fun}_l(E, \mathcal{T})$ is isomorphic to $\mathcal{G}(E)$ in this case. More generally, the following result can be established.

10.12 Proposition.

(a) *A function F is u.s.c. if and only if the set $\mathbf{X}^{\triangle}(F, t)$ is closed for every $t \in \mathcal{T}$.*

(a′) *A function F is l.s.c. if and only if the set $\mathbf{X}^{\triangledown}(F, t)$ is open for every $t \in \mathcal{T}$.*

Proof. “only if”: Assume that F is u.s.c. We show that the complement of $\mathbf{X}^{\triangle}(F, t)$ is open. Let $x \notin \mathbf{X}^{\triangle}(F, t)$; then $F(x) \not\geq t$, and so there is a neighbourhood V of x such that $F(y) \not\geq t$ for $y \in V$. In other words, $V \subseteq [\mathbf{X}^{\triangle}(F, t)]^c$.

The if-statement is proved similarly. □

10.13 Theorem. *Let E be a topological space and $\mathcal{T}$ a complete lattice.*

(a) *If F_i is u.s.c. for every i in some arbitrary index set I, then the pointwise infimum $\bigwedge_{i \in I} F_i$ given by $(\bigwedge_{i \in I} F_i)(x) = \bigwedge\{F_i(x) \mid i \in I\}$ is u.s.c. as well. The set $\mathrm{Fun}_u(E, \mathcal{T})$ with the pointwise ordering defines a complete lattice with the same infimum as $\mathrm{Fun}(E, \mathcal{T})$.*

(a′) *If F_i is l.s.c. for every i in some arbitrary index set I, then the pointwise supremum $\bigvee_{i \in I} F_i$ given by $(\bigvee_{i \in I} F_i)(x) = \bigvee\{F_i(x) \mid i \in I\}$ is l.s.c. as well. The set $\mathrm{Fun}_l(E, \mathcal{T})$ with the pointwise ordering defines a complete lattice with the same supremum as $\mathrm{Fun}(E, \mathcal{T})$.*

Proof. Assume that F_i is u.s.c. for every i in some index set I. Proposition 10.10 states that $\mathbf{X}^{\triangle}(\cdot, t)$ is an erosion, that is, $\mathbf{X}^{\triangle}(\bigwedge_{i \in I} F_i, t) = \bigcap_{i \in I} \mathbf{X}^{\triangle}(F_i, t)$, and the right-hand side defines a closed set. Now it follows from the previous proposition that $\bigvee_{i \in I} F_i$ is u.s.c. The function which is identically I is the greatest element of $\mathrm{Fun}_u(E, \mathcal{T})$. Thus it follows from Proposition 2.12 that $\mathrm{Fun}_u(E, \mathcal{T})$ is a complete lattice. □

One can also give a direct proof of this theorem using the very definition of u.s.c. and l.s.c. functions.

The following result contains an alternative representation of the lattice of u.s.c. functions (and dually, l.s.c. functions); compare Theorem 10.9.

10.14 Theorem. *Let E be a topological space and $\mathcal{T}$ a complete lattice.*

(a) *The mapping $\mathbf{X}^{\triangle}$ defines an isomorphism between the complete lattices $\mathrm{Fun}_u(E, \mathcal{T})$ and $\mathcal{F}(E)_{\triangle}\mathcal{T}$ with inverse $\mathbf{F}^{\triangle}$.*

(a′) *The mapping $\mathbf{X}^{\triangledown}$ defines an isomorphism between the complete lattices $\mathrm{Fun}_l(E, \mathcal{T})$ and $\mathcal{G}(E)_{\triangledown}\mathcal{T}$ with inverse $\mathbf{F}^{\triangledown}$.*

Proof. It has been shown that $\mathbf{X}^{\triangle}(F, t) \in \mathcal{F}(E)$ if F is u.s.c. Since the infimum in $\mathcal{F}(E)$ coincides with the intersection, Proposition 10.8 means that $\mathbf{X}^{\triangle}(F)$ satisfies property (10.12). To show that $\mathbf{F}^{\triangle}$ maps $\mathcal{F}(E)_{\triangle}\mathcal{T}$ into $\mathrm{Fun}_u(E, \mathcal{T})$, let $\mathbf{X} \in \mathcal{F}(E)_{\triangle}\mathcal{T}$ and $F = \mathbf{F}^{\triangle}(\mathbf{X})$. From Theorem 10.9 it follows that $\mathbf{X}^{\triangle}(F, t) = \mathbf{X}(t)$; therefore, this set is closed. Now Proposition 10.12 gives that F is u.s.c. The proof that $\mathbf{X}^{\triangle}$ and $\mathbf{F}^{\triangle}$ are each other's inverses is completely analogous to the proof of Theorem 10.9. □

10.15 Remark. There exists yet a third interesting topological representation of $\mathrm{Fun}_u(E, \mathcal{T})$ for the case $\mathcal{T} = \overline{\mathbb{R}}$, described in Vervaat (1988). Supply $\overline{\mathbb{R}}$ with the (non-Hausdorff) topology whose open sets are $\varnothing$, $\overline{\mathbb{R}}$, and $\{t \in \overline{\mathbb{R}} \mid t > t_0\}$, where t_0 ranges over $\overline{\mathbb{R}}$. One can easily prove that this defines a topology. If $E \times \overline{\mathbb{R}}$ is the product space supplied with the product topology, then the complete lattice $\mathcal{F}(E \times \overline{\mathbb{R}})$ is isomorphic with $\mathrm{Fun}_u(E, \overline{\mathbb{R}})$. Denoting elements of $\mathcal{F}(E \times \overline{\mathbb{R}})$ by U, the isomorphism is characterized by the mapping from $\mathcal{F}(E \times \overline{\mathbb{R}})$ into $\mathrm{Fun}_u(E, \overline{\mathbb{R}})$ given by

$$[\mathcal{F}(U)](x) = \bigvee \{t \in \overline{\mathbb{R}} \mid (x, t) \in U\},$$

and the mapping from $\mathrm{Fun}_u(E, \overline{\mathbb{R}})$ into $\mathcal{F}(E \times \overline{\mathbb{R}})$ given by

$$\mathcal{U}(F) = \{(x, t) \in E \times \overline{\mathbb{R}} \mid t \leq F(x)\}.$$

We will encounter the mapping $\mathcal{U}$ again in Section 11.6 under the name *umbra*.

10.6. Extension of lattice operators

The main goal of this section is to describe how to construct increasing operators on $\mathcal{L}\diamond\mathcal{T}$, $\mathcal{L}_\vartriangle\mathcal{T}$, and $\mathcal{L}_\triangledown\mathcal{T}$, given a family of increasing operators on the smaller lattice $\mathcal{L}$. As an important special case we consider the situation where the family consists of a single operator. Throughout this section we make the assumption that $\mathcal{T}$ is a strongly admissible complete lattice.

Suppose we are given a family $\{\psi_t \mid t \in \mathcal{T}\}$ of increasing operators on $\mathcal{L}$ which is decreasing with respect to t, that is, $\psi_t \leq \psi_s$ if $s \leq t$ (henceforth, we shall speak of a *decreasing family*). We can define an increasing operator $\psi^\diamond$ on $\mathcal{L}\diamond\mathcal{T}$ as follows:

$$(\psi^\diamond(\mathbf{X}))(t) = \psi_t(\mathbf{X}(t)), \qquad \text{for } \mathbf{X} \in \mathcal{L}\diamond\mathcal{T}. \tag{10.24}$$

Before we define extensions to $\mathcal{L}_\vartriangle\mathcal{T}$ and $\mathcal{L}_\triangledown\mathcal{T}$ as well, we prove the following lemma.

10.16 Lemma. *Assume that $\mathcal{T}$ is a strongly admissible complete lattice. Let $\{\psi_t \mid t \in \mathcal{T}\}$ be a decreasing family of increasing operators on $\mathcal{L}$, and let the operator $\psi^\diamond$ on $\mathcal{L}\diamond\mathcal{T}$ be given by* (10.24). *Then*

$$\check{\mathbf{I}}\psi^\diamond = \check{\mathbf{I}}\psi^\diamond\check{\mathbf{I}} = \check{\mathbf{I}}\psi^\diamond\hat{\mathbf{I}}, \tag{10.25}$$

$$\hat{\mathbf{I}}\psi^\diamond = \hat{\mathbf{I}}\psi^\diamond\check{\mathbf{I}} = \hat{\mathbf{I}}\psi^\diamond\hat{\mathbf{I}}. \tag{10.26}$$

Proof. We prove only the relations in (10.25). Since $\check{\mathbf{I}} \leq \mathrm{id}$ and $\hat{\mathbf{I}} \geq \mathrm{id}$ it follows immediately that

$$\check{\mathbf{I}}\psi^\diamond\check{\mathbf{I}} \leq \check{\mathbf{I}}\psi^\diamond \leq \check{\mathbf{I}}\psi^\diamond\hat{\mathbf{I}}.$$

Let $\mathbf{X} \in \mathcal{L}\diamond\mathcal{T}$; then

$$[\psi^\diamond(\check{\mathbf{I}}\mathbf{X})](t) = \psi_t\Big(\bigvee_{s \gg t} \mathbf{X}(s)\Big) \geq \bigvee_{s \gg t} \psi_s(\mathbf{X}(s)) = [\check{\mathbf{I}}\psi^\diamond(\mathbf{X})](t),$$

which gives $\psi^\diamond\check{\mathbf{I}} \geq \check{\mathbf{I}}\psi^\diamond$. Similarly, one shows that $\psi^\diamond\hat{\mathbf{I}} \leq \hat{\mathbf{I}}\psi^\diamond$. Thus, using (10.16) and (10.17), it follows that

$$\check{\mathbf{I}}\psi^\diamond\check{\mathbf{I}} \geq \check{\mathbf{I}}^2\psi^\diamond = \check{\mathbf{I}}\psi^\diamond,$$

and that

$$\check{\mathbf{I}}\psi^\diamond\hat{\mathbf{I}} \leq \check{\mathbf{I}}\hat{\mathbf{I}}\psi^\diamond = \check{\mathbf{I}}\psi^\diamond.$$

This finishes the proof. □

$$\begin{array}{ccc} \mathcal{L}_{\vartriangle}\mathcal{T} & \xrightarrow{\psi^{\vartriangle}} & \mathcal{L}_{\vartriangle}\mathcal{T} \\ \hat{\mathbf{I}}\uparrow\downarrow\check{\mathbf{I}} & & \hat{\mathbf{I}}\uparrow\downarrow\check{\mathbf{I}} \\ \mathcal{L}_{\triangledown}\mathcal{T} & \xrightarrow{\psi^{\triangledown}} & \mathcal{L}_{\triangledown}\mathcal{T} \end{array}$$

Figure 10.5 Intertwining diagram for the operators $\psi^{\vartriangle}$ and $\psi^{\triangledown}$.

Define the operators $\psi^{\vartriangle} \in \mathcal{O}^{+}(\mathcal{L}_{\vartriangle}\mathcal{T})$ and $\psi^{\triangledown} \in \mathcal{O}^{+}(\mathcal{L}_{\triangledown}\mathcal{T})$, respectively, by

$$\psi^{\vartriangle} = \hat{\mathbf{I}}\psi^{\diamond}, \tag{10.27}$$

that is,

$$\psi^{\vartriangle}(\mathbf{X})(t) = \bigwedge_{s \ll t} \psi_s(\mathbf{X}(s)), \tag{10.28}$$

and

$$\psi^{\triangledown} = \check{\mathbf{I}}\psi^{\diamond}, \tag{10.29}$$

that is,

$$\psi^{\triangledown}(\mathbf{X})(t) = \bigvee_{s \gg t} \psi_s(\mathbf{X}(s)). \tag{10.30}$$

We say that the family $\{\psi_t \mid t \in \mathcal{T}\}$ generates $\psi^{\vartriangle}$ and $\psi^{\triangledown}$. In Corollary 10.7 we have seen that the complete lattices $\mathcal{L}_{\vartriangle}\mathcal{T}$ and $\mathcal{L}_{\triangledown}\mathcal{T}$ are isomorphic. The given definitions are compatible with this isomorphism as is clearly shown by the intertwining relations

$$\check{\mathbf{I}}\psi^{\vartriangle} = \psi^{\triangledown}\check{\mathbf{I}} \qquad \text{on } \mathcal{L}_{\vartriangle}\mathcal{T}, \tag{10.31}$$

$$\hat{\mathbf{I}}\psi^{\triangledown} = \psi^{\vartriangle}\hat{\mathbf{I}} \qquad \text{on } \mathcal{L}_{\triangledown}\mathcal{T}; \tag{10.32}$$

see also the diagram in Fig. 10.5.

To prove, e.g., the first relation we use the identities (10.25) and (10.27)–(10.29) and Proposition 10.6. We get

$$\check{\mathbf{I}}\psi^{\vartriangle} = \check{\mathbf{I}}\hat{\mathbf{I}}\psi^{\diamond} = \check{\mathbf{I}}\psi^{\diamond} = \check{\mathbf{I}}\psi^{\diamond}\check{\mathbf{I}} = \psi^{\triangledown}\check{\mathbf{I}}.$$

We show that the given procedure for the construction of increasing operators on $\mathcal{L}_{\vartriangle}\mathcal{T}$ and $\mathcal{L}_{\triangledown}\mathcal{T}$ is compatible with the formation of suprema, infima, and composites.

10.17 Theorem. *Let $\mathcal{T}$ be a strongly admissible complete lattice. Assume that for every i in the index set I, $\{\psi_{i,t} \mid t \in \mathcal{T}\}$ is a decreasing family in $\mathcal{O}^+(\mathcal{L})$ which generates $\psi_i^\vartriangle$ and $\psi_i^\triangledown$.*

(a) *The family $\{\bigvee_{i\in I} \psi_{i,t} \mid t \in \mathcal{T}\}$ generates the operator $\hat{\mathbf{I}}(\bigvee_{i\in I} \psi_i^\vartriangle)$ on $\mathcal{L}_\vartriangle\mathcal{T}$ and the operator $\bigvee_{i\in I} \psi_i^\triangledown$ on $\mathcal{L}_\triangledown\mathcal{T}$ (where $\bigvee$ denotes the supremum in $\mathcal{L}\diamond\mathcal{T}$).*

(a′) *The family $\{\bigwedge_{i\in I} \psi_{i,t} \mid t \in \mathcal{T}\}$ generates the operator $\bigwedge_{i\in I} \psi_i^\vartriangle$ on $\mathcal{L}_\vartriangle\mathcal{T}$ and the operator $\check{\mathbf{I}}(\bigwedge_{i\in I} \psi_i^\triangledown)$ on $\mathcal{L}_\triangledown\mathcal{T}$ (where $\bigwedge$ denotes the infimum in $\mathcal{L}\diamond\mathcal{T}$).*

(b) *Let $\{\phi_t \mid t \in \mathcal{T}\}$ and $\{\psi_t \mid t \in \mathcal{T}\}$ be decreasing families in $\mathcal{O}^+(\mathcal{L})$ and let $\phi^\vartriangle, \phi^\triangledown$ and $\psi^\vartriangle, \psi^\triangledown$ be defined in the usual way. Then $\psi^\vartriangle\phi^\vartriangle$ and $\psi^\triangledown\phi^\triangledown$ are generated by the family $\{\psi_t\phi_t \mid t \in \mathcal{T}\}$.*

Proof. (a): Define $\psi_t = \bigvee_{i\in I} \psi_{i,t}$ for $t \in \mathcal{T}$. Let $\psi^\diamond$ be the operator on $\mathcal{L}\diamond\mathcal{T}$ generated by the family $\{\psi_t \mid t \in \mathcal{T}\}$, and let, for $i \in I$, $\psi_i^\diamond$ be the operator generated by $\{\psi_{i,t} \mid t \in \mathcal{T}\}$. It is obvious that $\psi^\diamond = \bigvee_{i\in I} \psi_i^\diamond$. Since $\check{\mathbf{I}}$ is a dilation it commutes with suprema, and so we find that

$$\psi^\triangledown = \check{\mathbf{I}}\psi^\diamond = \check{\mathbf{I}}(\bigvee_{i\in I} \psi_i^\diamond) = \bigvee_{i\in I} \check{\mathbf{I}}\psi_i^\diamond = \bigvee_{i\in I} \psi_i^\triangledown.$$

To compute $\psi^\vartriangle$ we use the intertwining relation (10.31) and the fact that $\check{\mathbf{I}}$ distributes over suprema:

$$\check{\mathbf{I}}\psi^\vartriangle = \psi^\triangledown\check{\mathbf{I}} = \bigvee_{i\in I} \psi_i^\triangledown\check{\mathbf{I}} = \bigvee_{i\in I} \check{\mathbf{I}}\psi_i^\vartriangle = \check{\mathbf{I}}(\bigvee_{i\in I} \psi_i^\vartriangle).$$

Now we apply $\hat{\mathbf{I}}$ on both sides and use that $\hat{\mathbf{I}}\check{\mathbf{I}} = \hat{\mathbf{I}}$ (cf. (10.16)) and that $\hat{\mathbf{I}}$ coincides with the identity operator on $\mathcal{L}_\vartriangle\mathcal{T}$. This gives the desired result.

(b): We prove only the first relation. Let $\phi^\vartriangle$, $\psi^\vartriangle$ be defined in the usual way. It is apparent that $\psi^\diamond\phi^\diamond$ is generated by the family $\{\psi_t\phi_t \mid t \in \mathcal{T}\}$. The operator on $\mathcal{L}_\vartriangle\mathcal{T}$ associated with this family is

$$\hat{\mathbf{I}}(\psi^\diamond\phi^\diamond) = (\hat{\mathbf{I}}\psi^\diamond\hat{\mathbf{I}})\phi^\diamond = (\hat{\mathbf{I}}\psi^\diamond)(\hat{\mathbf{I}}\phi^\diamond) = \psi^\vartriangle\phi^\vartriangle,$$

where we have used (10.26). This concludes the proof. □

10.18 Corollary. *Let $\mathcal{T}$ be a strongly admissible complete lattice, and let $\{\psi_t \mid t \in \mathcal{T}\}$ be a decreasing family in $\mathcal{O}^+(\mathcal{L})$.*

(a) *If every ψ_t is a dilation, then $\psi^\triangledown$ is a dilation as well.*

(a′) *If every ψ_t is an erosion, then $\psi^\vartriangle$ is an erosion as well.*

(b) *If every ψ_t is idempotent, then $\psi^\vartriangle$ and $\psi^\triangledown$ are idempotent as well.*

(c) *If every ψ_t is (anti-)extensive, then $\psi^\vartriangle$ and $\psi^\triangledown$ are (anti-)extensive as well.*

(d) *If every ψ_t is an opening, then $\psi^\vartriangle$ and $\psi^\triangledown$ are openings as well.*
(d′) *If every ψ_t is a closing, then $\psi^\vartriangle$ and $\psi^\triangledown$ are closings as well.*

Proof. (a): This follows easily if one uses that the supremum on $\mathcal{L}_\triangledown\mathcal{T}$ coincides with the supremum in $\mathcal{L}\diamond\mathcal{T}$ and that $\check{\mathbf{I}}$ is a dilation on $\mathcal{L}\diamond\mathcal{T}$.

(b): This is a straightforward consequence of Theorem 10.17(b).

(c) and (d): Easy. □

The given construction applies in particular to the case where the family of operators on $\mathcal{L}$ contains only one element ψ. Here the operators $\psi^\vartriangle$ and $\psi^\triangledown$ are, respectively, given by

$$\psi^\vartriangle(\mathbf{X})(t) = \bigwedge_{s \ll t} \psi(\mathbf{X}(s)) \tag{10.33}$$

and

$$\psi^\triangledown(\mathbf{X})(t) = \bigvee_{s \gg t} \psi(\mathbf{X}(s)), \tag{10.34}$$

and we say that $\psi^\vartriangle, \psi^\triangledown$ are generated by ψ. In this case we denote the extension of ψ to $\mathcal{L}\diamond\mathcal{T}$ also by ψ. The next result is a strong version of Corollary 10.18(a), (a′).

10.19 Proposition. *Assume that $\mathcal{T}$ is a strongly admissible complete lattice. Let (ε, δ) be an adjunction on $\mathcal{L}$; then $(\varepsilon^\vartriangle, \delta^\vartriangle)$ is an adjunction on $\mathcal{L}_\vartriangle\mathcal{T}$ and $(\varepsilon^\triangledown, \delta^\triangledown)$ is an adjunction on $\mathcal{L}_\triangledown\mathcal{T}$.*

Proof. Assume that (ε, δ) is an adjunction on $\mathcal{L}$. We prove that $(\varepsilon^\vartriangle, \delta^\vartriangle)$ is an adjunction on $\mathcal{L}_\vartriangle\mathcal{T}$. The second statement is proved analogously. We must show that $\delta^\vartriangle(\mathbf{Y}) \le \mathbf{X} \iff \mathbf{Y} \le \varepsilon^\vartriangle(\mathbf{X})$ for $\mathbf{X}, \mathbf{Y} \in \mathcal{L}_\vartriangle\mathcal{T}$. We prove only $\Rightarrow$; the reverse implication is proved similarly. The inequality $\delta^\vartriangle(\mathbf{Y}) \le \mathbf{X}$ is equivalent to $\hat{\mathbf{I}}\delta(\mathbf{Y}) \le \mathbf{X}$. Applying $\check{\mathbf{I}}$ on both sides gives $\check{\mathbf{I}}\hat{\mathbf{I}}\delta(\mathbf{Y}) \le \check{\mathbf{I}}\mathbf{X}$. Using that $\check{\mathbf{I}}\hat{\mathbf{I}} = \check{\mathbf{I}}$ and that $(\hat{\mathbf{I}}, \check{\mathbf{I}})$ is an adjunction on $\mathcal{L}\diamond\mathcal{T}$ (Proposition 10.6), we find $\delta(\mathbf{Y}) \le \hat{\mathbf{I}}\check{\mathbf{I}}\mathbf{X} = \hat{\mathbf{I}}\mathbf{X}$. Then, since (ε, δ) is obviously an adjunction on $\mathcal{L}\diamond\mathcal{T}$, $\mathbf{Y} \le \varepsilon(\hat{\mathbf{I}}\mathbf{X})$. Apply $\hat{\mathbf{I}}$ on both sides and use that $\mathbf{Y}$ is invariant under $\hat{\mathbf{I}}$ and that $\hat{\mathbf{I}}\varepsilon\hat{\mathbf{I}} = \hat{\mathbf{I}}\varepsilon = \varepsilon^\vartriangle$ (see (10.26)). This shows the result. □

Let $a : \mathcal{T} \to \mathcal{T}$ and $\mathbf{X} \in \mathcal{L}^\mathcal{T}$; denote by $\mathbf{X}a$ the composition

$$(\mathbf{X}a)(t) = \mathbf{X}(a(t)).$$

For later use we prove the following result.

10.20 Proposition. *Assume that $\mathcal{T}$ is a strongly admissible complete lattice. Let $\psi^\vartriangle, \psi^\triangledown$ be generated by the increasing operator ψ and let a be an automorphism on $\mathcal{T}$.*

(a) *If $\mathbf{X} \in \mathcal{L}_\vartriangle\mathcal{T}$, then $\mathbf{X}a \in \mathcal{L}_\vartriangle\mathcal{T}$ and $\psi^\vartriangle(\mathbf{X}a) = [\psi^\vartriangle(\mathbf{X})]a$.*

(a′) *If $\mathbf{X} \in \mathcal{L}_\triangledown\mathcal{T}$, then $\mathbf{X}a \in \mathcal{L}_\triangledown\mathcal{T}$ and $\psi^\triangledown(\mathbf{X}a) = [\psi^\triangledown(\mathbf{X})]a$.*

Proof. Under the given assumptions, that $\mathbf{X}a \in \mathcal{L}_\vartriangle\mathcal{T}$ if $\mathbf{X} \in \mathcal{L}_\vartriangle\mathcal{T}$ follows immediately from the fact that the binary relations $\ll$ and $\gg$ are preserved by every automorphism on $\mathcal{T}$. Furthermore,

$$[\psi^\vartriangle(\mathbf{X}a)](t) = \bigwedge_{s \ll t} \psi\big(\mathbf{X}(a(s))\big) = \bigwedge_{\sigma \ll a(t)} \psi(\mathbf{X}(\sigma)) = (\psi^\vartriangle\mathbf{X})(a(t));$$

here we have substituted $\sigma = a(s)$ in the third expression and used that $s \ll t$ if and only if $\sigma \ll a(t)$. This gives the desired relation. □

10.7. Lattices with negation

Recall that a mapping ν on a complete lattice $\mathcal{L}$ is called a negation if ν is a dual automorphism and $\nu^2 = \mathsf{id}$.

10.21 Examples.

(a) $\mathcal{T} = \overline{\mathbb{R}} \times \overline{\mathbb{R}}$ with the product ordering. The mappings $\nu(t_1, t_2) = (-t_1, -t_2)$ and $\nu(t_1, t_2) = (-t_2, -t_1)$ define negations on $\mathcal{T}$.

(b) $\mathcal{T} = \{0, 1\} \times \{0, 1\} \times \{0, 1\}$ with the product ordering. Every negation on $\mathcal{T}$ satisfies $\nu(0, 0, 0) = (1, 1, 1)$ and $\nu(1, 1, 1) = (0, 0, 0)$. It is completely determined by assigning to $(1, 0, 0)$ one of the vectors $(1, 1, 0)$, $(1, 0, 1)$, or $(0, 1, 1)$. For example, if $\nu(1, 0, 0) = (1, 1, 0)$, then $\nu(1, 1, 0) = (1, 0, 0)$ because $\nu^2 = \mathsf{id}$. Using that ν is a dual automorphism, one derives that

$$(1, 1, 0) = \nu(1, 0, 0) = \nu((1, 1, 0) \wedge (1, 0, 1)) = (1, 0, 0) \vee \nu(1, 0, 1).$$

It is not difficult to check that $\nu(1, 0, 1)$ cannot equal $(1, 1, 0)$; thus one finds that $\nu(1, 0, 1) = (0, 1, 0)$. Therefore, $\nu(0, 1, 0) = (1, 0, 1)$. Similarly, $\nu(0, 0, 1) = (0, 1, 1)$.

These examples show that for colour images, e.g., modelled by $\mathrm{Fun}(E, \mathcal{T})$, where $\mathcal{T} = \{0, 1, \ldots, N\} \times \{0, 1, \ldots, N\} \times \{0, 1, \ldots, N\}$, there exist many negations. The one which seems to make most sense in practice is given by $(r, g, b)^* = (N - r, N - g, N - b)$, where r, g, b represent the intensities of the red, green, and blue component, respectively.

10.22 Lemma. *Let $\mathcal{T}$ be a complete lattice with a negation $t \mapsto t^*$; then*

$$s \ll t \iff s^* \gg t^*, \tag{10.35}$$

for $s, t \in \mathcal{T}$.

The proof of this result is straightforward.

Throughout the remainder this section it will be assumed that both the lattice $\mathcal{L}$ and the lattice $\mathcal{T}$ have a negation, and we shall use the same notation for them: $X \mapsto X^*$ is the negation on $\mathcal{L}$ and $t \mapsto t^*$ the negation on $\mathcal{T}$. First we show how to construct negations on $\mathcal{L}_\vartriangle\mathcal{T}$ and $\mathcal{L}_\triangledown\mathcal{T}$ in this case. Define the mapping ν on $\mathcal{L}\diamond\mathcal{T}$ by

$$\nu(\mathbf{X})(t) = (\mathbf{X}(t^*))^*. \tag{10.36}$$

Note that ν preserves monotonicity properties of $\mathbf{X}$: if $\mathbf{X}$ is increasing (or decreasing) with respect to t, then so is $\nu(\mathbf{X})$. Furthermore, it is obvious that

$$\nu^2 = \mathrm{id}. \tag{10.37}$$

We prove the following result.

10.23 Theorem. *Assume that $\mathcal{T}$ is a strongly admissible complete lattice.*

(a) *If $\mathbf{X} \in \mathcal{L}\diamond\mathcal{T}$, then $\mathbf{X} \in \mathcal{L}_\vartriangle\mathcal{T}$ if and only if $\nu(\mathbf{X}) \in \mathcal{L}_\triangledown\mathcal{T}$.*

(b) *$\nu\check{\mathbf{I}} = \hat{\mathbf{I}}\nu$ on $\mathcal{L}\diamond\mathcal{T}$, and this mapping defines a negation on $\mathcal{L}_\vartriangle\mathcal{T}$.*

(b′) *$\nu\hat{\mathbf{I}} = \check{\mathbf{I}}\nu$ on $\mathcal{L}\diamond\mathcal{T}$, and this mapping defines a negation on $\mathcal{L}_\triangledown\mathcal{T}$.*

Proof. (a): Assume that $\mathbf{X} \in \mathcal{L}_\vartriangle\mathcal{T}$, and define $\mathbf{Y}(t) = \nu(\mathbf{X})(t) = (\mathbf{X}(t^*))^*$. If $t_i \in \mathcal{T}$ for $i \in I$, then

$$\mathbf{Y}(\bigwedge_{i\in I} t_i) = [\mathbf{X}(\bigvee_{i\in I} t_i^*)]^* = [\bigwedge_{i\in I} \mathbf{X}(t_i^*)]^* = \bigvee_{i\in I} \mathbf{X}(t_i^*)^* = \bigvee_{i\in I} \mathbf{Y}(t_i);$$

therefore, $\mathbf{Y} \in \mathcal{L}_\triangledown\mathcal{T}$. Here we have used that $\mathbf{X}$ satisfies relation (10.12).

(b): If $\mathbf{X} \in \mathcal{L}\diamond\mathcal{T}$, then

$$\begin{aligned}(\hat{\mathbf{I}}\nu)(\mathbf{X})(t) &= \bigwedge_{s\ll t} \nu(\mathbf{X})(s) = \bigwedge_{s\ll t} (\mathbf{X}(s^*))^* \\ &= [\bigvee_{s\ll t} \mathbf{X}(s^*)]^* = [\bigvee_{r^*\ll t} \mathbf{X}(r)]^* \\ &= [\bigvee_{r\gg t^*} \mathbf{X}(r)]^* = (\check{\mathbf{I}}\mathbf{X}(t^*))^* \\ &= \nu(\check{\mathbf{I}}\mathbf{X})(t),\end{aligned}$$

which proves the first assertion. Furthermore, it is obvious that $\nu\check{\mathbf{I}}$ defines a dual automorphism on $\mathcal{L}_\triangle\mathcal{T}$ and

$$(\nu\check{\mathbf{I}})(\nu\check{\mathbf{I}}) = \hat{\mathbf{I}}\nu\nu\check{\mathbf{I}} = \hat{\check{\mathbf{I}}}\check{\mathbf{I}} = \hat{\mathbf{I}}.$$

Here we have used $\nu^2 = \mathrm{id}$ and (10.16). Since $\hat{\mathbf{I}}$ coincides with the identity operator on $\mathcal{L}_\triangle\mathcal{T}$, the result is proved. □

Define negations $\nu^\triangle$ on $\mathcal{L}_\triangle\mathcal{T}$ and $\nu^\triangledown$ on $\mathcal{L}_\triangledown\mathcal{T}$ respectively by

$$\nu^\triangle = \nu\check{\mathbf{I}} = \hat{\mathbf{I}}\nu, \tag{10.38}$$

$$\nu^\triangledown = \nu\hat{\mathbf{I}} = \check{\mathbf{I}}\nu. \tag{10.39}$$

10.24 Theorem. *Let $\mathcal{T}$ be a strongly admissible complete lattice. Assume that $\{\psi_t \mid t \in \mathcal{T}\}$ is a decreasing family in $\mathcal{O}^+(\mathcal{L})$; then $\{(\psi_{t^*})^* \mid t \in \mathcal{T}\}$ is a decreasing family as well. Let $\psi_*^\diamond, \psi_*^\triangle, \psi_*^\triangledown$ be the operators on $\mathcal{L}\diamond\mathcal{T}$, $\mathcal{L}_\triangle\mathcal{T}$, and $\mathcal{L}_\triangledown\mathcal{T}$ generated by this family; then*

$$\psi_*^\diamond\nu = \nu\psi^\diamond, \tag{10.40}$$

$$\psi_*^\triangle\nu^\triangle = \nu^\triangle\psi^\triangle, \tag{10.41}$$

$$\psi_*^\triangledown\nu^\triangledown = \nu^\triangledown\psi^\triangledown. \tag{10.42}$$

Proof. To prove relation (10.40) observe that

$$\begin{aligned}(\psi_*^\diamond\nu)(\mathbf{X})(t) &= \psi_{t^*}^*(\nu(\mathbf{X})(t)) = \psi_{t^*}^*(\mathbf{X}(t^*)^*)\\ &= \left[\psi_{t^*}(\mathbf{X}(t^*))\right]^* = (\nu\psi^\diamond)(\mathbf{X})(t).\end{aligned}$$

Next we prove relation (10.41):

$$\begin{aligned}\psi_*^\triangle\nu^\triangle &= \hat{\mathbf{I}}\psi_*^\diamond\hat{\mathbf{I}}\nu = \hat{\mathbf{I}}\psi_*^\diamond\nu\check{\mathbf{I}} = \hat{\mathbf{I}}\nu\psi^\diamond\check{\mathbf{I}}\\ &= \nu\check{\mathbf{I}}\psi^\diamond\check{\mathbf{I}} = \nu\check{\mathbf{I}}\hat{\mathbf{I}}\psi^\diamond\check{\mathbf{I}} = \nu^\triangle\psi^\triangle.\end{aligned}$$

Here we have used (10.38), (10.40), and (10.26). Now relation (10.42) follows by duality. □

10.8. Operators: from sets to functions

Theorem 10.9 states that the function lattice $\mathrm{Fun}(E, \mathcal{T})$ on the one hand and the lattices $\mathcal{P}(E)_\triangle\mathcal{T}$ and $\mathcal{P}(E)_\triangledown\mathcal{T}$ on the other are isomorphic. Furthermore, Section 10.6 shows how to build an operator on $\mathcal{L}_\triangle\mathcal{T}$ and

$$\begin{array}{ccc} \mathrm{Fun}(E,\mathcal{T}) & \xrightarrow{\Psi} & \mathrm{Fun}(E,\mathcal{T}) \\ \mathbf{X}^{\vartriangle}\big\downarrow\big\uparrow\mathbf{F}^{\vartriangle} & & \mathbf{X}^{\vartriangle}\big\downarrow\big\uparrow\mathbf{F}^{\vartriangle} \\ \mathcal{P}(E)_{\vartriangle}\mathcal{T} & \xrightarrow{\psi^{\vartriangle}} & \mathcal{P}(E)_{\vartriangle}\mathcal{T} \end{array}$$

Figure 10.6 Intertwining diagram for the operators on $\mathrm{Fun}(E, \mathcal{T})$ and $\mathcal{P}(E)_{\vartriangle}\mathcal{T}$.

$\mathcal{L}_{\triangledown}\mathcal{T}$ given a decreasing family of increasing operators on $\mathcal{L}$. Combining both results yields a systematic approach to building increasing operators on the function lattice $\mathrm{Fun}(E, \mathcal{T})$. Here, this approach will be worked out for an arbitrary strongly admissible complete lattice $\mathcal{T}$. In the next chapter the cases $\mathcal{T} = \overline{\mathbb{R}}$, $\overline{\mathbb{Z}}$, and $\{0, 1, \ldots, N\}$ will be considered in more detail.

Let E be an arbitrary set and $\mathcal{T}$ a strongly admissible complete lattice. In Theorem 10.9 it has been shown that $\mathrm{Fun}(E, \mathcal{T})$ is (lattice) isomorphic to $\mathcal{P}(E)_{\vartriangle}\mathcal{T}$ and $\mathcal{P}(E)_{\triangledown}\mathcal{T}$. By means of this isomorphism every operator $\psi^{\vartriangle}$ on $\mathcal{P}(E)_{\vartriangle}\mathcal{T}$ corresponds in a one–one manner to an operator Ψ on $\mathrm{Fun}(E, \mathcal{T})$. This interrelation is expressed by the intertwining diagram in Fig. 10.6.

If the operator $\psi^{\vartriangle}$ on $\mathcal{P}(E)_{\vartriangle}\mathcal{T}$ is given, then the operator Ψ on $\mathrm{Fun}(E, \mathcal{T})$ is obtained from

$$\Psi = \mathbf{F}^{\vartriangle}\psi^{\vartriangle}\mathbf{X}^{\vartriangle}. \tag{10.43}$$

Obviously, this construction method is compatible with suprema, infima, and compositions. Note that $\psi^{\vartriangle}$, Ψ are not required to be increasing. One can combine this correspondence between operators on $\mathrm{Fun}(E, \mathcal{T})$ and $\mathcal{P}(E)_{\vartriangle}\mathcal{T}$ with the construction method for operators on the lattice $\mathcal{L}_{\vartriangle}\mathcal{T}$ described in Section 10.6. Let $\{\psi_t \mid t \in \mathcal{T}\}$ be a decreasing family of increasing set operators, let $\psi^{\vartriangle}$ be the increasing operator on $\mathcal{P}(E)_{\vartriangle}\mathcal{T}$ generated by this family, and let Ψ be the operator on $\mathrm{Fun}(E, \mathcal{T})$ obtained via (10.43). A straightforward computation shows that

$$\Psi(F)(x) = \bigvee\{t \in \mathcal{T} \mid x \in \psi_t(\mathbf{X}^{\vartriangle}(F, t))\}. \tag{10.44}$$

Instead of the foregoing procedure, one can also exploit the isomorphism between $\mathrm{Fun}(E, \mathcal{T})$ and $\mathcal{P}(E)_{\triangledown}\mathcal{T}$ to extend set operators to function operators. It is clear that such an approach should lead to the same result. See also the diagram depicted in Fig. 10.7. This diagram shows, for example, that $\mathbf{F}^{\triangledown}\psi^{\triangledown}\mathbf{X}^{\triangledown} = \mathbf{F}^{\vartriangle}\psi^{\vartriangle}\hat{\mathbf{I}}\mathbf{X}^{\triangledown}$.

Analogously to (10.44) one obtains the identity:

$$\Psi(F)(x) = \bigwedge\{t \in \mathcal{T} \mid x \notin \psi_t(\mathbf{X}^{\triangledown}(F, t)). \tag{10.45}$$

$$\begin{array}{ccccccc}
\mathrm{Fun}(E,\mathcal{T}) & \xrightarrow{\mathbf{X}^{\nabla}} & \mathcal{P}(E)_{\nabla}\mathcal{T} & \xrightarrow{\psi^{\nabla}} & \mathcal{P}(E)_{\nabla}\mathcal{T} & \xrightarrow{\mathbf{F}^{\nabla}} & \mathrm{Fun}(E,\mathcal{T}) \\
\Big| & & \hat{\mathbf{I}}\big\downarrow\big\uparrow\check{\mathbf{I}} & & \hat{\mathbf{I}}\big\downarrow\big\uparrow\check{\mathbf{I}} & & \Big| \\
\mathrm{Fun}(E,\mathcal{T}) & \xrightarrow{\mathbf{X}^{\Delta}} & \mathcal{P}(E)_{\Delta}\mathcal{T} & \xrightarrow{\psi^{\Delta}} & \mathcal{P}(E)_{\Delta}\mathcal{T} & \xrightarrow{\mathbf{F}^{\Delta}} & \mathrm{Fun}(E,\mathcal{T})
\end{array}$$

Figure 10.7 Intertwining diagram for operators on $\mathcal{P}(E)_{\Delta}\mathcal{T}, \mathcal{P}(E)_{\nabla}\mathcal{T}$, and $\mathrm{Fun}(E, \mathcal{T})$.

We call Ψ the *semi-flat (function) operator generated by* the family $\{\psi_t \mid t \in \mathcal{T}\}$. If this family contains only one operator ψ, then Ψ is given by

$$\Psi(F)(x) = \bigvee\{t \in \mathcal{T} \mid x \in \psi(\mathbf{X}^{\Delta}(F, t))\}, \tag{10.46}$$

or alternatively,

$$\Psi(F)(x) = \bigwedge\{t \in \mathcal{T} \mid x \notin \psi(\mathbf{X}^{\nabla}(F, t))\}, \tag{10.47}$$

and is called the *flat operator generated by* ψ. The next result is an immediate consequence of Theorem 10.17.

10.25 Theorem. *Let $\mathcal{T}$ be a strongly admissible complete lattice. Assume that for every i in the index set I, $\{\psi_{i,t} \mid t \in \mathcal{T}\}$ is a decreasing family of increasing operators on $\mathcal{P}(E)$ which generates the operator Ψ_i on* $\mathrm{Fun}(E, \mathcal{T})$.

(a) *The family $\{\bigvee_{i\in I} \psi_{i,t} \mid t \in \mathcal{T}\}$ generates $\bigvee_{i\in I} \Psi_i$ on* $\mathrm{Fun}(E, \mathcal{T})$.
(a′) *The family $\{\bigwedge_{i\in I} \psi_{i,t} \mid t \in \mathcal{T}\}$ generates $\bigwedge_{i\in I} \Psi_i$ on* $\mathrm{Fun}(E, \mathcal{T})$.
(b) *Let $\{\phi_t \mid t \in \mathcal{T}\}$ and $\{\psi_t \mid t \in \mathcal{T}\}$ be decreasing families of increasing operators on $\mathcal{P}(E)$, and let Φ and Ψ be the operators generated on* $\mathrm{Fun}(E, \mathcal{T})$. *Then the composition $\Psi\Phi$ is generated by the family $\{\psi_t\phi_t \mid t \in \mathcal{T}\}$.*

Furthermore, Corollary 10.18 implies the following result.

10.26 Corollary. *Let $\mathcal{T}$ be a strongly admissible complete lattice, and let $\{\psi_t \mid t \in \mathcal{T}\}$ be a decreasing family of increasing operators on $\mathcal{P}(E)$ which generates the operator Ψ on* $\mathrm{Fun}(E, \mathcal{T})$.

(a) *If every ψ_t is a dilation, then Ψ is a dilation as well.*
(a′) *If every ψ_t is an erosion, then Ψ is an erosion as well.*
(b) *If every ψ_t is idempotent, then Ψ is idempotent as well.*
(c) *If every ψ_t is (anti-)extensive, then Ψ is (anti-)extensive as well.*
(d) *If every ψ_t is an opening, then Ψ is an opening as well.*
(d′) *If every ψ_t is a closing, then Ψ is a closing as well.*

If the grey-value lattice $\mathcal{T}$ has a negation $t \mapsto t^*$, then $F \mapsto F^*$, where $F^*(x) = (F(x))^*$, defines a negation on $\text{Fun}(E, \mathcal{T})$. Define the dual Ψ^* of a function operator Ψ as usual, i.e.,

$$\Psi^*(F) = (\Psi(F^*))^*.$$

10.27 Theorem. *Assume that $\mathcal{T}$ is a strongly admissible complete lattice with a negation. Let ψ be an increasing set operator, and let Ψ be the semi-flat function operator generated by the decreasing family $\{\psi_t \mid t \in \mathcal{T}\}$. Then Ψ^* is the semi-flat function operator generated by the decreasing family $\{(\psi_{t^*})^* \mid t \in \mathcal{T}\}$.*

Proof. It is easy to show that

$$\mathbf{X}^{\triangle}(F^*, t) = \left[\mathbf{X}^{\triangledown}(F, t^*)\right]^{c}.$$

If one uses this relation along with (10.44), one gets that

$$\begin{aligned}
\Psi^*(F)(x) &= \left[\Psi(F^*)(x)\right]^* \\
&= \left[\bigvee\{t \in \mathcal{T} \mid x \in \psi_t(\mathbf{X}^{\triangle}(F^*, t))\}\right]^* \\
&= \left[\bigvee\{t \in \mathcal{T} \mid x \in \psi_t\big((\mathbf{X}^{\triangledown}(F, t^*))^{c}\big)\}\right]^* \\
&= \bigwedge\{t^* \mid x \in \psi_t\big((\mathbf{X}^{\triangledown}(F, t^*))^{c}\big)\} \\
&= \bigwedge\{t \mid x \in \psi_{t^*}\big((\mathbf{X}^{\triangledown}(F, t))^{c}\big)\} \\
&= \bigwedge\{t \mid x \notin \left[\psi_{t^*}\big((\mathbf{X}^{\triangledown}(F, t))^{c}\big)\right]^{c}\} \\
&= \bigwedge\{t \mid x \notin (\psi_{t^*})^*\big(\mathbf{X}^{\triangledown}(F, t)\big)\}.
\end{aligned}$$

Now (10.45) gives the result. □

In particular, this result says that Ψ^* is a flat operator generated by ψ^* given that Ψ is a flat operator generated by ψ.

10.9. Bibliographical notes

A great deal of the material presented in this chapter appears here for the first time. However, extension of set operators to functions by means of thresholding is a well-known technique; in the next chapter it will be discussed in great detail. In Section 11.12 one finds further references to other literature in this area. In a recent paper Serra (1993) discusses flat extensions of set operators which are not necessarily increasing. In that paper the role of anamorphoses in the definition of flat operators is strongly emphasized. Although the basic idea is very similar, the overlap between (Serra, 1993) and the present chapter is rather small.

The concept of an u.s.c. function occurs at several places in the literature. We refer to Dal Masso (1993) for an analytic treatment, and to Gierz et al. (1980) for a lattice-theoretical discussion. The relevance of this class of functions in the context of mathematical morphology is apparent from the work of Matheron (1967). Finally, we point out one other possible application of functions mapping into a lattice, namely, as a mathematical model for *image sequences*.

CHAPTER ELEVEN

Morphology for grey-scale images

Henk J.A.M. Heijmans
Formerly Centre for Mathematics and Computer Science, Amsterdam, Netherlands

Contents

This chapter comprises a systematic discussion of morphological operators for grey-scale images. The previous chapter has presented a general and formal discussion of functions mapping a space $\mathbb{E}^d$ into a complete lattice $\mathcal{T}$ satisfying some admissibility condition. This admissibility condition is satisfied for all relevant grey-value lattices such as $\overline{\mathbb{R}}$, $\overline{\mathbb{Z}}$, $\{0, 1, \ldots, N\}$, and powers of these lattices such as $\overline{\mathbb{R}}^d$, etc. In the previous chapter it was explained how to extend increasing set operators to functions. The goal of the present chapter is different. The first part shows how to apply the abstract results of the previous chapter to the scalar case $\mathcal{T} = \overline{\mathbb{R}}$. This part is self-contained in the sense that the results can be understood without knowledge of the contents of the previous chapter. A number of the results presented here are, in fact, specializations of results obtained in the previous chapter. In these cases, the proof will be omitted. Unless stated otherwise, $\overline{\mathbb{R}}$ is the grey-value set and $\mathbb{E}^d$ (i.e., $\mathbb{R}^d$ or $\mathbb{Z}^d$) the domain space, and in such cases the notation $\mathrm{Fun}(\mathbb{E}^d)$ will be used instead of $\mathrm{Fun}(\mathbb{E}^d, \overline{\mathbb{R}})$.

Advances in Imaging and Electron Physics, Volume 216
ISSN 1076-5670
https://doi.org/10.1016/bs.aiep.2020.07.011

11.1. Functions and threshold sets

We briefly recall some basic facts from Section 4.6. For every function F, its negative F^* is defined by

$$F^*(x) = -F(x).$$

The horizontal translation of F by the vector $h \in \mathbb{E}^d$ is

$$F_h(x) = F(x - h);$$

the vertical translation by the scalar $v \in \mathbb{R}$ is

$$(F + v)(x) = F(x) + v.$$

Define the *threshold sets* $\mathbf{X}(F, t)$ of the function F by

$$\mathbf{X}(F, t) = \{x \in \mathbb{E}^d \mid F(x) \geq t\}. \tag{11.1}$$

These sets, which were denoted by $\mathbf{X}^{\triangle}(F, t)$ in the previous chapter, satisfy the continuity relation

$$\mathbf{X}(F, t) = \bigcap_{s<t} \mathbf{X}(F, s), \tag{11.2}$$

for every $t \in \overline{\mathbb{R}}$. Note that $\mathbf{X}(F, -\infty) = \mathbb{E}^d$. Let $\mathcal{P}(\mathbb{E}^d)_{\triangle}\overline{\mathbb{R}}$ denote the set of all mappings $\mathbf{X} : \overline{\mathbb{R}} \to \mathcal{P}(\mathbb{E}^d)$ that satisfy continuity relation (11.2). The attentive reader might object that this definition is different from the one presented in (10.12). It is straightforward to show, however, that both definitions coincide in this particular case. Define, for a mapping $\mathbf{X} : \overline{\mathbb{R}} \to \mathcal{P}(\mathbb{E}^d)$, the function $\mathbf{F}(\mathbf{X}) \in \mathrm{Fun}(\mathbb{E}^d)$ by

$$\mathbf{F}(\mathbf{X})(x) = \bigvee \{t \in \overline{\mathbb{R}} \mid x \in \mathbf{X}(t)\}. \tag{11.3}$$

The following result, which is a special case of Theorem 10.9, expresses that there is a one–one correspondence between elements of $\mathrm{Fun}(\mathbb{E}^d)$ and elements of $\mathcal{P}(\mathbb{E}^d)_{\triangle}\overline{\mathbb{R}}$.

11.1 Theorem. *The mapping $F \mapsto \mathbf{X}(F, \cdot)$ defines a lattice isomorphism between* $\mathrm{Fun}(\mathbb{E}^d)$ *and $\mathcal{P}(\mathbb{E}^d)_{\triangle}\overline{\mathbb{R}}$ with inverse $\mathbf{X} \mapsto \mathbf{F}(\mathbf{X})$.*

We summarize some basic properties of the given thresholding operation.

11.2 Proposition. *Suppose that* $F \in \mathrm{Fun}(\mathbb{E}^d)$ *and that* $F_i \in \mathrm{Fun}(\mathbb{E}^d)$ *for every* i *in some index set* I. *If* $h \in \mathbb{E}^d$ *and* $v \in \mathbb{R}$, *then*

(a) $\mathbf{X}(F_h, t) = [\mathbf{X}(F, t)]_h$,

(b) $\mathbf{X}(F + v, t) = \mathbf{X}(F, t - v)$,

(c) $\mathbf{X}(\bigwedge_{i\in I} F_i, t) = \bigcap_{i\in I} \mathbf{X}(F_i, t)$,

(d) $\mathbf{X}(\bigvee_{i\in I} F_i, t) = \bigcap_{s<t} \bigcup_{i\in I} \mathbf{X}(F_i, s)$.

Proof. The properties (a) and (b) are obvious. Furthermore, (c) follows from the results in the previous chapter. It remains to prove (d). It is obvious that $\mathbf{X}(\bigvee_{i\in I} F_i, t) \supseteq \bigcup_{i\in I} \mathbf{X}(F_i, t)$; this gives the inclusion $\supseteq$. To prove $\subseteq$ take $x \in \mathbf{X}(\bigvee_{i\in I} F_i, t)$ and $s < t$; then $\bigvee_{i\in I} F_i(x) \geq t$, and so there is an $i \in I$ such that $F_i(x) \geq s$; that is, $x \in \bigcup_{i\in I} \mathbf{X}(F_i, s)$. This implies the result. □

It follows from (c) that $F \mapsto \mathbf{X}(F, t)$ defines an erosion from $\mathrm{Fun}(\mathbb{E}^d)$ to $\mathcal{P}(\mathbb{E}^d)$ for every $t \in \overline{\mathbb{R}}$. The adjoint dilation $C(\cdot, t)$ is a mapping from $\mathcal{P}(\mathbb{E}^d)$ to $\mathrm{Fun}(\mathbb{E}^d)$ which is given by

$$C(X, t) = \begin{cases} t, & x \in X \\ -\infty, & x \notin X. \end{cases} \tag{11.4}$$

Note that $C(X, t)$ is a kind of characteristic function of X; this function was called *flat function* in Section 10.4 and was denoted by $C^{\triangle}(X, t)$. The formal proof that $(\mathbf{X}(\cdot, t), C(\cdot, t))$ is an adjunction between $\mathrm{Fun}(\mathbb{E}^d)$ and $\mathcal{P}(\mathbb{E}^d)$ can be found in Proposition 10.10.

11.2. Semi-flat function operators

This chapter is concerned with operators on $\mathrm{Fun}(\mathbb{E}^d)$. From now on we shall call these operators *function operators*, contrary to operators on $\mathcal{P}(\mathbb{E}^d)$, which will be called *set operators*. In Section 10.8 it has been explained how to construct an increasing function operator given a decreasing family of increasing set operators. Here we recall briefly the underlying method and the main properties. To a large extent, the approach is based on the following observation.

11.3 Lemma. *Let* $\mathbf{Y} : \mathbb{R} \to \mathcal{P}(\mathbb{E}^d)$ *be an arbitrary mapping; then* $\mathbf{X}(\cdot)$ *given by*

$$\mathbf{X}(t) = \bigcap_{s<t} \mathbf{Y}(s), \qquad t \in \overline{\mathbb{R}},$$

is an element of $\mathcal{P}(\mathbb{E}^d)_{\triangle}\overline{\mathbb{R}}$.

In Section 10.3 the operator mapping $\mathbf{Y}$ to $\mathbf{X}$ is denoted by $\hat{\mathbf{I}}$, and Proposition 10.6 lists several interesting properties of this operator. Note that in this lemma $\mathbf{Y}$ is not necessarily defined in $\pm\infty$; it is clear from the definition, however, that $\mathbf{X}(-\infty) = \mathbb{E}^d$ and $\mathbf{X}(+\infty) = \bigcap_{s<\infty} \mathbf{Y}(s)$.

If an operator, not necessarily increasing, acts on the threshold sets of a function, then the transformed sets do not constitute a decreasing family in general, and so they need not correspond to a function. This observation makes clear that the extension of set operators to functions by thresholding makes sense as long as one restricts attention to increasing operators.

11.4 Proposition. *Let $\{\psi_t \mid t \in \mathbb{R}\}$ be a family of increasing operators on $\mathcal{P}(\mathbb{E}^d)$ which is decreasing with respect to t, i.e., $\psi_t \le \psi_s$ if $t \ge s$. Let Ψ be the operator on* $\mathrm{Fun}(\mathbb{E}^d)$ *given by*

$$\Psi(F)(x) = \bigvee \{t \in \mathbb{R} \mid x \in \psi_t(\mathbf{X}(F, t))\}, \tag{11.5}$$

for $F \in \mathrm{Fun}(\mathbb{E}^d)$ and $x \in \mathbb{R}$; then Ψ is increasing and

$$\mathbf{X}(\Psi(F), t) = \bigcap_{s<t} \psi_s(\mathbf{X}(F, s)), \tag{11.6}$$

for $F \in \mathrm{Fun}(\mathbb{E}^d)$ and $t \in \overline{\mathbb{R}}$. If every ψ_t is translation invariant, then Ψ is an H-operator.

Proof. This result, with the exception of the statement about translation invariance, follows immediately from the abstract theory developed in the previous chapter; see, in particular, Section 10.8. If every ψ_s is translation invariant, then

$$\begin{aligned} x \in \psi_t(\mathbf{X}(F_h, t)) &\iff x \in \psi_t([\mathbf{X}(F, t)]_h) \\ &\iff x \in [\psi_t(\mathbf{X}(F, t))]_h \\ &\iff x - h \in \psi_t(\mathbf{X}(F, t)), \end{aligned}$$

and so it follows that $\Psi(F_h)(x) = \Psi(F)(x - h)$. This shows that Ψ is an H-operator. □

Ψ is a *semi-flat function operator* generated by the family $\{\psi_t \mid t \in \mathbb{R}\}$. The next three results are special cases of Theorem 10.25, Corollary 10.26, and Theorem 10.27, respectively.

11.5 Theorem. *Assume that for every i in the index set I, $\{\psi_{i,t} \mid t \in \overline{\mathbb{R}}\}$ is a decreasing family of increasing operators on $\mathcal{P}(\mathbb{E}^d)$ which generates the operator Ψ_i on* $\mathrm{Fun}(\mathbb{E}^d)$.

(a) *The family* $\{\bigvee_{i\in I}\psi_{i,t} \mid t\in\overline{\mathbb{R}}\}$ *generates* $\bigvee_{i\in I}\Psi_i$ *on* $\mathrm{Fun}(\mathbb{E}^d)$.
(a′) *The family* $\{\bigwedge_{i\in I}\psi_{i,t} \mid t\in\overline{\mathbb{R}}\}$ *generates* $\bigwedge_{i\in I}\Psi_i$ *on* $\mathrm{Fun}(\mathbb{E}^d)$.
(b) *Let* $\{\phi_t \mid t\in\overline{\mathbb{R}}\}$ *and* $\{\psi_t \mid t\in\overline{\mathbb{R}}\}$ *be decreasing families of increasing operators on* $\mathcal{P}(\mathbb{E}^d)$, *and let* Φ *and* Ψ *be the operators generated on* $\mathrm{Fun}(\mathbb{E}^d)$, *respectively; then the composition* $\Psi\Phi$ *is generated by the family* $\{\psi_t\phi_t \mid t\in\overline{\mathbb{R}}\}$.

11.6 Corollary. *Let* $\{\psi_t \mid t\in\overline{\mathbb{R}}\}$ *be a decreasing family of increasing operators on* $\mathcal{P}(\mathbb{E}^d)$ *which generates the operator* Ψ *on* $\mathrm{Fun}(\mathbb{E}^d)$.
(a) *If every* ψ_t *is a dilation, then* Ψ *is a dilation as well.*
(a′) *If every* ψ_t *is an erosion, then* Ψ *is an erosion as well*
(b) *If every* ψ_t *is idempotent, then* Ψ *is idempotent as well.*
(c) *If every* ψ_t *is (anti-)extensive, then* Ψ *is (anti-)extensive as well.*
(d) *If every* ψ_t *is an opening, then* Ψ *is an opening as well.*
(d′) *If every* ψ_t *is a closing, then* Ψ *is a closing as well.*

Given an operator Ψ on $\mathrm{Fun}(\mathbb{E}^d)$, the negative operator Ψ^* is given by

$$\Psi^*(F) = -\Psi(-F).$$

11.7 Theorem. *Suppose that the decreasing family* $\{\psi_t \mid t\in\mathbb{R}\}$ *generates the operator* Ψ *on* $\mathrm{Fun}(\mathbb{E}^d)$; *then* $\{(\psi_{-t})^* \mid t\in\mathbb{R}\}$ *is a decreasing family as well, and it generates the negative operator* Ψ^*.

11.3. Flat function operators

The results of the previous section apply in particular to the case where the generating family has only one member, say ψ. In this case the associated function operator Ψ is given by

$$\Psi(F)(x) = \bigvee\{t\in\overline{\mathbb{R}} \mid x\in\psi(\mathbf{X}(F,t))\}, \tag{11.7}$$

and the threshold sets of $\Psi(F)$ are given by

$$\mathbf{X}(\Psi(F),t) = \bigcap_{s<t}\psi(\mathbf{X}(F,s)). \tag{11.8}$$

11.8 Definition. An increasing operator Ψ on $\mathrm{Fun}(\mathbb{E}^d)$ is called a *flat operator* if there is an increasing set operator ψ such that (11.7) or, alternatively, (11.8) holds. In that case we say that ψ is the *generator* of Ψ.

11.9 Proposition.

(a) *Every increasing flat function operator* Ψ *has a unique generator* ψ *given by*

$$\psi(X) = \mathbf{X}(\Psi(C(X,t)),t), \tag{11.9}$$

where $C(X,t)$ *is given by* (11.4) *and* $t \in \mathbb{R} \cup \{\infty\}$ *is arbitrary.*

(b) *Consider a flat function operator* Ψ *generated by* ψ. *If* $\psi(\varnothing) = \varnothing$, *then*

$$\Psi(C(X,t)) = C(\psi(X),t), \tag{11.10}$$

for every $t \in \overline{\mathbb{R}}$ *and* $X \subseteq \mathbb{E}^d$.

Proof. (a): Relation (11.9) follows immediately if one substitutes $F = C(X,t)$ in (11.8) and uses that $\mathbf{X}(C(X,t),s) = X$ for $s < t$.

(b): To prove (11.10), we use (11.7). Let $t \in \overline{\mathbb{R}}$ and $X \subseteq \mathbb{E}^d$; then $\mathbf{X}(C(X,t),s) = X$ if $s \le t$. Therefore

$$\begin{aligned} \Psi(C(X,t))(x) &= \bigvee\{s \in \mathbb{R} \mid x \in \psi(\mathbf{X}(C(X,t),s))\} \\ &= \bigvee\{s \le t \mid x \in \psi(X)\}. \end{aligned}$$

This shows the result. □

Relation (11.10) means that Ψ maps a flat function onto a flat function (with the same height). The converse is not true: there exist increasing function operators which do have this property but are not flat. Consider, for instance, the operator Ψ given by

$$\Psi(F)(x) = \begin{cases} M_F, & \text{if } F(x) > -\infty, \\ -\infty, & \text{if } F(x) = -\infty; \end{cases}$$

here $M_F = \bigvee\{F(x) \mid x \in \mathbb{E}^d\}$. Note that Ψ is a T-closing. It is easy to check that

$$\Psi(C(X,t)) = C(X,t),$$

meaning that (11.10) holds with $\psi = \mathrm{id}$. The operator Ψ is not generated by id, however, for otherwise it would be the identity operator.

If the set operator ψ is $\downarrow$-continuous, relation (11.8) can be simplified by interchanging ψ with the intersection on the right-hand side. This means

$$\mathbf{X}(\Psi(F),t) = \psi(\mathbf{X}(F,t)), \qquad \text{if } \psi \text{ is } \downarrow\text{-continuous.} \tag{11.11}$$

11.10 Proposition. *Every increasing flat function operator* Ψ *is invariant under vertical translations, i.e.,*

$$\Psi(F+v)=\Psi(F)+v,$$

for $F\in \mathrm{Fun}(\mathbb{E}^d)$ *and* $v\in\mathbb{R}$. *In particular,* Ψ *is a T-operator iff its generator is translation invariant.*

Proof. The statement concerning the horizontal translation invariance is straightforward; see also Proposition 11.4. Suppose that ψ is the generator of Ψ; then

$$\begin{aligned}\mathbf{X}(\Psi(F+v),t)&=\bigcap_{s<t}\psi(\mathbf{X}(F+v,s))=\bigcap_{s<t}\psi(\mathbf{X}(F,s-v))\\&=\bigcap_{s<t-v}\psi(\mathbf{X}(F,s))=\mathbf{X}(\Psi(F),t-v)\\&=\mathbf{X}(\Psi(F)+v,t);\end{aligned}$$

this shows that Ψ is invariant under vertical translations. □

11.11 Remark. The following converse of Proposition 11.10 holds: if Ψ is a semi-flat function operator generated by the family $\{\psi_t \mid t\in\mathbb{R}\}$ and if Ψ is invariant under vertical translations, then all ψ_t are identical and Ψ is flat. In fact, for $v\in\mathbb{R}$,

$$\begin{aligned}\mathbf{X}(\Psi(F+v),t)&=\bigcap_{s<t}\psi_s(\mathbf{X}(F+v,s))\\&=\bigcap_{s<t}\psi_s(\mathbf{X}(F,s-v)).\end{aligned}$$

This expression equals

$$\begin{aligned}\mathbf{X}(\Psi(F)+v,t)&=\mathbf{X}(\Psi(F),t-v)\\&=\bigcap_{s<t-v}\psi_s(\mathbf{X}(F,s)).\end{aligned}$$

Substituting $F=C(X,\infty)$ in both expressions and using that $\mathbf{X}(F,s)=X$ for every s, one gets

$$\bigcap_{s<t}\psi_s(X)=\bigcap_{s<t-v}\psi_s(X).$$

From the fact that this relation holds for every $t,v\in\mathbb{R}$, one derives that $\psi_t(X)$ is independent of t.

Recall from Definition 4.32 that a mapping $a : \overline{\mathbb{R}} \to \overline{\mathbb{R}}$ is called a *strong anamorphosis* if it is continuous and strictly increasing. Alternatively, a is a strong anamorphosis iff it is a lattice isomorphism from $\overline{\mathbb{R}}$ to its range $\{a(t) \mid t \in \overline{\mathbb{R}}\}$. It is evident that $t \mapsto t + v$ is a strong anamorphosis for every scalar $v \in \mathbb{R}$. The next result is a straightforward generalization of Proposition 10.20.

11.12 Theorem. *Let Ψ be a flat operator generated by ψ, and assume that $a : \overline{\mathbb{R}} \to \overline{\mathbb{R}}$ is a strong anamorphosis and that*

(i) $\psi(\mathbb{E}^d) = \mathbb{E}^d$ *if* $a(-\infty) > -\infty$;

(ii) $\psi(\varnothing) = \varnothing$ *if* $a(\infty) < \infty$.

Then $\Psi(a(F)) = a(\Psi(F))$ *for* $F \in \mathrm{Fun}(\mathbb{E}^d)$.

Proof. Since a is continuous and strictly increasing, it has a well-defined inverse a^{-1} on the interval $(a(-\infty), a(\infty))$. For $t \in (a(-\infty), a(\infty))$,

$$\begin{aligned}
\mathbf{X}(a(\Psi(F)), t) &= \mathbf{X}(\Psi(F), a^{-1}(t)) = \bigwedge_{s < a^{-1}(t)} \psi(\mathbf{X}(F, s)) \\
&= \bigwedge_{s<t} \psi(\mathbf{X}(F, a^{-1}(s))) = \bigwedge_{s<t} \psi(\mathbf{X}(a(F), s)) \\
&= \mathbf{X}(\Psi(a(F)), t).
\end{aligned}$$

Suppose that $a(-\infty) > -\infty$ and $t \in (-\infty, a(-\infty))$; then $\mathbf{X}(a(\Psi(F)), t) = \mathbb{E}^d$. On the other hand, by (iii),

$$\mathbf{X}(\Psi(a(F)), t) = \bigwedge_{s<t} \psi(\mathbf{X}(a(F), s) = \bigwedge_{s<t} \psi(\mathbb{E}^d) = \mathbb{E}^d.$$

The proof for $t \in (a(\infty), \infty)$ in case that $a(\infty) < \infty$ follows by a similar argument. □

It is evident that every lattice automorphism a on $\overline{\mathbb{R}}$ is a strong anamorphosis with $a(-\infty) = -\infty$ and $a(\infty) = \infty$. We specialize Theorem 11.5 to flat operators.

11.13 Theorem. *Let Ψ_i be a flat function operator generated by ψ_i, for every i in some index set I.*

(a) *The operator* $\bigvee_{i \in I} \psi_i$ *generates* $\bigvee_{i \in I} \Psi_i$ *on* $\mathrm{Fun}(\mathbb{E}^d)$.

(a′) *The operator* $\bigwedge_{i \in I} \psi_i$ *generates* $\bigwedge_{i \in I} \Psi_i$ *on* $\mathrm{Fun}(\mathbb{E}^d)$.

(b) *Let Φ, Ψ be flat function operators generated by ϕ and ψ, respectively; then $\Psi\Phi$ is a flat operator generated by $\psi\phi$.*

This result implies that flat function operators are ordered in the same way as their generators; if ϕ, ψ are the generators of Φ and Ψ and if $\phi \le \psi$, then also $\Phi \le \Psi$. The next result follows from Corollary 11.6.

11.14 Corollary. *Let Ψ be a flat function operator generated by the increasing set operator ψ.*

(a) *If ψ is a dilation, then Ψ is a dilation as well.*
(a′) *If ψ is an erosion, then Ψ is an erosion as well.*
(b) *If ψ is idempotent, then Ψ is idempotent as well.*
(c) *If ψ is (anti-)extensive, then Ψ is (anti-)extensive as well.*
(d) *If ψ is an opening, then Ψ is an opening as well.*
(d′) *If ψ is a closing, then Ψ is a closing as well.*

A dilation on functions which is flat is called a *flat dilation.* Flat erosions, openings, closings, filters, etc., are defined in a similar way. In fact one can prove slightly stronger statements for adjunctions than those in (a) and (a′).

11.15 Proposition. *Let $(\mathcal{E}, \Delta)$ be an adjunction on* $\text{Fun}(\mathbb{E}^d)$. *Suppose that at least one of these operators is flat; then both operators are flat. If ε and δ are the respective generators of $\mathcal{E}$ and Δ, then (ε, δ) constitutes an adjunction on $\mathcal{P}(\mathbb{E}^d)$.*

Proof. Suppose, e.g., that $\mathcal{E}$ is a flat erosion, and let ε be the set erosion which generates $\mathcal{E}$. Let δ be the adjoint set dilation, and let Δ be the function dilation generated by δ. We show that $(\mathcal{E}, \Delta)$ is an adjunction. Since Δ is uniquely determined by $\mathcal{E}$, this implies the result.

Let $F, G \in \text{Fun}(\mathbb{E}^d)$ be such that $\Delta(F) \le G$. We show that $F \le \mathcal{E}(G)$. For $t \in \mathbb{R}$ one has $\mathbf{X}(\Delta(F), t) = \bigwedge_{s<t} \delta(\mathbf{X}(F, s)) \subseteq \mathbf{X}(G, t)$; hence $\delta(\mathbf{X}(F, t)) \subseteq \mathbf{X}(G, t)$. From the fact that (ε, δ) is an adjunction it follows that $\mathbf{X}(F, t) \subseteq \varepsilon(\mathbf{X}(G, t)) = \mathbf{X}(\mathcal{E}(G), t)$; here we have used (11.11). We conclude that $F \le \mathcal{E}(G)$.

Let $F \le \mathcal{E}(G)$; we show that $\Delta(F) \le G$. Given $s \in \mathbb{R}$, we have $\mathbf{X}(F, s) \subseteq \mathbf{X}(\mathcal{E}(G), s) = \varepsilon(\mathbf{X}(G, s))$ because ε is $\downarrow$ continuous. Since (ε, δ) is an adjunction, $\delta(\mathbf{X}(F, s)) \subseteq \mathbf{X}(G, s)$. Let $t \in \mathbb{R}$, and take the intersection over $s < t$; this leads to $\bigwedge_{s<t} \delta(\mathbf{X}(F, s)) \subseteq \bigwedge_{s<t} \mathbf{X}(G, s) = \mathbf{X}(G, t)$, or equivalently, $\mathbf{X}(\Delta(F), t) \subseteq \mathbf{X}(G, t)$. This implies that $\Delta(F) \le G$. □

Adjunctions on $\text{Fun}(\mathbb{E}^d)$ resulting from flat operators will be called *flat adjunctions.* The next result follows from Theorem 11.7.

11.16 Theorem. *Let Ψ be a flat function operator generated by ψ; then Ψ^* is a flat function operator generated by ψ^*.*

11.17 Example. Let ε_A and δ_A be the erosion and dilation by the structuring element A, i.e.,

$$\varepsilon_A(X) = X \ominus A \qquad \text{and} \qquad \delta_A(X) = X \oplus A.$$

We compute the flat erosion $\mathcal{E}_A$ and dilation Δ_A generated by ε_A and δ_A, respectively. It is easy to see that the function operator generated by the translation $X \mapsto X_h$ is $F \mapsto F_h$. This fact in combination with Theorem 11.13 shows that $\mathcal{E}_A$ and Δ_A are respectively given by

$$\mathcal{E}_A(X) = \bigwedge_{h \in A} F_{-h} \qquad \text{and} \qquad \delta_A(X) = \bigvee_{h \in A} F_h.$$

See also (4.135)–(4.136) and Fig. 4.13.

11.18 Theorem. (Representation theorem for flat T-operators)
Every increasing flat T-operator can be represented as a supremum of flat T-erosions or, alternatively, as an infimum of flat T-dilations.

Proof. Every flat T-operator is generated by a translation invariant set operator. This set operator can be represented as a supremum of translation invariant erosions and also as an infimum of translation invariant dilations; see Theorem 4.15. Now the result follows from Theorem 11.13 and Corollary 11.14. □

11.19 Example. (Top-hat operator)
In Section 4.6 we have discussed the top-hat operator. Let α be an opening on $\mathcal{P}(\mathbb{E}^d)$, and let $\mathcal{\alpha}$ be its flat extension to $\mathsf{Fun}(\mathbb{E}^d)$. The mapping $F \mapsto F - \mathcal{\alpha}(F)$ is called top-hat operator. We demonstrate that

$$\mathcal{\alpha}(F - \mathcal{\alpha}(F)) = 0 \qquad \text{on } \mathbb{E}^d.$$

Put $F' = F - \mathcal{\alpha}(F)$, and suppose that $\mathcal{\alpha}(F')(x) > 0$ for some $x \in \mathbb{E}^d$. Fix t with $0 < t < \mathcal{\alpha}(F')(x)$. By definition, $\mathcal{\alpha}(F')(x) = \bigvee\{s \mid x \in \alpha(X(F', s))\}$. Put $C = \alpha(X(F', t))$; then $C \neq \varnothing$ because $x \in C$. Furthermore, since $C \subseteq X(F', t)$, $F'(y) \geq t$ for $y \in C$. From $F = F' + \mathcal{\alpha}(F)$ we conclude that $F(y) \geq \mathcal{\alpha}(F)(y) + t$ for $y \in C$. Since C is invariant under α, this means that $\mathcal{\alpha}(F)(y) \geq \mathcal{\alpha}(F)(y) + t$, a contradiction. We conclude that $\mathcal{\alpha}(F') = 0$ on $\mathbb{E}^d$.

11.20 Example. (Dolby openings)
Let α_1, α_2 be two openings on $\mathcal{P}(\mathbb{E}^d)$ with $\alpha_1 \leq \alpha_2$, and let $\mathcal{\alpha}_1, \mathcal{\alpha}_2$ be their respective flat extensions to $\mathsf{Fun}(\mathbb{E}^d)$. For a given function F we apply $\mathcal{\alpha}_1$

to the original F and α_2 to the lower thresholded copy of F_c given by

$$F_c(x) = \begin{cases} F(x), & \text{if } F(x) \geq c, \\ -\infty, & \text{otherwise.} \end{cases}$$

Let α be given by

$$\alpha(F) = \alpha_1(F) \vee \alpha_2(F_c).$$

It is evident that α is increasing and anti-extensive. We prove that α is idempotent, and hence an opening. First, observe that $\alpha_1\alpha \geq \alpha_1\alpha_1 = \alpha_1$. From $\alpha_2(F_c)(x) = \bigvee\{t \mid x \in \alpha_2(\mathbf{X}(F_c, t))\}$ it follows readily that $\alpha_2(F_c)$ does not assume values between $-\infty$ and c, and so $\alpha_2(F_c)_c = \alpha_2(F_c)$. Consequently,

$$\alpha_2(\alpha(F)_c) \geq \alpha_2(\alpha_2(F_c)_c) = \alpha_2(\alpha_2(F_c)) = \alpha_2(F_c).$$

This means that $\alpha\alpha \geq \alpha$. Since the reverse inequality follows immediately from the anti-extensivity of α, it follows that α is idempotent. The opening α filters lower values of a function more severly than higher values. Such performance resembles that of the Dolby filter in audio signal processing which filters high frequencies more severely when their amplitude is small. The construction of a Dolby opening is illustrated in Fig. 11.1; here α_1, α_2 are openings by line segments with lengths $l_1 \geq l_2$, respectively.

Observe that one can construct *Dolby closings* in a similar way; in this case one has to apply a strong closing to the original function, a weak closing to an upper thresholded copy, and take the infimum of both functions. Finally, we point out that one may likewise build Dolby openings and closings using families containing more than two operators.

11.4. Flat operators and Boolean functions

In Section 4.6 it has been explained briefly how to define increasing T-operators by means of Boolean functions. The present section contains a formal treatment of this issue. Recall the definition of a finite window operator on $\mathcal{P}(\mathbb{E}^d)$ from Definition 4.29. By Theorem 4.30 every increasing finite window operator is of the form

$$\psi_b(X) = \{h \in \mathbb{E}^d \mid b(X(a_1 + h), X(a_2 + h), \ldots, X(a_n + h)) = 1\}, \quad (11.12)$$

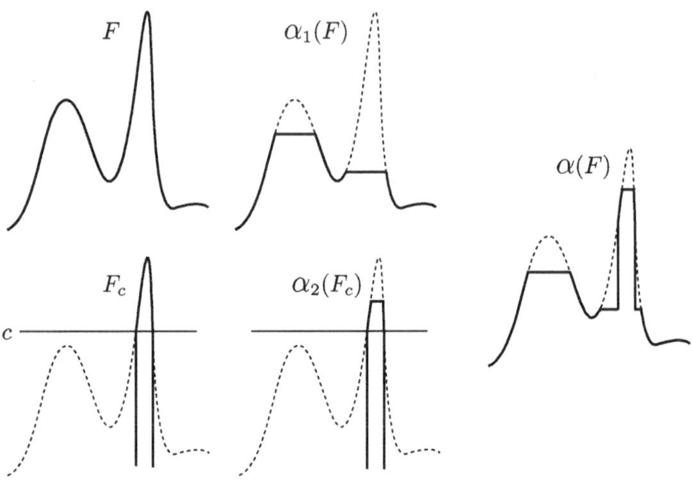

Figure 11.1 The Dolby opening $\alpha(F) = \alpha_1(F) \vee \alpha_2(F_c)$.

where $A = \{a_1, a_2, \ldots, a_n\}$ is a structuring element and b is an increasing Boolean function. There are two ways to generalize such operators to functions. First, one has the flat function operator Ψ_b generated by ψ_b. But alternatively, one can first extend b to $\overline{\mathbb{R}}^n$ and use this extension to define an operator on $\text{Fun}(\mathbb{E}^d)$. We show that both approaches lead to the same result. One can extend b to a function $b^\sim : \overline{\mathbb{R}}^n \to \overline{\mathbb{R}}$ in the following way:

$$b^\sim(t_1, t_2, \ldots, t_n) = \bigvee \{t \in \overline{\mathbb{R}} \mid b([t_1 \geq t], \ldots, [t_n \geq t]) = 1\}, \tag{11.13}$$

where $[t_1 \geq t]$ is a Boolean expression which equals 1 if $t_1 \geq t$ and 0 otherwise. It is easy to show that the following properties are satisfied:

$$(b_1 \cdot b_2)^\sim = b_1^\sim \wedge b_2^\sim,$$
$$(b_1 + b_2)^\sim = b_1^\sim \vee b_2^\sim,$$

where $\cdot$ and $+$ denote the logic AND and OR, respectively. For example, if $b(x_1, x_2, x_3) = x_1 x_2 + x_3$, then $b^\sim(t_1, t_2, t_3) = (t_1 \wedge t_2) \vee t_3$. Observe that this definition of $b^\sim$ corresponds to the definition in Section 4.6, where b was extended by replacing a product by an infimum, a sum by a supremum, a 0 by ∞ and a 1 by $-\infty$. Hereafter, a Boolean function and its extension to $\overline{\mathbb{R}}^n$ will be denoted by the same symbol. From the definition of Ψ_b, it follows that

$$\Psi_b(F)(x) = \bigvee \{t \in \overline{\mathbb{R}} \mid x \in \psi_b(\mathbf{X}(F, t))\}$$

$$= \bigvee \{t \in \overline{\mathbb{R}} \mid b([F(x+a_1) \geq t], \ldots, [F(x+a_n) \geq t]) = 1\}.$$

In combination with (11.13) this means that

$$\Psi_b(F)(x) = b(F(x+a_1), \ldots, F(x+a_n)). \tag{11.14}$$

Given a Boolean threshold function b with realization vector $(w_1, \ldots, w_n \mid s)$, assume that all entries are nonnegative integers. Thus,

$$b(u_1, \ldots, u_n) = \Big[\sum_{i=1}^{n} w_i u_i \geq s\Big].$$

In Section 2.4 it has been explained that one can compute $b(u_1, \ldots, u_n)$ by sorting the values u_i, counted w_i times, in decreasing order and picking the sth value. Using the expression given by (11.13), it is straightforward to show that for grey-values the same procedure can be followed. Let w be a function which maps $\mathbb{E}^d$ into $\mathbb{Z}_+$ and has finite domain, i.e., $w(h) = 0$ for all but finitely many vectors $h \in \mathbb{E}^d$. Consider the increasing set operator defined in (4.90):

$$X \star_s w = \{x \in \mathbb{E}^d \mid \sum_{h \in \mathbb{E}^d} w(h) X(x+h) \geq s\}.$$

Recall that this operator is a *weighted rank operator*. The flat function operator generated by this set operator is

$$(F \star_s w)(x) = \bigvee \{t \in \overline{\mathbb{R}} \mid \sum_{h \in \mathbb{E}^d} w(h) [F(x+h) \geq t] \geq s\},$$

and it is also called weighted rank operator. As explained previously, the value $(F \star_s w)(x)$ is most easily obtained by sorting the values $F(x+h)$, counted $w(h)$ times, in decreasing order and picking the sth value. If all nonzero weights are equal to 1, the resulting operator is called *rank operator* and is denoted by $\rho_{A,s}$; here A is the domain of w. Refer to Section 4.6 for further details.

11.5. H-operators

Recall that a function operator is called an H-operator if it commutes with horizontal translations. In Proposition 11.4 we have seen that a semi-flat operator generated by a family of translation invariant set operators is an

H-operator. The reader should have no trouble in verifying that the class of H-operators is closed under suprema, infima, compositions, and negation.

The main objective of this section is to give a complete characterization of H-adjunctions and a representation theorem for increasing H-operators in terms of H-dilations and H-erosions.

11.21 Proposition. *The pair of operators* $(\mathcal{E}, \Delta)$ *on* $\mathrm{Fun}(\mathbb{E}^d)$ *given by*

$$\Delta(F)(x) = \bigvee_{h \in \mathbb{E}^d} d_h(F(x-h)), \tag{11.15}$$

$$\mathcal{E}(F)(x) = \bigwedge_{h \in \mathbb{E}^d} e_h(F(x+h)), \tag{11.16}$$

is an H-adjunction if (e_h, d_h) *is an adjunction on* $\overline{\mathbb{R}}$ *for every* $h \in \mathbb{E}^d$. *Moreover, every H-adjunction is of this form.*

Proof. First we show that the pair $(\mathcal{E}, \Delta)$ given by (11.15)–(11.16) is an H-adjunction. It is easy to see that these operators are invariant under H-translations. It remains to show that $(\mathcal{E}, \Delta)$ satisfies the adjunction relation. Now

$$\begin{aligned}
\Delta(F) \le G &\iff \bigvee_{h \in \mathbb{E}^d} d_h(F(x-h)) \le G(x) \\
&\iff \forall h \in \mathbb{E}^d: \quad d_h(F(x-h)) \le G(x) \\
&\iff \forall h \in \mathbb{E}^d: \quad F(x-h) \le e_h(G(x)) \\
&\iff \forall h \in \mathbb{E}^d: \quad F(x) \le e_h(G(x+h)) \\
&\iff F(x) \le \bigwedge_{h \in \mathbb{E}^d} e_h(G(x+h)) \\
&\iff F \le \mathcal{E}(G),
\end{aligned}$$

which proves the result.

Assume that $(\mathcal{E}, \Delta)$ is an H-adjunction. Proposition 5.3 implies that for every $x, y \in \mathbb{E}^d$ there exists an adjunction $(\varepsilon_{y,x}, \delta_{x,y})$ on $\overline{\mathbb{R}}$ such that $\Delta(F)(x) = \bigvee_{x \in \mathbb{E}^d} \delta_{x,y}(F(x))$. But the H-invariance of Δ gives that $\delta_{x,y} = \delta_{x-y,0}$. Putting $d_h = \delta_{h,0}$, the result follows. □

Before we present any examples we give a complete description of adjunctions on $\overline{\mathbb{R}}$. The next result is trivial.

11.22 Lemma.

(a) *e is an erosion on $\overline{\mathbb{R}}$ if and only if e is increasing, continuous from the right, and satisfies $e(+\infty) = +\infty$.*

(a′) *d is a dilation on $\overline{\mathbb{R}}$ if d is increasing, continuous from the left, and satisfies $d(-\infty) = -\infty$.*

Proposition 3.14 implies that

$$e(t) = \bigvee\{s \in \overline{\mathbb{R}} \mid d(s) \le t\},$$
$$d(t) = \bigwedge\{s \in \overline{\mathbb{R}} \mid t \le e(s)\},$$

if (e, d) is an adjunction on $\overline{\mathbb{R}}$. It is obvious that the pair $d(t) = at + b$, $e(t) = (t - b)/a$, with $a > 0$ and $b \in \overline{\mathbb{R}}$, is an adjunction on $\overline{\mathbb{R}}$. In the case of ambiguous expressions one should use the convention that $s + t = -\infty$ if $s = -\infty$ or $t = -\infty$ and that $s - t = \infty$ if $s = \infty$ or $t = -\infty$. Note in particular that $\infty + (-\infty) = -\infty$, whereas $\infty - \infty = \infty$.

If one chooses $d_h(t) = t + G(h)$ and $e_h(t) = t - G(h)$ in Proposition 11.21, where $G: E \to \overline{\mathbb{R}}$ is an arbitrary function, then one arrives at the pair

$$\Delta(F)(x) = (F \oplus G)(x) = \bigvee_{h \in \mathbb{E}^d} [F(x - h) + G(h)],$$
$$\mathcal{E}(F)(x) = (F \ominus G)(x) = \bigwedge_{h \in \mathbb{E}^d} [F(x + h) - G(h)].$$

These operators are, respectively, the Minkowski addition and subtraction for functions with additive structuring function, previously discussed in Sections 4.6 and 5.7.1. Both operators are invariant under horizontal as well as vertical translations.

Alternatively, one can take $d_h(t) = A(h)t$ and $e_h(t) = t/A(h)$, where $A : E \to (0, \infty)$. In that case one obtains the Minkowski addition and subtraction with multiplicative structuring functions, discussed in Section 5.7.2. It is also possible to combine the additive and multiplicative approach by choosing $d_h(t) = A(h)t + G(h)$ and $e_h(t) = (t - G(h))/A(h)$. In this case,

$$\Delta(F)(x) = \bigvee_{h \in \mathbb{E}^d} [A(h)F(x - h) + G(h)],$$
$$\mathcal{E}(F)(x) = \bigwedge_{h \in \mathbb{E}^d} \frac{1}{A(h)}[F(x + h) - G(h)].$$

This section concludes with the following representation theorem.

11.23 Theorem. (Representation of H-operators)

(a) *Every increasing H-operator Ψ which satisfies $\Psi(O) = O$ can be represented as an infimum of H-dilations.*

(a′) *Every increasing H-operator Ψ which satisfies $\Psi(I) = I$ can be represented as a supremum of H-erosions.*

Proof. Let $\Psi(O) = O$, and let $\mathcal{D}$ be the set of all H-dilations Δ that dominate Ψ, that is, $\Delta \geq \Psi$. It is clear that $\bigwedge \mathcal{D} \geq \Psi$. To prove the reverse inequality, it suffices to show that for every $F \in \mathrm{Fun}(\mathbb{E}^d)$ there is a $\Delta \in \mathcal{D}$ such that

$$\Psi(F)(0) \geq \Delta(F)(0). \tag{$*$}$$

In fact, this implies that

$$\Psi(F)(0) \geq (\bigwedge \mathcal{D})(F)(0),$$

for every $F \in \mathrm{Fun}(\mathbb{E}^d)$, and hence that

$$\Psi(F)(x) = \Psi(F_{-x})(0) \geq (\bigwedge \mathcal{D})(F_{-x})(0) = (\bigwedge \mathcal{D})(F)(x).$$

This means that $\Psi \geq \bigwedge \mathcal{D}$.

To prove $(*)$, take $F \in \mathrm{Fun}(\mathbb{E}^d)$. Define for $h \in \mathbb{E}^d$ the mapping $d_h : \overline{\mathbb{R}} \to \overline{\mathbb{R}}$ by

$$d_h(t) = \begin{cases} -\infty, & \text{if } t = -\infty, \\ \Psi(F)(0), & \text{if } -\infty < t \leq F(-h), \\ \infty, & \text{if } t > F(-h). \end{cases}$$

Obviously, d_h is a dilation. Let Δ be the H-dilation given by (11.15). It follows immediately that

$$\Delta(F)(0) = \bigvee_{h \in \mathbb{E}^d} d_h(F(-h)) = \Psi(F)(0).$$

To finish the proof, one has to show that $\Delta \in \mathcal{D}$, i.e., $\Psi \leq \Delta$. Note first that $\Psi \leq \Delta$ if and only if $\Psi(F')(0) \leq \Delta(F')(0)$ for every function F'. Three cases can be distinguished:

(i) $F' = \mathcal{O}$: trivial.

(ii) $\mathcal{O} \neq F' \leq F$. Then

$$\Delta(F')(0) = \bigvee_{h \in \mathbb{E}^d} d_h(F'(-h)) = \Psi(F)(0) \geq \Psi(F')(0).$$

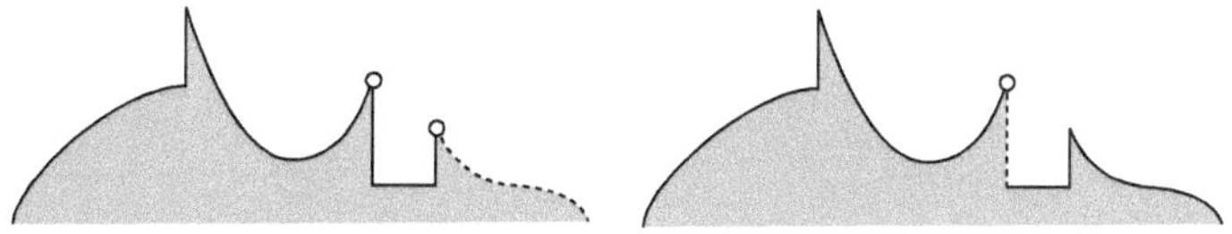

Figure 11.2 Pre-umbra (left) and umbra (right).

(iii) $F' \not\leq F$. Then $\Delta(F')(0) = \infty \geq \Psi(F')(0)$.
This concludes the proof. □

11.6. Umbra transform

Besides the extension of morphological operators from sets to functions by thresholding, there exists yet another procedure for extending set operators to functions. This procedure is based on the *umbra transform*. The current section presents a detailed discussion of this tool. It will become clear to the reader that the umbra approach combines well with the lattice framework.

11.24 Definition. A subset $U \subseteq \mathbb{E}^d \times \mathbb{R}$ is called a *pre-umbra* if $(x, t) \in U$ implies that $(x, s) \in U$ for every $s < t$. It is called an *umbra* if the reverse implication also holds, that is, if $(x, s) \in U$ for every $s < t$ implies that $(x, t) \in U$.

Both notions are illustrated in Fig. 11.2. Recall that the umbra transform has been used as a visualization tool in Section 4.6.

We denote the family of umbras on $\mathbb{E}^d$ by Umbra($\mathbb{E}^d$). Evidently, Umbra($\mathbb{E}^d$) is a subset of $\mathcal{P}(\mathbb{E}^d \times \mathbb{R})$ which is closed under intersections. Define, for a subset $X \subseteq \mathbb{E}^d \times \mathbb{R}$, the umbra $\mathcal{U}_s(X)$ as the smallest umbra that contains X; in other words, $\mathcal{U}_s(X)$ is the intersection of all umbras that contain X. Here the subscript "s" indicates that $\mathcal{U}_s$ returns the umbra of a set. Refer to Fig. 11.3 for an illustration.

A union of umbras is in general not a umbra. For instance, if U_n is the umbra $\{(0, t) \mid t \leq -\frac{1}{n}\}$, then $\bigcup_{n \geq 1} U_n = \{(0, t) \mid t < 0\}$, which is a pre-umbra but not an umbra since it does not contain the point $(0, 0)$. Nevertheless, one has the following result.

11.25 Proposition. *The set* Umbra($\mathbb{E}^d$) *with the inclusion ordering is a complete lattice. The infimum and supremum of a family U_i, $i \in I$, are given respectively*

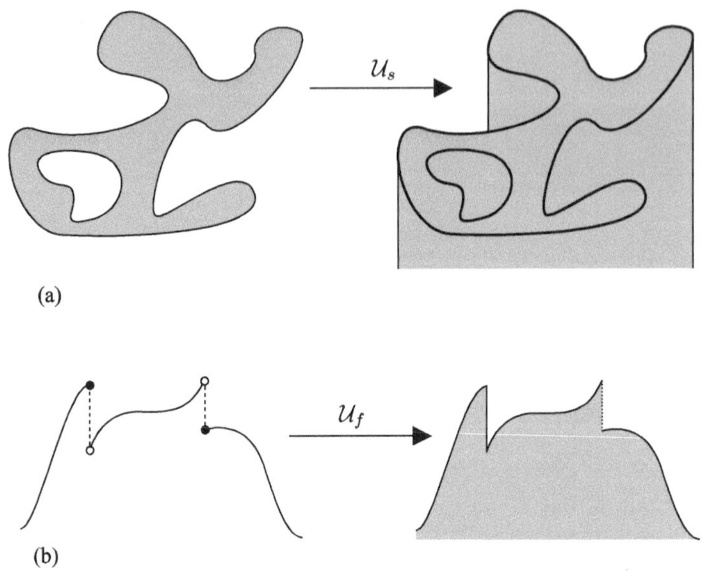

Figure 11.3 (a) The umbra $\mathcal{U}_S(X)$ of a set X and (b) the umbra $\mathcal{U}_f(F)$ of a function F.

by

$$\bigwedge_{i\in I} U_i = \bigcap_{i\in I} U_i,$$
$$\bigvee_{i\in I} U_i = \mathcal{U}_s(\bigcup_{i\in I} U_i).$$

To show that this lattice is isomorphic with $\mathrm{Fun}(\mathbb{E}^d)$, some further definitions are required. The umbra of a function F is defined by

$$\mathcal{U}_f(F) = \{(x, t) \mid t \leq F(x)\};$$

see Fig. 11.3 for an example. On the other hand, one can associate a function with every umbra in the following way:

$$\mathcal{F}_u(U)(x) = \bigvee \{t \in \mathbb{R} \mid (x, t) \in U\}.$$

From this point on we write $\mathcal{F}$ instead of $\mathcal{F}_u$, and $\mathcal{U}$ instead of $\mathcal{U}_f$.

11.26 Proposition. *The mapping $\mathcal{U}$ defines a lattice isomorphism between* $\mathrm{Fun}(\mathbb{E}^d)$ *and* $\mathrm{Umbra}(\mathbb{E}^d)$, *with inverse $\mathcal{F}$. Both mappings are invariant under horizontal and vertical translations.*

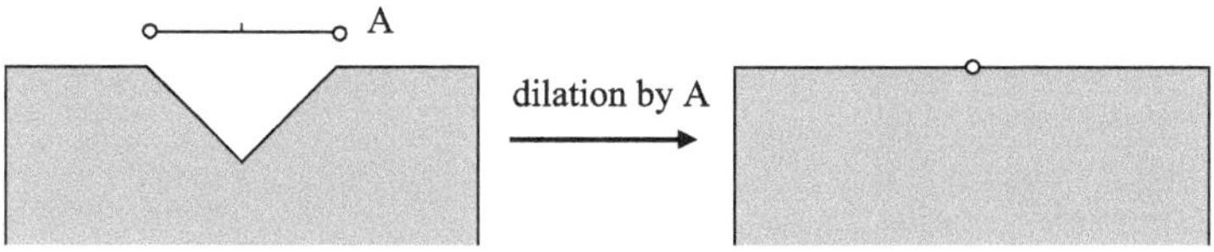

Figure 11.4 A dilation of an umbra need not be an umbra.

One can utilize this isomorphism, called *umbra transform*, to extend increasing operators on $\mathcal{P}(\mathbb{E}^d \times \mathbb{R})$ to $\text{Fun}(\mathbb{E}^d)$. But first we present a counterexample which shows there exist increasing translation invariant operators from $\mathcal{P}(\mathbb{E}^d \times \mathbb{R})$ into itself which do not map $\text{Umbra}(\mathbb{E}^d)$ into itself. Consider the dilation of $U \in \mathcal{P}(\mathbb{R} \times \mathbb{R})$ by the open line segment as depicted in Fig. 11.4. This dilation is a pre-umbra, but not an umbra since $(0, t) \in U \oplus A$ for every $t < 0$ but $(0, 0) \notin U \oplus A$. To circumvent such technical problems one may use the following result. Denote, for $X \subseteq \mathbb{E}^d \times \mathbb{R}$ and $v \in \mathbb{R}$, by X^v the vertical translate of X along the vector v.

11.27 Lemma. *Let* $U \subseteq \mathbb{E}^d \times \mathbb{R}$.

(a) *U is a pre-umbra if and only if $U \subseteq U^v$ for every $v > 0$.*

(b) *U is an umbra if and only if $U = \bigcap_{v>0} U^v$.*

(c) *If U is a pre-umbra, then $\mathcal{U}_s(U) = \bigcap_{v>0} U^v$.*

Proof. The proof of (a) is straightforward.

(b): The only if-statement is easy. To prove the if-statement, assume that $U = \bigcap_{v>0} U^v$. We show that U is an umbra. First, if $(x, t) \in U$, then $(x, t) \in U^v$ for every $v > 0$, showing that $(x, s) \in U$ for $s < t$. Conversely, $(x, s) \in U$ for $s < t$ is equivalent to $(x, t) \in U^v$ for $v > 0$ showing that $(x, t) \in U$.

(c): Assume that U is a pre-umbra, and let $\tilde{U} = \bigcap_{v>0} U^v$. That $\tilde{U}$ is an umbra follows readily from the observation that $\bigcap_{v>0} \tilde{U}^v = \bigcap_{v>0} U^v = \tilde{U}$. Assume that U_0 is an umbra such that $U \subseteq U_0$; then $\tilde{U} = \bigcap_{v>0} U^v \subseteq \bigcap_{v>0} U_0^v = U_0$; this proves that $\tilde{U}$ is the smallest umbra containing U. This means that $\tilde{U} = \mathcal{U}_s(U)$. □

Assume that $\psi : \mathcal{P}(\mathbb{E}^d \times \mathbb{R}) \to \mathcal{P}(\mathbb{E}^d \times \mathbb{R})$ is increasing and invariant under vertical translations. If U is a pre-umbra, then $U \subseteq U^v$ for $v > 0$, hence

$$\psi(U) \subseteq \psi(U^v) = [\psi(U)]^v,$$

meaning that $\psi(U)$ is a pre-umbra as well. The previous counterexample, however, shows that $\psi(U)$ is in general not an umbra, even if U is one. For that reason we define

$$\tilde{\psi}(X) = \bigcap_{v>0} [\psi(X)]^v. \tag{11.17}$$

Lemma 11.27 says that $\tilde{\psi}(U)$ is an umbra if U is an umbra. Since ψ is invariant under vertical translations, one may actually write

$$\tilde{\psi}(X) = \bigcap_{v>0} \psi(X^v). \tag{11.18}$$

11.28 Proposition. *Consider an operator ψ on $\mathcal{P}(\mathbb{E}^d \times \mathbb{R})$ which is increasing and invariant under vertical translations. The operator $\tilde{\psi}$ defined by* (11.18) *leaves* Umbra($\mathbb{E}^d$) *invariant.*

Observe that ψ coincides with $\tilde{\psi}$ on Umbra($\mathbb{E}^d$) if ψ is $\downarrow$-continuous, for in that case

$$\tilde{\psi}(U) = \bigcap_{v>0} \psi(U^v) = \psi(\bigcap_{v>0} U^v) = \psi(U);$$

here we have used that $U^v \downarrow U$ as $v \downarrow 0$ if U is an umbra.

11.29 Theorem. *Let ψ be an increasing operator on $\mathcal{P}(\mathbb{E}^d \times \mathbb{R})$ which is invariant under vertical translations, and let $\tilde{\psi}$ be given by* (11.18)*; then Ψ given by*

$$\Psi = \mathcal{F} \circ \tilde{\psi} \circ \mathcal{U} \tag{11.19}$$

defines an increasing operator on Fun($\mathbb{E}^d$) *which is invariant under vertical translations. Furthermore, Ψ is a T-operator if ψ is translation invariant.*

Consider, for example, Minkowski addition on Fun($\mathbb{E}^d$). Let $G \in$ Fun($\mathbb{E}^d$) be an arbitrary function, and let the operator ψ on $\mathcal{P}(\mathbb{E}^d \times \mathbb{R})$ be given by $\psi(X) = X \oplus \mathcal{U}(G)$. The operator Ψ given by (11.19) is the Minkowski addition $\Psi(F) = F \oplus G$. For using Proposition 11.25 and the translation invariance of $\mathcal{U}$, one gets

$$\begin{aligned}\mathcal{U}(F \oplus G) &= \mathcal{U}(\bigvee_{h\in\mathbb{E}^d} [F_h + G(h)]) = \mathcal{U}_s(\bigcup_{h\in\mathbb{E}^d} \mathcal{U}(F_h + G(h))) \\ &= \mathcal{U}_s(\bigcup_{h\in\mathbb{E}^d} [\mathcal{U}(F)]_h^{G(h)}) = \mathcal{U}_s(\mathcal{U}(F) \oplus \mathcal{U}(G)).\end{aligned}$$

From this identity the assertion follows immediately. An analogous statement can be derived for Minkowski subtraction. In this case, however, the proof is easier because erosion is $\downarrow$-continuous.

As a final remark we point out that the umbra representation can also be used to extend increasing set operators ψ on $\mathcal{P}(\mathbb{E}^d)$ to $\mathrm{Fun}(\mathbb{E}^d)$. In fact, with every such operator one can associate an increasing operator ψ' on $\mathcal{P}(\mathbb{E}^d \times \mathbb{R})$ which is invariant under vertical translations. Define, for $X \subseteq \mathbb{E}^d \times \mathbb{R}$, the threshold set $X(t) = \{x \mid (x, t) \in X\}$; let ψ' be the operator obtained by application of ψ to these threshold sets. In symbols,

$$\psi'(X) = \{(x, t) \in \mathbb{E}^d \times \mathbb{R} \mid x \in \psi(X(t))\}.$$

It is easy to check that ψ' is increasing and invariant under vertical translations. Using the foregoing results, one can transform ψ' into an operator on $\mathrm{Fun}(\mathbb{E}^d)$. We leave it as an exercise for the reader to show that this approach leads to the same function operator as the flat operator construction discussed in Section 11.3.

11.7. Grey-value set $\overline{\mathbb{Z}}$

The previous chapter has made clear to us that the lattice structure of the grey-value set is quite important. So far in this chapter it has been assumed that the grey-value set is $\overline{\mathbb{R}}$. Many of the results, however, carry over verbatim to the case where the grey-value set is $\overline{\mathbb{Z}}$. There are, however, some minor differences; these are discussed in this section. In Sections 11.8 and 11.9 the situation where the grey-value set is finite will be discussed.

The main difference between $\overline{\mathbb{Z}}$ and $\overline{\mathbb{R}}$ in the lattice-theoretical sense is that the finite points are not limit points; in symbols, $\bigvee\{s \in \overline{\mathbb{Z}} \mid s < t\} = t - 1 < t$ and $\bigwedge\{s \in \overline{\mathbb{Z}} \mid s > t\} = t + 1 > t$. The point $t = \infty$ is a lower limit point, however, and dually, $t = -\infty$ is an upper limit point. Note that this distinction between continuous and discrete grey-value sets shows up elegantly in Examples 10.2(a) and (b).

It turns out that in the present case the extension of set operators to function operators is rather simple. Assume that $\{\psi_t \mid t \in \mathbb{Z}\}$ is a family of increasing operators on $\mathcal{P}(\mathbb{E}^d)$ which is decreasing with respect to t. Define the increasing operator Ψ on $\mathrm{Fun}(E, \overline{\mathbb{Z}})$ as in (11.5), that is,

$$\Psi(F)(x) = \bigvee\{t \in \mathbb{Z} \mid x \in \psi_t(\mathbf{X}(F, t))\}.$$

The threshold sets of $\Psi(F)$ are given by

$$\mathbf{X}(\Psi(F), t) = \psi_t(\mathbf{X}(F, t)), \qquad t \in \mathbb{Z},$$
$$\mathbf{X}(\Psi(F), \infty) = \bigcap_{s<\infty} \psi_s(\mathbf{X}(F, s)).$$

The reader should compare these identities with (11.6). Besides this modification all the results of Section 11.2 and Section 11.3 carry over to the present situation almost verbatim. For instance, Ψ is a flat operator if there exists an increasing set operator ψ, called the generator of Ψ, such that

$$\mathbf{X}(\Psi(F), t) = \psi(\mathbf{X}(F, t)), \qquad t \in \mathbb{Z}.$$

Theorem 11.12, which states that flat function operators commute with strong anamorphoses, also extends easily to discrete grey-value sets; in fact, the continuity requirement on a can be omitted altogether in this case. Also the umbra approach simplifies considerably, for in this case the pre-umbra and the umbra coincide: a set $U \subseteq \mathbb{E}^d \times \mathbb{Z}$ is an umbra if $(x, t) \in U$ implies that $(x, s) \in U$ for every $s \leq t$. As a consequence, one does not need the intermediate operator $\tilde{\psi}$ given by (11.18). This leads to the following simplification of Proposition 11.28: every increasing operator on $\mathcal{P}(\mathbb{E}^d \times \mathbb{Z})$ which is invariant under vertical translations leaves $\text{Umbra}(\mathbb{E}^d)$ invariant.

11.8. Finite grey-value sets

In this and the next section it is assumed that the grey-value set $\mathcal{T}$ is finite, namely, $\mathcal{T} = \{0, 1, \ldots, N\}$. In this case the negative of a function F is given by

$$F^*(x) = N - F(x). \tag{11.20}$$

Given a family of increasing set operators $\psi_1 \geq \psi_2 \geq \cdots \geq \psi_N$, one can define the semi-flat function operator Ψ by

$$\Psi(F)(x) = \max\{1 \leq t \leq N \mid x \in \psi_t(\mathbf{X}(F, t))\}; \tag{11.21}$$

here $\Psi(F)(x) = 0$ if the set on the right-hand side is empty. This definition is equivalent to

$$\mathbf{X}(\Psi(F), t) = \psi_t(\mathbf{X}(F, t)), \qquad t = 1, 2, \ldots, N. \tag{11.22}$$

Most of the results of Section 11.2 and Section 11.3 remain valid in the finite case. For the convenience of the reader we will briefly recall the main results here and indicate the necessary modifications. First observe that, when $\psi_1, \psi_2, \ldots, \psi_N$ are translation invariant, Ψ is an H-operator.

The analogue of Theorem 11.7 concerning the negative operator becomes slightly different.

11.30 Proposition. *Let the function operator Ψ be generated by the family of set operators $\psi_1 \geq \psi_2 \geq \cdots \geq \psi_N$; then the negative operator Ψ^* is generated by the family $\psi_N^* \geq \psi_{N-1}^* \geq \cdots \geq \psi_1^*$.*

Proof. Use twice the fact that

$$\mathbf{X}(F^*, t) = \big(\mathbf{X}(F, N - t + 1)\big)^c.$$

For $t = 1, \ldots, N$ and $F \in \text{Fun}(\mathbb{E}^d, \mathcal{T})$,

$$\begin{aligned}\mathbf{X}(\Psi^*(F), t) &= \big(\mathbf{X}(\Psi(F^*), N - t + 1)\big)^c = \big(\psi_{N-t+1}(\mathbf{X}(F^*, N - t + 1))\big)^c \\ &= \big(\psi_{N-t+1}(\mathbf{X}(F, t)^c)\big)^c = \psi_{N-t+1}^*(\mathbf{X}(F, t));\end{aligned}$$

this proves the result. □

As in Theorem 11.5, the construction of semi-flat function operators is compatible with compositions, suprema, and infima. Furthermore, if every ψ_t is an erosion (dilation, etc.), then Ψ is again such; cf. Corollary 11.6. The operator Ψ is flat if there is an increasing set operator ψ such that

$$\mathbf{X}(\Psi(F), t) = \psi(\mathbf{X}(F, t)),$$

for $t = 1, 2, \ldots, N$ and $F \in \text{Fun}(\mathbb{E}^d, \mathcal{T})$. The generator ψ is uniquely determined by Ψ. Complements, compositions, suprema, and infima of flat operators are again flat; cf. Theorems 11.13 and 11.16. One can prove the following analogue of Theorem 11.12.

11.31 Theorem. *Let Ψ be a flat operator on $\text{Fun}(\mathbb{E}^d, \mathcal{T})$ generated by ψ, and let $a : \mathcal{T} \to \mathcal{T}$ be an increasing mapping. Assume moreover that*

(i) *$\psi(\mathbb{E}^d) = \mathbb{E}^d$ if $a(0) > 0$;*

(ii) *$\psi(\varnothing) = \varnothing$ if $a(N) < N$.*

Then $\Psi(a(F)) = a(\Psi(F))$, for every $F \in \text{Fun}(\mathbb{E}^d, \mathcal{T})$.

Proof. Suppose that the flat operator Ψ is generated by ψ; we show that $\mathbf{X}(a(\Psi(F)), t) = \mathbf{X}(\Psi(a(F)), t)$ for $t = 1, 2, \ldots, N$.

– Let $a(0) > 0$ and $t \leq a(0)$; hence $\psi(\mathbb{E}^d) = \mathbb{E}^d$. Then $\mathbf{X}(a(\Psi(F)), t) = \mathbb{E}^d$ and $\mathbf{X}(\Psi(a(F)), t) = \psi(\mathbf{X}(a(F), t)) = \psi(\mathbb{E}^d) = \mathbb{E}^d$.

– Let $a(N) < N$ and $t > a(N)$; hence $\psi(\varnothing) = \varnothing$. Then $\mathbf{X}(a(\Psi(F)), t) = \varnothing$, whereas $\mathbf{X}(\Psi(a(F)), t) = \psi(\mathbf{X}(a(F), t)) = \psi(\varnothing) = \varnothing$.

– Let $a(0) < t \leq a(N)$; define $b(t) = \min\{s \mid t \leq a(s)\}$. The following assertion is obvious: if $0 \leq s \leq N$ and $a(0) \leq t \leq a(N)$, then $t \leq a(s)$ iff $b(t) \leq s$. In other words, (a, b) is an adjunction between $\mathcal{T}$ and $a(\mathcal{T})$. So

$$\begin{aligned}\mathbf{X}(a(\Psi(F)), t) &= \mathbf{X}(\Psi(F), b(t)) = \psi(\mathbf{X}(F, b(t))\\ &= \psi(\mathbf{X}(a(F), t)) = \mathbf{X}(\Psi(a(F)), t).\end{aligned}$$

This conclude the proof. □

Assume that Ψ is a flat operator generated by ψ and that $\psi(\varnothing) = \varnothing$ and $\psi(\mathbb{E}^d) = \mathbb{E}^d$; then

$$\mathrm{Ran}(\Psi(F)) \subseteq \mathrm{Ran}(F), \tag{11.23}$$

for every function F. Here $\mathrm{Ran}(F) = \{F(x) \mid x \in \mathbb{E}^d\}$ denotes the range of F. To see this, choose a such that $\mathrm{Ran}(a) = \mathrm{Ran}(F)$ and $a(t) = t$ for $t \in \mathrm{Ran}(F)$, e.g., $a(t) = \min[\mathrm{Ran}(F) \cap \{t, t+1, \ldots, N\}]$ and $\max(\mathrm{Ran}(F))$ if this set is empty. By the last proposition, Ψ commutes with a; hence $\Psi(a(F)) = a(\Psi(F))$. But $a(F) = F$, and we obtain $\Psi(F) = a(\Psi(F))$; this proves the assertion.

11.32 Examples.

(a) We present an example which shows that condition (11.23) is not sufficient for Ψ being flat. Let $N \geq 2$ and $0 \leq p \leq N-2$. Define the increasing operator Ψ by

$$\Psi(F)(x) = \begin{cases} F(x), & \text{if } F(x) \leq p, \\ M_F, & \text{if } F(x) > p. \end{cases}$$

Here M_F denotes the maximum of F. Evidently, $\mathrm{Ran}(\Psi(F)) \subseteq \mathrm{Ran}(F)$. Suppose that Ψ were a flat operator generated by ψ. For $X \subseteq \mathbb{E}^d$, let $F = C(X, N)$ be the function which is N on X and 0 elsewhere. Obviously, $\Psi(F) = F$. From the assumption that Ψ is generated by ψ it follows that

$$X = \mathbf{X}(F, N) = \mathbf{X}(\Psi(F), N) = \psi(\mathbf{X}(F, N)) = \psi(X).$$

As this holds for every X, we conclude that $\psi = \mathrm{id}$. But then $\Psi = \mathrm{id}$ as well, a contradiction.

(b) Relation (11.23) does not hold for infinite grey-value sets such as $\overline{\mathbb{R}}$ and $\overline{\mathbb{Z}}$. In fact, let ψ be the increasing set operator

$$\psi(X) = \begin{cases} \varnothing, & \text{if } X = \varnothing, \\ \mathbb{E}^d, & \text{if } X \neq \varnothing. \end{cases}$$

If F is such that $\mathbf{X}(F, t) \neq \varnothing$ for every t, then $\Psi(F) = I$, which means $\Psi(F)(x) = \infty$ for every x. For example, this holds if there is a sequence x_n in $\mathbb{E}^d$ for which $F(x_n) \to \infty$. So there exist many functions F whose range contains only finite values and yet have $\infty \in \mathrm{Ran}(\Psi(F))$.

11.9. Finite grey-value sets and truncation

The finite set $\mathcal{T} = \{0, 1, \ldots, N\}$ is not closed under addition and subtraction and therefore the notion of a T-operator makes no sense when dealing with the function lattice $\mathrm{Fun}(\mathbb{E}^d, \mathcal{T})$. In particular, the dilation and erosion given by Minkowski function addition and subtraction, respectively, cannot be applied directly when $\mathcal{T}$ is finite. A solution to this problem is to truncate values below 0 and above N. To be specific, define for $t \in \mathbb{Z}$,

$$\lfloor t \rfloor = \begin{cases} 0, & \text{if } t < 0, \\ t, & \text{if } t \in \{0, 1, \ldots, N\}, \\ N, & \text{if } t > N. \end{cases}$$

Let G be a structuring function with domain $\mathrm{dom}(G) \subseteq \mathbb{E}^d$ and with values in $\overline{\mathbb{Z}}$. Define the dilation Δ_G by

$$\Delta_G(F)(x) = \bigvee_{h \in \mathrm{dom}(G)} \lfloor F(x-h) + G(h) \rfloor. \tag{11.24}$$

Indeed, the reader should have no difficulty in verifying that Δ_G is a dilation. Analogously, define the erosion $\mathcal{E}_G$ by

$$\mathcal{E}_G(F)(x) = \bigwedge_{h \in \mathrm{dom}(G)} \lfloor F(x+h) - G(h) \rfloor. \tag{11.25}$$

Unfortunately, the pair $\mathcal{E}_G, \Delta_G$ does in general not define an adjunction. This is nicely illustrated by Fig. 11.5, which shows two functions F_1 and F_2 on $\mathbb{Z}$ such that $\Delta_G(F_1) \leq F_2$ but $F_1 \not\leq \mathcal{E}_G(F_2)$.

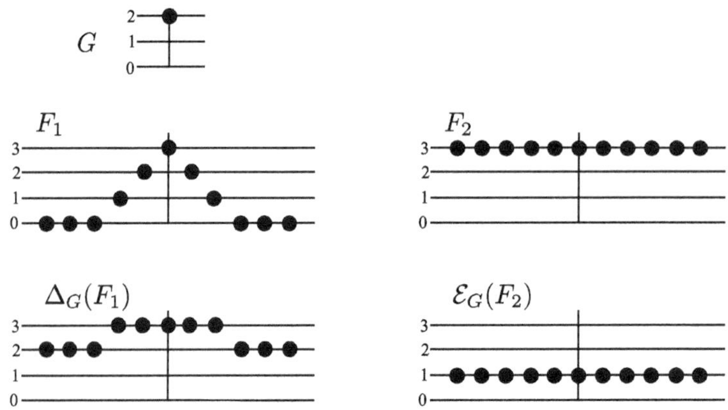

Figure 11.5 $\Delta_G(F_1) \leq F_2$ but $F_1 \not\leq \mathcal{E}_G(F_2)$.

In this example, $N = 3$ and G is the structuring function with domain $\{0\}$ and $G(0) = 2$. In fact one can show that the erosion $\mathcal{E}$ adjoint to Δ_G is given by

$$\mathcal{E}(F) = \begin{cases} O, & \text{if } F(x) < 2 \text{ for some } x \in \mathbb{Z} \\ F - 2, & \text{if } F(x) \geq 2 \text{ for every } x \in \mathbb{Z}. \end{cases}$$

Actually, if $G \leq 0$ everywhere, then the pair $(\mathcal{E}_G, \Delta_G)$ does define an adjunction. Note that in the expressions for $\Delta_G, \mathcal{E}_G$ just given the supremum and infimum are not taken over all $h \in \mathbb{E}^d$ but over the subset $\mathrm{dom}(G)$. In fact, one can extend G to the whole set $\mathbb{E}^d$ by putting $G(h) = -\infty$ for h outside the domain.

The difficulties that have been described can be overcome by utilizing the characterization of H-adjunctions on $\mathrm{Fun}(\mathbb{E}^d)$ given in Proposition 11.21. It is evident that this characterization also holds for the grey-value set $\mathcal{T}$ considered here. Thus, every H-adjunction $(\mathcal{E}, \Delta)$ on $\mathrm{Fun}(\mathbb{E}^d, \mathcal{T})$ has the form

$$\Delta(F)(x) = \bigvee_{h \in \mathbb{E}^d} d_h(F(x-h)), \tag{11.26}$$

$$\mathcal{E}(F)(x) = \bigwedge_{h \in \mathbb{E}^d} e_h(F(x+h)), \tag{11.27}$$

where (e_h, d_h) is an adjunction on $\mathcal{T}$ for every $h \in \mathbb{E}^d$. To avoid the problems with truncation, one has to assign the status of "absorbing barrier" to the grey-values 0 and N: if F takes the value 0 at a certain point, then this value cannot be changed by a vertical translation in the upward direction. Similarly, a vertical translation in downward direction cannot change points with grey-value N. We formalize this observation by introducing the modified addition $\dot{+}$ and subtraction $\dot{-}$. Define, for $v \in \mathbb{Z}$, the operation $t \mapsto t \dot{+} v$ on $\mathcal{T}$ by

$$\begin{cases} 0 \dot{+} v = 0, & \\ t \dot{+} v = 0, & \text{if } t > 0 \text{ and } t + v \le 0, \\ t \dot{+} v = t + v, & \text{if } t > 0 \text{ and } 0 \le t + v \le N, \\ t \dot{+} v = N, & \text{if } t > 0 \text{ and } t + v > N, \end{cases}$$

and the operation $t \mapsto t \dot{-} v$ by

$$\begin{cases} t \dot{-} v = 0, & \text{if } t < N \text{ and } t - v \le 0, \\ t \dot{-} v = t - v, & \text{if } t < N \text{ and } 0 \le t - v \le N, \\ t \dot{-} v = N, & \text{if } t < N \text{ and } t - v > N, \\ N \dot{-} v = N. & \end{cases}$$

For instance, if $N = 7$, then $(3 \dot{+} 5) \dot{-} 4 = 7$, $3 \dot{+} (5 \dot{-} 4) = 4$, and $(3 \dot{-} 4) \dot{+} 5 = 0$. Also, $0 \dot{+} 3 = 0$ and $3 \dot{+} 0 = 3$. Note in particular that $\dot{+}$ and $\dot{-}$ do not obey the commutative and associative laws satisfied by ordinary addition and subtraction.

11.33 Lemma. *The pair $e(t) = t \dot{-} v$, $d(t) = t \dot{+} v$ constitutes an adjunction on $\mathcal{T}$ for every $v \in \mathbb{Z}$.*

The proof of this result is straightforward. If $v = -\infty$, then (e, d) is the trivial adjunction (ι, o). Fig. 11.6 depicts the pair (e, d) where $N = 7$ and $v = 2$.

If one combines (11.26)–(11.27) with Lemma 11.33, one obtains the class of H-adjunctions of interest. Let G be a function with domain $\mathrm{dom}(G)$ which takes values in $\mathbb{Z}$. Let $d_h(t) = t \dot{+} G(h)$ for $h \in \mathrm{dom}(G)$ and $d_h \equiv 0$ otherwise. Dually, let $e_h(t) = t \dot{-} G(h)$ for $h \in \mathrm{dom}(G)$ and $e_h \equiv N$ otherwise. Now the pair

$$\Delta(F)(x) = (F \dot{\oplus} G)(x) = \bigvee_{h \in \mathrm{dom}(G)} (F(x-h) \dot{+} G(h)), \tag{11.28}$$

$$\mathcal{E}(F)(x) = (F \dot{\ominus} G)(x) = \bigwedge_{h \in \mathrm{dom}(G)} (F(x+h) \dot{-} G(h)) \tag{11.29}$$

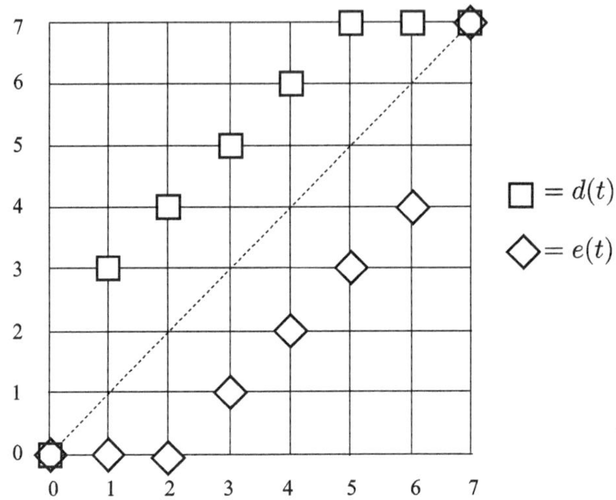

Figure 11.6 The adjunction $e(t) = t \dot{-} 2$, $d(t) = t \dot{+} 2$ on the grey-value set $\{0, 1, \ldots, 7\}$.

defines an H-adjunction on $\mathrm{Fun}(\mathbb{E}^d, \mathcal{T})$. In practice, G will only take values between $-N$ and N. If $G \leq 0$ everywhere, then $F \dot{\oplus} G$ and $F \dot{\ominus} G$ coincide with the "truncated" dilation and erosion given by (11.24) and (11.25), respectively.

11.34 Proposition.

(a) *Assume that G is a function which is nonnegative on its domain, and let Δ_G and $\mathcal{E}_G$ be given by (11.28) and (11.29); then*

$$\Delta_G(F \dot{+} v) = \Delta_G(F) \dot{+} v, \tag{11.30}$$

$$\mathcal{E}_G(F \dot{-} v) = \mathcal{E}_G(F) \dot{-} v, \tag{11.31}$$

if $v \geq 0$.

(b) *If Δ is an H-dilation on $\mathrm{Fun}(\mathbb{E}^d, \mathcal{T})$ satisfying $\Delta(F \dot{+} v) = \Delta(F) \dot{+} v$ for $F \in \mathrm{Fun}(\mathbb{E}^d, \mathcal{T})$ and $v \geq 0$, then Δ is of the form*

$$\Delta(F) = F \dot{\oplus} G,$$

for some nonnegative function G. Analogously, if $\mathcal{E}$ is an H-erosion on $\mathrm{Fun}(\mathbb{E}^d, \mathcal{T})$ satisfying $\mathcal{E}(F \dot{-} v) = \mathcal{E}(F) \dot{-} v$ for $F \in \mathrm{Fun}(\mathbb{E}^d, \mathcal{T})$ and $v \geq 0$, then $\mathcal{E}$ is of the form

$$\mathcal{E}(F) = F \dot{\ominus} G,$$

for some nonnegative function G.

Proof. The proof of (a) is straightforward.

(b): Let $f_{0,1}$ be the pulse function with altitude 1 at $x = 0$, i.e., $f_{0,1}(x) = 1$ if $x = 0$ and 0 otherwise. Define G by

$$\begin{cases} \mathrm{dom}(G) = \{x \mid \Delta(f_{0,1})(x) \geq 1\}, \\ G(x) = \Delta(f_{0,1})(x) - 1, \qquad \text{for } x \in \mathrm{dom}(G). \end{cases}$$

Since every $F = \bigvee_{h \in \mathbb{E}^d} f_{h,F(h)}$, it suffices to show that $\Delta(f_{h,v}) = f_{h,v} \dot{\oplus} G$ for $h \in \mathbb{E}^d$ and $v \in \mathcal{T}$. Because of the horizontal translation invariance we may restrict ourselves to the case $h = 0$. The result is trivial for $v = 0$, since $f_{0,0} = O$. So it remains to show that $\Delta(f_{0,v}) = f_{0,v} \dot{\oplus} G$ for $v = 1, \ldots, N$. Now

$$\Delta(f_{0,v}) = \Delta(f_{0,1} \dot{+} (v - 1)) = \Delta(f_{0,1}) \dot{+} (v - 1),$$

which means that

$$\Delta(f_{0,v})(x) = \Delta(f_{0,1})(x) \dot{+} (v - 1) = \begin{cases} (G(x) + 1) \dot{+} (v - 1), & x \in \mathrm{dom}(G), \\ 0, & x \notin \mathrm{dom}(G). \end{cases}$$

On the other hand,

$$(f_{0,v} \dot{\oplus} G)(x) = \bigvee_{h \in \mathrm{dom}(G)} f_{0,v}(x - h) \dot{+} G(h) = \begin{cases} v \dot{+} G(x), & x \in \mathrm{dom}(G), \\ 0, & x \notin \mathrm{dom}(G). \end{cases}$$

Since $(G(x) + 1) \dot{+} (v - 1) = v \dot{+} G(x)$ if $v \geq 1$ and $G(x) \geq 0$, the two expressions are equal, and the result follows. □

Refer to Fig. 11.7 for an example.

11.35 Theorem.

(a) *Let Ψ be an H-operator on* $\mathrm{Fun}(\mathbb{E}^d, \mathcal{T})$ *which satisfies $\Psi(I) = I$ and*

$$\Psi(F \dot{-} v) = \Psi(F) \dot{-} v,$$

for every $v \geq 0$; then Ψ can be written as a supremum of erosions of the form (11.29).

(a′) *If Ψ is an H-operator which satisfies $\Psi(O) = O$ and*

$$\Psi(F \dot{+} v) = \Psi(F) \dot{+} v,$$

for every $v \geq 0$, then Ψ can be written as an infimum of dilations of the form (11.28).

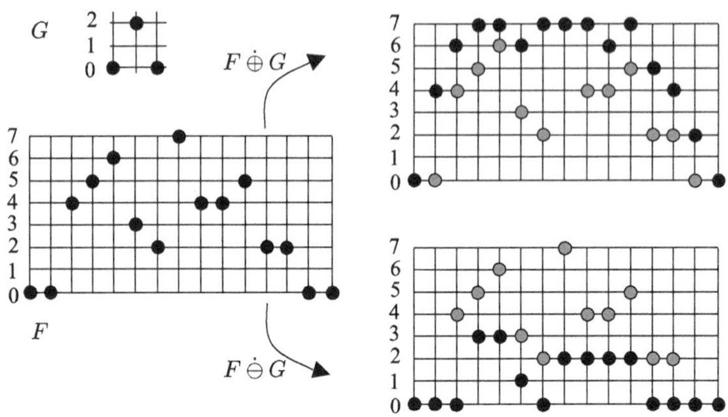

Figure 11.7 $F \dot{\oplus} G$ and $F \dot{\ominus} G$. The grey dots at the right represent the original function F.

Proof. We introduce the following notation. If G is a nonnegative function with domain $\mathrm{dom}(G)$, then $\hat{G} \in \mathrm{Fun}(\mathbb{E}^d)$ is defined by

$$\hat{G}(x) = \begin{cases} 0, & \text{if } x \notin \mathrm{dom}(G), \\ \min\{G(x)+1, N\}, & \text{if } x \in \mathrm{dom}(G). \end{cases}$$

Let Ψ be an H-operator satisfying $\Psi(I) = I$. The kernel $\mathcal{V}(\Psi)$ is defined as the set of all functions G with $\Psi(\hat{G})(0) \geq 1$. We prove that

$$\Psi(F) = \bigvee_{G \in \mathcal{V}(\Psi)} F \dot{\ominus} G.$$

"$\leq$": Let $\Psi(F)(x) \geq t$ for some $t \geq 1$. We show that $(F \dot{\ominus} G)(x) \geq t$ for some $G \in \mathcal{V}(\Psi)$. Let G be the function with domain $\mathrm{dom}(G) = \{h \mid F(x+h) \geq t\}$ given by $G(h) = F(x+h) \dot{-} t$. It is easy to check that $\hat{G} = F_x \dot{-} (t-1)$. Now $\Psi(F)(x) \geq t$ implies that $\Psi(F_x \dot{-} (t-1))(0) \geq 1$; hence $G \in \mathcal{V}(\Psi)$. But

$$(F \dot{\ominus} G)(x) = \bigwedge_{h \in \mathrm{dom}(G)} [F(x+h) \dot{-} G(h)] \geq t,$$

as follows immediately from the definition of G.

"$\geq$": Let $t \geq 1$, and suppose that $(F \dot{\ominus} G)(x) \geq t$ for some $G \in \mathcal{V}(\Psi)$; then $F(x+h) \dot{-} G(h) \geq t$ for all $h \in \mathrm{dom}(G)$. Since $t \neq 0$, however, this implies that $F_x(h) \dot{-} t \geq G(h)$ for every $h \in \mathrm{dom}(G)$, and thus $F_x \dot{-} (t-1) \geq \hat{G}$. But this means that

$$\Psi(F_x \dot{-} (t-1))(0) \geq \Psi(\hat{G})(0) \geq 1,$$

or equivalently, that $\Psi(F)(x) \dot{-} (t-1) \geq 1$. Now it follows that $\Psi(F)(x) \geq 1 \dot{+} (t-1) = t$, and the proof is completed. □

11.36 Example. (Annular opening)
Recall the following facts from Example 6.24. If $G : \mathbb{E}^d \to \overline{\mathbb{R}}$ is a function with the properties

(i) dom(G) is symmetric,
(ii) $G(x) + G(-x) \geq 0$ for $x \in \text{dom}(G)$,

then the operator $F \mapsto F \wedge (F \oplus G)$ is a T-opening, called annular opening. It is also possible to define annular openings when $\mathcal{T} = \{0, 1, \ldots, N\}$. For that purpose the dilations $F \mapsto F \dot{\oplus} G$ introduced previously will be utilized. First we prove the following result.

Assume that for every $h \in \mathbb{E}^d$, d_h is a dilation on $\mathcal{T}$ such that one of the following assumptions is satisfied:

(i) *d_h and d_{-h} are both identically zero;*
(ii) *$d_{-h} \circ d_h \geq \text{id}$.*

If $\Delta(F)(x) = \bigvee_{h \in \mathbb{E}^d} d_h(F(x-h))$, then $\text{id} \wedge \Delta$ is an H-opening.

Proof. First, observe that every dilation d on $\{0, 1, \ldots, N\}$ has the property that $d(s \wedge t) = d(s) \wedge d(t)$ for $s, t = 0, 1, \ldots, N$. We must show that the operator $\text{id} \wedge \Delta$ is idempotent. Then, since it is anti-extensive, it is an opening. Evidently, $(\text{id} \wedge \Delta)^2 \leq \text{id} \wedge \Delta$. To prove the converse, it suffices to show that $\Delta(\text{id} \wedge \Delta) \geq \text{id} \wedge \Delta$, since then $(\text{id} \wedge \Delta)^2 = \text{id} \wedge \Delta \wedge \Delta(\text{id} \wedge \Delta) \geq \text{id} \wedge \Delta$. Define $H = \{h \in \mathbb{E}^d \mid d_h \text{ is not identically } 0\}$. Let $F \in \text{Fun}(\mathbb{E}^d)$ and $x \in \mathbb{E}^d$; then

$$
\begin{aligned}
\Delta(\text{id} \wedge \Delta)(F)(x) &= \bigvee_{h \in H} d_h\Big(\bigvee_{h' \in H} [d_{h'}(F(x-h-h')) \wedge F(x-h)]\Big) \\
&= \bigvee_{h \in H} \bigvee_{h' \in H} \big[d_h d_{h'}(F(x-h-h')) \wedge d_h(F(x-h))\big] \\
&\geq \bigvee_{h \in H} \big[d_h d_{-h}(F(x)) \wedge d_h(F(x-h))\big] \\
&\geq \bigvee_{h \in H} \big[F(x) \wedge d_h(F(x-h))\big] \\
&= F(x) \wedge \Delta(F)(x).
\end{aligned}
$$

This completes the proof. □

Let G be a function with domain $\text{dom}(G) \subseteq \mathbb{E}^d$ which takes values in $\mathbb{Z}$. If $d_h(t) = t \dot{+} G(h)$, then the foregoing assumptions are satisfied if G satisfies the following two conditions.

(i) dom(G) is symmetric;
(ii) $G(h) \geq 0$ for $h \in \text{dom}(G)$.

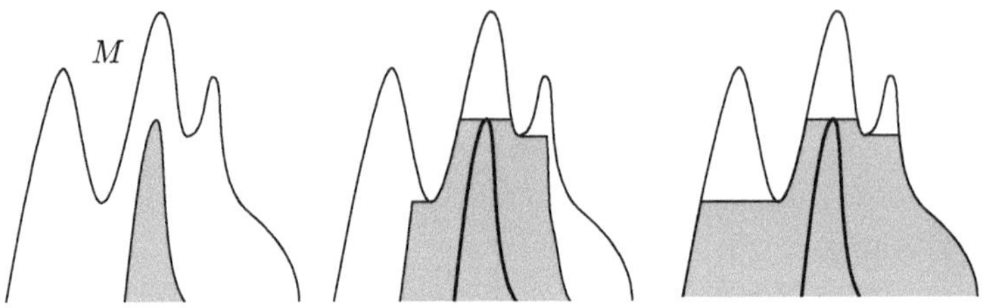

Figure 11.8 Geodesic dilation and reconstruction of a function.

To show that the second condition is necessary, suppose that $G(h) < 0$ for some $h \in \mathrm{dom}(G)$. Let $t = \min\{-G(h), N\}$; then

$$d_{-h}(t) = (t \dot{+} G(h)) \dot{+} G(-h) = 0 \dot{+} G(-h) = 0 \not\geq t,$$

and therefore $d_{-h} \circ d_h \not\geq \mathrm{id}$.

11.10. Geodesic and conditional operators

We assume throughout this section that the domain space is $\mathbb{R}^d$ and that the grey-value set is $\overline{\mathbb{R}}$. The flat operator construction can be used to extend the geodesic and conditional operators of Section 3.4 and Section 9.5 to grey-scale functions. First, consider the family of geodesic dilations. Let M be a mask function and $F \leq M$; fix $r > 0$, and choose $t \in \overline{\mathbb{R}}$. It is evident that the family $\delta^r(\mathbf{X}(F, t) \mid \mathbf{X}(M, t))$ is decreasing in t; here δ^r, $r \geq 0$, is the family of metric dilations introduced in Section 9.3. One can use this family to compute a function $\Delta^r(F \mid M)$, namely,

$$\Delta^r(F \mid M)(x) = \bigvee \{t \in \overline{\mathbb{R}} \mid x \in \delta^r(\mathbf{X}(F, t) \mid \mathbf{X}(M, t))\}. \tag{11.32}$$

It is also possible to formulate this in terms of semi-flat operators. In fact, it is easy to see that $\Delta^r(\cdot \mid M)$ is the semi-flat function operator generated by the family of set operators $\delta^r(\cdot \mid \mathbf{X}(M, t))$, $t \in \overline{\mathbb{R}}$. Note that this family is decreasing with respect to t for every fixed r. An illustration of a geodesic dilation is given in Fig. 11.8.

Proposition 9.46 says that

$$\delta^r(\cdot \mid \mathbf{X}(M, t))\delta^s(\cdot \mid \mathbf{X}(M, t)) = \delta^{r+s}(\cdot \mid \mathbf{X}(M, t)),$$

for $r, s \geq 0$. In combination with Theorem 11.5(b) this leads to the following result.

11.37 Proposition. *The geodesic dilations $\Delta^r(\cdot \mid M)$, $r \geq 0$, satisfy the semi-group property*

$$\Delta^r(\cdot \mid M)\Delta^s(\cdot \mid M) = \Delta^{r+s}(\cdot \mid M), \qquad r \geq 0. \tag{11.33}$$

The family $\varepsilon^r(\cdot \mid \mathbf{X}(M, t))$, $t \in \overline{\mathbb{R}}$, is in general not decreasing with respect to t; hence it is not possible to construct a geodesic erosion for functions using this family. Refer to Section 11.12 for an alternative definition of geodesic erosions for functions.

The geodesic reconstruction $\rho(\cdot \mid M)$ is defined analogously to (9.44), that is,

$$\rho(F \mid M) = \bigvee_{r \geq 0} \Delta^r(F \mid M);$$

see Fig. 11.8 for an illustration. Note that this definition makes sense because the family $\Delta^r(F \mid M)$ increases with r. We have noticed that $\Delta^r(\cdot \mid M)$ is the semi-flat function operator generated by the family of set operators $\delta^r(\cdot \mid \mathbf{X}(M, t))$, $t \in \overline{\mathbb{R}}$. From Theorem 11.5(a) it follows that $\rho(\cdot \mid M)$ is also a semi-flat operator, generated by the family $\rho(\cdot \mid \mathbf{X}(M, t))$. Recall from Section 9.5 that $\rho(X \mid M)$ is the set of all points in M for which there exists a rectifiable path to a point in X. A combination of these facts gives the following result.

11.38 Proposition. *If F, M are functions such that $F \leq M$, then $\rho(F \mid M)(x)$ equals the supremum over all $t \in \overline{\mathbb{R}}$ for which there exists a rectifiable path in $\mathbf{X}(M, t)$ which connects x with a point in $\mathbf{X}(F, t)$.*

In a similar way, it is possible to define conditional operators for grey-scale functions. These operators have been introduced in Section 3.4 for complete lattices and have been discussed for binary images in Section 9.5. We do not wish to give any details here, but content ourselves with writing down explicit expressions. The conditional dilation of $F \leq M$ by the structuring function G is

$$\Delta_G(F \mid\leq M) = (F \oplus G) \wedge M. \tag{11.34}$$

As the grey-level functions do not constitute a Boolean lattice, it is not possible to use the abstract results of Section 3.4 to derive an expression

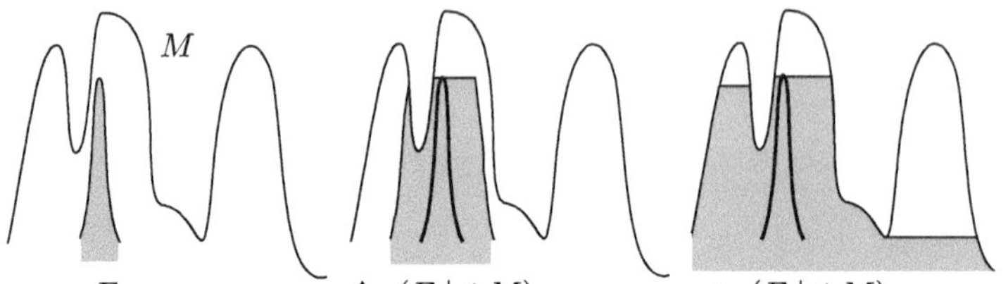

Figure 11.9 A function F, a mask function M, the conditional dilation $\Delta_A(F \mid\leq M)$, and the reconstruction $\rho_A(F \mid\leq M)$; here A is a line segment.

for the adjoint erosion; refer to Section 11.12 for some further comments about conditional erosions for grey-scale functions.

If G is a flat structuring function with domain A, then we will write $\Delta_A(\cdot \mid\leq M)$ instead of $\Delta_G(\cdot \mid\leq M)$. It is evident that $\Delta_A(\cdot \mid\leq M)$ is the semi-flat extension of the family of conditional dilations $\delta_t(X) = (X \oplus A) \cap M_t$, where $M_t = \mathbf{X}(M, t)$. The conditional dilation $\Delta_A(\cdot \mid\leq M)$ is illustrated in Fig. 11.9.

If $G(0) \geq 0$, then $\Delta_G(\cdot \mid\leq M)$ is extensive, and therefore

$$\Delta_G^n(F \mid\leq M) \leq \Delta_G^{n+1}(F \mid\leq M) \leq M, \qquad n \geq 1,$$

for every function $F \leq M$. In this case we define the *conditional reconstruction* $\rho_G(\cdot \mid\leq M)$ by

$$\rho_G(F \mid\leq M) = \bigvee_{n\geq 1} \Delta_G^n(F \mid\leq M). \tag{11.35}$$

When G is a flat structuring element with domain A containing the origin, then the conditional reconstruction is denoted by $\rho_A(\cdot \mid\leq M)$.

11.39 Example. (Dome extraction)

It is easy to see that the mapping $F \mapsto F - \rho(F - v \mid F)$, where $v > 0$, extracts *domes* of altitude v of the function F; see Fig. 11.10.

In the remainder of this example we restrict ourselves to the discrete case; i.e., we assume that the domain space is $\mathbb{Z}^2$ and that the grey-value lattice is $\overline{\mathbb{Z}}$. In this case, the geodesic reconstruction is replaced by a conditional reconstruction with a flat structuring element A. In the case of 8-connectivity one chooses for A the 3×3 square; in the case of 4-connectivity A is the rhombus. A *regional maximum* R of a function F is a connected component of $\mathbb{Z}^2$ such that

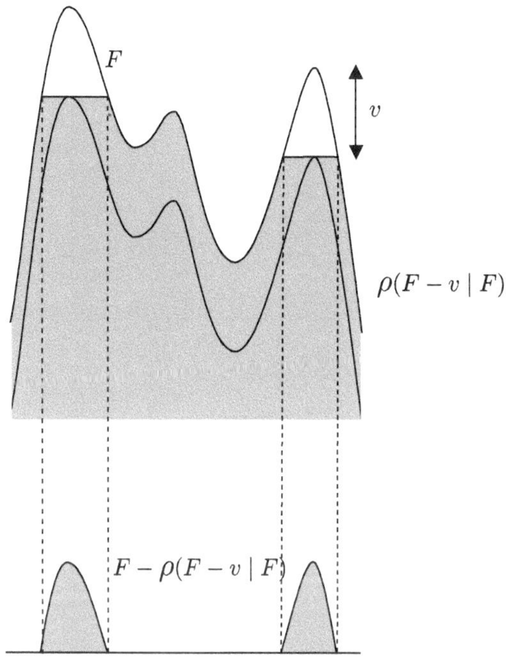

Figure 11.10 The function which maps F onto $F - \rho(F - v \mid F)$ contains the domes of altitude v of the function F.

- F is constant on R;
- if $r \in R$ and $x \notin R$ is a neighbour of r, then $F(x) < F(r)$.

It is easy to show that the domain of $F - \rho_A(F - 1 \mid\leq F)$ is the union of all regional maxima of F.

11.11. Granulometries

In Section 6.7 we have presented a formal theory of granulometries on complete lattices. Special attention has been given to the class of $(\mathbb{T}, \mathcal{R})$-Minkowski granulometries. In Section 9.6 the abstract results have been applied to the lattice $\mathcal{P}(\mathbb{R}^d)$. In this case a granulometry is called a Minkowski granulometry if the openings involved are translation invariant and compatible under scalings. If, moreover, every opening is assumed to be of structural type, then it follows that the structuring element is convex, assuming that it is compact.

In this section the abstract results of Section 6.7 are applied to the lattice of grey-scale images considered here. Throughout this section we assume that $\mathbb{R}^d$ is the domain space and that $\overline{\mathbb{R}}$ is the grey-value lattice.

11.40 Definition. A *granulometry* on $\text{Fun}(\mathbb{R}^d)$ is a one-parameter family of openings $\{\alpha_r \mid r > 0\}$, or briefly, $\{\alpha_r\}$, such that

$$\alpha_s \leq \alpha_r, \qquad \text{if } s \geq r. \tag{11.36}$$

From Theorem 3.24 it follows that a family of operators α_r which satisfies (11.36) also satisfies the semigroup property

$$\alpha_r \alpha_s = \alpha_s \alpha_r = \alpha_s, \qquad s \geq r. \tag{11.37}$$

In fact, (11.36) is equivalent to (11.37) and also to

$$\text{Inv}(\alpha_s) \subseteq \text{Inv}(\alpha_r), \qquad s \geq r.$$

A granulometry $\{\alpha_r\}$ is called a *T-granulometry* if every opening α_r is a T-opening; an *H-granulometry* is defined analogously. A general method for building grey-scale granulometries is derived from the (semi-) flat operator construction discussed in Sections 11.2–11.3.

11.41 Proposition.

(a) *Let, for every $t \in \overline{\mathbb{R}}$, $\{\alpha_{t,r}\}$ be a granulometry on $\mathcal{P}(\mathbb{R}^d)$, and let $\alpha_{s,r} \leq \alpha_{t,r}$ if $t \leq s$. Let α_r be the semi-flat operator generated by the family $\{\alpha_{t,r} \mid t \in \overline{\mathbb{R}}\}$. Then $\{\alpha_r\}$ is a granulometry on* $\text{Fun}(\mathbb{R}^d)$. *If every opening $\alpha_{t,r}$ is translation invariant, then $\{\alpha_r\}$ is an H-granulometry.*

(b) *Let $\{\alpha_r\}$ be a granulometry on $\mathcal{P}(\mathbb{R}^d)$, and let α_r be the flat operator generated by α_r; then $\{\alpha_r\}$ is a granulometry on* $\text{Fun}(\mathbb{R}^d)$. *If every α_r is translation invariant, then $\{\alpha_r\}$ is a T-granulometry.*

The granulometries in (a) and (b) are, respectively, called *semi-flat granulometries* and *flat granulometries*.

11.42 Example. Another way to build grey-scale granulometries is the following. Let $\{\alpha_r\}$ be a granulometry on $\mathcal{P}(\mathbb{R}^d)$, take $F \in \text{Fun}(\mathbb{R}^d)$, and define $\alpha_r(F)$ as the restriction of F to the opened domain $\alpha_r(\text{dom}(F))$; then $\{\alpha_r\}$ is a granulometry. If every opening α_r is translation invariant, then $\{\alpha_r\}$ is a T-granulometry.

Alternatively, one may replace $\text{dom}(F)$ by an arbitrary threshold set $\mathbf{X}(F, t)$. Vertical translation invariance is lost in this way, however, and one may only conclude that $\{\alpha_r\}$ is a H-granulometry if every α_r is translation invariant.

Throughout the remainder of this section we deal with Minkowski granulometries. As outlined in Section 6.7 the definition of a Minkowski granulometry requires a group of translations and a group of multiplications on $\text{Fun}(\mathbb{R}^d)$ which are compatible in the sense that $r\tau r^{-1}$ is a translation if r is a multiplication and τ a translation. In this section we deal exclusively with T-translations. As to the multiplication group there are two alternatives, namely, T-multiplication (or T-scaling)

$$(r \cdot F)(x) = rF(x/r),$$

and H-multiplication (or H-scaling)

$$(r \,\bullet\, F)(x) = F(x/r);$$

cf. (4.100) and (4.101). Both multiplications are compatible with T-translations. In fact,

$$r \cdot \big[(\frac{1}{r} \cdot F)_h + v\big] = F_{rh} + rv$$

and

$$r \,\bullet\, \big[(\frac{1}{r} \,\bullet\, F)_h + v\big] = F_{rh} + v,$$

for $r > 0$, $h \in \mathbb{R}^d$, $v \in \mathbb{R}$, and $F \in \text{Fun}(\mathbb{R}^d)$.

11.43 Definition. A granulometry $\{\alpha_r\}$ on $\text{Fun}(\mathbb{R}^d)$ is called a *(T,T)-granulometry* if every α_r is a T-opening and if the family is compatible with T-multiplications, that is,

$$\alpha_r(r \cdot F) = r \cdot \alpha_1(F), \qquad r > 0, \tag{11.38}$$

for $F \in \text{Fun}(\mathbb{R}^d)$. It is called a *(T,H)-granulometry* if every α_r is a T-opening and if the family is compatible with H-multiplications, that is,

$$\alpha_r(r \,\bullet\, F) = r \,\bullet\, \alpha_1(F), \qquad r > 0, \tag{11.39}$$

for $F \in \text{Fun}(\mathbb{R}^d)$.

Note that this terminology is more or less in conformity with Definition 6.38.

11.11.1 (T,T)-Minkowski granulometries

Let $F \circ G$ denote the T-opening of F by G as given in (4.138), that is,

$$F \circ G = \bigvee \{G_h + v \mid h \in \mathbb{R}^d, v \in \mathbb{R} \text{ and } G_h + v \leq F\}.$$

From Theorem 6.40 it follows that $\{\alpha_r\}$ is a (T,T)-Minkowski granulometry if and only if there exists a family $\mathcal{G} \subseteq \mathrm{Fun}(\mathbb{R}^d)$ such that

$$\alpha_r(F) = \bigvee_{s \geq r} \bigvee_{G \in \mathcal{G}} F \circ (s \cdot G). \tag{11.40}$$

The invariance domain $\mathrm{Inv}(\alpha_r)$ is the closure of $r \cdot \mathcal{G}$ under T-translations, suprema, and T-multiplications with scalars ≥ 1. As usual, $\{\alpha_r\}$ is called a *structural granulometry* if every α_r is a structural opening. A structural (T,T)-Minkowski granulometry must be of the form $\alpha_r(F) = F \circ (r \cdot G)$ where the structuring function G is such that

$$r \cdot G \text{ is } G\text{-open for } r \geq 1; \tag{11.41}$$

cf. (6.55).

11.44 Proposition. *Let G be a function with a convex umbra; then*

$$r \cdot G \oplus s \cdot G = (r + s) \cdot G, \qquad r, s > 0, \tag{11.42}$$

and (11.41) *holds.*

Proof. Assume that (11.42) holds; then

$$r \cdot G \circ G = \big(((r-1) \cdot G \oplus G) \ominus G\big) \oplus G = (r-1) \cdot G \oplus G = r \cdot G,$$

and hence (11.41) holds as well. Now suppose that G has a convex umbra. We show that (11.42) holds. In Section 11.6 we have seen that

$$\mathcal{U}(r \cdot G \oplus s \cdot G) = \mathcal{U}_s\big(\mathcal{U}(r \cdot G) \oplus \mathcal{U}(s \cdot G)\big).$$

Using $\mathcal{U}(r \cdot G) = r\mathcal{U}(G)$, we obtain $\mathcal{U}(r \cdot G) \oplus \mathcal{U}(s \cdot G) = r\mathcal{U}(G) \oplus s\mathcal{U}(G)$. Proposition 9.2 gives that this expression equals $(r+s)\mathcal{U}(G) = \mathcal{U}((r+s) \cdot G)$. This implies (11.42). □

11.45 Corollary. *Let $G \in \mathrm{Fun}(\mathbb{R}^d)$ have a convex umbra; then the openings $\alpha_r(F) = F \circ (r \cdot G)$ define a (T,T)-Minkowski granulometry.*

11.11.2 (T,H)-Minkowski granulometries

Before we give a characterization of (structural) (T,H)-Minkowski granulometries, we recall Example 11.42. Let $\{\alpha_r\}$ be a granulometry on $\mathcal{P}(\mathbb{R}^d)$, and let $\mathcal{A}_r(F)$ be the restriction of F to the domain $\alpha_r(\text{dom}(F))$. If $\{\alpha_r\}$ is of Minkowski type (in the sense of Section 9.6), then the resulting grey-scale granulometry $\{\mathcal{A}_r\}$ is of Minkowski type in the (T,T)-sense as well as in the (T,H)-sense. This follows from the observation that

$$\text{dom}(r \cdot F) = \text{dom}(r \,\bullet\, F) = r\,\text{dom}(F).$$

The details are left to the reader.

The flat operator construction provides a powerful method for obtaining (T,H)-Minkowski granulometries.

11.46 Proposition. *Let $\{\alpha_r\}$ be a Minkowski granulometry on $\mathcal{P}(\mathbb{R}^d)$, and let $\mathcal{A}_r$ be the flat function operator generated by α_r; then $\{\mathcal{A}_r\}$ defines a (T,H)-Minkowski granulometry on* Fun$(\mathbb{R}^d)$.

Proof. It is sufficient to show that $\mathcal{A}_r(F) = r \,\bullet\, \mathcal{A}_1(r^{-1} \,\bullet\, F)$. Using (11.7) and the identity $\mathbf{X}(r \,\bullet\, F, t) = r\mathbf{X}(F, t)$, one gets that

$$\begin{aligned}
\mathcal{A}_r(F)(x) &= \bigvee\{t \in \overline{\mathbb{R}} \mid x \in \alpha_r(\mathbf{X}(F,t))\} \\
&= \bigvee\{t \in \overline{\mathbb{R}} \mid x \in r\alpha_1(\tfrac{1}{r}\mathbf{X}(F,t))\} \\
&= \bigvee\{t \in \overline{\mathbb{R}} \mid \tfrac{1}{r}x \in \alpha_1(\mathbf{X}(\tfrac{1}{r} \,\bullet\, F, t))\} \\
&= \mathcal{A}_1(\tfrac{1}{r} \,\bullet\, F)(\tfrac{x}{r}) \\
&= \big(r \,\bullet\, \mathcal{A}_1(\tfrac{1}{r} \,\bullet\, F)\big)(x).
\end{aligned}$$

This concludes the proof. □

As before, it follows from Theorem 6.40 that $\{\mathcal{A}_r\}$ is a (T,H)-Minkowski granulometry if and only if there is a family $\mathcal{G} \subseteq \text{Fun}(\mathbb{R}^d)$ such that

$$\mathcal{A}_r(F) = \bigvee_{s \geq r} \bigvee_{G \in \mathcal{G}} F \circ (s \,\bullet\, G). \tag{11.43}$$

Clearly, every structural (T,H)-Minkowski granulometry $\{\mathcal{A}_r\}$ is of the form $\mathcal{A}_r(F) = F \circ (r \,\bullet\, G)$, where

$$r \,\bullet\, G \text{ is } G\text{-open for } r \geq 1. \tag{11.44}$$

Presently we shall, under some assumptions on G, derive a complete characterization of the functions G which obey this condition.

11.47 Definition. A function G is *compact* if it is u.s.c. and has a compact domain.

The next result, the proof of which will be given later, forms the basis for a characterization of (T,H)-Minkowski granulometries.

11.48 Proposition. *If G is compact, then* (11.44) *is satisfied if and only if* $\operatorname{dom}(G)$ *is convex and G is constant on* $\operatorname{dom}(G)$.

Since the opening by a structuring function is insensitive to vertical translations of this function, one may assume without loss of generality that $G = 0$ on its domain.

11.49 Corollary. *Every structural (T,H)-Minkowski granulometry $\{\alpha_r\}$ which uses a compact structuring function is flat. More specifically,*

$$\alpha_r(F) = F \circ r \centerdot G,$$

where $\operatorname{dom}(G)$ *is compact and convex and $G = 0$ on* $\operatorname{dom}(G)$.

We show by means of two examples that compactness of the domain as well as upper semi-continuity of the structuring function G is crucial. The continuous function $G : \mathbb{R} \to \mathbb{R}$ given by $G(x) = -|x|$ (see Fig. 11.11(a)) satisfies (11.44) but is not flat. Note that the domain of this function is unbounded. On the other hand, the function

$$G(x) = \begin{cases} 0, & x = 0, \\ 1, & 0 < x \le 1, \\ -\infty, & \text{elsewhere.} \end{cases}$$

has a compact domain but is not u.s.c.; see Fig. 11.11(b). It is easy to check that this function too satisfies (11.44).

The remainder of this subsection is devoted to the proof of Proposition 11.48. Note that the if-statement follows immediately from the theory in Section 9.6; hence it suffices to prove the only if-statement.

11.50 Lemma. *Assume that G is a compact function and that $r \centerdot G$ is G-open for $r \ge 1$; then* $\operatorname{dom}(G)$ *is convex.*

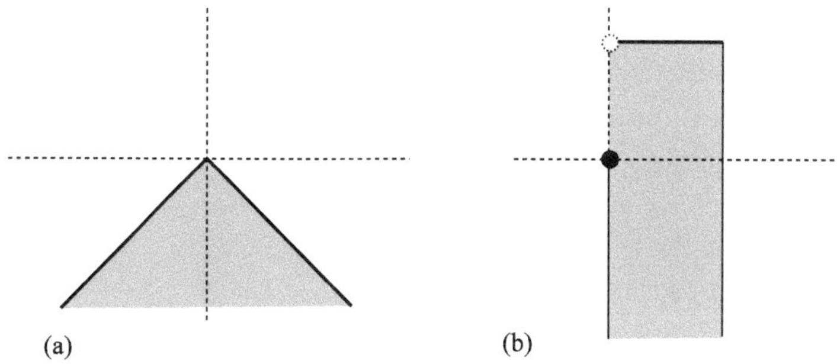

(a) (b)

Figure 11.11 (a) A function with noncompact domain and (b) a function which is not u.s.c.; both satisfy (11.44), but for neither (a) or (b) does the statement of Proposition 11.48 hold; see also the text preceding.

Proof. The mapping $\mathrm{dom}(\cdot) : \mathrm{Fun}(\mathbb{R}^d) \to \mathcal{P}(\mathbb{R}^d)$ is a dilation in the sense that

$$\mathrm{dom}(\bigvee_{i\in I} F_i) = \bigcup_{i\in I} \mathrm{dom}(F_i),$$

for every collection F_i, $i \in I$, in $\mathrm{Fun}(\mathbb{R}^d)$. Assume that G is a compact function such that $r \centerdot G$ is G-open; then

$$r \centerdot G = \bigvee \{G_h + v \mid h \in \mathbb{R}^d,\ v \in \mathbb{R} \text{ and } G_h + v \leq r \centerdot G\}.$$

Applying $\mathrm{dom}(\cdot)$ on both sides and using that

$$\mathrm{dom}(r \centerdot G) = r\,\mathrm{dom}(G) \qquad \text{and} \qquad \mathrm{dom}(G_h + v) = (\mathrm{dom}(G))_h,$$

one finds that

$$rD = \bigcup \{D_h \mid h \in \mathbb{R}^d,\ v \in \mathbb{R} \text{ and } G_h + v \leq r \centerdot G\},$$

where $D = \mathrm{dom}(G)$. Evidently, $G_h + v \leq r \centerdot G$ is a stronger condition than $\mathrm{dom}(G_h + v) \subseteq \mathrm{dom}(r \centerdot G)$, i.e., $D_h \subset rD$. This means that

$$rD \subseteq \bigcup \{D_h \mid h \in \mathbb{R}^d \text{ and } D_h \subseteq rD\} = rD \circ D.$$

The other inclusion is obvious, and one gets that $rD \circ D = rD$, for $r \geq 1$. Furthermore, D is compact, and one can apply Theorem 9.17, which gives that $D = \mathrm{dom}(G)$ is convex. □

Before we proceed, we point out that the key idea in the proof of Proposition 11.48 is to show that a compact function G which satisfies

(11.44) assumes its maximum at the extreme points of its domain. Then, using Zorn's Lemma 5.25 it can be shown that the maximum is attained at the convex hull of the extreme points. But since, by the previous lemma, the domain is convex, the Krein–Milman Theorem 9.21 gives that G is constant.

We start with the following alternative characterization of the extreme points $E(D)$ of a convex set D.

11.51 Lemma. *Let $D \subseteq \mathbb{R}^d$ be convex; then $e \in E(D)$ if and only if for every $r \geq 1$ the only solution x of*

$$D_{re-x} \subseteq rD \tag{11.45}$$

inside D is $x = e$.

Proof. "only if": Let $e \in E(D)$; assume that $x \neq e$ and $D_{re-x} \subseteq rD$. Then $y = \frac{1}{r}(e + re - x) \in D$, and thus $e = \frac{1}{r+1}x + \frac{r}{r+1}y$, meaning that $e \notin E(D)$, a contradiction.

"if": Assume that $e \notin E(D)$. We show that there exists an $r \geq 1$ and an $x \neq e$ such that (11.45) holds. There exist $u, v \in D$ such that $e = ru + (1 - r)v$ for some $r \in (0, 1)$. Take $s \in (0, r)$ and $x = su + (1 - s)v$; we show that x solves (11.45) if $r = (1 - s)/(1 - r) > 1$. Note first that

$$e + \frac{1}{r}(u - x) = u.$$

Take $y \in D$; using the convexity of D we find

$$\begin{aligned}\frac{1}{r}(y + re - x) &= e + \frac{1}{r}(y - x) = e + \frac{1}{r}(u - x) + \frac{1}{r}(y - u)\\ &= u + \frac{1}{r}(y - u) = \frac{1}{r}y + (1 - \frac{1}{r})u \in D.\end{aligned}$$

This proves the result. □

11.52 Lemma. *Let G be a function such that*

(i) *the domain $D = \mathrm{dom}(G)$ is convex;*

(ii) *$r \,.\, G$ is G-open for $r \geq 1$.*

If $x \in D$ with $G(x) \geq t$ and $e \in E(D)$, then $G(y) \geq t$ for every $y \in (e, x]$.

Proof. Let $e \in E(D)$ and $r \geq 1$. For every $\epsilon > 0$ there exists a $y \in \mathbb{R}^d$ and $v \in \mathbb{R}$ such that

(a) $G_y + v \leq r \,.\, G$;

(b) $G_y(re) + v > (r \,.\, G)(re) - \epsilon = G(e) - \epsilon$.

From (a) it follows that $D_y \subseteq rD$ and from (b) that $re \in D_y$. So there exists an $x \in D$ such that $re = x + y$, i.e., $y = re - x$. This means that

$$D_{re-x} \subseteq rD.$$

Now Lemma 11.51 implies that $x = e$, and thus $y = (r-1)e$. Substitution in (b) gives $G(e) + v > G(e) - \epsilon$, that is, $v > -\epsilon$. In combination with (a), this means that

$$G_{(r-1)e} - \epsilon \leq r \,.\, G.$$

Since this inequality holds for every $\epsilon > 0$,

$$G_{(r-1)e} \leq r \,.\, G.$$

Now let $G(x) \geq t$; substituting $y = x + (r-1)e$ in the last inequality gives

$$t \leq G(x) \leq G(\frac{1}{r}x + (1 - \frac{1}{r})e).$$

This proves the assertion. □

Now we are ready to complete the proof of the only if-statement of Proposition 11.48.

Proof of Proposition 11.48. Assume that $r \,.\, G$ is G-open for every $r \geq 1$. Lemma 11.50 implies that the domain $D = \mathrm{dom}(G)$ is convex. Let $t = \bigvee_{x \in D} G(x)$ be the supremum of G. Since G is compact, the value t is attained at some point in the domain, and therefore $\mathbf{X}(G,t) \neq \varnothing$. Note in fact that $G(x) = t$ for $x \in \mathbf{X}(G,t)$. We conclude from Lemma 11.52 that $[x, e) \subseteq \mathbf{X}(G,t)$ if $x \in \mathbf{X}(G,t)$ and $e \in E(D)$. Since G is u.s.c. it follows that $G(e) = t$, and therefore $E(D) \subset \mathbf{X}(G,t)$. Let $\mathcal{H}$ be the poset of all convex subsets of $\mathbf{X}(G,t)$. If $\mathcal{C}$ is a chain in $\mathcal{H}$, then $\bigcup \mathcal{C}$ is an upper bound. Thus $\mathcal{H}$ has a maximal element, which we denote by M. Obviously, M is a closed set, since otherwise $M \subseteq \overline{M} \subseteq \mathbf{X}(G,t)$ and $\overline{M}$ convex would contradict the maximality of M. Assume that $e \notin M$ for some $e \in E(D)$; then the convex hull M' of $\{e\} \cup M$ contains all segments $[e, x]$ where $x \in M$. Thus Lemma 11.52 gives that $M' \subseteq \mathbf{X}(G,t)$. But this means that M is not maximal, a contradiction. Therefore, $E(D) \subseteq M$. Since M is closed and convex, $\overline{\mathrm{co}}(E(D)) \subseteq M$. The Krein–Milman Theorem 9.21 gives that

$D = \overline{\text{co}}(E(D))$, and thus we get that $M = D \subseteq \mathbf{X}(G, t)$. This concludes the proof. □

11.12. Bibliographical notes

The literature on mathematical morphology for grey-scale images gives the impression that its development started much later than that of binary morphology. This is only partially true. Already Matheron (1967) had developed a theory of random u.s.c. functions. But Matheron's book (Matheron, 1975) does not contain this general theory; it restricts itself to random closed sets. To the best of our knowledge the first external publications concerning grey-scale morphology appeared only a decade after Matheron's pioneering work. But as we wrote in Section 4.7 we do not have the ambition to study in depth the history of grey-scale morphology.

The main references with regard to this chapter are the first book by Serra (1982), the work of Sternberg (1982, 1986), Maragos (1989b), a not so well-known paper by Janowitz (1986), our paper (Heijmans, 1991b), and some recent work by Serra (1993). Sections 11.1–11.3, 11.5, and 11.7–11.9 follow closely our account on grey-scale morphology in Heijmans (1991b); see also Heijmans (1994a).

The construction of flat operators by thresholding can also be found in Serra (1982, Chapter XII). Furthermore, this method is discussed extensively by Maragos and Schafer (1987a) and by Maragos (1989b). In all these discussions, however, the authors restrict themselves to u.s.c. functions. Maragos and Schafer (1987a), and Maragos and Ziff (1990) put great emphasis on the *superposition threshold principle*; by this they mean that the set operator ψ and the function operator Ψ are related by the formula

$$\psi(\mathbf{X}(F, t)) = \mathbf{X}(\Psi(F), t). \tag{11.46}$$

They call such operators *FSP operators*, where "FSP" stands for "Function-Set-Processing". To obey this condition they have to restrict themselves to u.s.c. operators; see Maragos (1989b); Maragos and Schafer (1987a) and Maragos (1985) for precise results. Furthermore, we point out the analogy between their results and the observation made in (11.11). In this chapter, upper semi-continuity is not required because we use (11.6), not (11.46).

Janowitz (1986) defines a class of operators which corresponds to our flat operators; note, however, that Janowitz calls them *flat filters*. Janowitz, however, was unaware of the developments in mathematical morphology, which might explain why his work remained unnoticed. In the same period, Wendt et al. (1986) studied a class of image transforms based on increasing Boolean functions, which they called *stack filters*. Essentially, stack filters and flat operators are the same objects. The observation that flat operators are compatible with anamorphoses was made by several authors, e.g., Serra (1982), Janowitz (1986), and Heijmans (1991b). Theorems 11.12 and 11.31 can be found in the last-mentioned reference. A variant of Theorem 11.31 as well as the inclusion in (11.23) can also be found in the paper by Janowitz (1986).

The top-hat operator is due to Meyer (1978); note, however, that this operator is usually called top-hat transform. The Dolby opening in the context of mathematical morphology is introduced in a recent paper by Serra (1993). A discussion of the literature concerning the construction of morphological operators from Boolean functions can be found in Section 4.7.

The umbra transform originates from work by Sternberg (1982, 1986). Other treatments can be found in a paper by Haralick et al. (1987) and the book of Giardina and Dougherty (1988, Chapter 6). The present discussion, which is extracted from Heijmans (1993), combines the original umbra approach with the complete lattice framework and avoids some of the pitfalls inherent in grey-scale morphology. Without giving any details, we point out that these pitfalls stem from the deviant position taken by the grey-values $\pm\infty$. This problem is discussed in great detail by Ronse (1990b). The account in Sections 11.8–11.9, dealing with finite grey-value lattices, is extracted from Heijmans (1991b); the annular opening in Example 11.36, however, appears in Heijmans (1994a). In the latter paper we also prove the following result.

Threshold Decomposition Theorem. *Let $\mathcal{T} = \{0, 1, \ldots, N\}$, and let G be a structuring function which assumes values between $-N$ and N. Then*

$$(F \dot{\oplus} G)(x) = \bigvee_{-N \le j \le N} \left[\sum_{i=1}^{N} \big(\mathbf{X}(F, i) \oplus \mathbf{X}(G, j)\big)(x) \dot{+} j \right],$$

$$(F \dot{\ominus} G)(x) = \bigwedge_{-N \le j \le N} \left[\sum_{i=1}^{N} \big(\mathbf{X}(F, i) \ominus \mathbf{X}(G, j)\big)(x) \dot{-} j \right],$$

for $F \in \mathrm{Fun}(\mathbb{E}^d, \mathcal{T})$ and $x \in \mathbb{E}^d$.

This decomposition theorem is closely related to a decomposition theorem by Shih and Mitchell (1992).

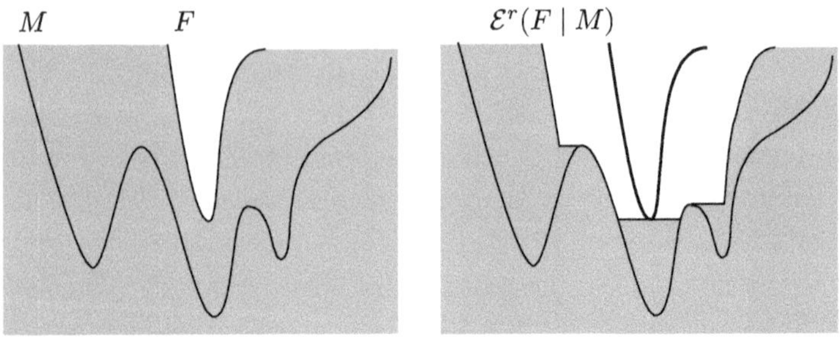

Figure 11.12 Geodesic erosion of a function F with respect to the mask function M; here $M \le F$.

The account in Section 11.10 on geodesic and conditional function operators is based mainly on work by members of the Fontainebleau school; we mention in particular the Ph.D. theses by Beucher (1990), Grimaud (1991), and Vincent (1990). Using negations, it is possible to define a geodesic erosion for functions as follows: suppose that M is a given mask function, and take $F \ge M$. Then $-F \le -M$ and using the results in Section 11.10 we can compute the geodesic dilation $\Delta^r(-F \mid -M)$. Now the operator $\mathcal{E}^r(\cdot \mid M)$ given by

$$\mathcal{E}^r(F \mid M) = -\Delta^r(-F \mid -M)$$

defines a *geodesic erosion*. An illustration can be found in Fig. 11.12.

Because $\mathrm{Fun}(\mathbb{E}^d)$ is not a Boolean lattice, one cannot find a nice expression for the (upper) conditional erosion $\mathcal{E}_G(\cdot \mid\le M)$ which is adjoint to the dilation given by (11.34). In the literature, one usually discusses the lower conditional erosion given by

$$\mathcal{E}_G(F \mid\ge M) = (F \ominus G) \vee M, \qquad \text{for } F \ge M;$$

see, e.g., Beucher (1990); Grimaud (1991), as well as Serra and Vincent (1992). Geodesic and conditional operators for grey-scale images are enormously important for practical applications of mathematical morphology, in particular, for segmentation purposes; see the overview by Beucher and Meyer (1993). Efficient algorithms for grey-scale reconstructions are discussed in a paper by Vincent (1993); see also the book of Schmitt and Vincent (2010). Finally, the reader may refer to two papers by Bleau et al. (1992a, 1992b), which also deal with geodesic operators for grey-scale images and their implementations.

Maragos (1989a) was among the first to generalize granulometries to grey-scale functions. The results on (T,T)-Minkowski granulometries have been obtained by Dougherty (1992). The discussion on (T,H)-Minkowski granulometries is extracted from a recent paper by Kraus et al. (1993). This paper discusses also (H,T)- and (H,H)-granulometries.

CHAPTER TWELVE

Morphological filters

Henk J.A.M. Heijmans
Formerly Centre for Mathematics and Computer Science, Amsterdam, Netherlands

Contents

This chapter presents a comprehensive treatment of the theory of morphological filters. Recall that a morphological filter is an increasing idempotent operator. Openings and closings, which have been studied in great detail in Chapter 6, are particular subclasses of filters. In this chapter the emphasis will lie on more general classes of filters. The simple observation that the class of filters is not closed under composition, supremum and infimum motivates the definition of other classes of lattice operators, such as overfilters, inf-overfilters, inf-filters, strong filters, and their dual notions.

In contrast to Chapter 5 and Chapter 6 the operators in this chapter are not assumed to be invariant under some automorphism group $\mathbb{T}$. The reason is that such a restriction merely would make the theory less transparent without leading to essentially stronger results. If he or she wishes, the reader may investigate the effects of adding $\mathbb{T}$-invariance.

12.1. Filters, overfilters, etc.

Throughout this chapter $\mathcal{L}$ is an arbitrary complete lattice unless stated otherwise. Recall that an increasing operator on $\mathcal{L}$ is called a *morphological filter*, or briefly *filter*, if it is idempotent, i.e.,

$$\psi^2 = \psi.$$

Advances in Imaging and Electron Physics, Volume 216
ISSN 1076-5670
https://doi.org/10.1016/bs.aiep.2020.07.012

One of the major challenges in low-level image processing is the restoration of distorted images. Morphological operators have proved quite successful for this purpose. If one uses an arbitrary operator, however, one can never be sure whether a second or third pass of the operator is necessary. On the other hand, when the operator is idempotent one knows in advance that a second pass does not have any further effect. Therefore, idempotence seems a sensible requirement for operators used to suppress noise.

A filter which is anti-extensive is called an *opening*. Dually, a filter which is extensive is called a *closing*. Example 3.29 shows that in general the composition of two openings is not an opening. In particular, this means that the class of filters is not closed under composition. As for openings (Theorem 3.24), however, one gets additional results if one filter lies below the other. Recall that $\mathrm{Inv}(\psi)$ is the invariance domain of ψ. In the literature, one usually calls elements which are invariant under a certain operator the *roots* of the operator.

12.1 Proposition. (Composition of filters)
Given two filters ϕ, ψ on $\mathcal{L}$ with $\phi \le \psi$; then

(a) $\phi \le \phi\psi\phi \le \left\{ \begin{matrix} \psi\phi \\ \phi\psi \end{matrix} \right\} \le \psi\phi\psi \le \psi;$

(b) $\psi\phi,\ \phi\psi,\ \psi\phi\psi,\ \phi\psi\phi$ *are filters;*

(c) $\mathrm{Inv}(\psi\phi\psi) = \mathrm{Inv}(\psi\phi)$ *and* $\mathrm{Inv}(\phi\psi\phi) = \mathrm{Inv}(\phi\psi)$.

Proof. (a): As $\phi \le \psi$, it follows that $\psi\phi\psi \le \psi\psi\psi = \psi$ and $\psi\phi = \psi\phi\phi \le \psi\phi\psi$. The other inequalities are proved analogously.

(b): That $\psi\phi$ is a filter follows from observing that $\psi\phi\psi\phi \ge \psi\phi\phi\phi = \psi\phi$ and $\psi\phi\psi\phi \le \psi\psi\psi\phi = \psi\phi$. The other assertions have similar proofs.

(c): Assume that $\psi\phi(X) = X$. Then

$$\psi\phi\psi(X) = \psi\phi\psi\psi\phi(X) = (\psi\phi)^2(X) = \psi\phi(X) = X.$$

This shows that $\mathrm{Inv}(\psi\phi) \subseteq \mathrm{Inv}(\psi\phi\psi)$. The other inclusions are proved in a similar way. □

That ϕ, ψ are filters does not imply that $\psi \vee \phi$ and $\psi \wedge \phi$ are filters as well. For a counterexample one may refer to Example 3.29, which shows that the infimum of two openings is not idempotent in general. Dually, the supremum of two closings is not a closing in general. If ϕ, ψ are filters, however, then $(\phi \vee \psi)^2 \ge \phi^2 = \phi$ and $(\phi \vee \psi)^2 \ge \psi^2 = \psi$; hence $(\phi \vee \psi)^2 \ge \phi \vee \psi$. Dually, $(\phi \wedge \psi)^2 \le \phi \wedge \psi$. The operators $\phi \vee \psi$ and

$\phi \wedge \psi$ are called *overfilter* and *underfilter*, respectively; a formal definition follows. These simple observations make clear that a comprehensive study of morphological filters must envisage other classes of operators, such as underfilters and overfilters. This fact was first realized by Matheron (1988), and in essence, this chapter comprises an overview of Matheron's pioneering work in this field.

12.2 Definition. An increasing operator on the complete lattice $\mathcal{L}$ is called

(a) an *overfilter* if $\psi^2 \geq \psi$;
(a′) an *underfilter* if $\psi^2 \leq \psi$;
(b) an *inf-overfilter* if $\psi(\mathrm{id} \wedge \psi) = \psi$;
(b′) a *sup-underfilter* if $\psi(\mathrm{id} \vee \psi) = \psi$;
(c) an *inf-filter* if ψ is a filter which satisfies $\psi(\mathrm{id} \wedge \psi) = \psi$;
(c′) a *sup-filter* if ψ is a filter which satisfies $\psi(\mathrm{id} \vee \psi) = \psi$;
(d) a *strong filter* if ψ is both a sup-filter and an inf-filter.

Several comments are in order. First, note that every inf-overfilter is an overfilter, since $\psi(\mathrm{id} \wedge \psi) \leq \psi^2$. To show that an increasing operator is an inf-overfilter it suffices to demonstrate that $\psi(\mathrm{id} \wedge \psi) \geq \psi$, as the reverse inequality holds trivially. Further, every inf-filter is both a filter and an inf-overfilter. One can easily establish the following properties.

12.3 Proposition.

(a) *Every extensive operator is an inf-overfilter.*
(a′) *Every anti-extensive operator is a sup-underfilter.*
(b) *If ψ is an overfilter, then ψ^n is an overfilter for every $n \geq 1$.*
(b′) *If ψ is an underfilter, then ψ^n is an underfilter for every $n \geq 1$.*
(c) *If ψ is an inf-overfilter, then ψ^n is an inf-overfilter for every $n \geq 1$.*
(c′) *If ψ is a sup-underfilter, then ψ^n is a sup-underfilter for every $n \geq 1$.*
(d) *If ψ is an inf-overfilter, then $\mathrm{id} \wedge \psi$ is an opening.*
(d′) *If ψ is a sup-underfilter, then $\mathrm{id} \vee \psi$ is a closing.*
(e) *Every opening and every closing is a strong filter.*

Proof. (a), (b), and (e) are obvious, and (d) has been proved in Theorem 6.26. It remains to prove (c). If ψ is an inf-overfilter, then $\psi^n(\mathrm{id} \wedge \psi^n) \geq \psi^n(\mathrm{id} \wedge \psi) = \psi^{n-1}\psi(\mathrm{id} \wedge \psi) = \psi^{n-1}\psi = \psi^n$. Here we have used that ψ^n is an overfilter, so $\psi^n \geq \psi$. □

An important class of inf-overfilters was discussed in Section 6.6. Assume that (ε, δ) and (ε', δ') are adjunctions between two complete lattices $\mathcal{L}$ and $\mathcal{M}$, that $\delta' \geq \delta$, and hence $\varepsilon' \leq \varepsilon$. Then $\delta'\varepsilon$ is an inf-overfilter on $\mathcal{L}$

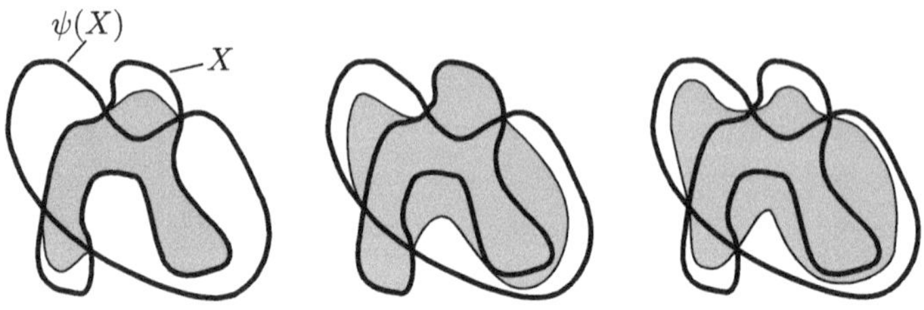

Figure 12.1 Geometrical characterization of (from left to right) inf-overfilters, sup-underfilters, and strong filters; see Proposition 12.6. The shaded region represents Y.

and $\varepsilon\delta'$ is a sup-underfilter on $\mathcal{M}$. Proposition 6.28 provides two methods of building inf-overfilters.

12.4 Proposition.

(a) *The set of overfilters is closed under suprema.*

(a′) *The set of underfilters is closed under infima.*

(b) *The set of inf-overfilters is closed under suprema.*

(b′) *The set of sup-underfilters is closed under infima.*

Proof. (a): If ψ_i are overfilters for $i \in I$, then $(\bigvee_{i\in I}\psi_i)(\bigvee_{i\in I}\psi_i) \geq \psi_i\psi_i \geq \psi_i$, and therefore $(\bigvee_{i\in I}\psi_i)(\bigvee_{i\in I}\psi_i) \geq \bigvee_{i\in I}\psi_i$, showing that $\bigvee_{i\in I}\psi_i$ is an overfilter.

(b): See the proof of Proposition 6.27. □

12.5 Proposition. *Let $\phi \leq \psi$ be filters on $\mathcal{L}$; then*

(a) *if ψ is a sup-filter, then $\phi\psi$ and $\psi\phi\psi$ are sup-filters;*

(a′) *if ϕ is an inf-filter, then $\psi\phi$ and $\phi\psi\phi$ are inf-filters;*

(b) *if $\psi\phi$ is a sup-filter, then $\phi\psi\phi$ is a sup-filter;*

(b′) *if $\phi\psi$ is an inf-filter, then $\psi\phi\psi$ is an inf-filter.*

Proof. (a): If ψ is a sup-filter, then

$$\phi\psi(\mathrm{id} \vee \phi\psi) \leq \phi\psi(\mathrm{id} \vee \psi^2) = \phi\psi(\mathrm{id} \vee \psi) = \phi\psi,$$

meaning that $\phi\psi$ is a sup-filter. Analogously, one shows that $\psi\phi\psi$ is a sup-filter.

(b): Analogous proof. □

There exist useful geometrical characterizations of inf-overfilters, sup-underfilters, and strong filters. These are illustrated in Fig. 12.1.

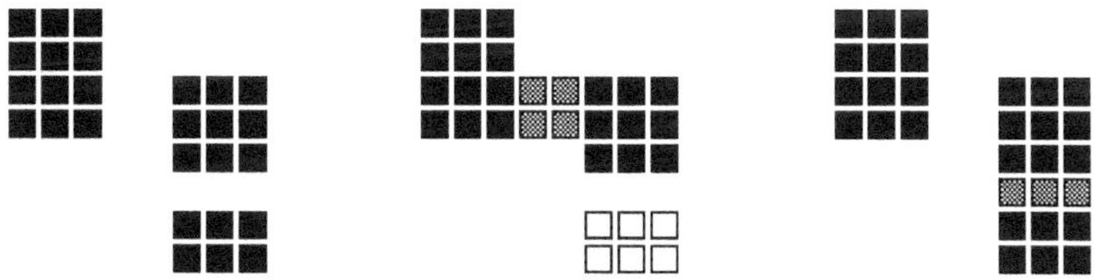

Figure 12.2 $(X \bullet A) \circ A \not\subseteq (X \circ A) \bullet A$ and $(X \circ A) \bullet A \not\subseteq (X \bullet A) \circ A$. From left to right: X, $(X \circ A) \bullet A$, and $(X \bullet A) \circ A$; here A is the 3×3 square.

12.6 Proposition. *Let ψ be an increasing operator.*

(a) *The following assertions are equivalent:*

- **(i)** *ψ is an inf-overfilter;*
- **(ii)** $X \wedge \psi(X) \le Y \le X \Rightarrow \psi(X) = \psi(Y)$, *for* $X, Y \in \mathcal{L}$.

(a′) *The following assertions are equivalent:*

- **(i)** *ψ is a sup-underfilter;*
- **(ii)** $X \le Y \le X \vee \psi(X) \Rightarrow \psi(X) = \psi(Y)$, *for* $X, Y \in \mathcal{L}$.

(b) *The following assertions are equivalent:*

- **(i)** *ψ is a strong filter;*
- **(ii)** $X \wedge \psi(X) \le Y \le X \vee \psi(X) \Rightarrow \psi(X) = \psi(Y)$, *for* $X, Y \in \mathcal{L}$.

Proof. We prove only (a); the proof of (b) is similar.

(i) $\Rightarrow$ (ii): Assume that ψ is an inf-overfilter and that $X \wedge \psi(X) \le Y \le X$. Then $\psi(X) = \psi(X \wedge \psi(X)) \le \psi(Y) \le \psi(X)$; hence $\psi(Y) = \psi(X)$.

(ii) $\Rightarrow$ (i): Take $Y = X \wedge \psi(X)$; then the left-hand side of (ii) is obeyed, which implies that $\psi(Y) = \psi(X \wedge \psi(X)) = \psi(X)$. Therefore, $\psi(\mathrm{id} \wedge \psi) = \psi$ and ψ is an inf-overfilter. □

12.7 Example. (Composition of openings and closings)
Consider an arbitrary opening α and closing β; then $\alpha \le \mathrm{id} \le \beta$, and by Proposition 12.1,

$$\alpha \le \alpha\beta\alpha \le \begin{Bmatrix} \beta\alpha \\ \alpha\beta \end{Bmatrix} \le \beta\alpha\beta \le \beta$$

are filters. Moreover, Proposition 12.5 gives that $\alpha\beta$ and $\beta\alpha\beta$ are sup-filters and that $\beta\alpha$ and $\alpha\beta\alpha$ are inf-filters. It is not true in general that $\beta\alpha \le \alpha\beta$. For example, Fig. 12.2 illustrates the case where $\alpha(X) = X \circ A$, $\beta(X) = X \bullet A$ on $\mathcal{P}(\mathbb{Z}^2)$ with A the 3×3 square. This example shows that $\beta\alpha \not\le \alpha\beta$ and $\alpha\beta \not\le \beta\alpha$.

Suppose, however, that α and β are such that $\beta\alpha \le \alpha\beta$. Then $\beta\alpha = \beta\alpha\alpha \le \alpha\beta\alpha \le \beta\beta\alpha = \beta\alpha$, i.e., $\alpha\beta\alpha = \beta\alpha$ and, by a similar argument, $\beta\alpha\beta = \alpha\beta$. Furthermore, the following identities hold in this case:

$$\mathrm{Inv}(\alpha\beta) = \mathrm{Inv}(\beta\alpha) = \mathrm{Inv}(\alpha) \cap \mathrm{Inv}(\beta).$$

We show that $\mathrm{Inv}(\alpha\beta) = \mathrm{Inv}(\alpha) \cap \mathrm{Inv}(\beta)$. The inclusion $\supseteq$ is obvious. Suppose that $X \in \mathrm{Inv}(\alpha\beta)$, i.e., $\alpha\beta(X) = X$. Then $\beta(X) = \beta(\alpha\beta(X)) = \alpha\beta(X) = X$, and thus $\alpha(X) = \alpha(\beta(X)) = X$, meaning that $X \in \mathrm{Inv}(\alpha) \cap \mathrm{Inv}(\beta)$. A class of openings and closings for which the inequality $\beta\alpha \le \alpha\beta$ holds is the openings and closings by a line segment.

Section 12.6 examines a general class of filters, the alternating sequential filters, which are compositions of openings and closings.

Throughout the remainder of this section it is assumed that $\mathcal{L}$ has a negation. The negative operator of ψ is denoted by ψ^*.

12.8 Proposition. *Consider a complete lattice $\mathcal{L}$ with a negation and an operator ψ on $\mathcal{L}$.*

(a) *ψ is an (inf-) overfilter iff ψ^* is an (sup-) underfilter.*
(b) *ψ is a filter iff ψ^* is a filter.*
(c) *ψ is an inf-filter iff ψ^* is a sup-filter.*
(d) *ψ is a strong filter iff ψ^* is a strong filter.*

Proof. The proof of this result follows immediately from the following two facts (see Proposition 3.2): for two arbitrary operators ϕ, ψ on $\mathcal{L}$ we have $(\psi\phi)^* = \psi^*\phi^*$, and $\phi \le \psi$ iff $\phi^* \ge \psi^*$.

(a): Suppose ψ is an overfilter, that is, $\psi^2 \ge \psi$. Then $(\psi^*)^2 = (\psi^2)^* \le \psi^*$, and so ψ^* is an underfilter. Furthermore, if ψ is an inf-overfilter, then $\psi(\mathrm{id} \wedge \psi) = \psi$, and therefore $\psi^*(\mathrm{id} \vee \psi^*) = \psi^*$, expressing that ψ^* is a sup-underfilter.

(b)–(d): Follow by similar arguments. □

12.9 Definition. Let $\mathcal{L}$ be a complete lattice with a negation. If ψ is a filter on $\mathcal{L}$ and $\psi^* = \psi$, then ψ is called a *self-dual filter*.

A self-dual filter has the desirable property that it treats the foreground and the background of an image identically. Whenever appropriate, it will be indicated in this chapter how to design self-dual filters.

It is obvious that openings and closings are not self-dual. Unfortunately, the composites $\alpha\beta$ and $\beta\alpha$ are not self-dual either, even if α and β are complementary operators, that is, $\beta = \alpha^*$. Similar remarks hold for the composites $\alpha\beta\alpha$ and $\beta\alpha\beta$.

12.2. Lattice of filters

In the previous section it was explained that the supremum of a family of filters is in general not a filter but an overfilter. The next result expresses that the overfilters are indeed the right class to consider.

12.10 Proposition.

(a) *Every overfilter ψ is the supremum of all filters $\leq \psi$.*
(a′) *Every underfilter ψ is the infimum of all filters $\geq \psi$.*
(b) *Every inf-overfilter ψ is the supremum of all strong filters $\leq \psi$.*
(b′) *Every sup-underfilter ψ is the infimum of all strong filters $\geq \psi$.*

Proof. (a): Let ψ be an overfilter. For $A \in \mathcal{L}$ we define a filter ϕ_A such that $\phi_A \leq \psi$ and $\phi_A(A) = \psi(A)$. It is clear that $\psi = \bigvee_{A \in \mathcal{L}} \phi_A$ in that case. Let

$$\phi_A(X) = \begin{cases} \psi(A), & \text{if } X \geq A \text{ or } X \geq \psi(A), \\ O, & \text{otherwise.} \end{cases}$$

It is easy to demonstrate that ϕ_A is a filter. We show that $\phi_A \leq \psi$. If $X \geq A$, then $\phi_A(X) = \psi(A) \leq \psi(X)$, since ψ is increasing. If $X \geq \psi(A)$, then $\phi_A(X) = \psi(A) \leq \psi^2(A) \leq \psi(X)$, where we used that ψ is an overfilter. For other X, $\phi_A(X) = O \leq \psi(X)$.

(b): We give a similar proof as in (a). Given an inf-overfilter ψ and $A \in \mathcal{L}$, we look for a strong filter ϕ_A such that $\phi_A \leq \psi$ and $\phi_A(A) = \psi(A)$. Define

$$\phi_A(X) = \begin{cases} \psi(A), & \text{if } X \geq A \wedge \psi(A), \\ O, & \text{otherwise.} \end{cases}$$

It is obvious that $\phi_A(A) = \psi(A)$. The proof that ϕ_A is a strong filter consists of two parts: (1) ψ is a sup-underfilter, and (2) ψ is an inf-overfilter.

(1): If $X \geq A \wedge \psi(A)$, then $\phi_A(X) = \psi(A)$. Therefore

$$X \vee \phi_A(X) \geq (A \wedge \psi(A)) \vee \psi(A) = \psi(A) \geq A \wedge \psi(A).$$

This yields $\phi_A(X \vee \phi_A(X)) = \psi(A) = \phi_A(X)$. If $X \not\geq A \wedge \psi(A)$, then $\phi_A(X) = O$, and so $\phi_A(X \vee \phi_A(X)) = \phi_A(X) = O$.

(2): We must show that $\phi_A(\mathrm{id} \wedge \phi_A) = \phi_A$. If $X \geq A \wedge \psi(A)$, then $\phi_A(X) = \psi(A)$, so $X \wedge \phi_A(X) \geq A \wedge \psi(A)$. This leads to $\phi_A(X \wedge \phi_A(X)) = \psi(A) = \phi_A(X)$. For $X \not\geq A \wedge \psi(A)$ the statement is trivial. It remains to show that $\phi_A \leq \psi$. If $X \geq A \wedge \psi(A)$, then $\phi_A(X) = \psi(A) = \psi(A \wedge \psi(A)) \geq$

$\psi(X)$; here we have used that ψ is an inf-overfilter. For other X the result is trivial. □

We introduce the notation $\mathsf{F}(\mathcal{L})$ for the filters and $\mathbb{F}(\mathcal{L})$ for the strong filters on $\mathcal{L}$. Using the notation of Section 2.1, $\langle \mathsf{F}(\mathcal{L}) \mid \vee \rangle$ denotes the smallest set in $\mathcal{O}^+(\mathcal{L})$ that contains all filters and is sup-closed; from Propositions 12.4(a) and 12.10(a) we infer that this class comprises all overfilters. Similarly, $\langle \mathbb{F}(\mathcal{L}) \mid \vee \rangle$ are the inf-overfilters.

Define the operator A on $\mathcal{O}^+(\mathcal{L})$ as follows:

$$\mathsf{A}\psi = \bigvee \{\phi \in \mathsf{F}(\mathcal{L}) \mid \phi \le \psi\}. \tag{12.1}$$

From the theory developed in Section 6.1 it follows that A is an opening on $\mathcal{O}^+(\mathcal{L})$ with invariance domain the set of overfilters on $\mathcal{L}$. Dually, define the operator B on $\mathcal{O}^+(\mathcal{L})$ by

$$\mathsf{B}\psi = \bigwedge \{\phi \in \mathsf{F}(\mathcal{L}) \mid \phi \ge \psi\}. \tag{12.2}$$

B is a closing on $\mathcal{O}^+(\mathcal{L})$ with invariance domain the set of underfilters.

12.11 Remark. Given an increasing operator ψ on $\mathcal{L}$, define $\mathcal{C}_\psi$ as the smallest class in $\mathcal{O}^+(\mathcal{L})$ that contains ψ and is closed under infima and self-composition; the latter means that $\phi \in \mathcal{C}_\psi$ implies that $\phi^n \in \mathcal{C}_\psi$ for $n \ge 1$. We show that

$$\mathsf{A}\psi = \bigwedge \mathcal{C}_\psi.$$

To do so, we put $\xi = \bigwedge \mathcal{C}_\psi$. The set $\{\phi \in \mathcal{O}^+(\mathcal{L}) \mid \mathsf{A}\psi \le \phi\}$ is closed under infima. Furthermore, if $\phi \ge \mathsf{A}\psi$, then $\phi^2 \ge (\mathsf{A}\psi)^2 \ge \mathsf{A}\psi$, because $\mathsf{A}\psi$ is an overfilter. This implies that this set is also closed under self-composition. Finally, it contains ψ, for $\mathsf{A}\psi \le \psi$. As $\mathcal{C}_\psi$ is the smallest set with these properties, it follows that

$$\mathcal{C}_\psi \subseteq \{\phi \in \mathcal{O}^+(\mathcal{L}) \mid \mathsf{A}\psi \le \phi\},$$

and hence that

$$\xi = \bigwedge \mathcal{C}_\psi \ge \bigwedge \{\phi \in \mathcal{O}^+(\mathcal{L}) \mid \mathsf{A}\psi \le \phi\} = \mathsf{A}\psi.$$

It remains to prove $\xi \le \mathsf{A}\psi$. As $\xi \in \mathcal{C}_\psi$, also $\xi^2 \in \mathcal{C}_\psi$, and so $\xi^2 \ge \xi$. This implies that ξ is an overfilter. Furthermore, $\xi \le \psi$, and since $\mathsf{A}\psi$ is the largest such overfilter, $\mathsf{A}\psi \ge \xi$.

12.12 Proposition.

(a) *If ψ is an underfilter, then $\mathsf{A}\psi$ is a filter.*

(a′) *If ψ is an overfilter, then $\mathsf{B}\psi$ is a filter.*

(b) *If $\psi \le \mathrm{id}$, then $\mathsf{A}\psi$ is an opening.*

(b′) *If $\psi \ge \mathrm{id}$, then $\mathsf{B}\psi$ is a closing.*

Proof. (a): Let ψ be an underfilter. From $\mathsf{A}\psi \le \psi$ it follows that $(\mathsf{A}\psi)^2 \le \psi^2 \le \psi$. Therefore, $(\mathsf{A}\psi)^2$ is an overfilter $\le \psi$, and, since $\mathsf{A}\psi$ is the largest overfilter with this property, $(\mathsf{A}\psi)^2 \le \mathsf{A}\psi$, in other words, $\mathsf{A}\psi$ is an underfilter. Therefore, $\mathsf{A}\psi$ is a filter.

(b): If $\psi \le \mathrm{id}$, then ψ is an underfilter, and (a) yields that $\mathsf{A}\psi$ is a filter. Furthermore, $\mathsf{A}\psi \le \psi \le \mathrm{id}$, and so $\mathsf{A}\psi$ is an opening. □

A consequence of this proposition is that for a given operator ψ, both $\mathsf{AB}\psi$ and $\mathsf{BA}\psi$ are filters.

Before we continue we recall some facts from Section 6.1. If ψ is an increasing operator on $\mathcal{L}$, then $\mathrm{Inv}(\mathrm{id} \wedge \psi)$ is sup-closed and there is a unique opening $\check{\psi}$ with

$$\mathrm{Inv}(\check{\psi}) = \mathrm{Inv}(\mathrm{id} \wedge \psi). \tag{12.3}$$

The operator $\check{\psi}$ is called the *lower envelope* of ψ. As $\check{\psi}$ maps into $\mathrm{Inv}(\check{\psi}) = \mathrm{Inv}(\mathrm{id} \wedge \psi)$, it follows that $\check{\psi} = (\mathrm{id} \wedge \psi)\check{\psi} \le \psi\check{\psi} \le \psi$, that is,

$$\check{\psi} \le \psi.$$

We point out that this also follows from the results of Section 6.1 where it was shown that the mapping $\psi \mapsto \check{\psi}$ is an opening on $\mathcal{O}^+(\mathcal{L})$.

Dually, there is a unique closing $\hat{\psi}$, called the *upper envelope* of ψ, determined by

$$\mathrm{Inv}(\hat{\psi}) = \mathrm{Inv}(\mathrm{id} \vee \psi). \tag{12.4}$$

The following two results establish the connection between the operators A and B on the one hand and the lower and upper envelope on the other.

12.13 Proposition. *Let ψ be an increasing operator on $\mathcal{L}$.*

(a) $\check{\psi} = \mathsf{A}(\mathrm{id} \wedge \psi)$*; in particular, $\check{\psi}$ is the largest opening $\le \psi$.*

(a′) $\hat{\psi} = \mathsf{B}(\mathrm{id} \vee \psi)$*; in particular, $\hat{\psi}$ is the smallest closing $\ge \psi$.*

Proof. Put $\alpha = \mathsf{A}(\mathrm{id} \wedge \psi)$. From Proposition 12.12(b) it follows that α is an opening, and from the very definition of A it is clear that α is the largest opening $\leq \mathsf{id} \wedge \psi$. We show that $\alpha = \check{\psi}$. As $\check{\psi}$ is an opening $\leq \psi$, we get that $\check{\psi} \leq \alpha$. To prove $\check{\psi} \geq \alpha$ it suffices to show that $\mathrm{Inv}(\alpha) \subseteq \mathrm{Inv}(\check{\psi}) = \mathrm{Inv}(\mathsf{id} \wedge \psi)$ (cf. Theorem 3.24). If $X \in \mathrm{Inv}(\alpha)$, then $\psi(X) \geq \alpha(X) = X$, in other words, $X \in \mathrm{Inv}(\mathsf{id} \wedge \psi)$. □

The next result expresses the mappings A and B in terms of the lower and upper envelope.

12.14 Proposition. *For every increasing operator* ψ*:*
(a) $\mathsf{A}\psi = \check{\psi}\psi$ *and* $\mathrm{Inv}(\mathsf{A}\psi) = \mathrm{Inv}(\psi)$*;*
(a′) $\mathsf{B}\psi = \hat{\psi}\psi$ *and* $\mathrm{Inv}(\mathsf{B}\psi) = \mathrm{Inv}(\psi)$*.*

Proof. Evidently, $\check{\psi}\psi \leq \psi$ and $\check{\psi}\psi\check{\psi}\psi \geq \check{\psi}\check{\psi}\check{\psi}\psi = \check{\psi}\psi$, meaning that $\check{\psi}\psi$ is an overfilter $\leq \psi$. Furthermore, if ϕ is an overfilter $\leq \psi$, then $\phi^2 \geq \phi$. In particular, $(\mathsf{id} \wedge \phi)\phi = \phi$, and so ϕ maps into $\mathrm{Inv}(\mathsf{id} \wedge \phi) = \mathrm{Inv}(\check{\phi})$. Therefore, $\check{\phi}\phi = \phi$. But $\check{\psi}\psi \geq \check{\phi}\phi$ (since $\psi \geq \phi$), and we get $\check{\psi}\psi \geq \phi$. We conclude that $\check{\psi}\psi$ is the largest overfilter $\leq \psi$, that is, $\mathsf{A}\psi = \check{\psi}\psi$.

We show that $\mathrm{Inv}(\mathsf{A}\psi) = \mathrm{Inv}(\psi)$. If $X \in \mathrm{Inv}(\psi)$, then $\psi(X) = X$, and so $\check{\psi}(X) = X$. But this yields $(\mathsf{A}\psi)(X) = \check{\psi}\psi(X) = X$, i.e., $X \in \mathrm{Inv}(\mathsf{A}\psi)$. Conversely, if $\check{\psi}\psi(X) = X$, then $\psi(X) \geq X$, which implies $\psi(\psi(X)) \geq \psi(X)$; hence $\psi(X) \in \mathrm{Inv}(\mathsf{id} \wedge \psi) = \mathrm{Inv}(\check{\psi})$. In other words, $\check{\psi}\psi(X) = \psi(X) = X$; hence $X \in \mathrm{Inv}(\psi)$. □

The class of filters $\mathsf{F}(\mathcal{L})$ is a partially ordered set under the partial ordering $\leq$ of $\mathcal{O}^+(\mathcal{L})$. The lattice $(\mathsf{F}(\mathcal{L}), \leq)$ is not a sublattice of $(\mathcal{O}^+(\mathcal{L}), \leq)$, however, as the supremum (resp. infimum) of a collection of filters is only an overfilter (resp. underfilter). Using the mappings A, B, however, the following result can be established.

12.15 Theorem. *The set* $\mathsf{F}(\mathcal{L})$ *is a complete lattice under the partial ordering* $\leq$*. The supremum of the collection of filters* ψ_i*,* $i \in I$*, is given by* $\mathsf{B}(\bigvee_{i\in I} \psi_i)$ *and the infimum by* $\mathsf{A}(\bigwedge_{i\in I} \psi_i)$*. Here* $\bigvee, \bigwedge$ *denote, respectively, the pointwise supremum and infimum in* $\mathcal{O}^+(\mathcal{L})$*.*

Proof. Let ψ_i, $i \in I$, be a collection of filters. We show that $\psi = \mathsf{A}(\bigwedge_{i\in I} \psi_i)$ is the largest filter that is $\leq \psi_i$ for every $i \in I$. First observe that $\bigwedge_{i\in I} \psi_i$ is an underfilter, and hence that ψ is a filter by Proposition 12.12. It is obvious that $\psi \leq \psi_i$, $i \in I$. Let ϕ be a filter which is $\leq \psi_i$ for $i \in I$. Then $\phi \leq \bigwedge_{i\in I} \psi_i$, and thus $\phi = \mathsf{A}\phi \leq \mathsf{A}(\bigwedge_{i\in I} \psi_i) = \psi$. Therefore, ψ is indeed the infimum of the collection ψ_i, $i \in I$. The result for the supremum follows by duality. □

A	B
$\mathsf{A}\psi = \bigvee\{\phi \in \mathsf{F}(\mathcal{L}) \mid \phi \le \psi\}$	$\mathsf{B}\psi = \bigwedge\{\phi \in \mathsf{F}(\mathcal{L}) \mid \phi \ge \psi\}$
$\mathsf{A}\psi$ is overfilter	$\mathsf{B}\psi$ is underfilter
$\mathsf{A}\psi \le \psi$	$\mathsf{B}\psi \ge \psi$
ψ underfilter $\Rightarrow$ $\mathsf{A}\psi$ filter	ψ overfilter $\Rightarrow$ $\mathsf{B}\psi$ filter
$\psi \le \mathrm{id} \Rightarrow \mathsf{A}\psi$ opening	$\psi \ge \mathrm{id} \Rightarrow \mathsf{B}\psi$ closing
$\check{\psi} = \mathsf{A}(\mathrm{id} \wedge \psi)$	$\hat{\psi} = \mathsf{B}(\mathrm{id} \vee \psi)$
$\mathsf{A}\psi = \check{\psi}\psi$	$\mathsf{B}\psi = \hat{\psi}\psi$
ψ sup-underfilter $\Rightarrow$ $\mathsf{A}\psi$ sup-filter	ψ inf-overfilter $\Rightarrow$ $\mathsf{B}\psi$ inf-filter

Figure 12.3 Properties of A and B.

If α_i, $i \in I$, is a collection of openings on $\mathcal{L}$, then $\bigwedge_{i\in I}\alpha_i$ is not an opening in general. The operator $\alpha = \mathsf{A}(\bigwedge_{i\in I}\alpha_i)$ is one, however, and by the same argument as used in the preceding proof it follows that α is the largest opening that is $\le \alpha_i$ for $i \in I$. Furthermore, $\mathsf{B}(\bigvee_{i\in I}\alpha_i) = \bigvee_{i\in I}\alpha_i$ in this case, since $\bigvee_{i\in I}\alpha_i$ is an opening, and hence an underfilter. Thus we arrive at the following results.

12.16 Corollary.

(a) *The class of openings on $\mathcal{L}$ is a complete lattice. The supremum of a collection of openings α_i, $i \in I$, is given by $\bigvee_{i\in I}\alpha_i$ and the infimum by $\mathsf{A}(\bigwedge_{i\in I}\alpha_i)$.*

(a′) *The class of closings on $\mathcal{L}$ is a complete lattice. The supremum of a collection of closings β_i, $i \in I$, is given by $\mathsf{B}(\bigvee_{i\in I}\beta_i)$ and the infimum by $\bigwedge_{i\in I}\beta_i$.*

In fact, these results express that both the class of openings and closings constitute a complete sublattice (Definition 2.19) of $(\mathsf{F}(\mathcal{L}), \le)$. This, in turn, is a complete underlattice of $(\mathcal{O}^+(\mathcal{L}), \le)$.

12.17 Proposition.

(a) *If ψ is a sup-underfilter, then $\mathsf{A}\psi$ is a sup-filter.*

(a′) *If ψ is an inf-overfilter, then $\mathsf{B}\psi$ is an inf-filter.*

Proof. If ψ is a sup-underfilter, then $\mathsf{A}\psi$ is a filter by Proposition 12.12(a). To show that $\mathsf{A}\psi$ is a sup-underfilter we use that $\mathsf{A}\psi = \check{\psi}\psi$ (Proposition 12.14). Then

$$\mathsf{A}\psi(\mathrm{id} \vee \mathsf{A}\psi) = \check{\psi}\psi(\mathrm{id} \vee \check{\psi}\psi) \le \check{\psi}\psi(\mathrm{id} \vee \psi) = \check{\psi}\psi = \mathsf{A}\psi.$$

This finishes the proof. □

The table in Fig. 12.3 summarizes the main properties of A and B.

Assume that $\mathcal{L}$ has a negation. If ψ is an operator on $\mathcal{L}$, then

$$(\mathsf{A}\psi)^* = \bigwedge\{\phi^* \mid \phi \in \mathsf{F}(\mathcal{L}) \text{ and } \phi \le \psi\}.$$

Substituting $\eta = \phi^*$, one gets

$$(\mathsf{A}\psi)^* = \bigwedge\{\eta \mid \eta \in \mathsf{F}(\mathcal{L}) \text{ and } \eta \ge \psi^*\} = \mathsf{B}(\psi^*).$$

Dually, one has

$$(\mathsf{B}\psi)^* = \mathsf{A}(\psi^*).$$

This means, in particular, that

$$(\check{\psi})^* = (\psi^*)^{\hat{}} \qquad \text{and} \qquad (\hat{\psi})^* = (\psi^*)^{\check{}}. \tag{12.5}$$

Here we used that $\check{\psi} = \mathsf{A}(\mathrm{id} \wedge \psi)$. An alternative proof of these identities has been given in Section 6.2.

12.3. Lattice of strong filters

Many of the results formulated in the previous section, starting with the definitions of the mappings A and B, can be reformulated for the class of strong filters.

Define the operator $\mathbb{A}$ on $\mathcal{O}^+(\mathcal{L})$ by

$$\mathbb{A}\psi = \bigvee\{\phi \in \mathbb{F}(\mathcal{L}) \mid \phi \le \psi\}. \tag{12.6}$$

In other words, $\mathbb{A}$ is the opening on $\mathcal{O}^+(\mathcal{L})$ with invariance domain the set of inf-overfilters. Dually,

$$\mathbb{B}\psi = \bigwedge\{\phi \in \mathsf{F}(\mathcal{L}) \mid \psi \le \phi\}, \tag{12.7}$$

the closing on $\mathcal{O}^+(\mathcal{L})$ with invariance domain the set of sup-underfilters. It is evident that

$$\mathbb{A}\psi \le \mathsf{A}\psi \qquad \text{and} \qquad \mathbb{B}\psi \ge \mathsf{B}\psi, \tag{12.8}$$

for every increasing operator ψ.

12.18 Remark. Using similar arguments as in Remark 12.11, one can show that $\mathbb{A}\psi = \bigwedge \mathcal{C}_\psi$, where $\mathcal{C}_\psi$ is the smallest class in $\mathcal{O}^+(\mathcal{L})$ that contains ψ, is closed under infima and under the composition $\phi \mapsto \phi(\mathrm{id} \wedge \phi)$.

12.19 Proposition.

(a) *If ψ is an underfilter, then $\mathbb{A}\psi$ is an inf-filter.*

(a′) *If ψ is an overfilter, then $\mathbb{B}\psi$ is a sup-filter.*

Proof. Since $\mathbb{A}\psi$ is an inf-overfilter, $(\mathbb{A}\psi)^2$ is an inf-overfilter too, by Proposition 12.14(c). Furthermore, $(\mathbb{A}\psi)^2 \le \psi^2 \le \psi$; hence $(\mathbb{A}\psi)^2$ is an inf-overfilter $\le \psi$. As $\mathbb{A}\psi$ is the largest such inf-overfilter, it follows that $(\mathbb{A}\psi)^2 \le \mathbb{A}\psi$, and hence that $\mathbb{A}\psi$ is an underfilter. □

Proposition 12.12(b) states that $\mathsf{A}\psi$ is an opening, namely, $\check{\psi}$, if $\psi \le \mathrm{id}$. As every opening is an inf-overfilter, we conclude that $\mathbb{A}\psi = \check{\psi}$ too if $\psi \le \mathrm{id}$.

The next result, which extends Proposition 12.14, expresses $\mathbb{A}$ and $\mathbb{B}$ in terms of the lower and upper envelope.

12.20 Proposition. *For every increasing operator ψ:*

(a) $\mathbb{A}\psi = \psi\check{\psi}$ *and* $\mathrm{Inv}(\mathbb{A}\psi) = \mathrm{Inv}(\psi)$;

(a′) $\mathbb{B}\psi = \psi\hat{\psi}$ *and* $\mathrm{Inv}(\mathbb{B}\psi) = \mathrm{Inv}(\psi)$.

Proof. It is easy to show that $\psi\check{\psi}$ is an inf-overfilter $\le \psi$. Now let ϕ be an inf-overfilter $\le \psi$. From Proposition 12.3(d) we know that $\mathrm{id} \wedge \phi$ is an opening; hence $\check{\phi} = \mathrm{id} \wedge \phi$. This gives $\phi\check{\phi} = \phi(\mathrm{id} \wedge \phi) = \phi \le \psi\check{\psi}$. Thus $\psi\check{\psi}$ is the largest inf-overfilter $\le \psi$.

If $X \in \mathrm{Inv}(\psi)$, then $\psi(X) = X$, and so $\check{\psi}(X) = X$. This implies that $(\mathbb{A}\psi)(X) = \psi\check{\psi}(X) = \psi(X) = X$. Conversely, if $\psi\check{\psi}(X) = X$, then $\psi(X) \ge X$, leading to $X \in \mathrm{Inv}(\mathrm{id} \wedge \psi) = \mathrm{Inv}(\check{\psi})$. Therefore, $\psi\check{\psi}(X) = \psi(X) = X$, i.e., $X \in \mathrm{Inv}(\psi)$. □

12.21 Corollary. *An increasing operator ψ is a strong filter if and only if*

$$\psi = \hat{\psi}\check{\psi} = \check{\psi}\hat{\psi}.$$

Proof. "if": From Example 12.7 we infer that $\check{\psi}\hat{\psi}$ is a sup-filter and that $\hat{\psi}\check{\psi}$ is an inf-filter. This proves the assertion.

"only if": Suppose ψ is a strong filter. Then $\mathbb{A}\psi = \psi\check{\psi} = \psi$; hence $\psi \le \hat{\psi}\check{\psi}$. But also, $\psi = \mathsf{B}\psi = \hat{\psi}\psi \ge \hat{\psi}\check{\psi}$. This means that $\psi = \hat{\psi}\check{\psi}$. The other equality is shown by a similar argument. □

In order to formulate conditions on ψ which guarantee $\mathbb{A}\psi$ and $\mathbb{B}\psi$ are strong filters, we recall Definition 2.15. The lattice $\mathcal{L}$ is modular if

$$X \vee (Y \wedge Z) = (X \vee Y) \wedge Z \qquad \text{if } X \le Z.$$

Every distributive lattice is modular. Furthermore, modularity is a self-dual property in the sense of the duality principle: if $\mathcal{L}$ is modular, then the opposite lattice $\mathcal{L}'$ is such as well.

12.22 Lemma. *Let $\mathcal{L}$ be a modular complete lattice.*
(a) *If ψ is an inf-overfilter, then $\psi(\mathrm{id} \vee \psi)$ is an inf-overfilter as well.*
(a′) *If ψ is a sup-underfilter, then $\psi(\mathrm{id} \wedge \psi)$ is a sup-underfilter as well.*

Proof. Given an inf-overfilter ψ, we show that $\phi = \psi(\mathrm{id} \vee \psi)$ is an inf-overfilter, too. We must show that $\phi(\mathrm{id} \wedge \phi) \geq \phi$. Note first that, since $\phi \geq \psi$,

$$\psi \geq \psi(\mathrm{id} \wedge \phi) \geq \psi(\mathrm{id} \wedge \psi) = \psi;$$

hence $\psi = \psi(\mathrm{id} \wedge \phi)$. Then

$$\begin{aligned}\phi(\mathrm{id} \wedge \phi) &= \psi(\mathrm{id} \vee \psi)(\mathrm{id} \wedge \phi) = \psi[(\mathrm{id} \wedge \phi) \vee \psi(\mathrm{id} \wedge \phi)] \\ &= \psi[(\mathrm{id} \wedge \phi) \vee \psi] = \psi[(\mathrm{id} \vee \psi) \wedge \phi] \\ &= \psi[(\mathrm{id} \vee \psi) \wedge \psi(\mathrm{id} \vee \psi)] = \psi(\mathrm{id} \wedge \psi)(\mathrm{id} \vee \psi) \\ &= \psi(\mathrm{id} \vee \psi) = \phi.\end{aligned}$$

This proves the result. □

12.23 Proposition. *Let $\mathcal{L}$ be a modular complete lattice.*
(a) *If ψ is a sup-underfilter, then $\mathbb{A}\psi$ is a strong filter.*
(a′) *If ψ is an inf-overfilter, then $\mathbb{B}\psi$ is a strong filter.*

Proof. From Proposition 12.19(a) we know that $\mathbb{A}\psi = \psi\check{\psi}$ is an inf-filter. Thus it remains to show that $\mathbb{A}\psi$ satisfies $\mathbb{A}\psi(\mathrm{id} \vee \mathbb{A}\psi) \leq \mathbb{A}\psi$. Since $\mathbb{A}\psi$ is an inf-overfilter, it follows from the previous lemma that $\mathbb{A}\psi(\mathrm{id} \vee \mathbb{A}\psi)$ is an inf-overfilter as well. Furthermore,

$$\psi = \psi(\mathrm{id} \vee \psi) \geq \mathbb{A}\psi(\mathrm{id} \vee \mathbb{A}\psi) \geq \mathbb{A}\psi.$$

Therefore, $\mathbb{A}\psi(\mathrm{id} \vee \mathbb{A}\psi)$ is an inf-overfilter which is $\leq \psi$, and also $\geq \mathbb{A}\psi$. Since, by definition, $\mathbb{A}\psi$ is the largest inf-overfilter $\leq \psi$, we conclude that $\mathbb{A}\psi = \mathbb{A}\psi(\mathrm{id} \vee \mathbb{A}\psi)$, that is, $\mathbb{A}\psi$ is a sup-underfilter. Together with the fact that it is an inf-filter, this proves the result. □

The table in Fig. 12.4 summarizes the main properties of $\mathbb{A}$ and $\mathbb{B}$. We formulate the following analogue of Theorem 12.15.

$\mathbb{A}$	$\mathbb{B}$
$\mathbb{A}\psi = \bigvee\{\phi \in \mathbb{F}(\mathcal{L}) \mid \phi \le \psi\}$	$\mathbb{B}\psi = \bigwedge\{\phi \in \mathbb{F}(\mathcal{L}) \mid \phi \ge \psi\}$
$\mathbb{A}\psi$ is inf-overfilter	$\mathbb{B}\psi$ is sup-underfilter
$\mathbb{A}\psi \le \mathsf{A}\psi \le \psi$	$\mathbb{B}\psi \ge \mathsf{B}\psi \ge \psi$
ψ underfilter $\Rightarrow$ $\mathbb{A}\psi$ inf-filter	ψ overfilter $\Rightarrow$ $\mathbb{B}\psi$ sup-filter
$\psi \le \mathrm{id} \Rightarrow \mathbb{A}\psi = \mathsf{A}\psi$ opening	$\psi \ge \mathrm{id} \Rightarrow \mathbb{B}\psi = \mathsf{B}\psi$ closing
$\mathbb{A}\psi = \psi\check{\psi}$	$\mathbb{B}\psi = \psi\hat{\psi}$
ψ sup-underfilter $\Rightarrow$ $\mathbb{A}\psi$ strong filter (if $\mathcal{L}$ modular)	ψ inf-overfilter $\Rightarrow$ $\mathbb{B}\psi$ strong filter (if $\mathcal{L}$ modular)

Figure 12.4 Properties of $\mathbb{A}$ and $\mathbb{B}$.

12.24 Theorem. *Let $\mathcal{L}$ be a modular complete lattice. The set $\mathbb{F}(\mathcal{L})$ of strong filters is a complete lattice under the partial ordering of $\mathcal{O}^+(\mathcal{L})$. If ψ_i, $i \in I$, is an arbitrary collection of strong filters, then its supremum is given by $\mathbb{B}(\bigvee_{i\in I}\psi_i)$ and its infimum by $\mathbb{A}(\bigwedge_{i\in I}\psi_i)$.*

If $\mathcal{L}$ has a negation, then

$$(\mathbb{A}\psi)^* = \mathbb{B}(\psi^*) \qquad \text{and} \qquad (\mathbb{B}\psi)^* = \mathbb{A}(\psi^*).$$

12.4. Invariance domain

To a certain extent the performance of a filter is determined by the structure of its invariance domain. In this section we explore this structure in some detail. Unless stated otherwise $\mathcal{L}$ is a complete lattice and ψ an increasing operator. One can easily show that

$$\hat{\psi}(Y) \le X \iff Y \le X, \qquad \text{for } X \in \mathrm{Inv}(\mathrm{id} \vee \psi),\ Y \in \mathcal{L}; \tag{12.9}$$

$$Y \le \check{\psi}(X) \iff Y \le X, \qquad \text{for } X \in \mathcal{L},\ Y \in \mathrm{Inv}(\mathrm{id} \wedge \psi). \tag{12.10}$$

Note that $\mathrm{Inv}(\mathrm{id} \wedge \psi)$ is an underlattice of $\mathcal{L}$ with the same supremum as $\mathcal{L}$ but with a different infimum. A dual remark applies to $\mathrm{Inv}(\mathrm{id} \vee \psi)$. In particular, (12.9)–(12.10) imply the following lemmas.

12.25 Lemma. *The pair $(\check{\psi}, \hat{\psi})$ defines an adjunction between* $\mathrm{Inv}(\mathrm{id} \vee \psi)$ *and* $\mathrm{Inv}(\mathrm{id} \wedge \psi)$.

12.26 Lemma.

(a) $\check{\psi}$ *maps* $\mathrm{Inv}(\mathrm{id} \vee \psi)$ *into* $\mathrm{Inv}(\psi)$.

(a′) $\hat{\psi}$ *maps* $\mathrm{Inv}(\mathrm{id} \wedge \psi)$ *into* $\mathrm{Inv}(\psi)$.

Proof. Since $\hat{\psi}\check{\psi}$ is an opening on $\mathrm{Inv}(\mathrm{id} \vee \psi)$, we have $\hat{\psi}\check{\psi}(X) \leq X$ for $X \in \mathrm{Inv}(\mathrm{id} \vee \psi)$. This implies that $\hat{\psi}\hat{\psi}\check{\psi}(X) \leq X$, and hence that $\hat{\psi}\psi\check{\psi}(X) \leq X$. Since $\check{\psi}(X) \in \mathrm{Inv}(\mathrm{id} \wedge \psi)$, also $\psi\check{\psi}(X) \in \mathrm{Inv}(\mathrm{id} \wedge \psi)$, and from Lemma 12.25 we derive that $\psi\check{\psi}(X) \leq \check{\psi}(X)$. The reverse inequality is trivially satisfied, and we conclude that $\psi\check{\psi}(X) = \check{\psi}(X)$ if $X \in \mathrm{Inv}(\mathrm{id} \vee \psi)$. This proves the result. □

The Tarski Fixpoint Theorem 3.3 shows that the invariance domain of an increasing operator is nonempty. Now we shall prove a stronger statement.

12.27 Tarski Fixpoint Theorem (strong version).

(a) *The invariance domain of an increasing operator ψ on the complete lattice $\mathcal{L}$ is an underlattice of $\mathcal{L}$. The supremum and infimum of a family X_i, $i \in I$ are given by $\hat{\psi}(\bigvee_{i\in I} X_i)$ and $\check{\psi}(\bigwedge_{i\in I} X_i)$, respectively.*

(b) *If, moreover, ψ is a filter, then the supremum and infimum are given by $\psi(\bigvee_{i\in I} X_i)$ and $\psi(\bigwedge_{i\in I} X_i)$, respectively.*

Proof. (a): Theorem 3.3 states that $\mathrm{Inv}(\psi) \neq \varnothing$. Suppose $X_i \in \mathrm{Inv}(\psi)$ for $i \in I$. Then $X_i \in \mathrm{Inv}(\mathrm{id} \wedge \psi)$, and as this set is closed under suprema, $\bigvee_{i\in I} X_i \in \mathrm{Inv}(\mathrm{id} \wedge \psi)$ as well. Thus Lemma 12.26 gives $\hat{\psi}(\bigvee_{i\in I} X_i) \in \mathrm{Inv}(\psi)$. It is obvious that $X_i \leq \hat{\psi}(\bigvee_{i\in I} X_i)$ for $i \in I$. If $Y \in \mathrm{Inv}(\psi)$ is an upper bound of X_i, $i \in I$, then $Y \geq \bigvee_{i\in I} X_i$; hence $Y = \hat{\psi}(Y) \geq \hat{\psi}(\bigvee_{i\in I} X_i)$. This implies that $\hat{\psi}(\bigvee_{i\in I} X_i)$ is the least upper bound. A similar argument gives that $\check{\psi}(\bigwedge_{i\in I} X_i)$ is the greatest lower bound.

(b): Similar proof. □

The following example is due to Ronse (1994a).

12.28 Example. (The lattice of filters)
Consider the complete lattice $\mathcal{O}^+(\mathcal{L})$ of increasing operators on $\mathcal{L}$. Let S be the operator given by self-composition, that is, $S\psi = \psi^2$. It is apparent that S is an increasing operator and that $\mathrm{Inv}(\mathrm{id} \wedge S)$ consists of the overfilters. This implies $\check{S} = \mathsf{A}$. Dually, $\hat{S} = \mathsf{B}$. Now the first part of the Tarski Fixpoint Theorem says that $\mathrm{Inv}(S)$, the filters on $\mathcal{L}$, constitute a complete underlattice of $\mathcal{O}^+(\mathcal{L})$. The supremum and infimum of a collection of filters ψ_i, $i \in I$, are given by $\mathsf{B}(\bigvee_{i\in I} \psi_i)$ and $\mathsf{A}(\bigwedge_{i\in I} \psi_i)$, respectively. This is exactly the content of Theorem 12.15.

12.29 Theorem. *Let ψ be an increasing operator; then $\hat{\psi}\check{\psi}$ is an inf-filter and $\check{\psi}\hat{\psi}$ is a sup-filter, and both operators have invariance domain* $\mathrm{Inv}(\psi)$.

If, in addition, ψ is a filter, then

$$\psi\check{\psi} = \hat{\psi}\check{\psi} \le \psi \le \check{\psi}\hat{\psi} = \psi\hat{\psi}.$$

Proof. That $\hat{\psi}\check{\psi}$ is an inf-filter follows from Example 12.7. Furthermore, Lemma 12.26 implies that $\hat{\psi}\check{\psi}$ maps into $\mathrm{Inv}(\psi)$. With this observation it becomes straightforward to show that $\mathrm{Inv}(\hat{\psi}\check{\psi}) = \mathrm{Inv}(\psi)$. The statement for $\check{\psi}\hat{\psi}$ follows by the duality principle.

Next, assume that ψ is a filter. It is obvious that $\psi\check{\psi} \le \hat{\psi}\check{\psi}$. To show the reverse inequality, we use $\hat{\psi}\psi = \psi$; note that this relation holds since ψ maps into $\mathrm{Inv}(\psi) \subseteq \mathrm{Inv}(\mathrm{id} \wedge \psi)$. Now $\psi\check{\psi} = \hat{\psi}\psi\check{\psi} \ge \hat{\psi}\check{\psi}\check{\psi} = \hat{\psi}\check{\psi}$. That $\psi\check{\psi} \le \psi$ is trivial, as $\check{\psi} \le \mathrm{id}$. □

12.30 Corollary.

(a) *ψ is an inf-filter iff $\psi = \beta\alpha$ for some closing β and opening α.*

(a′) *ψ is a sup-filter iff $\psi = \alpha\beta$ for some opening α and closing β.*

Proof. That every operator $\beta\alpha$ is an inf-filter follows from Example 12.7. On the other hand, if ψ is an inf-filter, then $\psi = \psi(\mathrm{id} \wedge \psi)$. But $\mathrm{id} \wedge \psi$ is an opening by Proposition 12.3(d); hence $\check{\psi} = \mathrm{id} \wedge \psi$, yielding that $\psi = \psi\check{\psi}$. Using Theorem 12.29(b), we find that $\psi = \psi\check{\psi} = \hat{\psi}\check{\psi}$. This concludes the proof. □

The Tarski Fixpoint Theorem states that the invariance domain of a filter is a complete underlattice. Now we will show that the converse also holds: with any complete underlattice $\mathcal{M}$ can be associated a filter with invariance domain $\mathcal{M}$. Suppose we are given a complete underlattice $\mathcal{M}$ with supremum $\sqcup$ and infimum $\sqcap$. Define the operators $\psi_\circ, \psi^\circ$ on $\mathcal{L}$ by

$$\psi_\circ(X) = \sqcup\{Y \in \mathcal{M} \mid Y \le X\},$$
$$\psi^\circ(X) = \sqcap\{Y \in \mathcal{M} \mid X \le Y\}.$$

It is evident that $\psi_\circ, \psi^\circ$ are increasing operators and that

$$\psi_\circ \le \psi^\circ.$$

Since both operators map into $\mathcal{M}$ and leave $\mathcal{M}$ invariant, it follows that they are idempotent. Let α, β be the opening and closing, respectively, given by

$$\alpha(X) = \bigvee\{Y \in \mathcal{M} \mid Y \le X\},$$
$$\beta(X) = \bigwedge\{Y \in \mathcal{M} \mid X \le Y\}.$$

Using the notation of Chapter 6, this means $\alpha = \alpha(\mathcal{M})$ and $\beta = \beta(\mathcal{M})$. We prove the following result.

12.31 Proposition. *Let $\mathcal{M}, \psi_\circ, \psi^\circ, \alpha, \beta$ be as before.*

(a) *$\psi_\circ, \psi^\circ$ are filters with invariance domain $\mathcal{M}$ and $\alpha \le \psi_\circ \le \psi^\circ \le \beta$.*

(b) *$\psi_\circ = \beta\alpha$ and $\psi^\circ = \alpha\beta$; in particular, $\psi_\circ$ is an inf-filter and ψ° is a sup-filter.*
Let ψ be an arbitrary filter with invariance domain $\mathcal{M}$; then:

(c) *$\psi_\circ = \psi\alpha$, $\psi^\circ = \psi\beta$ and $\psi_\circ \le \psi \le \psi^\circ$.*

(d) *For every family $Y_i \in \mathcal{M}, i \in I$,*

$$\bigsqcup_{i\in I} Y_i = \psi\Big(\bigvee_{i\in I} Y_i\Big), \qquad \bigsqcap_{i\in I} Y_i = \psi\Big(\bigwedge_{i\in I} Y_i\Big).$$

Proof. (a): Most of this part already has been demonstrated. That $\alpha \le \psi_\circ$ follows from the observation that the supremum with respect to $\mathcal{M}$ is greater than or equal to the supremum with respect to $\mathcal{L}$.

(b): It is obvious that $\alpha\psi_\circ = \beta\psi_\circ = \psi_\circ$. Furthermore,

$$\psi_\circ\alpha(X) = \sqcup\{Y \in \mathcal{M} \mid Y \le \alpha(X)\} = \sqcup\{Y \in \mathcal{M} \mid Y \le X\} = \psi_\circ(X);$$

here we have used that $Y \le \alpha(X)$ iff $Y \le X$ for $Y \in \mathcal{M}$. Similarly, it is true that $\psi_\circ\beta = \psi_\circ$. This gives

$$\beta\alpha \le \beta\psi_\circ = \psi_\circ = \psi_\circ\alpha \le \beta\alpha;$$

hence $\psi_\circ = \beta\alpha$. For ψ° the result follows by similar arguments.

(c): Let ψ be a filter with invariance domain $\mathcal{M}$. If $Y \in \mathcal{M}$ and $Y \le X$, then $Y = \psi(Y) \le \psi(X)$. This implies immediately that $\psi_\circ(X) \le \psi(X)$. Dually, it follows that $\psi \le \psi^\circ$. Then $\psi_\circ = \psi_\circ\alpha \le \psi\alpha \le \beta\alpha = \psi_\circ$, and therefore $\psi_\circ = \psi\alpha$. That $\psi^\circ = \psi\beta$ follows by duality.

(d): First note that $\psi(\bigvee_{i\in I} Y_i) \in \mathcal{M}$ and that $Y_i = \psi(Y_i) \le \psi(\bigvee_{i\in I} Y_i)$ for $i \in I$. If $Y_i \le M$, $i \in I$, for some $M \in \mathcal{M}$, then $\bigvee_{i\in I} Y_i \le M$, and therefore $\psi(\bigvee_{i\in I} Y_i) \le \psi(M) = M$. This indeed implies that $\bigsqcup_{i\in I} Y_i = \psi(\bigvee_{i\in I} Y_i)$. □

12.32 Example. (Convex sets)
Consider the underlattice in $\mathcal{L} = \mathcal{P}(\mathbb{R}^2)$ consisting of all convex subsets, i.e., $\mathcal{M} = \mathcal{C}(\mathbb{R}^2)$; cf. Example 2.16. The infimum in $\mathcal{M}$ is the ordinary set intersection, whereas the supremum is the convex hull of the set union. This means that

$$\psi_\circ(X) = \mathrm{co}\Big(\bigcup\{Y \in \mathcal{C}(\mathbb{R}^2) \mid Y \subseteq X\}\Big) = \mathrm{co}(X),$$

Figure 12.5 The chessboard pattern.

the convex hull of X. To prove this, one uses that every singleton is convex. Furthermore,

$$\psi^\circ(X) = \bigcap\{Y \in \mathcal{C}(\mathbb{R}^2) \mid X \subseteq Y\} = \mathrm{co}(X).$$

Thus $\psi_\circ = \psi^\circ = \mathrm{co}(\cdot)$. This means in particular that the closing $\mathrm{co}(\cdot)$ is the only filter on $\mathcal{P}(\mathbb{R}^2)$ that has the convex sets as invariance domain.

This section concludes with some remarks about the invariance domain of a self-dual filter. Assume that $\mathcal{L}$ is a negation. For a subset $\mathcal{M} \subseteq \mathcal{L}$ we put $\mathcal{M}^* = \{X^* \mid X \in \mathcal{M}\}$. It is evident that

$$\mathrm{Inv}(\psi^*) = (\mathrm{Inv}(\psi))^*, \tag{12.11}$$

for every self-dual operator ψ (not necessarily increasing or idempotent). This yields in particular that

$$X \in \mathrm{Inv}(\psi) \iff X^* \in \mathrm{Inv}(\psi),$$

if ψ is a self-dual operator.

12.33 Example. (The chessboard pattern and self-dual filters) Consider the square grid in $\mathbb{Z}^2$. Let C be the chessboard pattern shown in Fig. 12.5.

It is evident that C^c coincides with every translate C_h if $h = (h_1, h_2)$ is such that $h_1 + h_2$ is odd. Assume that ψ is a self-dual translation invariant operator on $\mathcal{P}(\mathbb{Z}^2)$. Then

$$[\psi(C)]^c = \psi^*(C^c) = \psi(C_h) = [\psi(C)]_h, \qquad \text{if } h_1 + h_2 \text{ is odd.}$$

This allows only four possibilities for $\psi(C)$, namely, $\varnothing$, $\mathbb{Z}^2$, C, and C^c. If we assume in addition that ψ is idempotent, then we can exclude $\psi(C) = C^c$.

For if $\psi(C) = C^c$, then

$$C^c = \psi(C) = \psi^2(C) = \psi(C^c) = [\psi^*(C)]^c = [\psi(C)]^c = C,$$

a contradiction. Thus, every translation invariant self-dual filter on $\mathcal{P}(\mathbb{Z}^2)$ maps the chessboard pattern C onto $\varnothing$, $\mathbb{Z}^2$, or C.

Further results on self-dual filters are derived in Section 13.9.

12.5. The middle filter

Suppose we are given two increasing operators $\phi \le \psi$; is it possible to construct a filter with invariance domain $\{X \mid \phi(X) \le X \le \psi(X)\}$? Observe that the opening $\check{\psi}$ has invariance domain $\{X \in \mathcal{L} \mid X \le \psi(X)\}$ and that the closing $\hat{\phi}$ has invariance domain $\{X \in \mathcal{L} \mid \phi(X) \le X\}$. So we are looking for a filter with invariance domain $\mathrm{Inv}(\check{\psi}) \cap \mathrm{Inv}(\hat{\phi})$. We start with the following lemma.

12.34 Lemma. *Let ϕ be an overfilter, ψ an underfilter and $\phi \le \psi$. Then $\mathcal{M} = \{X \in \mathcal{L} \mid \phi(X) \le X \le \psi(X)\}$ is a complete underlattice of $\mathcal{L}$ with supremum $\sqcup_{i\in I} Y_i = \hat{\phi}(\bigvee_{i\in I} Y_i)$ and infimum $\sqcap_{i\in I} Y_i = \check{\psi}(\bigwedge_{i\in I} Y_i)$, for every family $Y_i \in \mathcal{M}$, $i \in I$.*

Proof. Let $Y_i \in \mathcal{M}$ for $i \in I$. We show that $\sqcap_{i\in I} Y_i \in \mathcal{M}$. Note that $\check{\phi}\phi = \phi$ because ϕ is an inf-overfilter. Thus it follows that

$$\begin{aligned}
\phi(\sqcap_{i\in I} Y_i) &\le \phi(\bigwedge_{i\in I} Y_i) = \check{\phi}\phi(\bigwedge_{i\in I} Y_i) \\
&\le \check{\psi}\phi(\bigwedge_{i\in I} Y_i) \le \check{\psi}(\bigwedge_{i\in I} \phi(Y_i)) \\
&\le \check{\psi}(\bigwedge_{i\in I} Y_i) = \sqcap_{i\in I} Y_i.
\end{aligned}$$

On the other hand,

$$\begin{aligned}
\psi(\sqcap_{i\in I} Y_i) &= \psi\check{\psi}(\bigwedge_{i\in I} Y_i) \ge \check{\psi}\check{\psi}(\bigwedge_{i\in I} Y_i) \\
&= \check{\psi}(\bigwedge_{i\in I} Y_i) = \sqcap_{i\in I} Y_i.
\end{aligned}$$

This proves the assertion.

Next we show that if $Y \in \mathcal{M}$ and $Y \le Y_i$ for $i \in I$, then $Y \le \sqcap_{i\in I} Y_i$; this means that $\sqcap_{i\in I} Y_i$ is the greatest lower bound in $\mathcal{M}$ of the Y_i. If $Y \le Y_i$

for $i \in I$, then $Y \leq \bigwedge_{i \in I} Y_i$. Since $\psi(Y) \geq Y$, we have $\check{\psi}(Y) = Y$; hence $Y = \check{\psi}(Y) \leq \check{\psi}(\bigwedge_{i \in I} Y_i) = \sqcap_{i \in I} Y_i$. The result for the supremum follows by duality. □

Proposition 12.31 states that with every complete underlattice $\mathcal{M}$ of $\mathcal{L}$ there can be associated at least one morphological filter which has $\mathcal{M}$ as its invariance domain. The main result of this section gives the existence of a strong filter with invariance domain $\mathcal{M} = \{X \in \mathcal{L} \mid \phi(X) \leq X \leq \psi(X)\}$ if $\mathcal{L}$ is modular, ϕ an inf-overfilter, and ψ a sup-underfilter.

12.35 Theorem. *Consider a modular complete lattice $\mathcal{L}$, an inf-overfilter ϕ and a sup-underfilter ψ such that $\phi \leq \psi$. There exists a unique strong filter ω which satisfies $\phi \leq \omega \leq \psi$ and* $\mathrm{Inv}(\omega) = \{X \mid \phi(X) \leq X \leq \psi(X)\}$. *This filter has the following properties:*

$$\omega = \check{\psi}\hat{\phi} = \hat{\phi}\check{\psi} = (\mathrm{id} \wedge \psi\check{\psi}) \vee \phi\hat{\phi};$$

$$\check{\omega} = \check{\psi} \qquad \textit{and} \qquad \hat{\omega} = \hat{\phi};$$

$$\phi \leq \phi\hat{\phi} \leq \omega \leq \psi\check{\psi} \leq \psi.$$

Proof. From $\phi \leq \psi$, it follows that $\phi = \mathbb{A}\phi \leq \mathbb{A}\psi$, and hence that $\mathbb{B}\phi \leq \mathbb{B}\mathbb{A}\psi = \mathbb{A}\psi$; the latter identity follows from the fact that $\mathbb{A}\psi$ is a strong filter (Proposition 12.23). Since $\check{\psi} = \mathbb{A}(\mathrm{id} \wedge \psi) \leq \mathbb{A}\psi$ and $\mathrm{id} \wedge \mathbb{A}\psi$ is an opening, we conclude that $\check{\psi} = \mathrm{id} \wedge \mathbb{A}\psi$. Dually, $\hat{\phi} = \mathrm{id} \vee \mathbb{B}\phi$. Now,

$$\begin{aligned} \check{\psi}\hat{\phi} &= (\mathrm{id} \wedge \mathbb{A}\psi)(\mathrm{id} \vee \mathbb{B}\phi) \\ &= (\mathrm{id} \vee \mathbb{B}\phi) \wedge \mathbb{A}\psi(\mathrm{id} \vee \mathbb{B}\phi). \end{aligned}$$

Observe that

$$\mathbb{A}\psi \leq \mathbb{A}\psi(\mathrm{id} \vee \mathbb{B}\phi) \leq \mathbb{A}\psi(\mathrm{id} \vee \mathbb{A}\psi) = \mathbb{A}\psi;$$

here we have used that $\mathbb{A}\psi$ is a sup-underfilter. Therefore, using the modularity of $\mathcal{L}$,

$$\check{\psi}\hat{\phi} = (\mathrm{id} \vee \mathbb{B}\phi) \wedge \mathbb{A}\psi = (\mathrm{id} \wedge \mathbb{A}\psi) \vee \mathbb{B}\phi.$$

That $(\mathrm{id} \wedge \mathbb{A}\psi) \vee \mathbb{B}\phi$ equals $\hat{\phi}\check{\psi}$ can be shown using similar arguments as before. Thus we find $\check{\psi}\hat{\phi} = \hat{\phi}\check{\psi} = (\mathrm{id} \wedge \psi\check{\psi}) \vee \phi\hat{\phi}$, which, by Example 12.7, is both a sup-filter and an inf-filter; therefore, it is a strong filter. We put $\omega = \check{\psi}\hat{\phi}$ and conclude from Example 12.7 that

$$\mathrm{Inv}(\omega) = \mathrm{Inv}(\check{\psi}) \cap \mathrm{Inv}(\hat{\phi}) = \{X \in \mathcal{L} \mid \phi(X) \leq X \leq \psi(X)\}.$$

Furthermore,

$$\begin{aligned}\check{\omega} &= \mathrm{id} \wedge \omega = \mathrm{id} \wedge [(\mathrm{id} \wedge \mathbb{A}\psi) \vee \mathbb{B}\phi] \\ &= (\mathrm{id} \wedge \mathbb{A}\psi) \vee (\mathrm{id} \wedge \mathbb{B}\phi) = \mathrm{id} \wedge \mathbb{A}\psi \\ &= \check{\psi}.\end{aligned}$$

Similarly, $\hat{\omega} = \hat{\phi}$. Since $\omega = (\mathrm{id} \wedge \mathbb{A}\psi) \vee \mathbb{B}\phi$, we get $\omega \geq \mathbb{B}\phi = \phi\hat{\phi} \geq \phi$; the other estimates for ω follow by duality.

Assume that λ is another strong filter with $\phi \leq \lambda \leq \psi$ and $\mathrm{Inv}(\lambda) = \{X \mid \phi(X) \leq X \leq \psi(X)\}$. It is obvious that $\lambda\omega = \omega$, since ω maps into $\mathrm{Inv}(\omega) = \mathrm{Inv}(\lambda)$. On the other hand, by Corollary 12.21,

$$\lambda\omega = \hat{\lambda}\check{\lambda}\check{\psi}\hat{\phi} = \hat{\lambda}\check{\lambda}\hat{\phi} = \check{\lambda}\hat{\lambda}\hat{\phi} = \check{\lambda}\hat{\lambda} = \lambda,$$

where we used $\check{\lambda}\check{\psi} = \check{\lambda}$ (since $\check{\lambda} \leq \check{\psi}$) and $\hat{\lambda}\hat{\phi} = \hat{\lambda}$ (since $\hat{\lambda} \geq \hat{\phi}$). This implies $\lambda = \omega$. □

Note that $(\mathrm{id} \wedge \psi\check{\psi}) \vee \phi\hat{\phi} = (\mathrm{id} \vee \phi\hat{\phi}) \wedge \psi\check{\psi}$ is the centre of the operators $\phi\hat{\phi} \leq \psi\check{\psi}$; cf. Section 3.5.

The following result shows how one can obtain middle filters which are self-dual, assuming that the underlying lattice has a negation.

12.36 Proposition. *Let $\mathcal{L}$ be a complete lattice with a negation. If ψ is a sup-underfilter which satisfies $\psi \geq \psi^*$, then the middle filter $\omega = \check{\psi}(\psi^*)^\wedge$ is a self-dual strong filter.*

Proof. Observe that under the given assumptions ψ^* is an inf-overfilter. Therefore, by Theorem 12.35, $\omega = \check{\psi}(\psi^*)^\wedge$ is the middle filter. Furthermore,

$$\omega^* = (\check{\psi})^*\big((\psi^*)^\wedge\big)^* = (\psi^*)^\wedge\check{\psi} = \omega,$$

where we have used (12.5). □

Now the question remains how to find sup-underfilters ψ with $\psi \geq \psi^*$. An important class of operators with this property is constituted by the filters $\beta\alpha\beta$, where α is an opening and β its negative closing. We know from Example 12.7 that $\beta\alpha\beta$ is a sup-filter and that $\beta\alpha\beta \geq \alpha\beta\alpha = (\beta\alpha\beta)^*$. Example 13.47 will show that iteration of $\gamma = (\mathrm{id} \wedge \beta\alpha\beta) \vee \alpha\beta\alpha$ yields the middle filter, assuming that the underlying lattice is modular. Note that in this case the operator γ is self-dual.

12.6. Alternating sequential filters

This section discusses a family of filters obtained by composing openings and closings; cf. Example 12.7. These filters have become known as *alternating sequential filters* or *AS-filters*. Before introducing this new class of filters we present some preliminary results. Let ψ_n, $n = 1, \dots, N$, be a sequence of filters which satisfies one of the following two conditions:

(C1) $\psi_n\psi_m \le \psi_n \le \psi_m\psi_n, \qquad 1 \le m \le n \le N;$

(C2) $\psi_m\psi_n \le \psi_n \le \psi_n\psi_m, \qquad 1 \le m \le n \le N.$

Observe that (C2) is the dual of (C1) in the sense of the Duality Principle.

12.37 Lemma. *Consider the filters $\psi_1, \psi_2, \dots, \psi_N$ such that* (C1) *or* (C2) *holds. Let $n, i_1, i_2, \dots, i_k$ be indices such that $1 \le i_1, i_2, \dots, i_k \le n \le N$; then*

$$\psi_n\psi_{i_k}\psi_{i_{k-1}}\cdots\psi_{i_1}\psi_n = \psi_n. \tag{12.12}$$

In particular, $\psi_n\psi_{i_k}\psi_{i_{k-1}}\cdots\psi_{i_1}$ is a filter.

Proof. Assume, without loss of generality, that (C1) holds. Then $\psi_n\psi_{i_k} \le \psi_n$; hence $\psi_n\psi_{i_k}\psi_{i_{k-1}} \le \psi_n\psi_{i_{k-1}} \le \psi_n$. Repeating this argument one finds $\psi_n\psi_{i_k}\cdots\psi_{i_1} \le \psi_n$, which yields $\le$ in (12.12). On the other hand, $\psi_{i_1}\psi_n \ge \psi_n$, and so $\psi_{i_2}\psi_{i_1}\psi_n \ge \psi_{i_2}\psi_n \ge \psi_n$, and we find $\psi_{i_k}\cdots\psi_{i_1}\psi_n \ge \psi_n$, and therefore $\ge$ in (12.12) also holds. This proves the result. □

We introduce the following notation: if $\psi_1, \psi_2, \dots, \psi_N$ are operators and $1 \le n \le N$, then

$$(\psi)_n = \psi_n\psi_{n-1}\cdots\psi_1. \tag{12.13}$$

The following result is a straightforward consequence of Lemma 12.37.

12.38 Proposition. *Consider the filters $\psi_1, \psi_2, \dots, \psi_N$.*

(a) *If* (C1) *holds, then $(\psi)_n$ is a filter and $(\psi)_n \le \psi_n$.*

(a′) *If* (C2) *holds, then $(\psi)_n$ is a filter and $(\psi)_n \ge \psi_n$.*

If α_n is a collection of openings such that $\alpha_1 \ge \alpha_2 \ge \cdots \ge \alpha_N$, then $(\alpha)_n = \alpha_n$ since $\alpha_n\alpha_m = \alpha_m\alpha_n = \alpha_n$ if $n \ge m$. In general, one cannot prove this semigroup property for the filters $(\psi)_n$ in Proposition 12.38; yet the following *absorption laws* can be established.

12.39 Proposition. *Consider the filters $\psi_1, \psi_2, \dots, \psi_N$.*

(a) *If* (C1) *holds, then $(\psi)_n(\psi)_m = (\psi)_n \le (\psi)_m(\psi)_n$, $n \ge m$.*

(a′) *If* (C2) *holds, then $(\psi)_n(\psi)_m = (\psi)_n \ge (\psi)_m(\psi)_n$, $n \ge m$.*

Proof. Assume that (C1) holds. Then

$$\begin{aligned}(\psi)_n(\psi)_m &= \psi_n\psi_{n-1}\cdots\psi_{m+1}\psi_m\psi_{m-1}\cdots\psi_1\psi_m\psi_{m-1}\cdots\psi_1\\ &= \psi_n\psi_{n-1}\psi_{m+1}\psi_m\psi_{m-1}\cdots\psi_1\\ &= (\psi)_n.\end{aligned}$$

Here we have used that, in view of Lemma 12.37, $\psi_m\psi_{m-1}\cdots\psi_1\psi_m = \psi_m$.

We show that $(\psi)_m(\psi)_n \geq (\psi)_n$. Since $\psi_1\psi_n \geq \psi_n$, also $\psi_2\psi_1\psi_n \geq \psi_2\psi_n \geq \psi_n$, and we infer that $(\psi)_m\psi_n \geq \psi_n$. But this implies that

$$(\psi)_m(\psi)_n = (\psi)_m\psi_n\psi_{n-1}\cdots\psi_1 \geq \psi_n\psi_{n-1}\cdots\psi_1 = (\psi)_n.$$

This concludes the proof. □

A general method for obtaining a collection of filters which satisfies (C1), or dually (C2), is by means of openings and closings. Let $\alpha_1, \alpha_2, \ldots, \alpha_N$ be a family of openings with the semigroup property $\alpha_n\alpha_m = \alpha_n$ if $n \geq m$. Furthermore, let $\beta_1, \beta_2, \ldots, \beta_N$ be a family of closings with the semigroup property $\beta_n\beta_m = \beta_n$ if $n \geq m$. Although in practice the families α_n and β_n are often chosen dual with respect to each other there is no need to do so. We show that $\alpha_n\beta_n$ is a collection of filters which satisfies (C1). Namely, if $n \geq m$,

$$\alpha_n\beta_n\alpha_m\beta_m \leq \alpha_n\beta_n\beta_m = \alpha_n\beta_n$$

and

$$\alpha_m\beta_m\alpha_n\beta_n \geq \alpha_m\alpha_n\beta_n = \alpha_n\beta_n.$$

Dually, $\beta_n\alpha_n$ is a collection of filters which satisfies (C2). From Proposition 12.38 it follows that $(\alpha_n\beta_n)(\alpha_{n-1}\beta_{n-1})\cdots(\alpha_1\beta_1)$ is a filter; following the previous notational convention, this filter is denoted by $(\alpha\beta)_n$:

$$(\alpha\beta)_n = (\alpha_n\beta_n)(\alpha_{n-1}\beta_{n-1})\cdots(\alpha_1\beta_1). \tag{12.14}$$

From Proposition 12.39 it follows that $(\alpha\beta)_n$ obeys the absorption laws

$$(\alpha\beta)_n(\alpha\beta)_m = (\alpha\beta)_n \leq (\alpha\beta)_m(\alpha\beta)_n, \qquad n \geq m.$$

Dually, $(\beta\alpha)_n$ is the filter given by

$$(\beta\alpha)_n = (\beta_n\alpha_n)(\beta_{n-1}\alpha_{n-1})\cdots(\beta_1\alpha_1). \tag{12.15}$$

It obeys the absorption laws

$$(\beta\alpha)_n(\beta\alpha)_m = (\beta\alpha)_n \geq (\beta\alpha)_m(\beta\alpha)_n, \qquad n \geq m.$$

As both filters are constructed by applying openings and closings alternately, they are called *alternating sequential filters* or *AS-filters*. In practice, the α_n and β_n use larger and larger structuring elements as n increases. For an example the reader may refer to Example 12.40.

The filters $\beta_n\alpha_n\beta_n$ and $\alpha_n\beta_n\alpha_n$ do not satisfy (C1) or (C2) in general. Yet both families can be used to define AS-filters. To see this, consider the operator $(\beta\alpha\beta)_n$; note that results for $(\alpha\beta\alpha)_n$ follow by duality. First observe

$$\begin{aligned}(\beta_n\alpha_n\beta_n)(\beta_{n-1}\alpha_{n-1}\beta_{n-1})(\beta_{n-2}\alpha_{n-2}\beta_{n-2})\cdots(\beta_1\alpha_1\beta_1)\\ = \beta_n(\alpha_n\beta_n\beta_{n-1})(\alpha_{n-1}\beta_{n-1}\beta_{n-2})\cdots(\alpha_2\beta_2\beta_1)(\alpha_1\beta_1)\\ = \beta_n(\alpha_n\beta_n)(\alpha_{n-1}\beta_{n-1})\cdots(\alpha_1\beta_1).\end{aligned}$$

This leads to

$$(\beta\alpha\beta)_n = \beta_n(\alpha\beta)_n. \tag{12.16}$$

Then

$$\begin{aligned}(\beta\alpha\beta)_n(\beta\alpha\beta)_n &= \beta_n(\alpha\beta)_n\beta_n(\alpha\beta)_n \geq \beta_n(\alpha\beta)_n(\alpha\beta)_n\\ &= \beta_n(\alpha\beta)_n = (\beta\alpha\beta)_n.\end{aligned}$$

To show the reverse inequality, note that

$$(\beta\alpha\beta)_n \leq \beta_n\beta_{n-1}\cdots\beta_1 = \beta_n;$$

hence

$$(\beta\alpha\beta)_n(\beta\alpha\beta)_n \leq \beta_n(\beta\alpha\beta)_n = (\beta\alpha\beta)_n.$$

Thus $(\beta\alpha\beta)_n$ is a filter. We prove the absorption laws

$$(\beta\alpha\beta)_n(\beta\alpha\beta)_m = (\beta\alpha\beta)_n \geq (\beta\alpha\beta)_m(\beta\alpha\beta)_n, \qquad n \geq m. \tag{12.17}$$

First,

$$\begin{aligned}(\beta\alpha\beta)_n(\beta\alpha\beta)_m &= \beta_n(\alpha\beta)_n\beta_m(\alpha\beta)_m\\ &\geq \beta_n(\alpha\beta)_n(\alpha\beta)_m\\ &= \beta_n(\alpha\beta)_n = (\beta\alpha\beta)_n.\end{aligned}$$

To prove $\le$ observe that

$$\begin{aligned}(\beta\alpha\beta)_n &\le (\beta_n\alpha_n\beta_n)\cdots(\beta_{m+1}\alpha_{m+1}\beta_{m+1})\beta_m\cdots\beta_1\\ &= (\beta_n\alpha_n\beta_n)\cdots(\beta_{m+1}\alpha_{m+1}\beta_{m+1})\beta_m.\end{aligned}$$

Thus

$$\begin{aligned}(\beta\alpha\beta)_n(\beta\alpha\beta)_m &\le (\beta_n\alpha_n\beta_n)\cdots(\beta_{m+1}\alpha_{m+1}\beta_{m+1})\beta_m(\beta\alpha\beta)_m\\ &= (\beta\alpha\beta)_n.\end{aligned}$$

This yields the left equality in (12.17). To prove the inequality on the right,

$$\begin{aligned}(\beta\alpha\beta)_m(\beta\alpha\beta)_n &\le \beta_m\beta_{m-1}\cdots\beta_1(\beta\alpha\beta)_n\\ &= \beta_m(\beta\alpha\beta)_n = (\beta\alpha\beta)_n.\end{aligned}$$

Dually, one shows that $(\alpha\beta\alpha)_n$ is a filter which obeys

$$(\alpha\beta\alpha)_n = \alpha_n(\beta\alpha)_n, \tag{12.18}$$

and the absorption laws

$$(\alpha\beta\alpha)_n(\alpha\beta\alpha)_m = (\alpha\beta\alpha)_n \le (\alpha\beta\alpha)_m(\alpha\beta\alpha)_n, \qquad n \ge m. \tag{12.19}$$

Finally, we point out the following inequalities:

$$(\alpha\beta\alpha)_n \le \begin{Bmatrix}(\alpha\beta)_n\\ (\beta\alpha)_n\end{Bmatrix} \le (\beta\alpha\beta)_n. \tag{12.20}$$

These inequalities are almost trivial, e.g.,

$$(\beta\alpha\beta)_n = (\beta_n\alpha_n\beta_n)\cdots(\beta_1\alpha_1\beta_1) \ge (\alpha_n\beta_n)\cdots(\alpha_1\beta_1) = (\alpha\beta)_n.$$

It is easy to see that AS-filters are not self-dual (presuming that $\mathcal{L}$ has a negation). Even in the case that the families α_n and β_n are complementary, one gets "only" $((\alpha\beta)_n)^* = (\beta\alpha)_n \neq (\alpha\beta)_n$.

12.40 Example. Let α_n be the opening by the $(2n+1)\times(2n+1)$ square, and let β_n be the negative closing. Then α_n satisfies the semigroup property $\alpha_n\alpha_m = \alpha_m\alpha_n = \alpha_n$, $n \ge m$. The alternating sequential filter $(\beta\alpha)_n$ is illustrated in Fig. 12.6.

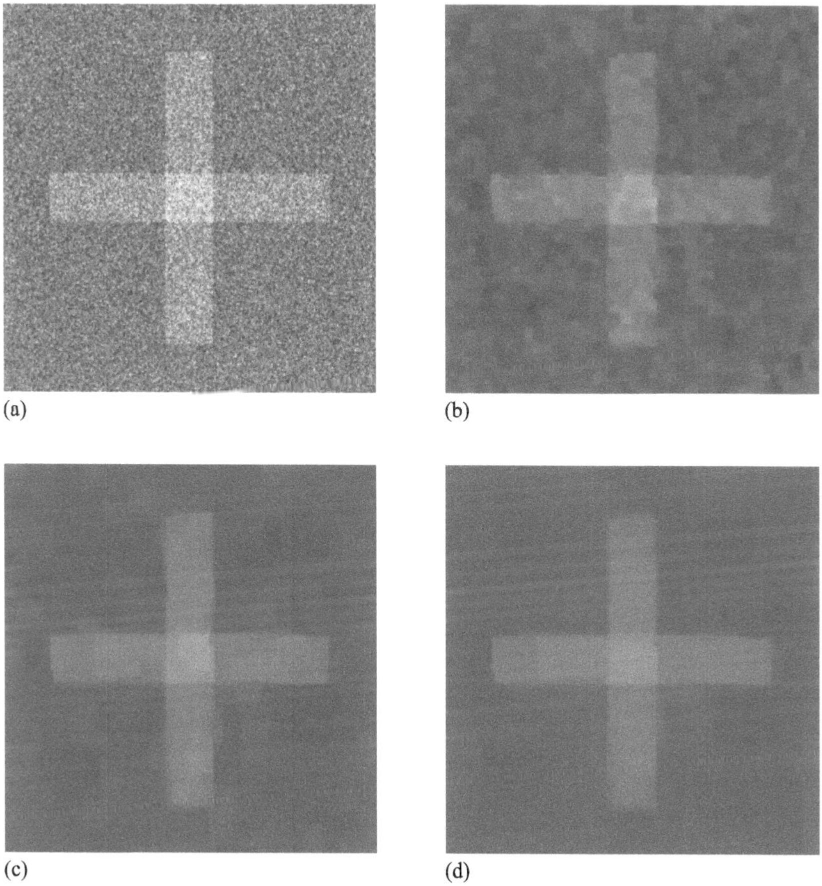

Figure 12.6 Alternating sequential filter. (a) Original image X; (b) $(\beta\alpha)_2(X)$; (c) $(\beta\alpha)_5(X)$; (d) $(\beta\alpha)_{10}(X)$. Here α_n, β_n are respectively the opening and closing by the $(2n+1) \times (2n+1)$ square.

12.7. Bibliographical notes

Most of the results presented in this chapter originate from the work of Matheron and Serra; see Serra (1988), in particular, Chapters 6 and 8. Note, however, that our notation is substantially different. The tutorial paper by Serra and Vincent (1992) comprises a nice introduction into the theory of morphological filters. In a recent paper, Ronse (1994a) discusses some links between the algebraic theory of morphological filters and (generalizations of) the Tarski Fixpoint Theorem. Furthermore, he points out some related results from theoretical computer science.

Alternating sequential filters are an invention of Sternberg (1986). A systematic theory, however, was first developed by Serra (1988). These filters have been shown to do a rather good job when used for noise filtering. Schonfeld and Goutsias (1991) show that AS-filters are optimal for restoration of noisy images.

CHAPTER THIRTEEN

Filtering and iteration

Henk J.A.M. Heijmans
Formerly Centre for Mathematics and Computer Science, Amsterdam, Netherlands

Contents

The previous chapter was concerned with a detailed investigation of algebraic properties of filters and derived notions such as overfilters, inf-overfilters, and strong filters. The present chapter explains how one can construct morphological filters by iteration of an arbitrary increasing operator. It is shown that one has to impose continuity conditions on the underlying operator. An important class of morphological operators which satisfy these continuity requirements are the finite window operators; every translation invariant operator that uses only finite structuring elements belongs to this class. The results can be used to construct the middle filter between an inf-overfilter and a sup-underfilter. The final section describes a method for the construction of filters which are self-dual.

13.1. Order convergence

In this section we explore a notion of sequential convergence on an arbitrary complete lattice $\mathcal{L}$ which is based on the partial ordering. As motivation consider the real line $\mathbb{R}$. A sequence x_n in $\mathbb{R}$ converges to the point x iff the following two conditions hold:

Advances in Imaging and Electron Physics, Volume 216
ISSN 1076-5670
https://doi.org/10.1016/bs.aiep.2020.07.013

(i) for every $\epsilon > 0$, there is an $N \geq 1$ such that $x_n \leq x + \epsilon$ for $n \geq N$;
(ii) for every $\epsilon > 0$, there is an $N \geq 1$ such that $x_n \geq x - \epsilon$ for $n \geq N$.

Instead of (i) and (ii) we can write "$\limsup x_n \leq x$" and "$\liminf x_n \geq x$", respectively; recall that $\limsup x_n = \bigwedge_{n\geq 1} \bigvee_{k\geq n} x_k$ and $\liminf x_n = \bigvee_{n\geq 1} \bigwedge_{k\geq n} x_k$. In other words, the Euclidean topology of the real line is determined completely by its ordering. One can mimic the definitions of lim sup and lim inf to arrive at a notion of convergence on an arbitrary complete lattice $\mathcal{L}$.

Given a sequence X_n in $\mathcal{L}$, define

$$\liminf X_n = \bigvee_{n\geq 1} \bigwedge_{k\geq n} X_k,$$
$$\limsup X_n = \bigwedge_{n\geq 1} \bigvee_{k\geq n} X_k.$$

It is obvious that $\bigwedge_{k\geq n} X_k \leq \bigvee_{k\geq m} X_k$ for every $n, m \geq 1$. This means that $\bigvee_{n\geq 1} \bigwedge_{k\geq n} X_k \leq \bigwedge_{m\geq 1} \bigvee_{k\geq m} X_k$. In other words:

$$\liminf X_n \leq \limsup X_n. \tag{13.1}$$

13.1 Definition. Given a sequence X_n in the complete lattice $\mathcal{L}$ and an element $X \in \mathcal{L}$, we say that X_n *(order-) converges* to X, written $X_n \to X$, if $\liminf X_n = \limsup X_n = X$.

Throughout this chapter we restrict attention to sequential convergence; in the final section we point out some relations with related concepts in the literature.

Let X_n, Y_n be sequences in $\mathcal{L}$ such that $X_n \leq Y_n$ for every $n \geq 1$. It is obvious that

$$\liminf X_n \leq \liminf Y_n \qquad \text{and} \qquad \limsup X_n \leq \limsup Y_n.$$

Furthermore, it is easy to check that

$$\liminf(X_n \wedge Y_n) \leq \liminf X_n \wedge \liminf Y_n, \tag{13.2}$$
$$\liminf(X_n \vee Y_n) \geq \liminf X_n \vee \liminf Y_n, \tag{13.3}$$

for all sequences X_n, Y_n. Similar relations hold for lim sup. In general, the reverse inequalities do not hold. But one can prove the following results.

13.2 Proposition. *Given two sequences X_n, Y_n in a complete lattice for which the infinite distributivity laws hold; then*

$$\liminf(X_n \wedge Y_n) = \liminf X_n \wedge \liminf Y_n, \tag{13.4}$$

$$\limsup(X_n \vee Y_n) = \limsup X_n \vee \limsup Y_n. \tag{13.5}$$

Proof. We prove (13.4). Put $\overline{X}_n = \bigwedge_{k\geq n} X_k$ and $\overline{Y}_n = \bigwedge_{k\geq n} Y_k$. It is evident that

$$\liminf(X_n \wedge Y_n) = \bigvee_{m\geq n} (\overline{X}_m \wedge \overline{Y}_m),$$

for every $n \geq 1$. Since $\overline{X}_n$, $\overline{Y}_n$ are increasing with n,

$$\begin{aligned}
\liminf(X_n \wedge Y_n) &= \bigvee_{n\geq 1} \bigvee_{m\geq n} (\overline{X}_m \wedge \overline{Y}_m) \\
&\geq \bigvee_{n\geq 1} \bigvee_{m\geq n} (\overline{X}_n \wedge \overline{Y}_m) \\
&= \bigvee_{n\geq 1} \bigvee_{m\geq 1} (\overline{X}_n \wedge \overline{Y}_m) \\
&= \bigvee_{n\geq 1} (\overline{X}_n \wedge \bigvee_{m>1} \overline{Y}_m) \\
&= (\bigvee_{n\geq 1} \overline{X}_n) \wedge (\bigvee_{m\geq 1} \overline{Y}_m) \\
&= \liminf X_n \wedge \liminf Y_n;
\end{aligned}$$

here we have used (2.5) twice. The reverse inequality is given in (13.2). □

Recall the following facts from Section 3.1. The notation $X_n \downarrow X$ means that X_n is a decreasing sequence ($X_n \leq X_{n-1}$) and $\bigwedge_{n\geq 1} X_n = X$. Dually, $X_n \uparrow X$ means that X_n is an increasing sequence ($X_n \geq X_{n-1}$) and $\bigvee_{n\geq 1} X_n = X$.

13.3 Proposition. *Consider a sequence X_n in a complete lattice $\mathcal{L}$. If $X_n \downarrow X$ or $X_n \uparrow X$, then $X_n \to X$.*

Proof. Assume that $X_n \downarrow X$; then $\bigvee_{k\geq n} X_k = X_n$. This implies immediately that $\limsup X_n = \bigwedge_{n\geq 1} X_n = X$. Furthermore, $\bigwedge_{k\geq n} X_k = \bigwedge_{k\geq 1} X_k$ for every $n \geq 1$, which means that $\liminf X_n = X$. Therefore, $X_n \to X$. The proof for $X_n \uparrow X$ is analogous. □

13.4 Proposition. *If $\mathcal{L}$ is a complete lattice with a negation, then*

$$\liminf X_n^* = (\limsup X_n)^*, \tag{13.6}$$
$$\limsup X_n^* = (\liminf X_n)^*, \tag{13.7}$$

for every sequence X_n in $\mathcal{L}$. In particular,

$$X_n \to X \qquad \textit{if and only if} \qquad X_n^* \to X^*.$$

The proof of this result is straightforward.

13.5 Examples.

(a) For a sequence X_n in $\mathcal{P}(E)$ one derives easily that

$$\liminf X_n = \{x \in E \mid x \in X_n \text{ eventually}\},$$
$$\limsup X_n = \{x \in E \mid x \in X_{n_k} \text{ for some subsequence } n_k\}.$$

If, e.g., $X_n = A$ if n is odd and $X_n = B$ if n is even, then $\liminf X_n = A \cap B$ and $\limsup X_n = A \cup B$.

(b) Consider the complete lattice of closed convex subsets in $\mathbb{R}^d$; choose two sets A, B such that $A \cap B = \varnothing$. Define $X_n = A$ if n is odd and $X_n = B$ if n is even. Then $\liminf X_n = \varnothing$ and $\limsup X_n = \overline{\text{co}}(A \cup B)$.

(c) On $\text{Fun}(E, \mathcal{T})$, with $\mathcal{T}$ a complete lattice, $F_n \to F$ if and only if $F_n(x) \to F(x)$ in $\mathcal{T}$ for every $x \in E$.

The complete lattice of closed sets of a topological space will be examined in Section 13.3.

13.2. Order continuity

Using the notions lim inf and lim sup, one can define lower and upper semi-continuity of operators between complete lattices.

13.6 Definition. Given two complete lattices $\mathcal{L}$ and $\mathcal{M}$ and an operator $\psi : \mathcal{L} \to \mathcal{M}$, we say that ψ is *$\downarrow$-continuous* if $X_n \to X$ in $\mathcal{L}$ implies that $\limsup \psi(X_n) \le \psi(X)$. Dually, we say that ψ is *$\uparrow$-continuous*, if $X_n \to X$ in $\mathcal{L}$ implies that $\psi(X) \le \liminf \psi(X_n)$. If ψ is both $\downarrow$-continuous and $\uparrow$-continuous, then we say that ψ is *$\updownarrow$-continuous*.

$\downarrow$- and $\uparrow$-continuity are dual notions in the sense of the Duality Principle: if $\psi : \mathcal{L} \to \mathcal{M}$ is $\downarrow$-continuous, then it is $\uparrow$-continuous as an operator between the dual lattices $\mathcal{L}'$ and $\mathcal{M}'$ and vice versa.

In Definition 3.7 an increasing operator ψ has been called $\downarrow$-continuous if $X_n \downarrow X$ implies that $\psi(X_n) \downarrow \psi(X)$, for every decreasing sequence X_n. We show that for increasing operators the conditions in Definition 3.7 and Definition 13.6 are equivalent.

13.7 Proposition. *Let ψ be an increasing operator between the complete lattices $\mathcal{L}$ and $\mathcal{M}$.*

(a) *The following statements are equivalent:*
 (i) *ψ is $\downarrow$-continuous;*
 (ii) *$X_n \downarrow X$ implies that $\psi(X_n) \downarrow \psi(X)$ for every sequence X_n;*
 (iii) $\limsup \psi(X_n) \leq \psi(\limsup X_n)$ *for every sequence X_n.*

(a′) *The following statements are equivalent:*
 (i) *ψ is $\uparrow$-continuous;*
 (ii) *$X_n \uparrow X$ implies that $\psi(X_n) \uparrow \psi(X)$ for every sequence X_n;*
 (iii) $\liminf \psi(X_n) \geq \psi(\liminf X_n)$ *for every sequence X_n.*

Proof. (i) $\Rightarrow$ (ii): Assume that ψ is $\downarrow$-continuous and that $X_n \downarrow X$. By Proposition 13.3 $X_n \to X$, and therefore $\limsup \psi(X_n) \leq \psi(X)$. As $\psi(X_n)$ is decreasing, $\limsup \psi(X_n) = \bigwedge_{n\geq 1} \psi(X_n)$, and so it follows that $\bigwedge_{n\geq 1} \psi(X_n) \leq \psi(X)$. But the reverse inequality holds trivially, and we conclude that $\psi(X_n) \downarrow \psi(X)$.

(ii) $\Rightarrow$ (iii): If X_n is an arbitrary sequence in $\mathcal{L}$, then $Y_n = \bigvee_{k\geq n} X_k$ is decreasing; moreover, $Y_n \downarrow \limsup X_n$. This implies that

$$\psi(\limsup X_n) = \bigwedge_{n\geq 1} \psi(Y_n).$$

However, $\psi(X_k) \leq \psi(Y_n)$ for $k \geq n$, and so $\bigvee_{k\geq n} \psi(X_k) \leq \psi(Y_n)$. Combining both facts gives that

$$\psi(\limsup X_n) \geq \bigwedge_{n\geq 1} \bigvee_{k\geq n} \psi(X_k) = \limsup \psi(X_n).$$

(iii) $\Rightarrow$ (i): Obvious. □

Consider, as an example, the operator $X \mapsto X \vee A$ on a complete lattice $\mathcal{L}$; here A is a fixed element of $\mathcal{L}$. From (13.3) it follows that $\liminf(X_n \vee A) \geq X \vee A$ if $X_n \to X$; this implies that $X \mapsto X \vee A$ is $\uparrow$-continuous. Dually, the operator $X \mapsto X \wedge A$ is $\downarrow$-continuous. Furthermore, we derive from Proposition 13.2 that both operators are $\updownarrow$-continuous if $\mathcal{L}$ obeys the infinite distributivity laws.

In the remainder of this section we investigate the class of $\downarrow$-continuous operators and the dual class of $\uparrow$-continuous operators.

13.8 Theorem.

(a) *Every erosion is $\downarrow$-continuous.*

(a′) *Every dilation is $\uparrow$-continuous.*

(b) *Every automorphism is $\updownarrow$-continuous.*

(c) *Every dual automorphism (e.g., every negation) is $\updownarrow$-continuous.*

This result is obvious. For instance, that erosions are $\downarrow$-continuous follows immediately from the fact that such operators act distributively over infima.

13.9 Proposition.

(a) *The infimum of an arbitrary collection of $\downarrow$-continuous operators is $\downarrow$-continuous.*

(a′) *The supremum of an arbitrary collection of $\uparrow$-continuous operators is $\uparrow$-continuous.*

Proof. Assume that ψ_i is $\downarrow$-continuous for every i in the index set I, and let $\psi = \bigwedge_{i\in I} \psi_i$. We show that $\limsup \psi(X_n) \leq \psi(X)$ if $X_n \to X$. Since ψ_i is $\downarrow$-continuous, $\limsup \psi_i(X_n) \leq \psi_i(X)$, for $i \in I$. Therefore

$$\limsup \bigwedge_{i\in I} \psi_i(X_n) \leq \psi_i(X),$$

for $i \in I$. Taking the infimum on the right-hand side, we arrive at

$$\limsup \bigwedge_{i\in I} \psi_i(X_n) \leq \bigwedge_{i\in I} \psi_i(X),$$

which was to be shown. □

13.10 Proposition. *Let $\psi_1, \psi_2, \ldots, \psi_p$ be operators from $\mathcal{L}$ into $\mathcal{M}$.*

(a) *Assume that $\mathcal{M}$ is semi-atomic. If every ψ_i is $\downarrow$-continuous, then $\psi_1 \vee \psi_2 \vee \cdots \vee \psi_p$ is $\downarrow$-continuous as well.*

(a′) *Assume that $\mathcal{M}$ is dual-semi-atomic. If every ψ_i is $\uparrow$-continuous, then $\psi_1 \wedge \psi_2 \wedge \cdots \wedge \psi_p$ is $\uparrow$-continuous as well.*

Proof. Let ψ_i be $\downarrow$-continuous for $i = 1, 2, \ldots, p$; we show that $\psi = \psi_1 \vee \psi_2 \vee \cdots \vee \psi_p$ is $\downarrow$-continuous as well. Assume that $X_n \to X$; we prove that $\limsup \psi(X_n) \leq \psi(X)$. If for a given semi-atom $A \leq \limsup \psi(X_n)$ it can be shown that $A \leq \psi(X)$, then the assertion follows immediately. For such A,

$$A \leq \bigvee_{k\geq n} \bigvee_{i=1}^{p} \psi_i(X_k) = \bigvee_{i=1}^{p} \bigvee_{k\geq n} \psi_i(X_k),$$

for every $n \geq 1$. As A is a semi-atom, it follows that $A \leq \bigvee_{k \geq n} \psi_{i(n)}(X_k)$ for some $i(n)$ between 1 and p. At least one of the values i between 1 and p will be assumed infinitely often by $i(n)$ if n ranges over the positive integers. For this i, we have $A \leq \bigvee_{k \geq n} \psi_i(X_k)$, for every $n \geq 1$. Thus,

$$A \leq \limsup \psi_i(X_n) \leq \psi_i(X) \leq \psi(X).$$

This finishes the proof. □

Observe that this proposition is valid for $\mathcal{M} = \mathcal{P}(E)$ as well as for $\mathcal{M} = \text{Fun}(E, \mathcal{T})$, where $\mathcal{T}$ is a complete chain.

13.11 Proposition. *Let ϕ, ψ be operators between the complete lattices $\mathcal{L}$ and $\mathcal{M}$.*

(a) *Under either of the following two assumptions the composition $\psi\phi$ is $\downarrow$-continuous:*
  (i) *ϕ is $\updownarrow$-continuous and ψ is $\downarrow$-continuous;*
  (ii) *ϕ is $\downarrow$-continuous and ψ is increasing and $\downarrow$-continuous.*

(a′) *Under either of the following two assumptions the composition $\psi\phi$ is $\uparrow$-continuous:*
  (i) *ϕ is $\updownarrow$-continuous and ψ is $\uparrow$-continuous;*
  (ii) *ϕ is $\uparrow$-continuous and ψ is increasing and $\uparrow$-continuous.*

Proof. (i): If $X_n \to X$, then $\phi(X_n) \to \phi(X)$, and thus $\limsup \psi(\phi(X_n)) \leq \psi(\phi(X))$. This means that $\psi\phi$ is $\downarrow$-continuous.

(ii): If $X_n \to X$, then $\limsup \phi(X_n) \leq \phi(X)$. Since ψ is increasing it follows that $\limsup \psi\phi(X_n) \leq \psi(\limsup \phi(X_n)) \leq \psi\phi(X)$. Therefore, $\psi\phi$ is $\downarrow$-continuous. □

Next we consider complete lattices with a negation. Theorem 13.8(c) states that every negation is $\updownarrow$-continuous.

13.12 Proposition. *Let ψ be an operator between the complete lattices $\mathcal{L}$ and $\mathcal{M}$.*

(a) *Assume that $\mathcal{M}$ has a negation. The operator ψ is $\downarrow$-continuous iff the operator $X \mapsto (\psi(X))^*$ is $\uparrow$-continuous.*

(b) *Assume that both $\mathcal{L}$ and $\mathcal{M}$ have a negation. Then ψ is $\downarrow$-continuous if and only if ψ^* is $\uparrow$-continuous.*

Proof. We use Proposition 13.4.

(a): Assume that ψ is $\downarrow$-continuous. If $X_n \to X$, then $\liminf(\psi(X_n))^* = (\limsup \psi(X_n))^* \geq (\psi(X))^*$, since $\limsup \psi(X_n) \leq \psi(X)$. Therefore, $X \mapsto (\psi(X))^*$ is $\uparrow$-continuous. The converse statement has an analogous proof.

(b): Assume that ψ is $\downarrow$-continuous. If $X_n \to X$, then $X_n^* \to X^*$, and so $\limsup \psi(X_n^*) \le \psi(X^*)$. Now

$$\begin{aligned}\liminf \psi^*(X_n) &= \liminf(\psi(X_n^*))^* = \left[\limsup \psi(X_n^*)\right]^* \\ &\ge (\psi(X^*))^* = \psi^*(X).\end{aligned}$$

This means that ψ^* is $\uparrow$-continuous. The if-statement is proved analogously. This concludes the proof. □

13.3. Relation with the hit-or-miss topology

Consider a locally compact Hausdorff space E with a countable basis. If he or she prefers, the reader may assume throughout this section that $E = \mathbb{R}^d$. Chapter 7 contains a comprehensive discussion of the hit-or-miss topology on the closed sets $\mathcal{F}(E)$. It also discusses upper and lower semi-continuous mappings from some topological space S into $\mathcal{F}(E)$. This section shows that convergence and continuity in the sense of the hit-or-miss topology are closely related to the corresponding notions of order convergence and order continuity defined in this chapter. A first result in this direction is Proposition 7.47, which says that an increasing operator on $\mathcal{F}(\mathbb{R}^d)$ is u.s.c. in the sense of the hit-or-miss topology if and only if it is $\downarrow$-continuous with respect to order convergence; see also Proposition 13.14.

To start, we point out the connection between $\limsup X_n$ and $\liminf X_n$ of a sequence X_n in $\mathcal{F}(E)$ on the one hand and, $\overline{\lim}\, X_n$ and $\underline{\lim}\, X_n$ on the other.

13.13 Proposition. *Given a sequence X_n in $\mathcal{F}(E)$; then*

$$\limsup X_n = \overline{\lim}\, X_n, \tag{13.8}$$

$$\liminf X_n \subseteq \underline{\lim}\, X_n. \tag{13.9}$$

In particular,

$$X_n \to X \Rightarrow X_n \xrightarrow{\mathcal{F}} X. \tag{13.10}$$

Proof. The first relation is implied by (7.15). To prove the second, recall from Section 7.4 that $\underline{\lim}\, X_n$ is the largest closed set that satisfies the criterion "if $X \cap G \ne \varnothing$, then $X_n \cap G \ne \varnothing$ eventually, for every open set G". Therefore, if we can show that $\liminf X_n$ satisfies this criterion, we may conclude that $\liminf X_n \subseteq \underline{\lim}\ X_n$. Assume that G is open and that

$$G \cap \liminf X_n = G \cap \overline{\bigcup_{n\geq 1} \bigcap_{k\geq n} X_k} \neq \varnothing.$$

This implies

$$G \cap \bigcup_{n\geq 1} \bigcap_{k\geq n} X_k = \bigcup_{n\geq 1} \bigcap_{k\geq n} (G \cap X_k) \neq \varnothing.$$

Thus there exists an $n \geq 1$ such that $X_k \cap G \neq \varnothing$ for $k \geq n$. This proves (13.9).

Finally, the implication in (13.10) is an immediate consequence of the previous two relations. In fact, if $X_n \to X$, then

$$\overline{\lim}\ X_n = \limsup X_n = X = \liminf X_n \subseteq \underline{\lim}\ X_n.$$

But $\underline{\lim}\ X_n \subseteq \overline{\lim}\ X_n$ is obvious, and the proof is finished. □

We show by means of an example that the inclusion in (13.9) may be strict, and hence that the reverse implication in (13.10) needs not hold. Let X_n be the circle in $\mathbb{R}^2$ centred at the origin with radius $1 - 1/n$, and let X be the circle with radius 1. It is obvious that $\overline{\lim}\ X_n = \limsup X_n = X$ and that $\underline{\lim}\ X_n = X$. However, $\liminf X_n = \varnothing$ as $\bigcap_{k\geq n} X_k = \varnothing$ for every $n \geq 1$.

The next result expresses the relation between both (semi-) continuity notions.

13.14 Proposition. *Let ψ be an arbitrary operator on $\mathcal{F}(E)$.*
(a) *If ψ is u.s.c., then ψ is $\downarrow$-continuous.*
(b) *If ψ is increasing and $\downarrow$-continuous, then ψ is u.s.c.*
(c) *If ψ is increasing and l.s.c., then ψ is $\uparrow$-continuous.*

Proof. (a): Assume that ψ is u.s.c. and that $X_n \to X$. Then, by (13.10), $X_n \xrightarrow{\mathcal{F}} X$, and therefore $\overline{\lim}\ \psi(X_n) \leq \psi(X)$ because ψ is u.s.c. But $\overline{\lim}\ \psi(X_n)$ and $\limsup \psi(X_n)$ coincide by (13.8), and we conclude that $\limsup \psi(X_n) \leq \psi(X)$. Therefore ψ is $\downarrow$-continuous.

(b): Follows from Propositions 7.39 and 13.7.

(c): Suppose that ψ is increasing and l.s.c.; furthermore, let $X_n \uparrow X$. By Proposition 13.7(a′), the proof is complete if we can show that $\psi(X_n) \uparrow \psi(X)$. Using Proposition 7.28(b), we get $X = \overline{\bigcup_{n\geq 1} X_n}$ and $X_n \xrightarrow{\mathcal{F}} X$. The same result gives $\psi(X_n) \xrightarrow{\mathcal{F}} \overline{\bigcup_{n\geq 1} \psi(X_n)}$. As ψ is l.s.c., we get $\psi(X) \subseteq$

$\underline{\lim}\, \psi(X_n) = \overline{\bigcup_{n\geq 1} \psi(X_n)}$. Since X_n is an increasing sequence and $X_n \subseteq X$, the reverse inclusion holds trivially. This implies that $\psi(X) = \overline{\bigcup_{n\geq 1} \psi(X_n)}$, which means that $\psi(X_n) \uparrow \psi(X)$. □

13.4. Translation invariant set operators

The main goal of this and the next section is to construct a large class of morphological operators on $\mathcal{P}(\mathbb{E}^d)$ or $\mathrm{Fun}(\mathbb{E}^d, \mathcal{T})$ which are $\updownarrow$-continuous. The next result shows that flat function operators inherit continuity properties from the corresponding set operators.

13.15 Proposition. *Let Ψ be an increasing flat function operator on* $\mathrm{Fun}(\mathbb{E}^d, \overline{\mathbb{R}})$ *generated by the set operator ψ.*

(a) *If ψ is $\downarrow$-continuous, then Ψ is $\downarrow$-continuous as well and*

$$\mathbf{X}(\Psi(F), t) = \psi(\mathbf{X}(F, t)), \tag{13.11}$$

for every function F and $t \in \mathbb{R}$.

(b) *If ψ is $\uparrow$-continuous, then Ψ is $\uparrow$-continuous as well.*

Proof. (a): Assume that ψ is $\downarrow$-continuous. Relation (13.11) is exactly (11.11). To prove that Ψ is $\downarrow$-continuous, assume that $F_n \downarrow F$; we must show that $\Psi(F_n) \downarrow \Psi(F)$. By Proposition 11.2(c) and (11.8),

$$\begin{aligned}
\mathbf{X}(\bigwedge_{n\geq 1} \Psi(F_n), t) &= \bigcap_{n\geq 1} \mathbf{X}(\Psi(F_n), t) \\
&= \bigcap_{n\geq 1} \bigcap_{s<t} \psi(\mathbf{X}(F_n, s)) \\
&= \bigcap_{s<t} \bigcap_{n\geq 1} \psi(\mathbf{X}(F_n, s)).
\end{aligned}$$

Since $\mathbf{X}(F_n, s) \downarrow \mathbf{X}(F, s)$ for every s and ψ is $\downarrow$-continuous,

$$\begin{aligned}
\mathbf{X}(\bigwedge_{n\geq 1} \Psi(F_n), t) &= \bigcap_{s<t} \psi(\bigcap_{n\geq 1} \mathbf{X}(F_n, s)) \\
&= \bigcap_{s<t} \psi(\mathbf{X}(F, s)) \\
&= \mathbf{X}(\Psi(F), t).
\end{aligned}$$

As this holds for every t, we conclude that $\bigwedge_{n\geq 1} \Psi(F_n) = \Psi(F)$. Thus Ψ is $\downarrow$-continuous.

(b): We use twice the observation that an operator is $\downarrow$-continuous iff the complemented operator is $\uparrow$-continuous. Thus, if ψ is $\uparrow$-continuous, then ψ^* is $\downarrow$-continuous. Combining (a) with Theorem 11.16, it follows that Ψ^* is $\downarrow$-continuous. Therefore, Ψ is $\uparrow$-continuous. □

It is easy to see that this result is also valid for functions with values in $\overline{\mathbb{Z}}$ or $\{0, 1, \ldots, N\}$.

Let X_n be a sequence in $\mathcal{P}(\mathbb{E}^d)$; recall the following expressions for $\limsup X_n$ and $\liminf X_n$ from Example 13.5(a):

$$\limsup X_n = \{x \in \mathbb{E}^d \mid x \in X_{n_k} \text{ for some subsequence } n_k\}, \tag{13.12}$$

$$\liminf X_n = \{x \in \mathbb{E}^d \mid x \in X_n \text{ eventually}\}. \tag{13.13}$$

We are particularly interested in continuity properties of translation invariant operators on $\mathcal{P}(\mathbb{E}^d)$ such as Minkowski addition and subtraction. Theorem 13.8 states that every dilation is $\uparrow$-continuous and that every erosion is $\downarrow$-continuous. For finite structuring elements, however, one can prove a much stronger result.

13.16 Proposition. *Let A be a finite structuring element in $\mathbb{E}^d$. The dilation $X \mapsto X \oplus A$, the erosion $X \mapsto X \ominus A$, the closing $X \mapsto X \bullet A$ and the opening $X \mapsto X \circ A$ are $\updownarrow$-continuous.*

Proof. First consider dilation. Theorem 13.8 gives that this operator is $\uparrow$-continuous. Obviously, every translation operator $X \mapsto X_h$ is an automorphism; hence $\updownarrow$-continuous by Theorem 13.8(b). Since dilation by a finite structuring element can be written as a finite union of translations, Proposition 13.10 gives that such a dilation is $\downarrow$-continuous; hence it is $\updownarrow$-continuous. The result for the erosion follows by duality. The proof for the opening and the closing can easily be established by applying Proposition 13.11. □

We present an example which shows that the finiteness condition is essential.

13.17 Examples.

(a) Let $A \subseteq \mathbb{E}^d$ be a structuring element which contains infinitely many points. Let $a_n \in A$ be distinct points, and put $X_n = \{-a_n, -a_{n+1}, \ldots\}$. It is obvious that $X_n \downarrow \varnothing$; hence $X_n \to \varnothing$ by Proposition 13.3. Obviously, $0 \in X_n \oplus A$ for every n, and so $0 \in \limsup(X_n \oplus A)$. However, $\varnothing \oplus A = \varnothing$, which shows that the operator $X \mapsto X \oplus A$ is not $\downarrow$-continuous.

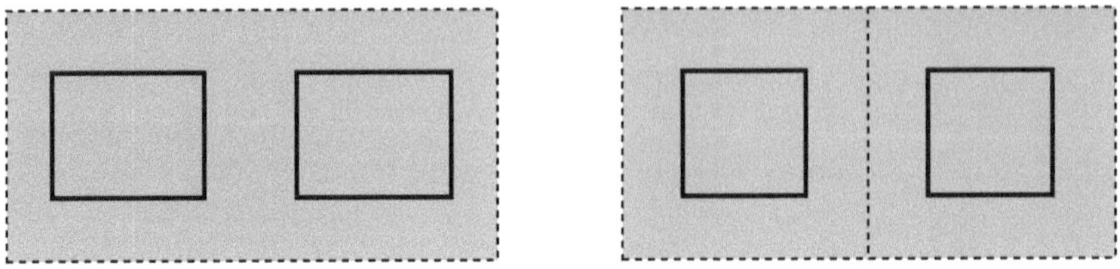

Figure 13.1 Dilation by an infinite structuring element is not $\downarrow$-continuous; see Example 13.17(b).

(b) In Fig. 13.1 we depict another example. Here X_n consists of two rectangles with width $1 + (1 - d_n)/2$ and height 1, which lie at distance d_n; here d_n increases toward 1 as $n \to \infty$. Assume that the locations of the left border of the left rectangle and the right border of the right rectangle do not vary with n.

Evidently, $X_n \downarrow X$ where X consists of two squares with sides 1 at distance 1. If the structuring element A is the open unit square, then $\bigcap_{n\geq 1} X_n \oplus A$ is the closed rectangle with width 4 and height 2, whereas $X \oplus A$ is the union of two open rectangles with width 2 and height 2.

Next we consider the hit-or-miss operator, the thinning and the thickening; cf. Section 4.2.

13.18 Proposition. *Consider two structuring elements $A, B \subseteq \mathbb{E}^d$.*

(a) *The hit-or-miss operator $X \mapsto X \circledast (A, B)$ is $\downarrow$-continuous. If both A and B are finite, then it is $\updownarrow$-continuous.*

(b) *The thickening operator $X \mapsto X \odot (A, B)$ is $\downarrow$-continuous. If both A and B are finite, then it is $\updownarrow$-continuous.*

(c) *The thinning operator $X \mapsto X \circledcirc (A, B)$ is $\uparrow$-continuous. If both A and B are finite, then it is $\updownarrow$-continuous.*

Although it is quite simple to give a direct proof of these results, we refer to the following section, which discusses finite window operators; the operators in Proposition 13.18 belong to this class. At the end of that section it will be pointed out that the results obtained there provide a proof of the proposition just stated.

Recall that the kernel of an operator ψ on $\mathcal{P}(\mathbb{E}^d)$ is defined by $\mathcal{V}(\psi) = \{A \subseteq \mathbb{E}^d \mid 0 \in \psi(A)\}$. The next result is the order convergence analogue of Proposition 7.45.

13.19 Proposition. *A translation invariant operator ψ on $\mathcal{P}(\mathbb{E}^d)$ is $\downarrow$-continuous if and only if $\mathcal{V}(\psi)$ is closed under* $\limsup$*, i.e., $X_n \in \mathcal{V}(\psi)$ for $n \geq 1$ implies that* $\limsup X_n \in \mathcal{V}(\psi)$.

Proof. "only if": We use that $\limsup \psi(X_n) \subseteq \psi(\limsup X_n)$; see Proposition 13.7. If $X_n \in \mathcal{V}(\psi)$ for every n, then $0 \in \psi(X_n)$, and therefore $0 \in \limsup \psi(X_n)$. But this implies that $0 \in \psi(\limsup X_n)$, and therefore $\limsup X_n \in \mathcal{V}(\psi)$.

"if": Let $X_n \to X$; we show that $\limsup \psi(X_n) \subseteq \psi(X)$. If $h \in \limsup \psi(X_n)$, then $h \in \psi(X_{n_k})$ for some subsequence n_k. But this means that $(X_{n_k})_{-h} \in \mathcal{V}(\psi)$. As $\mathcal{V}(\psi)$ is closed under lim sup, we get $X_{-h} = \limsup(X_{n_k})_{-h} \in \mathcal{V}(\psi)$, i.e., $h \in \psi(X)$. This concludes the proof. □

13.5. Finite window operators

A systematic treatment of $\updownarrow$-continuous operators is beyond the scope of this chapter. In forthcoming sections it will become clear in what sense (semi-) continuity properties of operators play a role in establishing convergence of sequences of iterates of a given operator ψ. The examples that will be presented are concerned with operators on $\mathcal{P}(\mathbb{E}^d)$ or their flat extensions to $\text{Fun}(\mathbb{E}^d)$. Proposition 13.15 shows that the flat extension of an increasing $\downarrow$-continuous ($\uparrow$-continuous) operator on $\mathcal{P}(E)$ to $\text{Fun}(E)$ is $\downarrow$-continuous ($\uparrow$-continuous), too. In the remainder of this section we concentrate on set operators.

13.20 Definition. Consider an operator ψ on $\mathcal{P}(E)$; assume that $W(h)$ is a finite subset of E for every $h \in E$. We say that ψ is a *finite window operator* with window W if

$$h \in \psi(X) \iff h \in \psi(X \cap Z),$$

for every $h \in E$, $X \subseteq E$ and $Z \supseteq W(h)$.

We make the following two observations. First, if ψ is increasing, then it suffices to assume that $h \in \psi(X)$ iff $h \in \psi(X \cap W(h))$ for every $h \in E$ and $X \subseteq E$. Second, if ψ is translation invariant ($E = \mathbb{Z}^d$ or $\mathbb{R}^d$), then we can take $W(h) = W_h$, where $W \subseteq E$ is finite.

13.21 Theorem. *Every finite window operator on $\mathcal{P}(E)$ is $\updownarrow$-continuous.*

Proof. First observe from (13.12)–(13.13) that $X_n \to X$ in $\mathcal{P}(E)$ implies $X_n \cap C = X \cap C$ eventually, for any finite set $C \subseteq E$.

Let ψ be a finite window operator with window W. We prove only that ψ is $\downarrow$-continuous; the $\uparrow$-continuity of ψ follows by similar arguments. Assume $X_n \to X$; it has to be demonstrated that $\limsup \psi(X_n) \subseteq \psi(X)$. Take

$h \in \limsup \psi(X_n) = \bigcap_{n\geq 1} \bigcup_{k\geq n} \psi(X_k)$. There is a subsequence n_k such that $h \in \psi(X_{n_k})$ for every k. This means that $h \in \psi(X_{n_k} \cap W(h))$. But $X_{n_k} \cap W(h) = X \cap W(h)$ for k sufficiently large. This implies that $h \in \psi(X \cap W(h))$. But then $h \in \psi(X)$, which concludes the proof. □

The remaining part of this section discusses finite window operators.

13.22 Proposition. *The operator ψ on $\mathcal{P}(E)$ is a finite window operator if and only if ψ^* is such as well. In that case, both operators have the same window.*

Proof. Suppose that ψ is a finite window operator with window W. Let $X \subseteq E$ and $Z \supseteq W(h)$; we show that

$$h \in \psi^*(X) \iff h \in \psi^*(X \cap Z).$$

Observe that for every $Y \subseteq E$,

$$\begin{aligned} h \in \psi^*(Y) &\iff h \notin \psi(Y^c) \iff h \notin \psi(Y^c \cap Z) \\ &\iff h \notin \psi((Y \cup Z^c)^c) \iff h \in \psi^*(Y \cup Z^c). \end{aligned}$$

Substitution of $Y = X \cap Z$ leads to

$$h \in \psi^*(X \cap Z) \iff h \in \psi^*((X \cap Z) \cup Z^c) = \psi^*(X \cup Z^c).$$

$h \in \psi^*(X \cup Z^c)$ is equivalent to $h \in \psi^*(X)$, however, and the assertion is proved. □

The following result is obvious.

13.23 Proposition. *Let, for every i in some index set I, ψ_i be a finite window operator on $\mathcal{P}(E)$ with window W_i. If $W(h) = \bigcup_{i\in I} W_i(h)$ is finite for every $h \in E$, then $\bigcap_{i\in I} \psi_i$ and $\bigcup_{i\in I} \psi_i$ are finite window operators with window W.*

For composites the following result holds.

13.24 Proposition. *Consider two finite window operators ϕ, ψ with windows U and V, respectively. Then $\psi\phi$ is a finite window operator with window $W(h) = \bigcup_{k\in V(h)} U(k)$.*

Proof. Given $h \in E$ and $Z \supseteq W(h)$, we show that

$$\phi(X) \cap V(h) = \phi(X \cap Z) \cap V(h), \tag{13.14}$$

for every set $X \subseteq E$. In fact, take $k \in V(h)$; since $W(h) \subseteq Z$, also $U(k) \subseteq Z$. Therefore, $k \in \phi(X)$ if and only if $k \in \phi(X \cap Z)$. In other words, $\phi(X) \cap \{k\} = \phi(X \cap Z) \cap \{k\}$. Taking the union over all $k \in V(h)$ leads to (13.14). Now,

$$\begin{aligned} h \in \psi(\phi(X)) &\Longleftrightarrow h \in \psi(\phi(X) \cap V(h)) \\ &\Longleftrightarrow h \in \psi(\phi(X \cap Z) \cap V(h)) \\ &\Longleftrightarrow h \in \psi(\phi(X \cap Z)); \end{aligned}$$

this concludes the proof. □

Assume $\mathbb{E}^d = \mathbb{R}^d$ or $\mathbb{Z}^d$ and $a \in \mathbb{E}^d$. The translation operator $X \mapsto X_a$ is a finite window operator with window $W(h) = \{h - a\}$. With the aid of Proposition 13.23 it follows that dilation or erosion by a finite structuring element is $\updownarrow$-continuous (cf. Proposition 13.16). Subsequently, Proposition 13.24 gives that opening or closing by a finite structuring element is $\updownarrow$-continuous as well. Furthermore, it is evident that the complement operator $X \mapsto X^c$ is a finite window operator with window $\{h\}$. This observation implies that hit-or-miss operators, as well as thickening and thinning operators, with finite structuring elements are $\updownarrow$-continuous. This provides a proof of Proposition 13.18.

13.25 Proposition. *Every increasing translation invariant finite window operator on $\mathcal{P}(\mathbb{E}^d)$ can be decomposed as a finite union of erosions with finite structuring elements or, alternatively, as a finite intersection of dilations with finite structuring elements.*

Proof. If we are able to prove the statement about erosions, then the one about dilations follows with the aid of Proposition 13.22. Let ψ be an increasing translation invariant finite window operator, and let W be the corresponding window. First, observe that one may assume $W(h) = W_h$, as ψ is translation invariant. Define $\mathcal{V}_0 = \{A \cap W \mid A \in \mathcal{V}(\psi)\}$. Since $0 \in \psi(A)$ iff $0 \in \psi(A \cap W)$, it follows that $\mathcal{V}_0 \subseteq \mathcal{V}(\psi)$. It is evident that

$$\psi(X) = \bigcup_{A \in \mathcal{V}(\psi)} X \ominus A = \bigcup_{A \in \mathcal{V}_0} X \ominus A.$$

Furthermore, every $A \in \mathcal{V}_0$ is finite since it is a subset of W and W is finite. Finally, note that $\mathcal{V}_0$ is finite since $\mathcal{V}_0 \subseteq \mathcal{P}(W)$. □

Stated differently, this result says that every increasing translation invariant finite window operator has a finite basis comprising finite structuring elements; cf. Section 7.6.

13.6. Iteration and idempotence

Throughout this section $\mathcal{L}$ is an arbitrary complete lattice unless stated otherwise. Given a sequence ψ_n of operators on $\mathcal{L}$, we say that ψ_n converges to the operator ψ on $\mathcal{L}$ if $\psi_n(X) \to \psi(X)$ for every $X \in \mathcal{L}$, and we denote this by $\psi_n \to \psi$.

13.26 Proposition. *Consider a sequence of operators ψ_n on the complete lattice $\mathcal{L}$ and an operator ψ such that $\psi_n \to \psi$.*

(a) *If τ is an automorphism on $\mathcal{L}$ and $\psi_n\tau = \tau\psi_n$ for every n, then $\psi\tau = \tau\psi$.*

(b) *If every ψ_n is increasing, then ψ is increasing, too.*

(c) *If ϕ, ϕ_n are operators such that $\phi_n \to \phi$ and if $\phi_n \leq \psi_n$ for every n, then $\phi \leq \psi$.*

(d) *If $\mathcal{L}$ has a negation, then $\psi_n^* \to \psi^*$.*

Proof. (a): For every $X \in \mathcal{L}$,

$$\begin{aligned}\psi(\tau(X)) &= \limsup \psi_n(\tau(X)) = \limsup \tau(\psi_n(X)) \\ &= \tau(\limsup \psi_n(X)) = \tau(\psi(X)).\end{aligned}$$

Here we have used that an automorphism distributes over infima and suprema.

(b): If $X \leq Y$, then $\psi_n(X) \leq \psi_n(Y)$. This implies that

$$\psi(X) = \limsup \psi_n(X) \leq \limsup \psi_n(Y) = \psi(Y),$$

and therefore ψ is increasing.

(c), (d): Obvious. □

As a corollary of (a), the limit of a sequence of translation invariant operators on $\mathcal{P}(\mathbb{E}^d)$ is translation invariant as well. A consequence of (c), e.g., is that the limit of a sequence of anti-extensive operators is anti-extensive as well.

This section deals exclusively with sequences of iterates of some fixed operator ψ. The limit of a sequence ψ^n, assuming it converges, is denoted by ψ^∞. The main question treated in this section is: "When is ψ^∞ idempotent?" Hereafter, when we speak of ψ^∞ we always assume implicitly that $\psi^n \to \psi^\infty$. We start with the following observation.

13.27 Lemma. *Let ψ be an operator on a complete lattice $\mathcal{L}$. If $\psi^n \to \psi^\infty$ and $\psi\psi^\infty = \psi^\infty$, then ψ^∞ is idempotent and* $\mathrm{Inv}(\psi^\infty) = \mathrm{Inv}(\psi)$.

Proof. From $\psi\psi^\infty = \psi^\infty$ one derives by induction that $\psi^n\psi^\infty = \psi^\infty$ for every $n \geq 1$. As $\psi^n\psi^\infty \to \psi^\infty\psi^\infty$, this implies that $\psi^\infty\psi^\infty = \psi^\infty$, and so ψ^∞ is idempotent.

We prove that $\mathrm{Inv}(\psi^\infty) = \mathrm{Inv}(\psi)$. First, if $\psi(X) = X$, then $\psi^n(X) = X$; hence $\psi^\infty(X) = X$. Conversely, if $\psi^\infty(X) = X$, then $X = \psi^\infty(X) = \psi\psi^\infty(X) = \psi(X)$. This concludes the proof. □

Unfortunately, the identity $\psi\psi^\infty = \psi^\infty$, or to be expressive, $\psi^{\infty+1} = \psi^\infty$, does not hold in general. The following is a counterexample; see also Example 13.33.

13.28 Example. This example discusses an increasing, extensive operator ψ for which ψ^∞ is not idempotent. Note that $\psi \geq \mathrm{id}$ implies that $\psi^{n+1} \geq \psi^n$, and so $\psi^n \to \psi^\infty = \bigvee_{n\geq 1}\psi^n$. Let $\mathbb{N}$ be the nonnegative integers and $\overline{\mathbb{N}} = \mathbb{N} \cup \{\infty\}$. Given $X \subseteq \overline{\mathbb{N}}$, define $X+1 = \{x+1 \mid x \in X\}$, where $\infty + 1 = \infty$. Define the operator ψ on $\mathcal{P}(\overline{\mathbb{N}})$ by

$$\begin{cases} \psi(X) = X \cup (X+1), & \text{if } X \neq \mathbb{N}, \\ \psi(\mathbb{N}) = \overline{\mathbb{N}}. \end{cases}$$

It follows immediately that ψ is increasing and extensive. Now $\psi(\{0\}) = \{0,1\}$, $\psi(\{0,1\}) = \{0,1,2\}$, etc. It follows immediately that $\psi^\infty(\{0\}) = \bigcup_{n\geq 1}\psi^n(\{0\}) = \mathbb{N}$. On the other hand, $\psi(\psi^\infty(\{0\})) = \psi(\mathbb{N}) = \overline{\mathbb{N}}$, which shows that $(\psi^\infty)^2 \neq \psi^\infty$.

In fact, this example shows that one has to take recourse to transfinite iteration to achieve idempotence. This was actually done by Maisonneuve (1982); see also Ronse (1994a). Although such a procedure is of great theoretical interest, it cannot be used in practice to obtain idempotent operators. For that reason we restrict attention to natural numbers here.

13.29 Lemma. *Let ψ be an operator on the complete lattice $\mathcal{L}$.*

(a) *If ψ is $\downarrow$-continuous and $\psi^n \to \psi^\infty$, then $\psi^\infty \leq \psi\psi^\infty$.*

(a′) *If ψ is $\uparrow$-continuous and $\psi^n \to \psi^\infty$, then $\psi^\infty \geq \psi\psi^\infty$.*

Proof. From the $\downarrow$-continuity of ψ and the fact that $\psi^n(X) \to \psi^\infty(X)$, it follows that

$$\limsup \psi(\psi^n(X)) \leq \psi\psi^\infty(X).$$

But the left-hand side equals $\psi^\infty(X)$, and the result is proved. □

13.30 Proposition. *Consider an operator ψ on the complete lattice $\mathcal{L}$.*

(a) *If ψ is $\downarrow$-continuous and anti-extensive, then $\psi^\infty = \bigwedge_{n\geq 1} \psi^n$ is idempotent.*
(a′) *If ψ is $\uparrow$-continuous and extensive, then $\psi^\infty = \bigvee_{n\geq 1} \psi^n$ is idempotent.*
(b) *If ψ is $\updownarrow$-continuous and $\psi^n \to \psi^\infty$, then ψ^∞ is idempotent.*

Proof. (a): Since ψ is anti-extensive, $\psi^{n+1} \leq \psi^n$, meaning that $\psi^\infty = \bigwedge_{n\geq 1} \psi^n$. From the previous lemma we obtain $\psi^\infty \leq \psi\psi^\infty$. The reverse inequality is a direct consequence of the anti-extensivity of ψ. Now Lemma 13.27 gives that ψ^∞ is idempotent.

(b): Combining (a) and (a′) in Lemma 13.29 yields that $\psi^\infty = \psi\psi^\infty$. Now the idempotence of ψ is a consequence of Lemma 13.27. □

So far the operators were not assumed to be increasing. From now on we deal with increasing operators exclusively.

13.31 Proposition.

(a) *If ψ is an underfilter, then ψ^∞ is an underfilter as well.*
(a′) *If ψ is an overfilter, then ψ^∞ is an overfilter as well.*
(b) *If ψ is a $\downarrow$-continuous underfilter, then $\psi^\infty = \mathsf{A}\psi$ is a $\downarrow$-continuous filter.*
(b′) *If ψ is a $\uparrow$-continuous overfilter, then $\psi^\infty = \mathsf{B}\psi$ is a $\uparrow$-continuous filter.*
(c) *If ψ is a sup-underfilter, then ψ^∞ is a sup-underfilter as well.*
(c′) *If ψ is an inf-overfilter, then ψ^∞ is an inf-overfilter as well.*
(d) *If ψ is a $\downarrow$-continuous sup-underfilter, then ψ^∞ is a $\downarrow$-continuous sup-filter.*
(d′) *If ψ is a $\uparrow$-continuous inf-overfilter, then ψ^∞ is a $\uparrow$-continuous inf-filter.*

Proof. (a): If ψ is an underfilter, then every ψ^n is an underfilter by Proposition 12.3(b′) and $\psi^{n+1} \leq \psi^n$. Thus, $\psi^\infty = \bigwedge_{n\geq 1} \psi^n$ and, since the class of underfilters is inf-closed, also an underfilter.

(b): It is evident that $\psi^n(X) \downarrow \psi^\infty(X) = \bigwedge_{n\geq 1} \psi^n(X)$, for every X. As ψ is $\downarrow$-continuous, it follows that $\psi(\psi^n(X)) \downarrow \psi(\psi^\infty(X))$. Since $\psi(\psi^n(X)) = \psi^{n+1}(X) \downarrow \psi^\infty(X)$, it follows that $\psi^\infty = \psi\psi^\infty$. Lemma 13.27 gives that ψ^∞ is idempotent. Since ψ^∞ is also increasing it is a filter. To obtain that $\psi^\infty = \mathsf{A}\psi$ it has to be demonstrated that every overfilter $\phi \leq \psi$ satisfies $\phi \leq \psi^\infty$. Let ϕ be any such overfilter; then $\phi \leq \phi^n \leq \psi^n$, and hence $\phi \leq \psi^\infty$. The $\downarrow$-continuity is a consequence of the $\downarrow$-continuity of every iterate ψ^n (Proposition 13.11) and the fact that the class of $\downarrow$-continuous operators is inf-closed (Proposition 13.9).

(c): Similar to (a).

(d): From (b) we get that ψ^∞ is a $\downarrow$-continuous filter. Furthermore, by Proposition 12.3(c′), ψ^n is a sup-underfilter. Since the class of sup-underfilters is inf-closed, ψ^∞ is a sup-underfilter as well. This yields the result. □

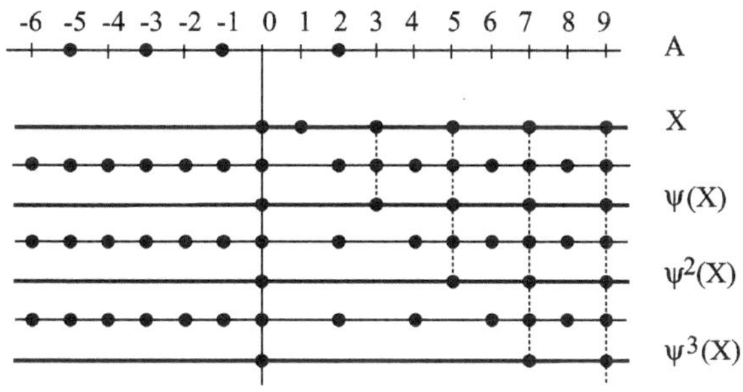

Figure 13.2 The operator ψ^∞ is not idempotent, for $\psi^\infty(X) = \{0\}$ but $(\psi^\infty)^2(X) = \varnothing$.

From (b), (b′), and the fact that $\check{\psi} = \mathsf{A}(\mathrm{id} \wedge \psi)$ and $\hat{\psi} = \mathsf{B}(\mathrm{id} \vee \psi)$ (cf. Proposition 12.13), it is clear how to compute the lower and upper envelope by iteration.

13.32 Corollary. *Suppose that ψ is an increasing operator on a complete lattice $\mathcal{L}$.*
(a) *If ψ is $\downarrow$-continuous, then $\check{\psi} = (\mathrm{id} \wedge \psi)^\infty$.*
(a′) *If ψ is $\uparrow$-continuous, then $\hat{\psi} = (\mathrm{id} \vee \psi)^\infty$.*

The following example shows that the continuity requirement is essential indeed.

13.33 Example. We construct an increasing, anti-extensive, translation invariant operator ψ on $\mathcal{P}(\mathbb{Z})$ for which ψ^∞ is not idempotent (hence not an opening). See Fig. 13.2.

Consider the structuring element $A = \{\ldots, -5, -3, -1, 2\}$, and define $\psi(X) = X \cap (X \oplus A)$. Note that by the infiniteness of A, the dilation $X \mapsto X \oplus A$ is not $\downarrow$-continuous; see Example 13.17. Take $X = \{0, 1, 3, 5, \ldots\}$; a straightforward computation shows that $\psi^n(X) = \{0, 2n+1, 2n+3, 2n+5, \ldots\}$. This implies that $\psi^\infty(X) = \bigwedge_{n \geq 1} \psi^n(X) = \{0\}$. However, $\psi\psi^\infty(X) = \psi(\{0\}) = \varnothing \neq \psi^\infty(X)$, and we conclude that ψ^∞ is not idempotent. In particular, $\check{\psi} \leq \psi^\infty$ but $\check{\psi} \neq \psi^\infty$.

13.34 Example. (Infimum of openings)
Consider the complete Boolean lattice $\mathcal{P}(\mathbb{Z}^2)$. Let A be the 2×2 square and B the rhombus. Let α_A, α_B be the structural openings by A and B, respectively. It is obvious that $\psi = \alpha_B \alpha_A$ is an increasing, anti-extensive operator; however, ψ is not idempotent. Since ψ is a finite window operator it is continuous, and therefore ψ^∞ is an opening. Fig. 13.3 illustrates the iteration of ψ. We show that $\mathrm{Inv}(\psi^\infty) = \mathrm{Inv}(\psi) = \mathrm{Inv}(\alpha_A) \cap \mathrm{Inv}(\alpha_B)$. The first relation follows from Lemma 13.27. The inclusion $\supseteq$ in the second

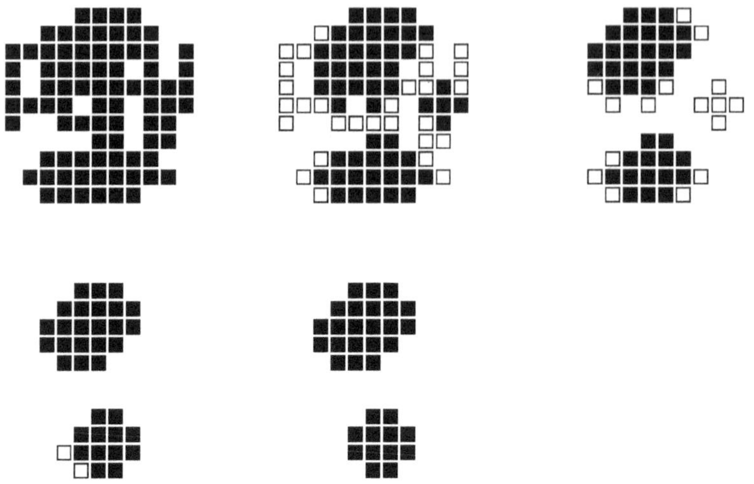

Figure 13.3 Iteration of $\alpha_B\alpha_A$, where A is the 2×2 square and B the rhombus; see Example 13.34. From left to right and top to bottom: the original image X and 4 iterates. The final object is not changed by subsequent iterations. The white pixels represent those points which are deleted by subsequent iterations.

relation is obvious. We show that $\mathrm{Inv}(\psi) \subseteq \mathrm{Inv}(\alpha_A) \cap \mathrm{Inv}(\alpha_B)$. Assume that $\psi(X) = X$. Since $\alpha_A \leq \mathrm{id}$ we may conclude that $X = \alpha_B\alpha_A(X) \subseteq \alpha_B(X)$, whence it follows that $\alpha_B(X) = X$. By a similar argument it follows that $\alpha_A(X) = X$.

Although $\mathrm{Inv}(\alpha_A)$ and $\mathrm{Inv}(\alpha_B)$ have a simple structure, a complete description of $\mathrm{Inv}(\alpha_A) \cap \mathrm{Inv}(\alpha_B)$ is cumbersome; some typical invariants are depicted in Fig. 13.3.

An alternative way to obtain the opening ψ^∞ is by iteration of $\phi = \alpha_A \wedge \alpha_B$. Namely,

$$\phi^2 \leq \psi \leq \phi,$$

from which it follows that $\phi^{2n} \leq \psi^n \leq \phi^n$ for every $n \geq 1$. Therefore, $\phi^\infty = \psi^\infty$.

It is easy to check that the observations made in the previous example can be generalized.

13.35 Proposition. *Let $\psi_1, \psi_2, \ldots, \psi_p$ be increasing operators on the complete lattice $\mathcal{L}$.*

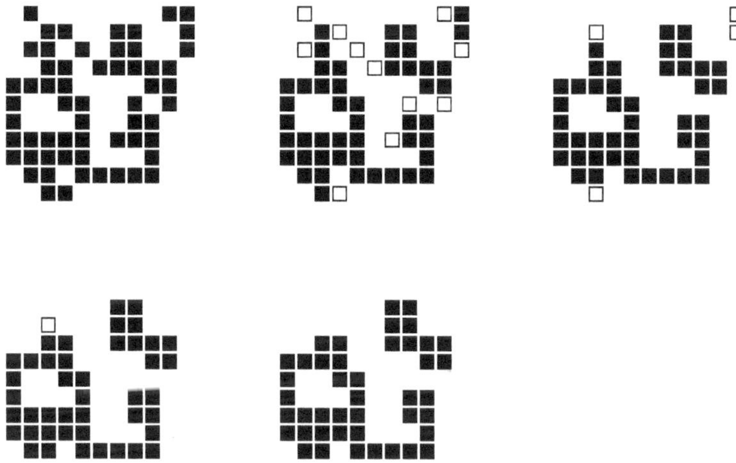

Figure 13.4 Iteration of id $\wedge \mu_A$, where A is the rhombus, yields the median opening; see Example 13.36. From left to right and top to bottom: the original image X and 4 iterates. The final object is not changed by subsequent iterations.

(a) *If every ψ_i is anti-extensive, then*

$$(\psi_p \psi_{p-1} \cdots \psi_1)^\infty = (\psi_p \wedge \psi_{p-1} \wedge \cdots \wedge \psi_1)^\infty$$

is an opening with invariance domain $\text{Inv}(\psi_1) \cap \text{Inv}(\psi_2) \cap \cdots \cap \text{Inv}(\psi_p)$.

(a′) *If every ψ_i is extensive, then*

$$(\psi_p \psi_{p-1} \cdots \psi_1)^\infty = (\psi_p \vee \psi_{p-1} \vee \cdots \vee \psi_1)^\infty$$

is a closing with invariance domain $\text{Inv}(\psi_1) \cup \text{Inv}(\psi_2) \cup \cdots \cup \text{Inv}(\psi_p)$.

13.36 Example. (Median opening)
Let A be the rhombus, and let μ_A be the median operator on $\mathcal{P}(\mathbb{Z}^2)$ associated with A. Since μ_A is a finite window operator, it is $\updownarrow$-continuous, and it follows that $(\text{id} \wedge \mu_A)^n$ converges to an opening; we call this opening the *median opening*.

Fig. 13.4 shows how the iterates $(\text{id} \wedge \mu_A)^n(X)$ decrease toward an invariant of the median opening.

13.7. Iteration of the centre operator

The centre $\gamma = \gamma(\phi, \psi)$ of two operators $\phi \leq \psi$ on a modular lattice is defined by

$$\gamma = (\mathrm{id} \wedge \psi) \vee \phi = (\mathrm{id} \vee \phi) \wedge \psi.$$

More generally, the centre of an arbitrary collection of operators ψ_i, $i \in I$, on $\mathcal{L}$ is defined by $\gamma = (\mathrm{id} \wedge \bigvee_{i\in I} \psi_i) \vee \bigwedge_{i\in I} \psi_i$; cf. Sections 3.5 and 3.6. In this section we investigate the sequence of iterates γ^n (with the convention that $\gamma^0 = \mathrm{id}$). We start with some general statements; later we specialize to the function lattice $\mathrm{Fun}(E, \mathcal{T})$, where $\mathcal{T}$ is a complete chain. In Section 13.8 it will be shown that, under certain conditions, the sequence γ^n converges toward the middle filter $\hat{\phi}\check{\psi} = \check{\psi}\hat{\phi}$, assuming that ϕ is an inf-overfilter and ψ a sup-underfilter. For the reader's convenience we recall the definition of the activity ordering from Definition 3.35: the operator ψ is *more active* than the operator ϕ, written $\phi \preccurlyeq \psi$, if

$$\mathrm{id} \wedge \psi \le \mathrm{id} \wedge \phi \qquad \text{and} \qquad \mathrm{id} \vee \psi \ge \mathrm{id} \vee \phi.$$

13.37 Proposition. *Let $\mathcal{L}$ be a complete modular lattice, ϕ an overfilter and ψ an underfilter such that $\phi \le \psi$. Furthermore, let $\gamma = (\mathrm{id} \wedge \psi) \vee \phi$ be the centre operator. Then*

$$\gamma^{n-1} \preccurlyeq \gamma^n \tag{13.15}$$

and

$$\mathrm{Inv}(\gamma^n) = \{X \in \mathcal{L} \mid \phi(X) \le X \le \psi(X)\}, \tag{13.16}$$

for every $n \ge 1$.

Proof. To establish (13.15) it is shown that

$$\gamma^n = (\mathrm{id} \wedge \psi\gamma^{n-1}) \vee \phi\gamma^{n-1} = (\mathrm{id} \vee \phi\gamma^{n-1}) \wedge \psi\gamma^{n-1}, \tag{13.17}$$

for every $n \ge 1$. Then, using the modularity of $\mathcal{L}$, we infer that

$$\mathrm{id} \wedge \gamma^n = \mathrm{id} \wedge \psi\gamma^{n-1}, \tag{13.18}$$

$$\mathrm{id} \vee \gamma^n = \mathrm{id} \vee \phi\gamma^{n-1}. \tag{13.19}$$

In particular, this gives

$$\mathrm{id} \wedge \gamma^n \le \mathrm{id} \wedge \psi^2\gamma^{n-2} \le \mathrm{id} \wedge \psi\gamma^{n-2} \le \mathrm{id} \wedge \gamma^{n-1};$$

here we used that $\gamma \le \psi$ and that ψ is an underfilter. Analogously, we get

$$\mathrm{id} \vee \gamma^n \ge \mathrm{id} \vee \gamma^{n-1}.$$

Thus, (13.15) follows.

We prove (13.17) by induction. For $n = 1$ the result is obvious. Assume that the result is true for a particular n. Then

$$\begin{aligned}\gamma^{n+1} &= \big((\mathrm{id} \wedge \psi) \vee \phi\big)\gamma^n \\ &= (\gamma^n \wedge \psi\gamma^n) \vee \phi\gamma^n \\ &= \Big[\big((\mathrm{id} \vee \phi\gamma^{n-1}) \wedge \psi\gamma^{n-1}\big) \wedge \psi\gamma^n\Big] \vee \phi\gamma^n.\end{aligned}$$

Since $\psi\gamma^{n-1} \geq \psi^2\gamma^{n-1} \geq \psi\gamma^n$, we get

$$\begin{aligned}\gamma^{n+1} &= \big[(\mathrm{id} \vee \phi\gamma^{n-1}) \wedge \psi\gamma^n\big] \vee \phi\gamma^n \\ &= (\mathrm{id} \vee \phi\gamma^{n-1} \vee \phi\gamma^n) \wedge \psi\gamma^n \\ &= (\mathrm{id} \vee \phi\gamma^n) \wedge \psi\gamma^n.\end{aligned}$$

Here we used the modularity of $\mathcal{L}$ and the fact that $\phi\gamma^{n-1} \leq \phi^2\gamma^{n-1} \leq \phi\gamma^n$.

It remains to prove (13.16). Assume $\gamma^n(X) = X$; then $\psi(X) \geq \psi^n(X) \geq \gamma^n(X) = X$; similarly, $\phi(X) \leq X$. Conversely, assume that $\phi(X) \leq X \leq \psi(X)$. Then $\gamma(X) = (X \wedge \psi(X)) \vee \phi(X) = X$; hence $\gamma^n(X) = X$. □

13.38 Definition. Let $\mathcal{L}$ be a complete lattice. The operator ψ is said to be *activity-extensive* if $\psi^n \preccurlyeq \psi^{n+1}$, for every $n \geq 0$.

Every increasing operator which is extensive or anti-extensive is activity-extensive. The converse is not true, however.

13.39 Remark.

(a) If, in addition to the assumptions of the previous proposition, ψ is an inf-filter, then $\psi = \psi(\mathrm{id} \wedge \psi) \leq \psi\gamma \leq \psi^2 \leq \psi$; this gives $\psi\gamma^n = \psi$. Thus $\gamma^n = (\mathrm{id} \wedge \psi) \vee \phi\gamma^{n-1}$. Dually, if ϕ is a sup-filter, one gets $\gamma^n = (\mathrm{id} \vee \phi) \wedge \psi\gamma^{n-1}$.

(b) Suppose that $\mathcal{L}$ is modular, ϕ is a sup-underfilter, ψ is an inf-overfilter, and $\phi \leq \psi$. Then the centre is a filter. In fact,

$$\begin{aligned}\gamma^2 &= (\gamma \wedge \psi\gamma) \vee \phi\gamma = \big[(\mathrm{id} \vee \phi) \wedge \psi \wedge \psi\gamma\big] \vee \phi\gamma \\ &= \big[(\mathrm{id} \vee \phi) \wedge \psi\big] \vee \phi\gamma = (\mathrm{id} \wedge \psi) \vee \phi \vee \phi\gamma \\ &= (\mathrm{id} \wedge \psi) \vee \phi = \gamma.\end{aligned}$$

Here we used that $\psi\gamma \geq \psi(\mathrm{id} \wedge \psi) = \psi$ and that $\phi\gamma \leq \phi(\mathrm{id} \vee \phi) = \phi$.

(c) Let ψ_i, $i \in I$, be a family of strong filters on the modular lattice $\mathcal{L}$, and let their centre γ be given by $\gamma = (\mathrm{id} \wedge \psi) \vee \phi$, where $\psi = \bigvee_{i \in I} \psi_i$ and

$\phi = \bigwedge_{i \in I} \psi_i$. We show that

$$\gamma = \hat{\phi}\check{\psi} = \check{\psi}\hat{\phi}.$$

In particular, this identity implies that γ is a strong filter. Observe that $\hat{\phi} = \mathrm{id} \vee \phi$ since ϕ is a sup-underfilter and that $\check{\psi} = \mathrm{id} \wedge \psi$ since ψ is an inf-overfilter. Note that γ is a filter by the remark in (b). First we show that $\psi = \psi(\mathrm{id} \vee \phi)$. Since every ψ_i is a strong filter,

$$\psi_i = \psi_i(\mathrm{id} \vee \psi_i) \geq \psi_i(\mathrm{id} \vee \phi) \geq \psi_i,$$

whence $\psi_i = \psi_i(\mathrm{id} \vee \phi)$. Taking the supremum over all i, we get $\psi = \psi(\mathrm{id} \vee \phi)$, as asserted. From this identity we obtain

$$\begin{aligned}\gamma &= (\mathrm{id} \vee \phi) \wedge \psi = (\mathrm{id} \vee \phi) \wedge \psi(\mathrm{id} \vee \phi) \\ &= (\mathrm{id} \wedge \psi)(\mathrm{id} \vee \phi) = \check{\psi}\hat{\phi}.\end{aligned}$$

The relation $\gamma = \hat{\phi}\check{\psi}$ follows by duality.

Suppose that ψ is an increasing operator on the modular lattice $\mathcal{L}$. We are interested in increasing operators π which are less active than ψ and which are activity-extensive. Recall from Proposition 13.37 that the centre $(\mathrm{id} \wedge \eta) \vee \xi$ of an overfilter ξ and underfilter η with $\xi \leq \eta$ is activity-extensive. Furthermore, this operator is less active than ψ if

$$\mathrm{id} \wedge \psi \leq \mathrm{id} \wedge \eta \qquad \text{and} \qquad \mathrm{id} \vee \psi \geq \mathrm{id} \vee \xi. \tag{13.20}$$

This observation forms the basis for the following result.

13.40 Proposition. *Consider a modular lattice $\mathcal{L}$ and an increasing operator ψ with lower envelope $\check{\psi}$ and upper envelope $\hat{\psi}$. If $\alpha \leq \check{\psi}$ is an opening and $\beta \geq \hat{\psi}$ is a closing, then*

$$\pi = (\mathrm{id} \wedge \psi\beta) \vee \psi\alpha \tag{13.21}$$

is activity-extensive and $\pi \preccurlyeq \psi$. Furthermore, if $\alpha' \leq \alpha$ and $\beta' \geq \beta$ are an opening and a closing, respectively, and if $\pi' = (\mathrm{id} \wedge \psi\beta') \vee \psi\alpha'$, then $\pi' \preccurlyeq \pi$.

Proof. It is obvious that $\mathrm{id} \wedge \psi \leq \mathrm{id} \wedge \psi\beta$ and that $\mathrm{id} \vee \psi \geq \mathrm{id} \vee \psi\alpha$. Furthermore, $(\psi\alpha)^2 = \psi\alpha\psi\alpha \geq \psi\alpha\check{\psi}\alpha \geq \psi\alpha^3 = \psi\alpha$, and therefore $\psi\alpha$ is an overfilter. Analogously, $\psi\beta$ is an underfilter. With the observations made previously this gives that π is activity-extensive and $\pi \preccurlyeq \psi$.

The proof that $\pi' \preccurlyeq \pi$ is straightforward. □

13.41 Remark. Under the hypotheses of the last proposition, the operator

$$\pi' = (\mathrm{id} \wedge \beta\psi) \vee \alpha\psi$$

is activity-extensive, too. It is also less active than ψ. It follows from Theorem 13.46 that π in (13.21) is more interesting because $\psi\alpha$ is an inf-overfilter whereas $\alpha\psi$ is only an overfilter.

The operator π in (13.21) satisfies

$$\mathrm{id} \wedge \alpha\beta \leq \pi \leq \mathrm{id} \vee \beta\alpha. \tag{13.22}$$

Since $\beta\alpha$ is a sup-filter, $\mathrm{id} \vee \beta\alpha$ is a closing. Similarly, one shows that $\mathrm{id} \wedge \alpha\beta$ is an opening.

There exists a nice interpretation of the relation

$$\psi^n \preccurlyeq \psi^{n+1}$$

when $\mathcal{L} = \mathrm{Fun}(E, \mathcal{T})$, with $\mathcal{T}$ an arbitrary complete lattice. Recall from Section 3.6 that $\preccurlyeq$ defines a partial ordering on $\mathrm{Fun}(E, \mathcal{T})$.

13.42 Definition. Assume that $\mathcal{T}$ is a complete lattice. The sequence F_n, $n \geq 0$, in $\mathrm{Fun}(E, \mathcal{T})$ is called *pointwise monotone* if the sequence $F_n(x)$ is either increasing or decreasing for every $x \in E$.

If $\mathcal{T} = \{0, 1\}$, in which case $\mathrm{Fun}(E, \mathcal{T})$ is isomorphic to $\mathcal{P}(E)$, the situation is described conveniently using terminology from systems theory. Let X_n be a sequence of subsets in E; say that a point $x \in E$ has state 0 (resp. 1) at time n if $x \notin X_n$ (resp. $x \in X_n$). The sequence X_n is pointwise monotone if every point $x \in E$ changes its state at at most one time instance. For example, $0, 0, \ldots, 0, 1, 1, \ldots$ is an admitted sequence as well as $1, 1, \ldots, 1, 0, 0, \ldots$. However, $0, 0, \ldots, 0, 1, 1, 0, 0, \ldots$ is a prohibited sequence.

The next result shows that the property $\psi^n \preccurlyeq \psi^{n+1}$ on $\mathrm{Fun}(E, \mathcal{T})$ can be interpreted in terms of pointwise monotonicity.

13.43 Proposition. *Let $\mathcal{T}$ be an complete lattice. The operator ψ on* $\mathrm{Fun}(E, \mathcal{T})$ *is activity-extensive if and only if the sequence $\psi^n(F)$ is pointwise monotone for every function F.*

Proof. We prove only the only if-statement; the proof of the if-statement is very similar. Given a function F, put $F_n = \psi^n(F)$ for $n \geq 0$, take $x \in E$, and assume that $n \geq 1$ is such that $F_0(x) = \cdots = F_{n-1}(x) < F_n(x)$. From $F_{n+1}(x) \vee F(x) \geq F_n(x) \vee F(x)$ we derive that $F_{n+1}(x) \geq F_n(x)$. Repeating this argument, we find that $F_n(x) \leq F_{n+1}(x) \leq F_{n+2}(x) \leq \cdots$, which proves the assertion. The case that $F_{n-1}(x) > F_n(x)$ is treated analogously. □

On $\mathcal{P}(E)$ the condition $\psi^n \preccurlyeq \psi^{n+1}$ means that $\psi^n(X) \cap X$ becomes smaller and that $\psi^n(X) \cap X^c$ becomes larger as n increases.

13.44 Proposition. *Let $\mathcal{T}$ be a strongly admissible complete lattice. An increasing operator ψ on $\mathcal{P}(E)$ is activity-extensive if and only if its flat extension to* Fun$(E, \mathcal{T})$ *is such as well.*

This result follows directly from the theory developed in Chapter 10. In fact, this theory says that the extension of operators on $\mathcal{P}(E)$ to Fun$(E, \mathcal{T})$ preserves the partial ordering. The admissibility condition for complete lattices was introduced in Chapter 10. It holds for all grey-value sets that are of practical value, such as $\overline{\mathbb{R}}$, $\overline{\mathbb{Z}}$ and $\{0, 1, \ldots, N\}$.

Assume that the modular lattice $\mathcal{L}$ has a negation. If ψ is an underfilter and $\psi \geq \psi^*$, then the centre $(\mathrm{id} \wedge \psi) \vee \psi^*$ is self-dual and activity-extensive. Refer to Proposition 13.49 for a more general result.

13.45 Examples.

(a) Assume that α is an opening and that β is a closing (not necessarily each other's negative) on a modular lattice. Using the observations made in Example 12.7, we distinguish at least three different ways to obtain an overfilter ϕ and an underfilter ψ such that $\phi \leq \psi$.

First, let $\phi = \alpha\beta\alpha$ and $\psi = \beta\alpha\beta$. This case will be investigated in more detail in Example 13.47.

Second, take $\phi = \alpha\beta\alpha$ and $\psi = \alpha\beta \wedge \beta\alpha$. Note that ϕ is an inf-filter whereas ψ is "only" an underfilter.

And finally, take $\phi = \alpha\beta \vee \beta\alpha$ and $\psi = \beta\alpha\beta$. Now ψ is a sup-filter, but ϕ is "only" an overfilter. The resulting centre operators (which are all activity-extensive) are illustrated in Fig. 13.5.

(b) Let α, β be the opening and closing on $\mathcal{P}(\mathbb{Z}^2)$ with the 2×2 square. Furthermore, let μ be the median operator associated with the rhombus structuring element. Then $\alpha \leq \mu \leq \beta$ (see Corollary 4.26) and Proposition 6.28(a) gives that $\mu\alpha$ is an inf-overfilter and that $\mu\beta$ is a sup-underfilter. Furthermore, $\mu\alpha \leq \mu\beta$. Let γ be the centre operator, i.e., $\gamma = (\mathrm{id} \wedge \mu\beta) \vee \mu\alpha$. The theory developed here gives that $\gamma^n \preccurlyeq \gamma^{n+1}$. Furthermore, $\gamma \preccurlyeq \mu$. See also Fig. 13.7 in Example 13.59.

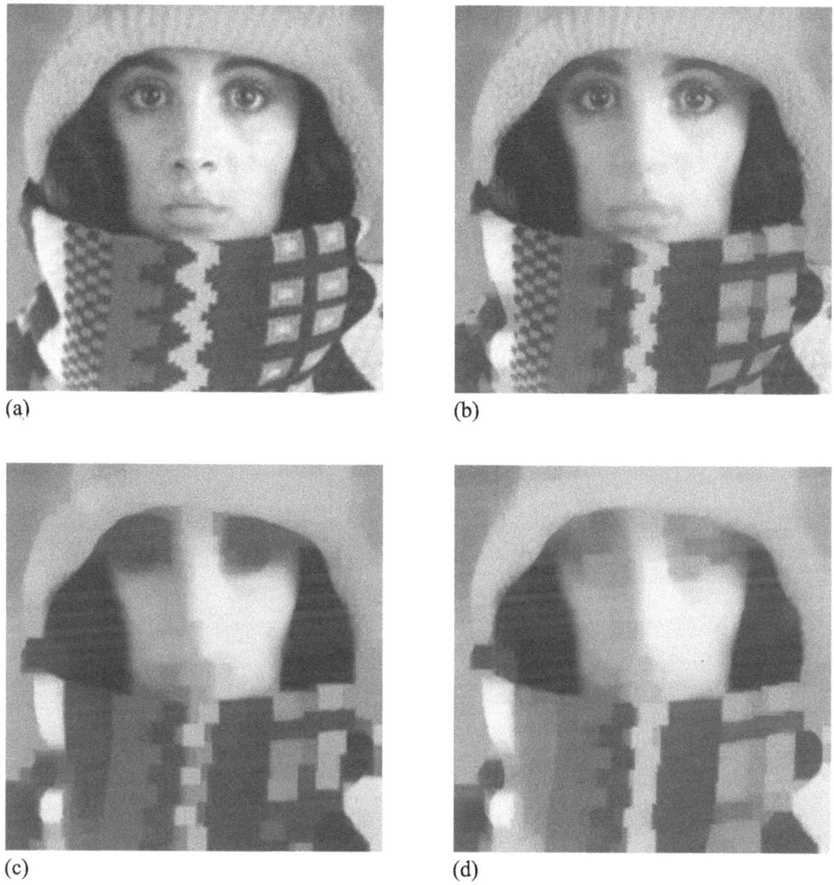

Figure 13.5 Let α, β be the opening and closing by a 5×5 square. Depicted are: (a) original image F; (b) centre of $\phi = \alpha\beta\alpha$ and $\psi = \beta\alpha\beta$ applied to F; (c) centre of $\phi = \alpha\beta\alpha$ and $\psi = \alpha\beta \wedge \beta\alpha$ applied to F; (d) centre of $\phi = \alpha\beta \vee \beta\alpha$ and $\psi = \beta\alpha\beta$ applied to F.

13.8. From centre operator to middle filter

In this section we point out the connection between the centre operator, introduced in Section 3.5 and studied in more detail in the previous section, and the middle filter studied in Section 12.5. We will deal exclusively with operators on $\mathrm{Fun}(E, \mathcal{T})$, where $\mathcal{T}$ is a complete chain. Observe that this lattice is modular.

Assume that ϕ is an overfilter and that ψ is an underfilter on $\mathrm{Fun}(E, \mathcal{T})$ such that $\phi \le \psi$. Let $\gamma = (\mathrm{id} \wedge \psi) \vee \phi$ be the centre operator. Then, by

Proposition 13.37,

$$\gamma^{n-1} \preccurlyeq \gamma^n, \qquad n \geq 1.$$

From Proposition 13.43 we know that the sequence $\gamma^n(F)$ is pointwise monotone for every function F. Therefore, this sequence converges (cf. Example 13.5(c)) to a limit, which we denote by $\gamma^\infty(F)$; in other words,

$$\gamma^n \to \gamma^\infty.$$

Proposition 13.30(b) implies that γ^∞ is a morphological filter if γ is $\updownarrow$-continuous. From Propositions 13.9–13.10 it follows that γ is $\updownarrow$-continuous if both ϕ and ψ are $\updownarrow$-continuous; here we use that Fun$(E, \mathcal{T})$ is both semi-atomic and dual-semi-atomic.

A lot more can be said if one assumes in addition that ϕ is an inf-overfilter and that ψ is a sup-underfilter; in this case γ^∞ is the strong middle filter $\check{\psi}\hat{\phi} = \hat{\phi}\check{\psi}$ given by Theorem 12.35. To prove this, some preparations are necessary. First observe that

$$\psi \leq \psi(\mathrm{id} \vee \phi) \leq \psi(\mathrm{id} \vee \psi) = \psi,$$

hence

$$\psi = \psi(\mathrm{id} \vee \phi). \tag{13.23}$$

Similarly,

$$\phi = \phi(\mathrm{id} \wedge \psi). \tag{13.24}$$

This yields, in particular, that

$$\gamma = (\mathrm{id} \wedge \psi) \vee \phi = (\mathrm{id} \wedge \psi) \vee \phi(\mathrm{id} \wedge \psi) = (\mathrm{id} \vee \phi)(\mathrm{id} \wedge \psi),$$

and similarly $\gamma = (\mathrm{id} \wedge \psi)(\mathrm{id} \vee \phi)$. Consequently,

$$\gamma^n = (\mathrm{id} \wedge \psi)^n(\mathrm{id} \vee \phi)^n = (\mathrm{id} \vee \phi)^n(\mathrm{id} \wedge \psi)^n, \tag{13.25}$$

for every $n \geq 1$. This implies

$$\gamma^n \geq (\mathrm{id} \vee \phi)^n \bigwedge_{k\geq 1} (\mathrm{id} \wedge \psi)^k.$$

If ψ is $\downarrow$-continuous, then, by Corollary 13.32(a), $\bigwedge_{k\geq 1}(\mathrm{id}\wedge\psi)^k=\check{\psi}$; hence

$$\gamma^n \geq (\mathrm{id}\vee\phi)^n\check{\psi}.$$

The sequence on the left-hand side converges toward γ^∞, whereas, by Corollary 13.32(a′), the sequence on the right-hand side converges toward $\hat{\phi}\check{\psi}$ if ϕ is $\uparrow$-continuous. Thus we conclude

$$\hat{\phi}\check{\psi}\leq\gamma^\infty.$$

A similar argument shows

$$\check{\psi}\hat{\phi}\geq\gamma^\infty.$$

But Theorem 12.35 states that, under the given assumptions, $\check{\psi}\hat{\phi}=\hat{\phi}\check{\psi}$ is the middle filter from which

$$\gamma^\infty=\check{\psi}\hat{\phi}=\hat{\phi}\check{\psi}.$$

The results are summarized in the following theorem.

13.46 Theorem. *Let $\mathcal{T}$ be a complete chain, let ϕ be a $\uparrow$-continuous inf-overfilter, and let ψ be a $\downarrow$-continuous sup-underfilter on* $\mathrm{Fun}(E,\mathcal{T})$ *such that $\phi\leq\psi$. Furthermore, let γ be the centre operator* $(\mathrm{id}\wedge\psi)\vee\phi$. *Then γ^n converges pointwise monotone to the middle filter $\omega=\check{\psi}\hat{\phi}=\hat{\phi}\check{\psi}$.*

Recall that activity-extensivity of γ means that $\mathrm{id}\wedge\gamma^n\leq\mathrm{id}\wedge\gamma^{n-1}$ and $\mathrm{id}\vee\gamma^n\geq\mathrm{id}\vee\gamma^{n-1}$. In fact, one can prove the following identities:

$$\mathrm{id}\wedge\gamma^n=(\mathrm{id}\wedge\psi)^n, \tag{13.26}$$

$$\mathrm{id}\vee\gamma^n=(\mathrm{id}\vee\phi)^n. \tag{13.27}$$

It is evident that these identities imply that γ is activity-extensive. To derive the first identity, we show first that

$$(\mathrm{id}\wedge\psi)^n=\mathrm{id}\wedge\psi(\mathrm{id}\wedge\psi)^{n-1},\qquad n\geq 1.$$

For $n=1$, this is obvious. If it holds for n, then

$$\begin{aligned}(\mathrm{id}\wedge\psi)^{n+1}&=(\mathrm{id}\wedge\psi)^n(\mathrm{id}\wedge\psi)=(\mathrm{id}\wedge\psi(\mathrm{id}\wedge\psi)^{n-1})(\mathrm{id}\wedge\psi)\\&=\mathrm{id}\wedge\psi\wedge\psi(\mathrm{id}\wedge\psi)^n=\mathrm{id}\wedge\psi(\mathrm{id}\wedge\psi)^n,\end{aligned}$$

where we have used that $\psi \geq \psi(\mathrm{id} \wedge \psi)^n$. Using relations (13.18) and (13.25), we get

$$\begin{aligned}\mathrm{id} \wedge \gamma^n &= \mathrm{id} \wedge \psi\gamma^{n-1} \\ &= \mathrm{id} \wedge \psi(\mathrm{id} \wedge \psi)^{n-1}(\mathrm{id} \vee \phi)^{n-1} \\ &= \mathrm{id} \wedge \psi(\mathrm{id} \vee \phi)^{n-1}(\mathrm{id} \wedge \psi)^{n-1}.\end{aligned}$$

Since $\psi(\mathrm{id} \vee \phi) = \psi$, by (13.23), we get $\psi(\mathrm{id} \vee \phi)^{n-1} = \psi$, whence

$$\mathrm{id} \wedge \gamma^n = \mathrm{id} \wedge \psi(\mathrm{id} \wedge \psi)^{n-1} = (\mathrm{id} \wedge \psi)^n,$$

as required. The identity in (13.27) is proved similarly.

For a first application of Theorem 13.46 refer to Example 13.45(b): observe that the operators μ, α, β in this example are finite window operators, and therefore $\updownarrow$-continuous. Therefore, γ^n converges to the middle filter $(\mu\alpha)^{\wedge}(\mu\beta)^{\vee}$, which is a self-dual strong filter with invariance domain $\{X \subseteq \mathbb{Z}^2 \mid \mu\alpha(X) \subseteq X \subseteq \mu\beta(X)\}$.

13.47 Example. Let $\mathcal{T}$ be a complete chain. If α is an opening and β a closing on $\mathrm{Fun}(E, \mathcal{T})$, then $\alpha\beta\alpha$ is an inf-filter, $\beta\alpha\beta$ a sup-filter and $\alpha\beta\alpha \leq \beta\alpha\beta$; see Example 12.7. The centre $(\mathrm{id} \wedge \beta\alpha\beta) \vee \alpha\beta\alpha$ converges pointwise monotone to the middle filter $(\alpha\beta\alpha)^{\wedge}(\beta\alpha\beta)^{\vee}$, assuming that α, β are $\updownarrow$-continuous. This convergence is illustrated in Fig. 13.6. Note that both the centre and the middle filter are self-dual if $\alpha = \beta^*$.

The operator $\psi\alpha$ in Proposition 13.40 is not just an overfilter but an inf-overfilter, and dually, $\psi\beta$ is a sup-underfilter; see also Section 6.6. Therefore, one can use this result in combination with Theorem 13.46 to construct strong filters. An application will be given in the following section.

13.9. Self-dual operators and filters

The interest in self-dual operators stems from the fact that such operators treat foreground and background of an image identically. A well-known self-dual operator is the median operator μ_A on $\mathcal{P}(\mathbb{E}^d)$, with A a structuring element containing an odd number of points; see Section 4.5.

13.48 Example. (Boolean functions and self-dual operators) Let ψ_b be the operator on $\mathcal{P}(\mathbb{E}^d)$ given by (4.82):

$$\psi_b(X) = \{h \in \mathbb{E}^d \mid b(X(a_1 + h), \ldots, X(a_n + h)) = 1\}.$$

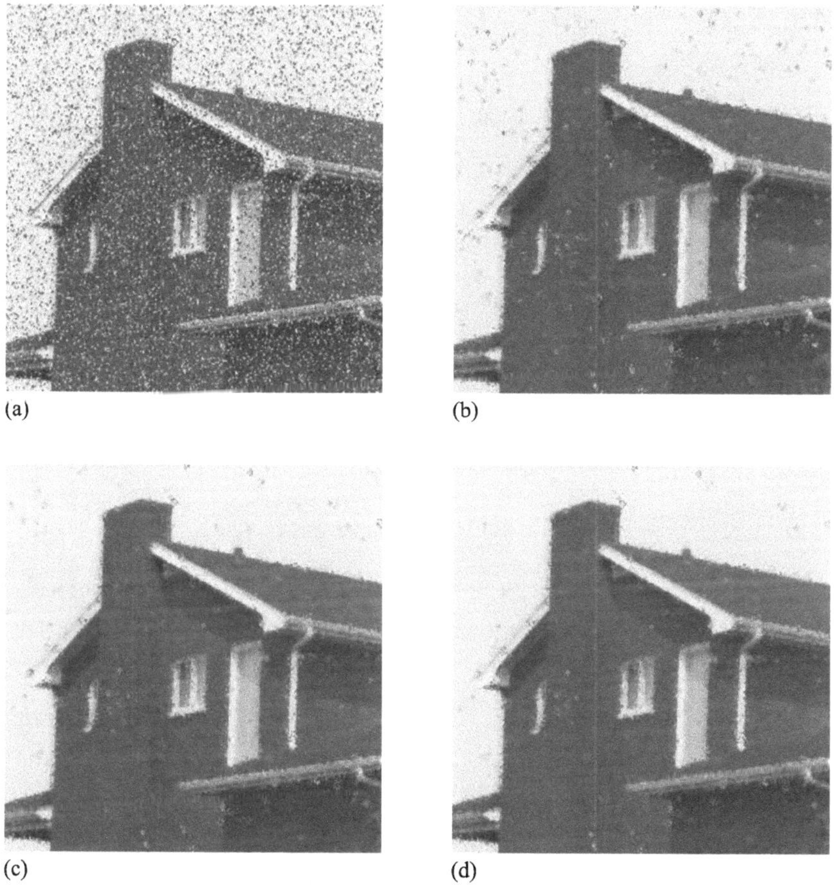

Figure 13.6 Iterations of the centre of $\alpha\beta\alpha$ and $\beta\alpha\beta$, where α and β are the opening and closing by the rhombus, respectively; see Example 13.47. (a) Original; (b) 1 iteration; (c) 2 iterations; (d) ∞ iterations.

Here $A \subseteq \mathbb{E}^d$ is a structuring element containing n points $a_1, a_2, \ldots, a_n$ and b is a Boolean function of n variables. The operator ψ_b is self-dual iff the Boolean function b is self-dual. Let, for example, A be the 3×3 square in $\mathbb{Z}^2$ centred at the origin. Let a_1 represent the origin, and let $a_2, \ldots, a_9$ be the 8 neighbours. If b is the Boolean threshold function with realization vector $(w, 1, 1, \ldots, 1 \mid s)$, then the negative b^* is the Boolean threshold function with realization vector $(w, 1, 1, \ldots, 1 \mid w + 9 - s)$. Thus, if w is odd and $2s = w + 9$, then $b^* = b$. Note that ψ_b is the median operator if $w = 1$ and $s = 5$.

The next result, the proof of which is obvious, is related to Example 3.44.

13.49 Proposition. *Let $\mathcal{L}$ be a complete modular lattice with negation, and let ψ be an increasing operator. The centre of $\psi \wedge \psi^*$ and $\psi \vee \psi^*$, given by*

$$\gamma = \Big(\mathrm{id} \wedge (\psi \vee \psi^*)\Big) \vee (\psi \wedge \psi^*),$$

is self-dual. Furthermore, $\gamma \preccurlyeq \psi$ and $\gamma \preccurlyeq \psi^*$.

13.50 Example. (Rank operators and self-duality)
Consider the lattice $\mathcal{P}(\mathbb{E}^d)$, and let $A \subseteq \mathbb{E}^d$ be a structuring element with n points. If $2s \geq n+1$, then the rank operator $\rho_{A,s}$ has negative $\rho_{A,n-s+1}$ and $\rho_{A,s} \leq \rho_{A,n-s+1}$; so in this case

$$\psi_s = (\mathrm{id} \wedge \rho_{A,n-s+1}) \vee \rho_{A,s}$$

is a self-dual operator. This operator extends the median operator: if n is odd and $s = (n+1)/2$, then $\psi_s = \mu_A$. Note that $\psi_s \preccurlyeq \psi_t$ if $s \geq t$ and $2s, 2t \geq n+1$ and that $\psi_n = \mathrm{id}$ if $0 \in A$.

In what follows we present a systematic method to construct increasing self-dual operators on $\mathcal{P}(E)$.

13.51 Definition. An anti-extensive operator $\sigma : \mathcal{P}(E) \to \mathcal{P}(E)$ is called a *switch operator* if it satisfies
(S1) $\sigma(Y) \cap X \subseteq \sigma(X)$ if $X \subseteq Y \subseteq E$;
(S2) $\sigma(X \cup \{h\}) \cap \sigma(X^c \cup \{h\}) = \varnothing$, for $h \in E$, $X \subseteq E$.

Before we explain these conditions, we present an example. Let μ be the median operator with the rhombus as structuring element. If a point $h \in X$ is surrounded by at least three 4-neighbours which are not contained in X, as in one of the following figures:

then $h \notin \mu(X)$. Define $\sigma(X)$ as the set of points which lie not in $\mu(X)$; this means that one of these four configurations centred at X lies entirely in X^c. Dually, if h is a background pixel, i.e., $h \in X^c$, then $h \in \mu(X)$ if at least three of its 4-neighbours belong to X. In that case $h \in \sigma(X^c)$. Thus application of σ to X yields all points in X that are deleted by μ and application of σ to X^c yields all points in X^c that are added by μ. In other words, $\sigma(X)$ and $\sigma(X^c)$ represent the pixels whose state switches from 1 to 0 and from 0 to 1, respectively.

Condition (S1) expresses that a point h in Y which switches by application of σ to Y also switches if σ is applied to a smaller set X which contains h. The condition (S2) guarantees that two complementary configurations centred at a point h cannot both force a switch of h; note that the fact that σ is anti-extensive means that the intersection in (S2) contains at most the point h.

13.52 Proposition. *To every increasing, self-dual operator ψ there corresponds a unique switch operator σ, and vice versa. The operators ψ and σ are interrelated by the following two inverse formulas:*

$$\psi(X) = \big(X \setminus \sigma(X)\big) \cup \sigma(X^c); \tag{13.28}$$

$$\sigma(X) = X \cap \psi(X^c). \tag{13.29}$$

Proof. First we show that the operator ψ given by (13.28) is an increasing self-dual operator when σ is a switch operator. To prove that ψ is increasing, let $X \subseteq Y$. If $h \in \psi(X)$, then either $h \in X \setminus \sigma(X)$ or $h \in \sigma(X^c)$. In the first case, $h \notin \sigma(Y)$ since $X \cap \sigma(Y) \subseteq \sigma(X)$, and therefore $h \in \psi(Y)$. Assume that $h \in \sigma(X^c)$. If $h \notin Y$, then, as $\sigma(X^c) \cap Y^c \subseteq \sigma(Y^c)$, also $h \in \sigma(Y^c)$, whence $h \in \psi(Y)$ follows. If $h \in Y$, then $h \in \psi(Y)$ if we can show that $h \notin \sigma(Y)$. To this purpose, observe that $Y^c \cup \{h\} \subseteq X^c$ and infer from (S1) that $\sigma(X^c) \cap (Y^c \cup \{h\}) \subseteq \sigma(Y^c \cup \{h\})$. This means that $h \in \sigma(Y^c \cup \{h\})$. But, then (S2) implies $h \notin \sigma(Y \cup \{h\}) = \sigma(Y)$. That ψ is self-dual follows by a straightforward calculation.

Next we show that σ given by (13.29) is a switch operator if ψ is an increasing self-dual operator. To prove that σ obeys (S1), let $X \subseteq Y$. Then $\sigma(Y) \cap X = Y \cap \psi(Y^c) \cap X \subseteq \psi(X^c) \cap X = \sigma(X)$. Here we have used that $Y^c \subseteq X^c$ and that ψ is increasing. Next we prove (S2). We must show that $h \notin \sigma(X \cup \{h\}) \cap \sigma(X^c \cup \{h\})$, for every $X \subseteq E$. Consider the case where $h \in X^c$; the case $h \in X$ is treated similarly. Now

$$\begin{aligned}\sigma(X \cup \{h\}) \cap \sigma(X^c) &= (X \cup \{h\}) \cap \psi((X \cup \{h\})^c) \cap X^c \cap \psi(X) \\ &\subseteq \psi(X^c) \cap \psi(X) \\ &= [\psi(X)]^c \cap \psi(X) \\ &= \varnothing.\end{aligned}$$

Here we have used that $\psi(X^c) = [\psi(X)]^c$ by the self-duality of ψ. Thus (S2) is proved.

The proof that (13.28) and (13.29) are inverse relations consists of two parts. As both parts are very similar, we prove only one of them, namely, if ψ is an increasing, self-dual operator, σ is given by (13.29) and $\psi'(X) = (X \setminus \sigma(X)) \cup \sigma(X^c)$, then $\psi' = \psi$. For

$$\begin{aligned}
\psi'(X) &= \big[X \cap (\sigma(X))^c\big] \cup \sigma(X^c) \\
&= \big[X \cap (X \cap \psi(X^c))^c\big] \cup \big[X^c \cap \psi(X)\big] \\
&= \big[X \cap (X^c \cup \psi(X))\big] \cup \big[X^c \cap \psi(X)\big] \\
&= \big[(X \cap X^c) \cup (X \cap \psi(X))\big] \cup \big[X^c \cap \psi(X)\big] \\
&= \big[X \cap \psi(X)\big] \cup \big[X^c \cap \psi(X)\big] \\
&= (X \cup X^c) \cap \psi(X) \\
&= \psi(X).
\end{aligned}$$

This conclude the proof. □

From now on this section deals exclusively with translation invariant operators on $\mathcal{P}(\mathbb{E}^d)$. The next result provides a complete characterization of translation invariant switch operators.

13.53 Proposition. *Let $\mathcal{A} \subseteq \mathcal{P}(\mathbb{E}^d)$ be a collection of structuring elements with the property that*

$$0 \notin A \text{ and } A \cap B \neq \varnothing \qquad \text{for } A, B \in \mathcal{A}; \tag{13.30}$$

then

$$\sigma_{\mathcal{A}}(X) = X \cap \Big[\bigcup_{A \in \mathcal{A}} X^c \ominus A\Big] \tag{13.31}$$

is a switch operator. Conversely, if σ is a switch operator and

$$\mathcal{A} = (\mathcal{V}(\sigma))^* = \{X^c \mid 0 \in \sigma(X)\}, \tag{13.32}$$

then $\mathcal{A}$ obeys condition (13.30) *and $\sigma = \sigma_{\mathcal{A}}$.*

Proof. We show that $\sigma_{\mathcal{A}}$ given by (13.31) is a switch operator. To prove (S1), assume that $X \subseteq Y$ and that $h \in \sigma(Y) \cap X$. Then $h \in Y^c \ominus A$ for some $A \in \mathcal{A}$. But this implies that $h \in X^c \ominus A$ as well. Since $h \in X$, we conclude $h \in \sigma(X)$. To prove (S2), assume that $h \in \sigma(X \cup \{h\})$ and that $h \in \sigma(X^c \cup \{h\})$. This means $A_h \subseteq (X \cup \{h\})^c$ and $B_h \subseteq (X^c \cup \{h\})^c$, for some $A, B \in \mathcal{A}$. Since $(X \cup \{h\})^c \cap (X^c \cup \{h\})^c = \varnothing$, however, this implies $A \cap B = \varnothing$, a contradiction.

Conversely, suppose we are given a switch operator σ; define $\mathcal{A}$ as in (13.32). We demonstrate that $\sigma = \sigma_{\mathcal{A}}$. First assume that $h \in \sigma(X)$; then $0 \in \sigma(X_{-h})$, and thus $X^c_{-h} \in \mathcal{A}$. Since $(X^c_{-h})_h \cap X = \varnothing$, we conclude that $h \in \sigma_{\mathcal{A}}(X)$. Conversely, assume that $h \in \sigma_{\mathcal{A}}(X)$. This means that $h \in X$ and $A_h \cap X = \varnothing$ for some $A \in \mathcal{A}$. Therefore, $X_{-h} \subseteq A^c$. From (S1) we derive that $\sigma(A^c) \cap X_{-h} \subseteq \sigma(X_{-h})$. But this implies that $0 \in \sigma(X_{-h})$, i.e., $h \in \sigma(X)$.

It remains to prove (13.30). First, if $A \in \mathcal{A}$, then $0 \in \sigma(A^c) \subseteq A^c$. This yields that $0 \notin A$. Second, assume that $A \cap B = \varnothing$; then $A \cup \{0\} \subseteq B^c$. Note that $0 \in \sigma(B^c) \subseteq B^c$, because σ is anti-extensive. From (S1) we conclude that

$$\sigma(B^c) \cap (A \cup \{0\}) \subseteq \sigma(A \cup \{0\});$$

it follows that $0 \in \sigma(A \cup \{0\})$. By definition, $0 \in \sigma(A^c) \subseteq A^c$. This implies that $A^c = A^c \cup \{0\}$, whence it follows that $0 \in \sigma(A^c \cup \{0\})$. Thus $0 \in \sigma(A \cup \{0\}) \cap \sigma(A^c \cup \{0\})$, which contradicts (S2). This proves that $A \cap B \neq \varnothing$ for $A, B \in \mathcal{A}$. □

If ψ is an increasing, self-dual operator and σ the associated switch operator given by (13.29), then the set $\mathcal{A}$ in (13.32) is

$$\mathcal{A} = \{A \in \mathcal{V}(\psi) \mid 0 \notin A\}. \tag{13.33}$$

Note that $A \cap B \neq \varnothing$ if $A, B \in \mathcal{V}(\psi)$. For assume that $A \cap B = \varnothing$; then $A \subseteq B^c$, and so $0 \in \psi(A) \subseteq \psi(B^c) = (\psi(B))^c$. However, $0 \in \psi(B)$, a contradiction.

Most of the structuring elements in (13.33) are redundant. For example, if $A, B \in \mathcal{V}(\psi)$ and $A \subseteq B$, then B can be omitted.

13.54 Examples.

(a) Let μ be the median operator on $\mathcal{P}(\mathbb{Z}^2)$ with the rhombus as structuring element. It is easy to verify that ψ is of the form (13.28), where $\sigma = \sigma_{\mathcal{A}}$ and $\mathcal{A}$ contains the following structuring elements:

In fact, it is obvious that every element in the kernel of μ that does not contain the origin must include one of these four configurations.

(b) Suppose that $\mathcal{A}$ contains only one element, namely,

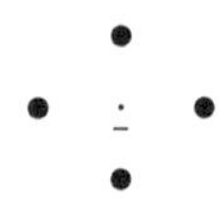

Thus a point in X is deleted by $\sigma_{\mathcal{A}}$ if all of its 4-neighbours lie in the complement of X. It follows that the associated self-dual operator is

$$\psi_4 = (\mathrm{id} \wedge \rho_{A,2}) \vee \rho_{A,4},$$

defined in Example 13.50, A being the rhombus structuring element.

13.55 Proposition. *Given a switch operator σ and an increasing extensive operator η; then $\sigma' = \mathrm{id} \wedge \sigma\eta$ is a switch operator as well, and $\sigma' \leq \sigma$. Let ψ, ψ' be the self-dual operators associated with σ, σ', respectively, by means of relation* (13.28). *Then*

$$\psi' = (\mathrm{id} \wedge \psi\eta) \vee \psi\eta^*. \tag{13.34}$$

In other words, ψ' is the centre of $\psi\eta^*$ and $\psi\eta$.

Proof. To prove (S1), assume that $X \subseteq Y$; then

$$\begin{aligned}\sigma'(Y) \cap X &= \sigma\eta(Y) \cap X = \sigma\eta(Y) \cap \eta(X) \cap X \\ &\subseteq \sigma\eta(X) \cap X = \sigma'(X).\end{aligned}$$

Here we have used that $\sigma\eta(Y) \cap \eta(X) \subseteq \sigma\eta(X)$ since $\eta(X) \subseteq \eta(Y)$.

It remains to prove (S2). Suppose there exist a point h and a set X such that

$$h \in (\sigma\eta \wedge \mathrm{id})(X \cup \{h\}) \cap (\sigma\eta \wedge \mathrm{id})(X^{\mathrm{c}} \cup \{h\}),$$

then $h \in \sigma\eta(X \cup \{h\})$ and $h \in \sigma\eta(X^{\mathrm{c}} \cup \{h\})$. Define $Y = \left[\eta(X \cup \{h\})\right]^{\mathrm{c}}$ and $Z = \eta(X^{\mathrm{c}} \cup \{h\})$; then $h \in \sigma(Y^{\mathrm{c}}) = \sigma(Y^{\mathrm{c}} \cup \{h\})$ since $h \in Y^{\mathrm{c}}$. Furthermore, $h \in Z$, whence $Z = Z \cup \{h\}$, and so $h \in \sigma(Z \cup \{h\})$. It is easy to see that $Y \subseteq Z$, and we get from (S1) that

$$\sigma(Z \cup \{h\}) \cap (Y \cup \{h\}) \subseteq \sigma(Y \cup \{h\}).$$

But this implies $h \in \sigma(Y \cup \{h\})$, and so $h \in \sigma(Y \cup \{h\}) \cap \sigma(Y^{\mathrm{c}} \cup \{h\})$, a contradiction.

Since $X \subseteq \eta(X)$ for every $X \subseteq E$, we get from (S1) that $\sigma(\eta(X)) \cap X \subseteq \sigma(X)$. In other words, $\sigma' = \sigma\eta \wedge \mathrm{id} \leq \sigma$. In particular, $\sigma(P) = \varnothing$ implies that $\sigma'(P) = \varnothing$.

The identity in (13.34) follows by a straightforward computation. □

13.56 Proposition. *Let σ, σ' be switch operators on $\mathcal{P}(E)$, and let ψ, ψ' be the corresponding self-dual operators given by* (13.28); *then*

$$\sigma \leq \sigma' \iff \psi \preccurlyeq \psi'.$$

Proof. For $X \subseteq E$,

$$\begin{aligned}\sigma(X) \subseteq \sigma'(X) &\iff X \cap \psi(X^c) \subseteq X \cap \psi'(X^c)\\ &\iff X \cap (\psi(X))^c \subseteq X \cap (\psi'(X))^c\\ &\iff X^c \cup \psi(X) \supseteq X^c \cup \psi'(X).\end{aligned}$$

Intersection with X leads to

$$X \cap \psi(X) \supseteq X \cap \psi'(X).$$

To prove that $X \cup \psi(X) \subseteq X \cup \psi'(X)$, one has to substitute X^c into the previous relation, take complements, and use that ψ is self-dual. This proves the result. □

Suppose that ψ is a self-dual operator on $\mathcal{P}(\mathbb{E}^d)$. If $\alpha \leq \check{\psi}$ is an opening and if $\beta = \alpha^*$ is the negative closing, then $\beta \geq (\check{\psi})^* = (\psi^*)^{\hat{}} = \hat{\psi}$; here we have used (12.5). The operator

$$\pi = (\mathrm{id} \wedge \psi\beta) \vee \psi\alpha \tag{13.35}$$

is self-dual and activity-extensive by Proposition 13.40. Furthermore,

$$\pi \preccurlyeq \psi,$$

and the larger the α, the more active the π. Theorem 13.46 gives that the sequence π^n converges to the middle filter ω of $\psi\alpha$ and $\psi\beta$, assuming that $\psi\beta$ and $\psi\alpha$ are $\downarrow$-continuous and $\uparrow$-continuous, respectively. This filter ω is strong and self-dual.

The most active ψ of this form is

$$\pi = (\mathrm{id} \wedge \psi\hat{\psi}) \vee \psi\check{\psi}. \tag{13.36}$$

In general it is not possible to find a closed expression for $\check{\psi}$. To get an opening $\alpha \leq \check{\psi}$ one must look for structuring elements $A \in \mathrm{Inv}(\check{\psi}) = \mathrm{Inv}(\mathrm{id} \wedge \psi)$. In fact, if $\mathcal{M} \subseteq \mathrm{Inv}(\check{\psi})$, then α given by

$$\alpha(X) = \bigcup_{A \in \mathcal{M}} X \circ A$$

satisfies $\alpha \leq \check{\psi}$. The next lemma explains how to find elements of $\mathrm{Inv}(\check{\psi})$ when the switch operator associated with ψ is known.

13.57 Lemma. *Let ψ be a self-dual increasing operator on $\mathcal{P}(\mathbb{E}^d)$, and let σ be the corresponding switch operator; then*

$$\mathrm{Inv}(\check{\psi}) = \{A \subseteq \mathbb{E}^d \mid \sigma(A) = \varnothing\}.$$

Proof. Using (13.29), we get $A \in \mathrm{Inv}(\check{\psi}) = \mathrm{Inv}(\mathrm{id} \wedge \psi)$ iff $A \subseteq \psi(A)$. This is equivalent to $A \cap (\psi(A))^c = \varnothing$. However, $(\psi(A))^c = \psi(A^c)$ since ψ is self-dual. This means that $A \in \mathrm{Inv}(\check{\psi})$ iff $\sigma(A) = A \cap \psi(A^c) = \varnothing$. □

The results obtained in this chapter suggest the following general construction method for self-dual (strong) filters.

13.58 Corollary. *Let σ be a switch operator on $\mathcal{P}(\mathbb{E}^d)$, and let ψ be the corresponding self-dual operator. Assume that ψ is continuous (this holds, e.g., if σ uses only finite structuring elements). Let $A_1, A_2, \ldots, A_p$ be finite structuring elements which satisfy $\sigma(A_j) = \varnothing$. If α is the opening $\alpha(X) = \bigcup_{j=1}^{p} X \circ A_j$ and if β is the negative closing, then the operator*

$$\pi = (\mathrm{id} \wedge \psi\beta) \vee \psi\alpha$$

is self-dual and activity-extensive; the sequence π^n converges to a strong self-dual filter.

We conclude with some examples.

13.59 Examples.

(a) Let μ be the median operator associated with the rhombus. The 2×2 square A is invariant under μ. Let α be the structural opening with A and $\beta = \alpha^*$. The foregoing theory gives that $\pi = (\mathrm{id} \wedge \mu\beta) \vee \mu\alpha$ is self-dual and activity-extensive and that $\pi \preccurlyeq \mu$. See also Example 13.45(b). The operator π and the median operator are illustrated in Fig. 13.7.

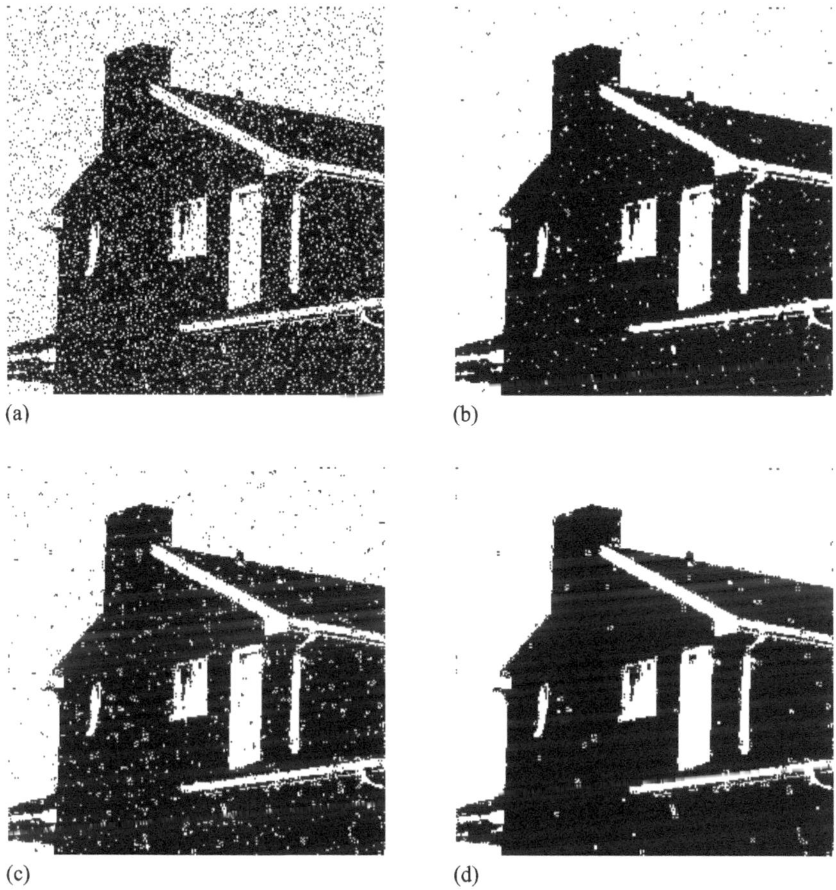

Figure 13.7 Iterates of $\pi = (\mathrm{id} \wedge \mu\beta) \vee \mu\alpha$, where μ is the median operator associated with the rhombus, α is the 2×2 opening and β the negative closing. (a) Original image; (b) median; (c) 1 iteration of π; (d) ∞ iterations of π.

(b) Some of the ideas in this section can be traced back to an interesting unpublished paper by Maisonneuve (1982). Here we briefly discuss an example from that paper. Let E be the hexagonal grid, and let $\mathcal{A}$ consist of the configurations

The operator ψ associated with $\sigma_{\mathcal{A}}$ resembles the median operator μ associated with the elementary hexagon. To obtain μ, however, one has to

add

 + rotations,

to the set $\mathcal{A}$. In particular, the operator ψ is less active than μ. It is easy to see that ψ is not activity-extensive. To obtain a modification of ψ which is activity-extensive, one has to find one or more structuring elements in $\mathrm{Inv}(\mathrm{id} \wedge \psi)$. It is evident that the triangles

$T_1 =$ , $T_2 =$ 

satisfy this property. If $\alpha_T = \alpha_{T_1} \vee \alpha_{T_2}$ and $\beta_T = \alpha_T^*$, then the operator $\pi = (\mathrm{id} \wedge \psi\beta_T) \vee \psi\alpha_T$ is increasing, self-dual, and activity-extensive. Since π is a finite window operator, it is $\updownarrow$-continuous. From Proposition 13.30(b) we get that $\omega = \pi^\infty$ is a filter, which we call the *Maisonneuve filter* after its inventor. In fact, Maisonneuve was able to show that ω is the middle filter between the inf-filter $\alpha_T\beta_T\alpha_T$ and the sup-filter $\beta_T\alpha_T\beta_T$.

(c) Our final example concerns a modification of the median operator μ associated with the 3×3 square. Let $\mathcal{A}$ consist of the structuring elements

 + rotations.

It is obvious that the self-dual operator ψ associated with the switch operator $\sigma_{\mathcal{A}}$ is less active than μ. To obtain a modification of ψ which is self-dual one has to find a structuring element C such that $\sigma_{\mathcal{A}}(C) = \varnothing$. It is easy to see that the 2×2 square is a solution.

13.10. Bibliographical notes

The notions lim sup and lim inf in complete lattices occur in several places in the literature, e.g., in Birkhoff (1967) and Gierz et al. (1980). A discussion of these notions for sets can be found in Kuratowski and Mostowski (1976, Section IV.2). A systematic discussion appeared in Heijmans and Serra (1992).

The construction of idempotent operators by (transfinite) iteration was first discussed by Maisonneuve (1982). The observation that one can restrict attention to sequential convergence if one imposes continuity conditions on the iterated operator is due to Heijmans (1989, 1990). The method is pursued further by Heijmans and Ronse (1990), who use it to construct openings. Heijmans and Serra (1992) extend these labours to operators which are not necessarily increasing.

The activity ordering is introduced by Serra (1988, Section 8.2) and studied in more detail by Meyer and Serra (1989a). The observation that the centre of an overfilter and an underfilter is activity-extensive was made by Serra (1988); see also Meyer and Serra (1989b). Maisonneuve (1982) showed that the middle filter can be obtained by transfinite iteration of the centre of an inf-overfilter and a sup-underfilter; cf. Theorem 13.46.

Proposition 13.15 is due to Heijmans (1991b). In the latter reference it is also shown that the converse of this proposition holds: if (13.11) holds, then ψ and Ψ are both $\downarrow$-continuous.

The counterexample 13.33 is due to Heijmans (1989). Example 13.47 is due to Maisonneuve (1982); see also Serra (1988, Section 8.7). For some further discussions on self-dual filters refer to Serra (1988, Chapter 8) and Meyer and Serra (1989b).

References

Aubin, J., & Frankowska, H. (1990). *Set-valued analysis*. Boston: Birkhäuser.

Baddeley, A. J. (1991). *Hausdorff metric for capacities* (Research report BS-R9127). CWI.

Baddeley, A. J., & Heijmans, H. J. A. M. (1992). *Incidence and lattice calculus with applications to stochastic geometry and image analysis* (Research report BS-R9213). CWI.

Banon, G. J. F., & Barrera, J. (1991). Minimal representation for translation-invariant set mappings by mathematical morphology. *SIAM Journal on Applied Mathematics*, *51*, 1782–1798.

Banon, G. J. F., & Barrera, J. (1993). Decomposition of mappings between complete lattices by mathematical morphology: Part I. General lattices. *Signal Processing*, *30*, 299–327.

Beer, G. (1985). On convergence of closed sets in a metric space and distance functions. *Bulletin of the Australian Mathematical Society*, *31*, 421–432.

Beer, G. (1987). Metric spaces with nice closed balls and distance functions for closed sets. *Bulletin of the Australian Mathematical Society*, *35*, 81–96.

Berger, M. (1989). *Geometry I*. Berlin: Springer-Verlag.

Beucher, S. (1990). *Segmentation d'Images et Morphologie Mathématique* (Ph.D. thesis). Fontainebleau: Ecole Nationale Supérieure des Mines de Paris.

Beucher, S., & Meyer, F. (1993). The morphological approach to segmentation: The watershed transformation. In E. R. Dougherty (Ed.), *Mathematical morphology in image processing* (pp. 433–481). New York: Marcel Dekker (Chapter 12).

Birkhoff, G. (1967). *Lattice theory* (3rd ed.). *American mathematical society colloquium publications: Vol. 25*. Providence, RI: American Mathematical Society.

Biswas, N. N. (1975). *Introduction to logic and switching theory*. New York: Gordon and Breach Science Publishers.

Bleau, A., de Guise, J., & LeBlanc, A. R. (1992a). A new set of fast algorithms for mathematical morphology: I. Idempotent geodesic transforms. *CVGIP. Image Understanding*, *56*, 178–209.

Bleau, A., de Guise, J., & LeBlanc, A.-R. (1992b). A new set of fast algorithms for mathematical morphology: II. Identification of topographic features on grayscale images. *CVGIP. Image Understanding*, *56*, 210–229.

Bloomberg, D., & Maragos, P. (1990). Generalized hit-miss operations. In *Image algebra and morphological image processing. SPIE proceedings: Vol. 1350* (pp. 116–128).

Blum, H. (1967). A transformation for extracting new descriptors of shape. In W. Wathen-Dunn (Ed.), *Models for the perception of speech and visual forms* (pp. 362–380). Amsterdam: MIT Press.

Blum, H. (1973). Biological shape and visual sciences (part I). *Journal of Theoretical Biology*, *38*, 205–287.

Blumenthal, L. M. (1953). *Theory and applications of distance geometry*. Oxford: Clarendon Press.

Blumenthal, L. M., & Menger, K. (1970). *Studies in geometry*. San Francisco: W. H. Freeman and Company.

Blyth, T. S., & Janowitz, M. F. (1972). *Residuation theory*. Oxford: Pergamon Press.

Borgefors, G. (1984). Distance transformations in arbitrary dimensions. *Computer Vision, Graphics, and Image Processing*, *27*, 321–345.

Borgefors, G. (1986). Distance transformations in digital images. *Computer Vision, Graphics, and Image Processing*, *34*, 344–371.

Bovik, A. C., Huang, T. S., & Munson, D. C. (1983). A generalization of median filtering using linear combinations of order statistics. *IEEE Transactions on Acoustics, Speech, and Signal Processing*, *31*, 1342–1349.

Busemann, H. (1955). *The geometry of geodesics*. New York: Academic Press.

Calabi, L., & Hartnett, W. E. (1968). Shape recognition, prairie fires, convex deficiencies and skeletons. *The American Mathematical Monthly*, *75*, 335–342.

Coster, M., & Chermant, J.-L. (1985). *Précis d'Analyse d'Images*. Paris: CNRS.

Crimmins, T. R., & Brown, W. R. (1985). Image algebra and automatic shape recognition. *IEEE Transactions on Aerospace and Electronic Systems*, *21*, 60–69.

Dal Masso, G. (1993). *An introduction to* Γ*-convergence*. Boston: Birkhäuser.

David, H. A. (1970). *Order statistics*. New York: Wiley.

Dorst, L., & Smeulders, A. W. M. (1987). Length estimators for digitized contours. *Computer Vision, Graphics, and Image Processing*, *40*, 311–333.

Dougherty, E. R. (1992). Euclidean gray-scale granulometries: Representation and umbra inducement. *Journal of Mathematical Imaging and Vision*, *1*, 7–21.

Dougherty, E. R. (Ed.). (1993). *Mathematical morphology in image processing*. New York: Marcel Dekker.

Dougherty, E. R., & Giardina, C. R. (1987a). Error bounds for morphologically derived measurements. *SIAM Journal on Applied Mathematics*, *47*, 425–440.

Dougherty, E. R., & Giardina, C. R. (1987b). *Image processing – continuous to discrete*. Englewood Cliffs, NJ: Prentice-Hall.

Dugundji, J. (1966). *Topology*. Boston: Allyn and Bacon.

Edelsbrunner, H., Kirkpatrick, D. G., & Seidel, R. (1983). On the shape of a set of points in the plane. *IEEE Transactions on Information Theory*, *29*, 551–559.

Engbers, E. A., van den Boomgaard, R., & Smeulders, A. W. M. (2001). Decomposition of separable concave structuring functions. *Journal of Mathematical Imaging and Vision*, *15*, 181–195.

Fell, J. M. G. (1962). A Hausdorff topology for the closed subsets of a locally compact non-Hausdorff space. *Proceedings of the American Mathematical Society*, *13*, 472–476.

Ghosh, P. K. (1994). On negative shape. In *Proceedings of the workshop "shape in picture"*. Springer-Verlag.

Giardina, C. R., & Dougherty, E. R. (1988). *Morphological methods in image and signal processing*. Englewood Cliffs, NJ: Prentice Hall.

Gierz, G., Hofmann, K. H., Keimel, K., Lawson, J. D., Mislove, M., & Scott, D. S. (1980). *A compendium of continuous lattices*. Berlin: Springer-Verlag.

Goetcherian, V. (1980). From binary to grey tone image processing using fuzzy logic concepts. *Pattern Recognition*, *12*, 7–15.

Goutsias, J. (1993). Modeling random objects: An introduction to random set theory. In R. M. Haralick (Ed.), *Mathematical morphology: Theory and hardware*. Oxford University Press. Publication abandoned.

Grimaud, M. (1991). *La Géodésie Numérique en Morphologie Mathématique. Application à la Detection Automatique de Microcalcifications en Mammographie Numérique* (Ph.D. thesis). Fontainebleau: Ecole Nationale Supérieure des Mines de Paris.

Haas, A., Matheron, G., & Serra, J. (1967a). Morphologie mathématique et granulométries en place. 1e Partie. *Annales Des Mines*, *XI*, 735–753.

Haas, A., Matheron, G., & Serra, J. (1967b). Morphologie mathématique et granulométries en place. 2e Partie. *Annales Des Mines*, *XII*, 767–782.

Hadwiger, H. (1950). Minkowskische Addition und Subtraktion beliebiger Punktmengen und die Theoreme von Erhard Schmidt. *Mathematische Zeitschrift*, *53*, 210–218.

Hadwiger, H. (1957). *Vorlesungen über Inhalt, Oberfläche und Isoperimetrie.* Berlin: Springer-Verlag.

Halmos, P. R. (1963). *Lectures on Boolean algebras.* Princeton: D. Van Nostrand Company, Inc.

Haralick, R. M., & Shapiro, L. G. (1992). *Computer and robot vision: Vol. I.* Reading, MA: Addison-Wesley.

Haralick, R. M., Sternberg, S. R., & Zhuang, X. (1987). Image analysis using mathematical morphology. *IEEE Transactions on Pattern Analysis and Machine Intelligence, 9*, 532–550.

Harrison, M. A. (1965). *Introduction to switching and automata theory.* New York: McGraw-Hill.

Hausdorff, F. (1962). *Set theory.* New York: Chelsea.

Heijmans, H. J. A. M. (1987). Mathematical morphology: An algebraic approach. *CWI Newsletter, 14*, 7–27.

Heijmans, H. J. A. M. (1989). Iteration of morphological transformations. *CWI Quarterly, 2*, 19–36.

Heijmans, H. J. A. M. (1990). Morphological filtering and iteration. In *Visual communications and image processing '90. SPIE proceedings: Vol. 1360* (pp. 166–175).

Heijmans, H. J. A. M. (1991a). Morphological discretization. In U. Eckhardt, A. Hübler, W. Nagel, & G. Werner (Eds.), *Geometrical problems of image processing* (pp. 99–106). Akademie Verlag.

Heijmans, H. J. A. M. (1991b). Theoretical aspects of gray-level morphology. *IEEE Transactions on Pattern Analysis and Machine Intelligence, 13*, 568–582.

Heijmans, H. J. A. M. (1992a). Discretization of morphological operators. *Journal of Visual Communication and Image Representation, 3*, 182–193.

Heijmans, H. J. A. M. (1992b). Mathematical morphology: A geometrical approach in image processing. *Nieuw Archief Voor Wiskunde, 10*, 237–276.

Heijmans, H. J. A. M. (1993). A note on the umbra transform in mathematical morphology. *Pattern Recognition Letters, 14*(11), 877–881.

Heijmans, H. J. A. M. (1994a). From binary to grey-level morphology. In R. M. Haralick (Ed.), *Mathematical morphology: Theory and hardware.* Oxford University Press. Publication abandoned.

Heijmans, H. J. A. M. (1994b). Mathematical morphology as a tool for shape description. In *Proceedings of the workshop "shape in picture".* Springer-Verlag (CWI Report BS-R9216).

Heijmans, H. J. A. M., Nacken, P., Toet, A., & Vincent, L. (1992). Graph morphology. *Journal of Visual Communication and Image Representation, 3*, 24–38.

Heijmans, H. J. A. M., & Ronse, C. (1990). The algebraic basis of mathematical morphology – part I: Dilations and erosions. *Computer Vision, Graphics, and Image Processing, 50*, 245–295.

Heijmans, H. J. A. M., & Serra, J. (1992). Convergence, continuity and iteration in mathematical morphology. *Journal of Visual Communication and Image Representation, 3*, 84–102.

Heijmans, H. J. A. M., & Vincent, L. (1993). Graph morphology in image analysis. In E. R. Dougherty (Ed.), *Mathematical morphology in image processing* (pp. 171–203). New York: Marcel Dekker (Chapter 6).

Herman, S. (1988). Feature-size dependent selective edge enhancement of X-ray images. In *SPIE meeting on medical imaging*.

Janowitz, M. F. (1986). A model for ordinal filtering of digital images. In E. J. Wegman, & D. J. DePriest (Eds.), *Statistical image processing and graphics*. New York: Marcel Dekker.

Justusson, B. I. (1981). Median filtering: Statistical properties. In T. S. Huang (Ed.), *Two-dimensional digital signal processing II: Transforms and median filters* (pp. 161–196). Berlin: Springer-Verlag.

Kendall, D. G. (1974). Foundations of a theory of random sets. In E. F. Harding, & D. G. Kendall (Eds.), *Stochastic geometry* (pp. 322–376). Chichester: John Wiley and Sons.

Khalimskii, E. D. (1969). On topologies of generalized segments. *Soviet Mathematics. Doklady, 10*, 1508–1511.

Klein, F. (1872). Vergleichende Betrachtungen über neuere geometrische Forschungen. *Ges. Math. Abh. I*, 460–497.

Klein, R. (1989). *Concrete and abstract Voronoi diagrams. Lecture notes in computer science: Vol. 400*. Berlin: Springer-Verlag.

Kong, T. Y., Kopperman, R., & Meyer, P. R. (1991). A topological approach to digital topology. *The American Mathematical Monthly, 98*(10), 901–917.

Kong, T. Y., & Rosenfeld, A. (1989). Digital topology: Introduction and survey. *Computer Vision, Graphics, and Image Processing, 48*, 357–393.

Kovalevsky, V. A. (1989). Finite topology as applied to image analysis. *Computer Vision, Graphics, and Image Processing, 46*, 141–161.

Kraus, E. J., Heijmans, H. J. A. M., & Dougherty, E. R. (1993). Gray-scale granulometries compatible with spatial scalings. *Signal Processing, 34*(1), 1–17 (CWI Report BS-R9212).

Kuratowski, K. (1972). *Introduction to set theory and topology* (2nd ed.). Oxford: Pergamon Press.

Kuratowski, K., & Mostowski, A. (1976). *Set theory*. Amsterdam: North Holland.

Lantuéjoul, C. (1978). *La Squelettisation et son Application aux Mesures Topologiques des Mosaiques Cristallines* (Ph.D. thesis). Paris: Ecole Nationale Supérieure des Mines de Paris.

Lantuéjoul, C., & Beucher, S. (1980). On the use of the geodesic metric in image analysis. *Journal of Microscopy*, *121*, 29–49.

Lantuéjoul, C., & Maisonneuve, F. (1984). Geodesic methods in quantitative image analysis. *Pattern Recognition*, *17*, 177–187.

Levialdi, S. (1971). Parallel pattern processing. *IEEE Transactions on Systems, Man and Cybernetics*, *1*, 292–296.

Maisonneuve, F. (1982). *Ordinaux transfinis et sur- (ou sous-) potentes* (Technical report N780). Ecole Nationale Supérieure des Mines de Paris.

Maragos, P. (1985). *Unified theory of translation-invariant systems with applications to morphological analysis and coding of images* (Ph.D. thesis). Atlanta: School of Electrical Engineering, Georgia Institute of Technology.

Maragos, P. (1986). Morphological skeleton representation and coding of binary images. *IEEE Transactions on Acoustics, Speech, and Signal Processing*, *34*, 1228–1244.

Maragos, P. (1987). Tutorial on advances in morphological image processing and analysis. *Optical Engineering*, *26*, 623–632.

Maragos, P. (1989a). Pattern spectrum and multiscale shape representation. *IEEE Transactions on Pattern Analysis and Machine Intelligence*, *11*, 701–716.

Maragos, P. (1989b). A representation theory for morphological image and signal processing. *IEEE Transactions on Pattern Analysis and Machine Intelligence*, *11*, 586–599.

Maragos, P. (1990). Affine morphology and affine signal models. In *Image algebra and morphological image processing. SPIE proceedings: Vol. 1350* (pp. 31–43).

Maragos, P., & Schafer, R. W. (1987a). Morphological filters – part I: Their set-theoretic analysis and relations to linear shift-invariant filters. *IEEE Transactions on Acoustics, Speech, and Signal Processing*, *35*, 1153–1169.

Maragos, P., & Schafer, R. W. (1987b). Morphological filters – part II: Their relations to median, order-statistics, and stack filters. *IEEE Transactions on Acoustics, Speech, and Signal Processing*, *35*, 1170–1184.

Maragos, P., & Ziff, R. D. (1990). Threshold superposition in morphological image analysis systems. *IEEE Transactions on Pattern Analysis and Machine Intelligence*, *12*, 498–504.

Marr, D. (1982). *Vision*. San Francisco: W. H. Freeman.

Marti, J. T. (1977). *Konvexe Analysis*. Basel: Birkhäuser.

Matheron, G. (1967). *Eléments pour une Théorie des Milieux Poreux*. Paris: Masson.

Matheron, G. (1974). *Les topologies finies* (CMM Report N-366). Ecole Nationale Supérieure des Mines de Paris.

Matheron, G. (1975). *Random sets and integral geometry*. New York; John Wiley and Sons.

Matheron, G. (1988). Filters and lattices. In J. Serra (Ed.), *Image analysis and mathematical morphology. Vol. 2: Theoretical advances*. London: Academic Press (Chapter 6).

Mattioli, J. (1993). *Problèmes Inverses et Relations Différentielles en Morphologie Mathématique* (Ph.D. thesis). Paris: Université Paris Dauphine.

Mattioli, J., & Schmitt, M. (1992). Inverse problems for granulometries by erosion. *Journal of Mathematical Imaging and Vision*, *2*, 217–232.

Melter, R. A., Rosenfeld, A., & Bhattacharya, P. (Eds.). (1991). *Vision geometry. Contemporary mathematics: Vol. 119*. Providence, RI: American Mathematical Society.

Menger, K. (1928). Untersuchungen über allgemeine Metrik I, II, III. *Mathematische Annalen*, *100*, 75–163.

Meyer, F. (1978). Contrast features extraction. In J. L. Chermant (Ed.), *Quantitative analysis of microstructures in materials science, biology and medecine* (pp. 374–380). Stuttgart: Riederer-Verlag.

Meyer, F. (1991). Skeletons and perceptual graphs. *Signal Processing*, *16*, 335–363.

Meyer, F., & Serra, J. (1989a). Contrasts and activity lattice. *Signal Processing*, *16*, 303–317.

Meyer, F., & Serra, J. (1989b). Filters: From theory to practice. In *Proceedings of the 5th European congress on stereology. Acta stereologica: Vol. 8* (pp. 503–508).

Michael, E. (1951). Topologies on spaces of subsets. *Transactions of the American Mathematical Society*, *71*, 152–182.

Miller, W. (1972). *Symmetry groups and their applications*. New York: Academic Press.

Minkowski, H. (1903). Volumen und Oberfläche. *Mathematische Annalen*, *57*, 447–495.

Motzkin, T. S. (1935a). Sur quelques propriétés caractéristique des ensembles convexes. *Rend. Reale Acad. Lincei, Classe Sci. Fis., Mat. Nat., 21*, 562–567.

Motzkin, T. S. (1935b). Sur quelques propriétés caractéristique des ensembles bornés non convexes. *Rend. Reale Acad. Lincei, Classe Sci. Fis., Mat. Nat., 21*, 773–779.

Muroga, S. (1971). *Threshold logic and its applications*. New York: Wiley-Interscience.

Nacken, P. (1993). *Chamfer metrics in mathematical morphology* (Report BS-R9309). CWI.

Niblack, C., Capson, D., & Gibbobs, P. (1990). Generating skeletons and centerlines from the medial axis transform. In *Proceedings of the tenth international conference on pattern recognition*.

Nodes, T. A., & Gallagher, N. C. (1982). Median filters: Some modifications and their properties. *IEEE Transactions on Acoustics, Speech, and Signal Processing, 30*, 739–746.

Pitas, I., & Venetsanopoulos, A. N. (1990). *Nonlinear digital filters*. Boston: Kluwer Academic Publishers.

Pratt, W. K. (1991). *Digital image processing* (2nd ed.). New York: John Wiley and Sons.

Preston Jr., K. (1983). Ξ-filters. *IEEE Transactions on Acoustics, Speech, and Signal Processing, 31*, 861–876.

Preston Jr., K., & Duff, M. J. B. (1984). *Modern cellular automata*. New York: Plenum Press.

Rinow, W. (1961). *Die Innere Geometrie der Metrischen Räume*. Berlin: Springer-Verlag.

Roerdink, J. B. T. M. (1989). *Mathematical morphology on homogeneous spaces – part I: The simply transitive case* (Research Report AM-R8924). CWI.

Roerdink, J. B. T. M. (1990a). *Mathematical morphology on homogeneous spaces – part II: The transitive case* (Research Report AM-R9006). CWI.

Roerdink, J. B. T. M. (1990b). Mathematical morphology on the sphere. In *Visual communications and image processing '90. SPIE proceedings: Vol. 1360* (pp. 263–271).

Roerdink, J. B. T. M. (1990c). On the construction of translation and rotation invariant morphological operators. In R. M. Haralick (Ed.), *Mathematical morphology: Theory and hardware*. Oxford University Press. Publication abandoned.

Roerdink, J. B. T. M. (1993). Mathematical morphology with non-commutative symmetry groups. In E. R. Dougherty (Ed.), *Mathematical*

morphology in image processing (pp. 205–254). New York: Marcel Dekker (Chapter 7).
Roerdink, J. B. T. M. (1994). Manifold shape: From differential geometry to mathematical morphology. In *Proceedings of the workshop "shape in picture"*. Springer-Verlag.
Roerdink, J. B. T. M., & Heijmans, H. J. A. M. (1988). Mathematical morphology for structures without translation-symmetry. *Signal Processing*, *15*, 271–277.
Ronse, C. (1988). *Extraction of narrow peaks and ridges in images by combination of local low rank and max filters: Implementation and applications to clinical angiography* (Working Document WD17). Brussels, Belgium: Philips Research Laboratory.
Ronse, C. (1989a). A bibliography on digital and computational convexity. *IEEE Transactions on Pattern Analysis and Machine Intelligence*, *11*, 181–190.
Ronse, C. (1989b). *Fourier analysis, mathematical morphology, and vision* (Working Document WD54). Brussels, Belgium: Philips Research Laboratory.
Ronse, C. (1990a). *Regular open or closed sets* (Working Document WD). Brussels, Belgium: Philips Research Laboratory.
Ronse, C. (1990b). Why mathematical morphology needs complete lattices. *Signal Processing*, *21*, 129–154.
Ronse, C. (1992). Toggles of openings, and a new family of idempotent operators on partially ordered sets. *Applicable Algebra in Engineering, Communication and Computing*, *3*, 99–128.
Ronse, C. (1994a). Lattice-theoretical fixpoint theorems in morphological image filtering. *Journal of Mathematical Imaging and Vision*, *4*, 19–41. Publication abandoned.
Ronse, C. (1994b). Openings: Main properties, and how to construct them. In R. M. Haralick (Ed.), *Mathematical morphology: Theory and hardware*. Oxford University Press.
Ronse, C., & Heijmans, H. J. A. M. (1991). The algebraic basis of mathematical morphology – part II: Openings and closings. *CVGIP: Image Understanding*, *54*, 74–97.
Rosenfeld, A., & Pfalz, J. L. (1966). Sequential operations in digital picture processing. *Journal of the Association for Computing Machinery*, *13*, 471–494.
Rosenfeld, A., & Pfalz, J. L. (1968). Distance functions on digital pictures. *Pattern Recognition*, *1*, 33–61.
Russ, J. C. (1992). *The image processing handbook*. Boca Raton: CRC Press, Inc.

Schmitt, M. (1989). *Des Algorithmes Morphologiques à l'Intelligence Artificielle* (Ph.D. thesis). Fontainebleau: Ecole Nationale Supérieure des Mines de Paris.

Schmitt, M., & Vincent, L. (2010). *Morphology: Algorithms and applications.* Cambridge University Press.

Schneider, R. (1993). *Convex bodies: The Brunn–Minkowski theory*. Cambridge: Cambridge University Press.

Schonfeld, D., & Goutsias, J. (1991). Optimal morphological pattern restoration from noisy binary images. *IEEE Transactions on Pattern Analysis and Machine Intelligence, 13*, 14–29.

Serra, J. (1982). *Image analysis and mathematical morphology*. London: Academic Press.

Serra, J. (Ed.). (1988). *Image analysis and mathematical morphology. II: Theoretical advances*. London: Academic Press.

Serra, J. (1991). *Skeleton decompositions* (Preprint).

Serra, J. (1993). Anamorphoses and function lattices (multivalued morphology). In E. R. Dougherty (Ed.), *Mathematical morphology in image processing* (pp. 483–523). New York: Marcel Dekker (Chapter 13).

Serra, J., & Vincent, L. (1992). An overview of morphological filtering. *IEEE Transactions on Circuits, Systems and Signal Processing, 11*, 47–108.

Shih, F. Y., & Mitchell, O. R. (1992). A mathematical morphology approach to Euclidean distance transformation. *IEEE Transactions on Image Processing, 1*, 197–204.

Song, J., & Delp, E. J. (1990). The analysis of morphological filters with multiple structuring elements. *Computer Vision, Graphics, and Image Processing, 50*, 308–328.

Sternberg, S. R. (1981). Parallel architecture for image processing. In *Proc. third international IEEE Compsac.*

Sternberg, S. R. (1982). Cellular computers and biomedical image processing. In J. Sklansky, & J. C. Bisconte (Eds.), *Biomedical images & computers. Lecture notes in medical informatics: Vol. 17* (pp. 294–319). Berlin: Springer-Verlag.

Sternberg, S. R. (1986). Grayscale morphology. *Computer Vision, Graphics, and Image Processing, 35*, 333–355.

Tukey, J. W. (1977). *Exploratory data analysis.* Reading, MA: Addison-Wesley.

Valentine, F. A. (1964). *Convex sets.* New York: McGraw-Hill.

van den Boomgaard, R. (1990). Threshold logic and mathematical morphology. In V. Cantoni, L. P. Cordella, S. Levianldi, & G. Sanniti di Baja (Eds.), *Progress in image analysis and processing*. Singapore: World Scientific.

van Lierop, M. (1987). *Digitisation functions in computer graphics* (Ph.D. thesis). Eindhoven: Technical University Eindhoven.

Vervaat, W. (1988). *Random upper semicontinuous functions and extremal processes* (Research report MS-R8801). CWI.

Verwer, B. J. H. (1991). *Distance transform: Metrics, algorithms and applications* (Ph.D. thesis). Delft University of Technology.

Vincent, L. (1989). Graphs and mathematical morphology. *Signal Processing*, *16*, 365–388.

Vincent, L. (1990). *Algorithmes Morphologiques a Base de Files d'Attente et de Lacets. Extension aux Graphes* (Ph.D. thesis). Fontainebleau: Ecole Nationale Supérieure des Mines de Paris.

Vincent, L. (1991). Morphological transformations of binary images with arbitrary structuring elements. *Signal Processing*, *22*, 3–23.

Vincent, L. (1993). Morphological grayscale reconstruction in image analysis: Efficient algorithms and applications. *IEEE Transactions on Image Processing*, *2*(2), 176–201.

Wendt, P. D., Coyle, E. J., & Callagher, N. C. (1986). Stack filters. *IEEE Transactions on Acoustics, Speech, and Signal Processing*, *34*, 898–911.

Wilson, S. S. (1992). Theory of matrix morphology. *IEEE Transactions on Pattern Analysis and Machine Intelligence*, *14*, 636–652.

Wilson, S. S. (1993). Training structuring elements in morphological networks. In E. R. Dougherty (Ed.), *Mathematical morphology in image processing* (pp. 1–41). New York: Marcel Dekker (Chapter 1).

Xu, J. (1991). Decomposition of convex polygonal morphological structuring elements into neighbourhood subsets. *IEEE Transactions on Pattern Analysis and Machine Intelligence*, *13*, 153–162.

Zhuang, X., & Haralick, R. M. (1986). Morphological structuring element decomposition. *Computer Vision, Graphics, and Image Processing*, *35*, 370–382.

Notation Index

$\text{card}(X)$: cardinality (number of elements) of set X [Section 3.1]
$\text{co}(X)$: convex hull of X [Example 2.16]
$\overline{\text{co}}(X)$: closed convex hull of X [Example 2.16]
$\text{conn}(X)$: connected components of X [Example 9.49(b)]
$\mathcal{C}(\mathbb{R}^2)$: convex sets in $\mathbb{R}^2$ [Example 2.16]
$\mathcal{C}_{\mathcal{F}}(\mathbb{R}^2)$: closed convex sets in $\mathbb{R}^2$ [Example 2.16]
γ: centre [Definition 3.35]
γ_x: connectivity opening [Section 9.7]
$d(x, y)$: distance between x and y [Section 7.2]
$d_k(x, y)$: $\{\Sigma_{i=1}^{d}|x_i - y_i|^k\}^{1/k}$ [Example 7.11]
$d_\infty(x, y)$: $\max_{1\le i\le d}|x_i - y_i|$ [Example 7.11]
d_E: geodesic metric [Section 9.2]
$d(h, X)$: distance between point h and set X [Section 7.2]
$\text{dom}(G)$: domain of G; $\text{dom}(G) = \{x \mid G(x) > -\infty\}$ [Section 4.6]
$D(X, Y)$: Hausdorff–Busemann metric, i.e., $\sup_{h\in E}|d(h, X) - d(h, Y)|e^{-d(o,h)}$ [Section 7.3]
$\hat{D}(X, Y)$: $\sup_{x\in X} d(x, Y)$ [Section 7.3]
$D_H(X, Y)$: Hausdorff metric, i.e., $\sup_{h\in E}|d(h, X) - d(h, Y)|$ [Section 7.3]
∂X: boundary of X [Section 7.1]
δ: dilation [Section 3.2]
$\check{\delta}$: reflection of δ [Section 3.2]
$\delta(X \mid\le M)$: (upper) conditional dilation [Section 3.4]
$\delta(X \mid\ge M)$: lower conditional dilation [Section 3.4]
δ^r: metric dilation [Section 9.3]
$\delta_A(X \mid\subseteq M)$: conditional dilation [Section 9.5]
$\delta^r(X \mid M)$: geodesic dilation of X in M [Section 9.5]
Δ_G: dilation by G [Section 4.6]
$\mathbf{\Delta}$: distance transform [Section 9.4]
$\mathbf{\Delta}^{\leftarrow}$: inverse distance transform [Section 9.4]
(E, d): metric space [Definition 7.10]
ε: erosion [Section 3.2]
(ε, δ): adjunction [Definition 3.10]
$\check{\varepsilon}$: reflection of ε [Section 3.2]
$\varepsilon(X \mid\le M)$: (upper) conditional erosion [Section 3.4]
$\varepsilon(X \mid\ge M)$: lower conditional erosion [Section 3.4]
$\varepsilon_A(X \mid\subseteq M)$: conditional erosion [Section 9.5]
$\varepsilon^r(X \mid M)$: geodesic erosion of X in M [Section 9.5]
$\mathcal{E}_G$: erosion by G [Section 4.6]
$f_{x,t}$: pulse function at x with height t [Example 2.38]
F_h: horizontal translation of function F over h [Section 4.6]
$F + v$: vertical translation of function F over v [Section 4.6]
$F \oplus G$: Minkowski sum of functions F and G [Section 4.6]
$F \ominus G$: Minkowski difference of functions F and G [Section 4.6]
$\mathcal{F}(\mathbb{R}^d)$: closed subsets of $\mathbb{R}^d$ [Example 2.9]
$\mathcal{F}^A$: $\{X \in \mathcal{F} \mid X \cap A = \varnothing\}$ [Section 7.4]
$\mathcal{F}_A$: $\{X \in \mathcal{F} \mid X \cap A \ne \varnothing\}$ [Section 7.4]

$\mathcal{F}'(E)$:	$\mathcal{F}(E) \setminus \{\varnothing\}$ [Section 7.4]
$\mathcal{F}^K_{G_1,G_2,\ldots,G_n}$:	$\mathcal{F}^K \cap \mathcal{F}_{G_1} \cap \cdots \cap \mathcal{F}_{G_n}$ [Section 7.4]
$\mathrm{Fun}(E,\mathcal{T})$:	functions mapping E into $\mathcal{T}$ [Example 2.10]
$\mathrm{Fun}_u(E,\mathcal{T})$:	upper semi-continuous functions [Section 10.5]
$\mathrm{Fun}_l(E,\mathcal{T})$:	lower semi-continuous functions [Section 10.5]
$\check{G}$:	reflection of G; $\check{G}(x) = G(-x)$ [Section 4.6]
$\mathcal{G}(\mathbb{R}^d)$:	open subsets of $\mathbb{R}^d$ [Example 2.9]
$\mathrm{id}_{\mathcal{L}}$, id:	identity operator [Section 3.1]
inf:	infimum [Section 2.1]
I:	unit element = greatest element [Section 2.1]
$\mathrm{Inv}(\psi)$:	invariance domain of ψ [Section 3.1]
$\mathrm{Inv}_+(\psi)$:	extensivity domain of ψ, i.e., $\{X \in \mathcal{L} \mid \psi(X) \geq X\}$ [Section 6.1]
$\mathrm{Inv}_-(\psi)$:	anti-extensivity domain of ψ, i.e., $\{X \in \mathcal{L} \mid \psi(X) \leq X\}$ [Section 6.1]
ι:	operator identically I [Section 3.1]
$\mathcal{K}^A$:	$\{X \in \mathcal{K} \mid X \cap A = \varnothing\}$ [Section 7.5]
$\mathcal{K}_A$:	$\{X \in \mathcal{K} \mid X \cap A \neq \varnothing\}$ [Section 7.5]
$\mathcal{K}^F_{G_1,G_2,\ldots,G_n}$:	$\mathcal{K}^F \cap \mathcal{K}_{G_1} \cap \cdots \cap \mathcal{K}_{G_n}$ [Section 7.5]
$\mathcal{K}'(E)$:	$\mathcal{K}(E) \setminus \{\varnothing\}$ [Section 7.5]
κ:	anti-centre [Definition 3.35]
$\lim X_n$:	limit of X_n [Section 7.3]
$\overline{\lim}\, X_n$:	upper limit of X_n [Section 7.3]
$\underline{\lim}\, X_n$:	lower limit of X_n [Section 7.3]
ℓ:	set of atoms [Section 2.2]
$\ell(A)$:	elements in $a \in \ell$ with $a \leq A$ [Section 5.2]
$\mathcal{L}$:	complete lattice [Definition 2.1]
$\mathcal{L}_1 \times \mathcal{L}_2$:	product lattice of $\mathcal{L}_1$ and $\mathcal{L}_2$ [Example 2.11]
$\mathcal{L}^d$:	power lattice [Example 2.11]
$\mathcal{L}_{(\leq M)}$:	elements in $\mathcal{L}$ which are $\leq M$ [Section 3.4]
$\mathcal{L}_{(\geq M)}$:	elements in $\mathcal{L}$ which are $\geq M$ [Section 3.4]
μ_A:	median operator [Section 4.5]
$\mathbb{N}$:	positive integers including 0 [Example 2.3]
ν:	complement operator or negation operator [Section 2.2]
$\nu(X \mid \leq M)$:	(upper) conditional complement [Section 3.4]
$\nu(X \mid \geq M)$:	lower conditional complement [Section 3.4]
o:	operator identically O [Section 3.1]
O:	zero element = least element [Section 2.1]
$\mathcal{O}(\mathcal{L},\mathcal{M})$:	operators between $\mathcal{L}$ and $\mathcal{M}$ [Section 3.1]
$\mathcal{O}(\mathcal{L})$:	operators on $\mathcal{L}$ [Section 3.1]
$\mathcal{O}^+(\mathcal{L},\mathcal{M})$:	increasing operators between $\mathcal{L}$ and $\mathcal{M}$ [Section 3.1]
$\mathcal{O}^+(\mathcal{L})$:	increasing operators on $\mathcal{L}$ [Section 3.1]
$\mathcal{O}_{\mathbb{T}}(\mathcal{L})$:	$\mathbb{T}$-invariant operators on $\mathcal{L}$ [Section 5.2]
$\mathcal{O}^+_{\mathbb{T}}(\mathcal{L})$:	$\mathbb{T}$-invariant increasing operators on $\mathcal{L}$ [Section 5.2]
ω:	middle filter [Section 12.5]
$p_B(x)$:	Minkowski functional [Section 9.2]
$\mathcal{P}(E)$:	power set of E; subsets of E [Example 2.7]
ψ^c:	$\psi^c(X) = [\psi(X)]^c$ [Section 3.5]
ψ^*:	$\psi^*(X) = [\psi(X^*)]^*$ [Section 3.1]

$X_n \xrightarrow{\mathcal{F}} X$:	X_n converges to X in hit-or-miss topology [Section 7.4]
$X_n \xrightarrow{\mathcal{K}} X$:	X_n converges to X in myope topology [Section 7.5]
$X_n \to X$:	order convergence [Section 13.1]
$\mathbf{X}(F, t)$:	threshold set [Section 11.1]
$\mathbb{Z}$:	integers [Example 2.7]
$\overline{\mathbb{Z}}$:	integers including $\pm\infty$ [Example 2.7]

Subject Index

C

H

I

FSC
www.fsc.org
MIX
Papier | Fördert
gute Waldnutzung
FSC® C083411

Kolibri 360 GmbH
Ferdinand-Jühlke-Straße 7
99095 Erfurt, Deutschland
produktsicherheit@kolibri360.de